Open Channel Hydraulics, River Hydraulic Structures and Fluvial Geomorphology

For Engineers, Geomorphologists and Physical Geographers

Open Channel Hydraulics, River Hydraulic Structures and Fluvial Geomorphology

For Engineers, Geomorphologists and Physical Geographers

Editors

Artur Radecki-Pawlik
Institute of Structural Mechanics
Faculty of Civil Engineering
Cracow University of Technology

Previously employed in:
Department of Hydraulics Engineering and Geotechnic
University of Agriculture in Krakow
Poland

Stefano Pagliara
DESTEC – Department of Energy, Engineering,
Systems, Land and Construction
University of Pisa
Italy

Jan Hradecký
Department of Physical Geography and Geoecology
Faculty of Science
University of Ostrava
Czech Republic

Associate Editor
Erik Hendrickson
Large Lakes Observatory (LLO)
University of Minnesota-Duluth
USA

CRC Press
Taylor & Francis Group
Boca Raton London New York

CRC Press is an imprint of the
Taylor & Francis Group, an **informa** business

A SCIENCE PUBLISHERS BOOK

Cover Illustrations

Background Image: The Tara River in Montenegro—picture taken by Artur Radecki-Pawlik (editor of the book)

Top image: Peterka type of ramp hydraulic structure on the Lubenka River in Southern Poland—Reproduced by kind courtesy of Karol Plesinski (author of Chapter 5)

CRC Press
Taylor & Francis Group
6000 Broken Sound Parkway NW, Suite 300
Boca Raton, FL 33487-2742

First issued in paperback 2020

© 2018 by Taylor & Francis Group, LLC
CRC Press is an imprint of Taylor & Francis Group, an Informa business

No claim to original U.S. Government works

ISBN-13: 978-1-4987-3082-2 (hbk)
ISBN-13: 978-0-367-78194-1 (pbk)

Visit the Taylor & Francis Web site at
http://www.taylorandfrancis.com

and the CRC Press Web site at
http://www.crcpress.com

Preface

One day, I was sitting in my office after giving a lecture to students for a course called "Integrated Watershed Management" I thought we have taught this course at the University of Agriculture in Cracow at the Environmental Faculty facility for our students and others from Erasmus for many years, but we do not have many teaching materials available for it. Also, we teach many other water resource courses about rivers, hydraulics, and fluvial processes; I was struck by the thought that maybe we could create a useful teaching material for our students by compiling information from all of these different water resource subjects at one place. While rivers are attractive and important enough by themselves, the question remained for me: How do we improve the materials for the course? Can we provide students with good informational material for a range of topics that would help me and other teachers/professors? I was then struck by another thought and a clear idea formed in my head: I need a good, well-written, and nicely presented textbook from very skilled specialists in water resource subjects from all over the world who have experience working in river catchments. I also wanted specialists who are as passionate, enthusiastic, and enjoy rivers and streams as much as I do. In the next couple of days following my new idea, I flew to Pisa, Italy (thankfully we have direct flights between Cracow and Pisa), where I met my colleague Stefano and told him about the textbook idea focused on streams, rivers, basics of hydraulics and fluvial geomorphology, river engineering, watershed management and other useful water resources subjects. Stefano liked the idea and the two of us decided we needed a good fluvial geomorphologist on board, so we called Jan (in Czech his name is Honza) from Ostrava, Czech Republic. Jan, as a dean of faculty and a very good professional in his field, immediately appreciated our idea about the textbook and agreed to work with us.

Whilst identifying chapter authors we decided we needed a good professional language editor to work with us, a native speaker but familiar with this field. It was a difficult task to find somebody like this but then I remembered that during one of courses we run in Czorsztyn, Poland for students from the University of Wisconsin–Stevens Point (our close friends from the United States, who have visited Poland for two week periods every summer for over 25 years), I had met Erik Hendrickson. I simply asked Erik for help, and after a short conversation, he agreed to help. He was a very big help! Thanks, Erik!

Anyway—as I said—we need a good textbook on rivers, fluvial geomorphology, river engineering and hydraulics for our European students, but also for teachers and professors. Also, we hope the students we teach from the University of Wisconsin–Stevens Point will be happy with our input—in addition to students from other continents. We have needed a nice, professional textbook as a valuable tool for all who want to understand water issues and problems. Now you know how this book began, and now, fortunately, this book is in your hands!

We invited brilliant specialists from around the world to work on this book. Nearly all continents are represented by these specialists here in this book. The people who contributed are world famous specialists in river-related problems, are enthusiastic and, very professional. Thank you, friends, for giving your time and for doing your best for our students, teachers, professors, and for the future of rivers.

So, finally, we have a product which is directed to fluvial geomorphologists, hydraulic and river engineers, physical geographers, geophysicists, and water resource managers. We hope you appreciate the subjects in this book and that the book is useful for you.

Once more, thank you to all the people who were involved in preparing this book. It is ready for classes, students, professors, and teachers. I believe we have created a very good textbook. So please, have fun and enjoy becoming a professional working with rivers, hydraulics, and in general, water resources.

Artur Radecki-Pawlik
Stefano Pagliara
Jan Hradecký
Erik Hendrickson

Contents

CHAPTER 1

Principles of Hydraulics of Open Channels

Stefano Pagliara* and Michele Palermo

INTRODUCTION

Open channel flow can be classified according to the characteristics of the channel. Namely, open channel flow can occur both in natural streams and rivers and in artificial channels, such as drainage channel systems, sewer pipes, concrete/earth channels, etc. According to the geometry of the channel, the water discharge can flow either in prismatic channels (generally characterized by a constant transversal cross-section geometry and longitudinal slope) or in non-prismatic channels (natural streams or rivers, in which both the transversal cross-section geometry and longitudinal bed slope vary). Furthermore, channel bed characteristics also contribute to distinguish the open channel flows, due to the roughness conditions and erosive process, which can contribute to the modification of the channel geometry. In other words, the channel bed can be smooth, rough, erodible, fixed, etc. Therefore, the flow characteristics modify accordingly. In general, two main distinctions can be done according to the flow characteristics variation with time and along the channel (Chow 1959; Citrini and Noseda 1987; De Marchi 1986; Henderson 1966). Considering the flow characteristics along the channel, uniform flow conditions (UF) occur when the discharge Q, the water depth y and the average flow velocity u at every cross-section do not vary. Namely, at the generic cross-section (located at the longitudinal distance x respect to a selected reference system), the following conditions should be verified:

$$\frac{\partial y}{\partial x} = \frac{\partial Q}{\partial x} = \frac{\partial u}{\partial x} = 0 \tag{1}$$

These analytical conditions occur only when the cross-section geometry do not vary. Therefore, if the water depth is constant, from Eq. (1) it can be easily deduced that the free surface should be parallel to the channel bed (see Fig. 1).

DESTEC – Department of Energy, Engineering, Systems, Land and Construction, University of Pisa, Via Gabba 22, 56122, Pisa, Italy.
E-mail: michele.palermo@ing.unipi.it
* Corresponding author: s.pagliara@ing.unipi.it

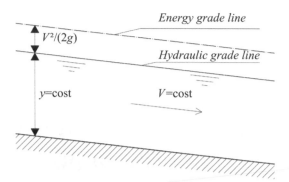

Fig. 1. Schematic representation of UF conditions.

In the case in which, the analytical conditions reported in Eq. (1) are not satisfied, that is, when Eq. (2) is valid, varied flow conditions take place.

$$\frac{\partial y}{\partial x} \neq \frac{\partial Q}{\partial x} \neq \frac{\partial u}{\partial x} \neq 0 \tag{2}$$

These last flow conditions can be further subdivided into rapidly varied flow (RVF), gradually varied flow (GVF), and spatially varied flow (SVF). GVF is characterized by a gradual variation of the water depth along the channel, resulting in a gradual variation of the streamlines curvature. For this flow typology, the energy losses occur along the channel and they are mainly due to the wall friction effects. RVF is characterized by significant streamlines curvature, because of the rapid variation of the water depth. Therefore, in this last case, energy losses are mainly concentrated in a relatively short channel length. Finally, SVF is characterized by a discontinuous flow, that is, the discharge varies along the channel. This is the typical case occurring in the presence of a tributary along the river. Figure 2 reports a diagram sketch illustrating the three mentioned flow conditions.

Nevertheless, open channel flow can be classified also according to its behavior with respect to time. Namely, two main distinctions can be done. The flow is steady when

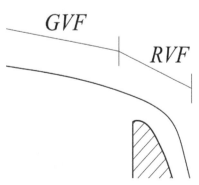

Fig. 2. Schematic representation of different flow regimes: UF, GVF and RVF.

its characteristics do not change with time, that is, this flow regime can be analytically described by the following Eq. (3):

$$\frac{\partial y}{\partial t} = \frac{\partial Q}{\partial t} = \frac{\partial u}{\partial t} = 0 \tag{3}$$

Conversely, when the analytical conditions expressed by Eq. (4) take place, for example, when the water depth varies, the flow conditions are termed unsteady.

$$\frac{\partial y}{\partial t} \neq \frac{\partial Q}{\partial t} \neq \frac{\partial u}{\partial t} \neq 0 \tag{4}$$

In synthesis, the following flow typologies can be distinguished:

a) Steady uniform flow
b) Steady varied flow
c) Unsteady uniform flow (rare condition)
d) Unsteady varied flow

Figure 3 reports several diagram sketches illustrating the flow conditions (a) Steady uniform flow, (b) Steady varied flow, and (d) Unsteady varied flow.

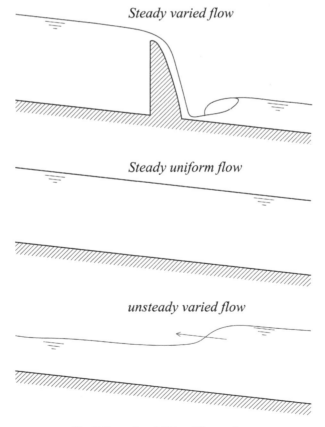

Steady varied flow

Steady uniform flow

unsteady varied flow

Fig. 3. Examples of different flow regimes.

Channel Geometric Characteristics

The main geometric characteristics of a channel, which are of interest for open channel flow, can be synthetized as follows:

- cross-section area A
- wetted perimeter P, that is, the perimeter of the cross-section area in contact with water
- hydraulic radius R_h, that is, $R_h = A/P$
- hydraulic depth D_h, that is, $D_h = A/W$, where W is the top width of the channel.

Pressure Distribution

In open channel flow, pressure distribution depends on the streamlines configuration. When they are almost straight and parallel, pressure distribution is hydrostatic, that is, there is a linear increase of the pressure with the depth. Nevertheless, when streamlines are significantly curved, pressure distribution is not more hydrostatic, due to the presence of a centrifugal force, resulting in a different pressure distribution, which could assume either larger or smaller values than the hydrostatic distribution, according to the curvature of the streamlines (convex or concave). Figure 4 represents a diagram sketch illustrating pressure deviation from hydrostatic distribution for different curvature of the streamlines.

In general, the variation of pressure distribution can be expressed as $p = k\gamma y$, where $k=1$ for hydrostatic distribution, and k either >1 or <1 when streamlines are concave or convex, and p is the pressure. Nevertheless, for practical applications, streamlines curvature is generally negligible for both UF and GVF conditions, thus resulting in a pressure distribution which can be considered hydrostatic. However, for RVF, the streamlines curvature is very prominent and the simplifying hypothesis valid for the previous mentioned flow regimes cannot be assumed.

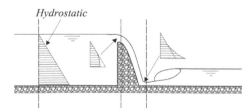

Fig. 4. Pressure distribution variation.

Velocity Distribution

Due to the free surface and friction effects exerted by channel walls, the velocity distribution is not uniformly distributed in a transversal cross section. Namely, it assumes different values (generally higher) in the center of the channel. Furthermore, the irregular shape of natural channels causes a random velocity distribution, resulting in the formation of both transversal and vertical velocity components and vortexes. Figure 5 shows some examples of transversal velocity distribution inside a channel, obtained by neglecting the transversal velocity components. It should be noted that the velocity distribution is characterized by a high gradient in correspondence with the walls. This causes a vertical velocity distribution which could be represented by the trend reported in Fig. 6. Namely, in general, a logarithmic profile can well represent the velocity distribution up to the level in which the velocity reaches its maximum. The velocity maximum is generally reached at a certain distance from the free water surface, which generally ranges between 0.05 and 0.25 y, where y is the water depth.

The maximum velocity does not occur in correspondence with the free water surface, because of the formation of superficial re-circulation flows, which are more significant when the ratio between water depth and transversal section width decreases. For practical applications, the average velocity can be generally estimated as the average value of the velocities measured at 0.2 y and 0.8 y. Other studies suggest that a preliminary estimation can be done considering the average cross section velocity equal to 0.7–0.8V_s, where V_s is the velocity of the water surface.

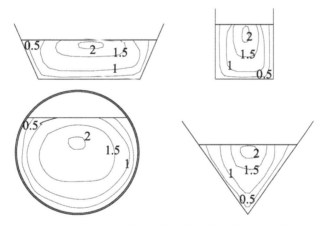

Fig. 5. Examples of velocity distributions in a channel transversal section.

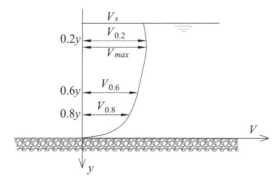

Fig. 6. Example of a vertical distribution of velocity.

Uniform Flow

In the past, several relationships were developed in order to estimate the main flow characteristics in open channels. Namely, several researchers developed different empirical (or semi-empirical) relationships valid in UF conditions, involving both flow and channel properties (i.e., average flow velocity, bed slope, roughness, water depth, hydraulic radius, discharge, cross-section area, etc.). In particular, Chezy, a French engineer, proposed the following equation (Chow 1959):

$$U = \chi \sqrt{R_m I_0} \tag{5}$$

where I_0 is the bed slope, R_m the hydraulic radius, U the average flow velocity in the section, and χ is a dimensional coefficient, accounting for wall roughness. Nevertheless, despite the simplicity of the expression proposed by Chezy, the determination of the correct value of the coefficient χ is quite complex. In fact, it depends on several factors, including the material constituting the channel bed (earth, gravels, rocks, etc.), channel walls characteristics (which could be made of concrete, gabions, etc.), vegetation present in the channel (which could vary during the years, due to the natural growth process), section shape and geometry, water depth, etc. Therefore, it appears evident that the main problem is the selection of an appropriate and effective coefficient in order to furnish a good estimation of the desired parameters. Generally, the estimation of the coefficient χ is made by adopting other empirical expressions proposed by Kutter (Eq. 6) and Bazin (Eq. 7). Both Kutter and Bazin (Chow 1959) assumed that χ is a function of the hydraulic radius R_m. In particular, Kutter proposed to estimate χ by using Eq. (6) in which γ is a dimensional parameter depending on the roughness of the walls of the channel (Chow 1959).

$$\chi = \frac{87}{1 + \dfrac{\gamma}{\sqrt{R_m}}} \tag{6}$$

Similarly, Kutter proposed Eq. (7) to estimate χ, in which he introduced the parameter m, having the same role of the parameter γ of the previous equation.

$$\chi = \frac{100}{1 + \dfrac{m}{\sqrt{R_m}}} \tag{7}$$

The values of these two parameters can be found in any specialized technical manual. Another fundamental empirical expression is proposed by Manning, an Irish engineer (Chow 1959). Namely, he proposed the following Eq. (8) in order to estimate the average water velocity U in the case of uniform flow conditions:

$$U = \frac{1}{n} R_m^{2/3} I_0^{1/2} \tag{8}$$

Also in this case, Manning introduced a coefficient n [$m^{-1/3}$s], accounting for channel bed and wall roughness properties. The values of the Manning's coefficient n can be found in any specialized manual. The previous equation is similar to that proposed by Gauckler-Strickler. Equation (9) represents the so called Gauckler-Strickler's equation, which is analytically identical to that proposed by Manning. The only difference is that in the Gauckler-Strickler's equation, the channel roughness is taken into account by the coefficient k_s [m$^{1/3}$s^{-1}], which is practically equal to n^{-1}.

$$U = k_s R_m^{2/3} I_0^{1/2} \tag{9}$$

By comparing Eqs. (5) and (8) it is easy to derive the following relationship between Manning's and Chezy's coefficients:

$$n = \frac{R_m^{1/6}}{\chi} \tag{10}$$

In addition, from Eq. (5) and Eq. (8), the flow discharge can be easily estimated by adopting the following Eq. (11) and Eq. (12), respectively.

$$Q = \chi A \sqrt{R_m I_0} \tag{11}$$

$$Q = \frac{1}{n} A R_m^{2/3} I_0^{1/2} \tag{12}$$

By analyzing both Eq. (11) and (12) and considering that the water cross-section area A and the hydraulic radius R_m can be expressed as function of the water depth y (i.e., $A=f(y)$ and $R_m=f(y)$), it is evident that Q can be expressed as only function of the variable y, in the case of UF conditions. In other words, there is a biunique relationship between the variable Q and the parameter y, that is, the stage-discharge relation. Generally, in a natural channel (e.g., river, stream, torrent, etc.), UF conditions are rarely achieved. Nevertheless, in a channel branch characterized by a quite regular shape, stable walls and bed, absence of any transversal hydraulic structure which can cause modifications of the water depths (e.g., dam, sills, check-dam, gate, etc.), and for steady flow conditions, it can be assumed that a quasi-steady uniform flow condition takes place, thus resulting in a biunique relationship between Q and y. This simplifying assumption allows researchers to establish a relationship between water depth and discharge in an opportunely selected section of the branch. Namely, in practical applications, a hydrometric station is located in the channel by which the water levels are measured. Generally, it is located close to the channel wall. Contemporarily, the water discharge is measured by conducting water velocity measurements in a sufficient number of positions inside the cross-section. In this way, several couples of water depth-discharge values can be measured, and interpolated by a curve in a diagram y-Q, thus furnish an analytical expression of the function $Q=f(y)$. The knowledge of this curve assumes a fundamental importance as the water discharge in a selected section, which is a parameter requiring a laborious estimation, can be easily derived by reading the water level in the same selected section.

Specific Energy

Let us consider a generic channel cross-section, in which the discharge is fixed and GVF flow regime occurs. Then, specific energy E is defined as follows:

$$E = y + \frac{\alpha V^2}{2g} = y + \frac{\alpha Q^2}{2gA^2} \tag{13}$$

in which y is the water depth, V is the average flow velocity, Q the water discharge and α Coriolis' coefficient (which can be assumed equal to 1 for usual practical applications). Namely, E represents the energy respect to the channel bed, whereas the total energy H is relative to a selected horizontal plane. Figure 7 shows a schematic diagram of both channel cross-section and channel profile along with the indication of the main geometric and hydraulic parameters.

Specific energy represents a local parameter of a water flow. For a generic cross-section, it is a function of both Q and y, as $A=f(y)$. Figure 8 shows the plot of Eq. (13).

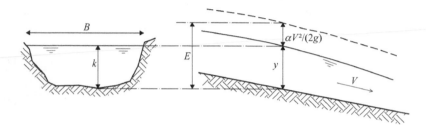

Fig. 7. Schematic diagram of a channel cross-section and profile.

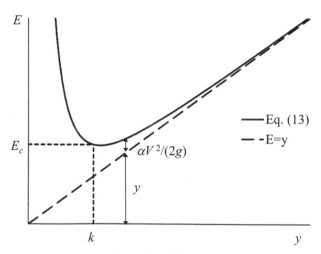

Fig. 8. Plot of Eq. (13).

If we consider Eq. (13), it is easy to observe that it tends to infinity for y tending to both 0 and infinity. Namely, it occurs that:

$$\lim_{y\to 0}\left(y+\frac{\alpha V^2}{2g}\right) = \lim_{y\to +\infty}\left(y+\frac{\alpha V^2}{2g}\right) = +\infty \qquad (14)$$

Nevertheless, it should also result that:

$$E>0 \qquad (15)$$

as the specific energy is a sum of two positive quantities. Combining both the deductions expressed by Eqs. (14) and (15), it can be noted that the function relationship $E(y)$ should have a minimum, which can be obtained by imposing the following analytical condition:

$$\frac{dE}{dy} = 0 \tag{16}$$

Therefore, considering both Eq. (13) and Eq. (16), one can easily derive the following expression:

$$\frac{dE}{dy} = 1 - \frac{\alpha Q^2}{gA^3} \frac{dA}{dy} = 0 \tag{17}$$

But, noting that

$$\frac{dA}{dy} = B \tag{18}$$

where B is the water surface width, it can be concluded that the condition for which the specific energy assumes the minimum value can be represented by the following equation:

$$\frac{\alpha Q^2}{gA^3} B = 1 \tag{19}$$

Which can be re-written in the most common way as follows:

$$\frac{A^3}{B} = \frac{\alpha Q^2}{g} \tag{20}$$

Furthermore, noting that $A=A(y)$ (as also $B=B(y)$), Eq. (20) furnishes the value of the variable y for which E is minimum. This last water depth is called critical depth and generally is indicated by the using the symbol k. It can be concluded that the critical depth is the water depth for which a fixed discharge flows with minimum specific energy. Consequently, the critical status is defined as the condition for which the water flows with a depth equal to the critical depth k. The average velocity at the critical status is called critical velocity V_c. V_c can be easily calculated by applying Eq. (20). In fact, noting that $Q=V_c A$, Eq. (20) can be re-written as follows:

$$\frac{A}{B} = \frac{\alpha Q^2}{gA^2} = \frac{\alpha V_c^2}{g} \tag{21}$$

Which leads to the following expression of V_c:

$$V_c = \left(\frac{g}{\alpha} \frac{A}{B} \right)^{0.5} \tag{22}$$

From Fig. 8, it can be easily observed that each value of the specific energy can be obtained with two different water depths y. Nevertheless, the minimum value of the specific energy E_c (critical energy) is obtained only when $y=k$. For $E<E_c$, Eq. (22) has no solutions, therefore the water flow is physically impossible, as there is not enough energy. The plot represented in Fig. 8 is obtained for a fixed Q value, therefore each

point of the curve represents the water depth y and specific energy E relative to the selected discharge. The point on the curve corresponding to both the minimum specific energy value E_c and to the critical depth k divides the curve into two parts. For $y<k$ we have that $V>V_c$, that is, supercritical flow conditions occurs. For $y>k$ we have that $V<V_c$, that is, subcritical flow conditions occur. It is evident that different curves similar to that sketched in Fig. 8 are obtained for each Q value. In particular, the difference between supercritical depth and subcritical depth decreases with Q, as shown in Fig. 9.

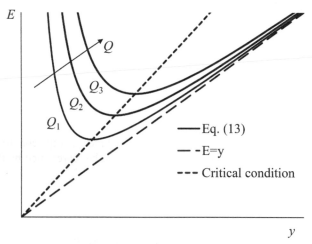

Fig. 9. Plot of Eq. (13) for different Q values.

All the previous deductions are based on the hypothesis that Q is kept constant. A similar analysis can be conducted considering that the specific energy E is constant. Therefore, in this case, the variation of discharge Q with y can be analyzed. Let's consider again Eq. (13). From Eq. (13), Eq. (23) can be easily obtained:

$$Q = A\sqrt{\frac{2g}{\alpha}(E-y)}$$

(23)

By observing Eq. (23), it results that there are two water depth conditions for which $Q=0$ m³/s. Namely $Q=0$ m³/s when either $y=0$ m or $y=E$. If $y=0$ m, then the liquid area $A=0$ m², therefore $Q=0$ m³/s. If $y=E$, then all the flow energy is due to the water depth, therefore $\alpha V^2/2g=0$ m, that is, $V=0$ m/s which implies that the discharge $Q=0$ m³/s. Based on these observations and on the obvious condition that the flow occurs when $Q>0$, it can be concluded that Eq. (23) should have a maximum for a certain y value. This maximum can be calculated using the following analytical relationship:

$$\frac{dQ}{dy} = 0$$

(24)

From the previous equations, the following analytical expression can be found:

$$\frac{dA}{dy}\sqrt{\frac{2g}{\alpha}(E-y)} - \frac{gA}{\alpha\sqrt{\frac{2g}{\alpha}(E-y)}} = 0$$

(25)

Considering that for Eq. (18), $dA/dy=B$, Eq. (25) can be re-written as follows:

$$B\sqrt{\frac{2g}{\alpha}(E-y)}-\frac{\sqrt{2g}A}{2\sqrt{\alpha(E-y)}}=0 \qquad (26)$$

Or, assuming $\alpha\approx1$, equivalently:

$$B\sqrt{(E-y)}-\frac{A}{2\sqrt{(E-y)}}=0 \qquad (27)$$

From which:

$$\frac{A}{2B}=E-y \qquad (28)$$

But, considering that

$$E_c=E_{min}=k+\frac{\alpha V_c^2}{2g} \qquad (29)$$

It is easy to observe that Eq. (29) can be re-written as follows:

$$E_c=E_{min}=k+\frac{A}{2B} \qquad (30)$$

therefore:

$$E_c-k=\frac{A}{2B} \qquad (31)$$

By comparing Eq. (28) and Eq. (31), it results that the condition represented by Eq. (28) occurs in correspondence with the critical flow condition, that is, the maximum discharge related to a certain specific energy occurs when $y=k$. Figure 10 shows the plot $Q=Q(y)$ of Eq. (23). Also in this case, for each discharge value, two water depth values can take place, that is, $y_A>k$ and $y_B<k$, where y_A and y_B are the water depth in subcritical and super critical flow conditions, respectively.

If the water discharge increases the difference between y_A and y_B decreases, up to when they coincide and become equal to k, for which the maximum discharge Q_{max} occurs.

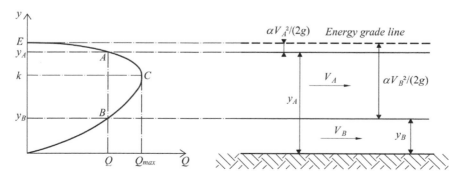

Fig. 10. Plot of Eq. (23).

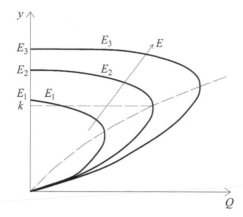

Fig. 11. Plot of Eq. (23) for different specific energies.

Froude Number

In order to classify the flow regimes (supercritical vs. subcritical) a non-dimensional parameter is adopted, that is, the Froude number F_r. F_r^2 represents the ratio between inertial forces F_i and forces due to gravity F_g, that is, the main forces acting in open channel flows (Eq. 32).

$$F_r^2 = \frac{F_i}{F_g} = \frac{\rho L^3 L T^{-2}}{\rho L^3 g} = \frac{V^2}{gL} \tag{32}$$

Where L represents a characteristic length, ρ water density, g acceleration due to gravity, and T the time. Rearranging Eq. (32), it is easy to obtain Eq. (33).

$$F_r = \frac{V}{\sqrt{gL}} \tag{33}$$

Therefore, the numerator of Froude number is the average velocity V, whereas the denominator is $(gL)^{0.5}$. For open channel flow, the characteristics length is assumed $L=D=A/B$, which is the hydraulic depth. Because of this assumption, Eq. (33) becomes:

$$F_r = \frac{V}{\sqrt{g\dfrac{A}{B}}} = \frac{V}{\sqrt{gD}} \tag{34}$$

If the channel cross-section is rectangular, $D=y$ and Eq. (34) is:

$$F_r = \frac{V}{\sqrt{gy}} \tag{35}$$

If the flow regime is critical, $y=k$ and $V=V_c$, i.e., Eq. (35) can be re-written as follows:

$$F_r = \frac{V_c}{\sqrt{gk}} \tag{36}$$

Assuming $\alpha=1$ and considering Eq. (22), it is easy to derive the following result:

$$F_r = \frac{V_c}{\sqrt{gk}} = \frac{\sqrt{gk}}{\sqrt{gk}} = 1 \tag{37}$$

Therefore, if the flow regime is critical, $F_r=1$. Let's synthetize the previous findings related to the critical flow regime:

1) for a fixed discharge, the specific energy is minimum;
2) for a fixed specific energy, the water discharge is maximum;
3) the Froude number is equal to 1.

Channel Bed Slope

The analysis of the flow regime can be extended to a discrete channel branch. Namely, in a prismatic channel branch, three different water depth conditions can occur, that is, the relationship between uniform flow depth and critical depth can be expressed by the following analytical conditions:

$$y_u > k \tag{38.1}$$

$$y_u < k \tag{38.2}$$

$$y_u = k \tag{38.3}$$

The previous equations represent three different flow regimes, that is, the channel slope is mild if Eq. (38.1) is valid, steep if Eq. (38.2) is valid, and critical if $y=k$.

Flow Surface Profiles in Steady Flow Conditions

Let's consider a GVF steady flow. Let's assume the following hypotheses: (1) channel bed slope less than 10%; (2) the differences of cross-section geometries are negligible; (3) incompressible fluid; (4) hydrostatic distribution of pressure; (5) uniform distribution of velocity; (6) the distributed losses can be expressed by using either Manning's or Chezy's formulas; (7) concentrated losses are negligible; (8) the discharge is constant in the channel branch. Let I_0 be the bed slope and j energy head slope. Therefore, if we consider a channel branch whose length is ds, the variation of channel bed and energy head level is equal to $I_0 ds$ and jds, respectively. Note that because of the above reported hypothesis 7, j is only related to distributed losses. Figure 12 illustrates a diagram sketch of the channel branch, along with the hydraulic and geometric parameters.

Let E be the energy head at the beginning of the channel branch. Therefore, at the end of the channel branch the energy head will be:

$$E + \frac{dE}{ds}ds \tag{39}$$

Therefore the energy variation is:

$$\frac{dE}{ds} = I_0 - j \tag{40}$$

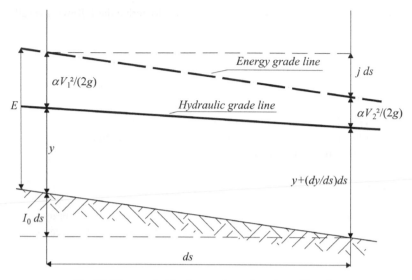

Fig. 12. Diagram sketch of energy losses in a channel branch.

Nevertheless, E can be expressed as follows:

$$E = y + \frac{\alpha V^2}{2g} = y + \frac{\alpha Q^2}{2gA^2}$$

(41)

By differentiating Eq. (41), it is easy to obtain the following analytical expression:

$$\frac{dE}{ds} = \frac{dy}{ds} - \frac{d}{ds}\left(\frac{\alpha Q^2}{2gA^2}\right) = \frac{dy}{ds} - \frac{\alpha Q^2}{gA^3}\frac{dA}{ds}$$

(42)

dA/ds can be re-written as follows:

$$\frac{dA}{ds} = \frac{\partial A}{\partial y}\frac{dy}{ds} + \frac{\partial A}{\partial s} = B\frac{dy}{ds} + \frac{\partial A}{\partial s}$$

(43)

According to the hypothesis 2, the variation of A along the channel branch abscissa s is negligible, that is, we have:

$$\frac{\partial A}{\partial s} = 0$$

(44)

Therefore, considering Eq. (44), Eq. (43) becomes:

$$\frac{dA}{ds} = \frac{\partial A}{\partial y}\frac{dy}{ds} = B\frac{dy}{ds}$$

(45)

Substituting Eq. (45) in Eq. (42), it is easy to derive the following:

$$\frac{dE}{ds} = \frac{dy}{ds} - \frac{\alpha Q^2}{gA^3}B\frac{dy}{ds}$$

(46)

Combining Eq. (40) and Eq. (46), we have:

$$I_0 - j = \frac{dy}{ds} - \frac{\alpha Q^2}{gA^3} B \frac{dy}{ds} \tag{47}$$

From which the following general expression can be derived:

$$\frac{I_0 - j}{1 - \frac{\alpha Q^2}{gA^3} B} = \frac{dy}{ds} \tag{48}$$

Furthermore, Eq. (34) can be re-arranged as follows:

$$F_r^2 = \frac{V^2}{g \frac{A}{B}} = \frac{Q^2}{gA^3} B \tag{49}$$

Therefore, combining Eqs. (48) and (49), the following Eq. (50) can be easily derived:

$$\frac{I_0 - j}{1 - F_r^2} = \frac{dy}{ds} \tag{50}$$

Equation (48) and Eq. (50) represent the general differential equations governing the water surface profile in GVF conditions. Nevertheless, E is function of the variables s and y, therefore the following Eq. (51) can be written:

$$\frac{dE}{ds} = \frac{dE}{dy} \frac{dy}{ds} \tag{51}$$

By substituting Eq. (51) in Eq. (40), we obtain:

$$\frac{dy}{ds} = \frac{I_0 - j}{\frac{dE}{dy}} \tag{52}$$

Water Surface Profiles for Mild Bed Slopes

In this paragraph, the flow conditions occurring in mild bed slope channels and the corresponding water surface profiles will be analyzed. Firstly, according to what stated above, a channel is characterized by a mild bed slope if the bed slope I_0 is less than the critical bed slope I_c (i.e., $I_0 < Ic$). For a mild bed slope channel, according to Eq. (38.1), the uniform flow depth is bigger than critical depth, that is, $y_u > k$. To simplify the analytical derivations, let's assume that the channel cross-section geometry is wide and rectangular, that is, $R_h \approx y$. Note, that the following considerations can be extended to all prismatic channels. Let's assume that the discharge Q is constant. Using the above proposed relationships, it is easy to evaluate both the values of uniform water depth y_u and critical water depth k. The knowledge of these two depths allows for an identification of three zones along the channel branch, delimited by two parallel lines whose distance from the channel bed is y_u and k, respectively. Let's term zone 1, the region in which the water depth y is $y > y_u > k$, zone 2 the region for which $y_u > y > k$, and zone 3 the region for which $y_u > k > y$. The water surface profiles occurring in the three zones are termed M1 (zone 1), M2 (zone 2) and M3 (zone 3), respectively. They will be analyzed in details in the following paragraphs. Figure 13 reports a sketch of the three mentioned profiles.

For a wide rectangular channel, it is easy to derive the following general differential equation, which will be useful in the following in order to illustrate the qualitative behavior of the water surface profiles for the general case of prismatic channels:

$$\frac{dy}{ds} = \frac{I_0 - \dfrac{I_0 y_u^{10/3}}{y^{10/3}}}{1 - \dfrac{k^3}{y^3}} = I_0 \frac{1 - \left(\dfrac{y_u}{y}\right)^{10/3}}{1 - \left(\dfrac{k}{y}\right)^3} \tag{53}$$

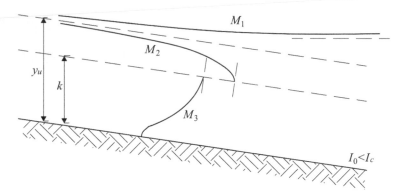

Fig. 13. Water surface profiles for mild bed slopes.

Water surface profile M1

This profile typology is characterized by the following condition:

$$y > y_u > k \tag{54}$$

In this case, we have:

$$\frac{dy}{ds} > 0 \tag{55}$$

Therefore, y is a monotonic increasing function of s. It means that water depth increases with channel abscissa (towards downstream). In particular, by considering Eq. (53), it is easy to estimate the values of the variable dy/ds at the limits of the existence interval in zone 1, that is, for $y \to y_u$ and $y \to \infty$:

$$\lim_{y \to y_u} \left[I_0 \frac{1 - \left(\dfrac{y_u}{y}\right)^{10/3}}{1 - \left(\dfrac{k}{y}\right)^3} \right] = 0 \tag{56}$$

and

$$\lim_{y \to \infty} \left[I_0 \frac{1-\left(\dfrac{y_u}{y}\right)^{10/3}}{1-\left(\dfrac{k}{y}\right)^3} \right] = I_0 \tag{57}$$

Equation (56) and Eq. (57) state that in the upstream part of the river branch, the water profile asymptotically tends to uniform flow depth y_u, whereas downstream it asymptotically tends to horizontal. Therefore, the M1 profile is characterized by a water surface increase towards downstream.

Water surface profile M2

This profile typology occurs when the water depth satisfies the following relationship; therefore, it occurs in zone 2:

$$y_u > y > k \tag{58}$$

From Eq. (53), it is easy to derive that the following analytical condition occurs in the case of M2 profile:

$$\frac{dy}{ds} < 0 \tag{59}$$

Namely, the water depth y is a monotonic decreasing function of the variable s. Therefore, the water depth decreases in the downstream direction. In particular, the extremes of this profile are the uniform water depth y_u and the critical depth k, respectively. Thus, in the upstream part, the water depth tends to the uniform flow depth y_u, whereas the limit for y tending to k is equal to infinity, that is, a vertical asymptotical behavior of the water surface profile occurs downstream, as shown by Eqs. (60) and (61):

$$\lim_{y \to y_u} \left[I_0 \frac{1-\left(\dfrac{y_u}{y}\right)^{10/3}}{1-\left(\dfrac{k}{y}\right)^3} \right] = 0 \tag{60}$$

and

$$\lim_{y \to k} \left[I_0 \frac{1-\left(\dfrac{y_u}{y}\right)^{10/3}}{1-\left(\dfrac{k}{y}\right)^3} \right] = \infty \tag{61}$$

From the two previous equations, it is evident that the water surface profile converges towards the uniform flow depth in the upstream part whereas it disposes vertically in correspondence with the critical water depth level in the downstream part. This is the typical behavior of the profile M2.

Water surface profile M3

This profile typology occurs when the water depth satisfies the following relationship, therefore it occurs in zone 3:

$$y_u > k > y \qquad (62)$$

From Eq. (53), it is easy to derive that the following analytical condition occurs in the case of M3 profile:

$$\frac{dy}{ds} > 0 \qquad (63)$$

Namely, the water depth y is a monotonic increasing function of the variable s. Therefore, the water depth increases towards downstream. In particular, the extremes of this profile are $y=0$ m and $y=k$, respectively. Thus, for y tending to k, the variable dy/ds tends to infinity, resulting in vertical asymptotical behavior of the water surface profile in the downstream part, as shown by Eq. (64):

$$\lim_{y \to k} \left[I_0 \frac{1 - \left(\dfrac{y_u}{y} \right)^{10/3}}{1 - \left(\dfrac{k}{y} \right)^3} \right] = \infty \qquad (64)$$

In the upstream part ($y=0$ m), the water surface profile assumes the typical configuration reported in Fig. 13.

Water Surface Profiles for Steep Bed Slopes

In this paragraph, we will analyze the water profile behavior occurring for steep bed slopes, that is, when the bed slope I_0 is bigger than the critical slope I_c ($I_0 > I_c$). In the case of the steep bed slope, according to the previous deductions, $y_u < k$. Therefore, in this case, the channel branch can be divided in three zones, termed zone 3 (where $0 < y < y_u$), zone 2 (where $y_u < y < k$), and zone 1 (where $y > k$). The three profiles occurring in the mentioned zones are named S1, S2, and S3, respectively. Figure 14 show the diagram sketch of a channel branch including the trends of the different water surface profiles.

Water surface profile S1

This profile typology occurs when the water depth satisfies the following relationship; therefore it occurs in zone 1:

$$y > k > y_u \qquad (65)$$

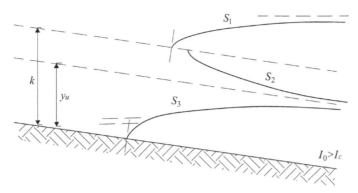

Fig. 14. Water surface profiles for steep bed slopes.

From Eq. (53), it is easy to derive that the following analytical condition occurs in the case of S1 profile:

$$\frac{dy}{ds} > 0 \tag{66}$$

Therefore, y is a monotonic increasing function of s. It means that water depth increases with channel abscissa s (towards downstream). In particular, by considering Eq. (53), it is easy to estimate the values of the variable dy/ds at the limits of the existence interval in zone 1, i.e., $y \to k$ and $y \to \infty$:

$$\lim_{y \to k} \left[I_0 \frac{1 - \left(\dfrac{y_u}{y}\right)^{10/3}}{1 - \left(\dfrac{k}{y}\right)^{3}} \right] = \infty \tag{67}$$

and

$$\lim_{y \to \infty} \left[I_0 \frac{1 - \left(\dfrac{y_u}{y}\right)^{10/3}}{1 - \left(\dfrac{k}{y}\right)^{3}} \right] = I_0 \tag{68}$$

In particular, in the upstream part, the water surface tends to dispose vertically in correspondence with the critical depth level (vertical asymptotical behavior), whereas it tends to the horizontal proceeding downstream.

Water surface profile S2

This profile typology occurs when the water depth satisfies the following relationship; therefore it occurs in zone 2:

$$k > y > y_u \tag{69}$$

From Eq. (53), it is easy to derive that the following analytical condition occurs in the case of S2 profile:

$$\frac{dy}{ds} < 0 \tag{70}$$

Therefore, y is a monotonic decreasing function of s. It means that water depth decreases with channel abscissa s (towards downstream). In particular, by considering Eq. (53), it is easy to estimate the values of the variable dy/ds at the limits of the existence interval in zone 2, that is,

$$\lim_{y \to k} \left[I_0 \frac{1 - \left(\dfrac{y_u}{y}\right)^{10/3}}{1 - \left(\dfrac{k}{y}\right)^3} \right] = \infty \tag{71}$$

and

$$\lim_{y \to y_u} \left[I_0 \frac{1 - \left(\dfrac{y_u}{y}\right)^{10/3}}{1 - \left(\dfrac{k}{y}\right)^3} \right] = 0 \tag{72}$$

In particular, in the upstream part, the water surface tends to dispose vertically in correspondence with the critical depth level (vertical asymptotical behavior), whereas it tends to the uniform flow downstream.

Water surface profile S3

This profile typology occurs when the water depth satisfies the following relationship; therefore it occurs in zone 3:

$$k > y_u > y \tag{73}$$

From Eq. (53), it is easy to derive that the following analytical condition occurs in the case of S3 profile:

$$\frac{dy}{ds} > 0 \tag{74}$$

Therefore, y is a monotonic increasing function of s. It means that hydraulic depth increases with channel abscissa s (towards downstream). In particular, by considering Eq. (53), it is easy to estimate the values of the variable dy/ds at the limits of the existence interval in zone 2, that is:

$$\lim_{y \to y_u} \left[I_0 \frac{1 - \left(\dfrac{y_u}{y}\right)^{10/3}}{1 - \left(\dfrac{k}{y}\right)^3} \right] = 0 \tag{75}$$

Namely, in the downstream part, the water surface tends to the uniform flow depth, whereas it assumes the typical configuration shown in Fig. 14 for y tending to 0 m.

Hydraulic Jump

In the previous paragraphs, water surface profiles relative to GVF conditions were analyzed. Namely, a water flow can gradually transit by the critical condition passing from subcritical to supercritical flow regime. For example, when the bed slope increases, two different profiles can occur: upstream of the slope variation a subcritical flow occurs which gradually transit to supercritical flow in correspondence with the channel slope variation. But the transition from supercritical to subcritical flow does not occur gradually. Namely, it is characterized by a sharp variation of the liquid profile. This is the typical configuration which can occur downstream of a spillway or a partially open sluice gate. The transition from supercritical to subcritical flow condition results in a horizontal axis vortex formation. The water surface appears very irregular and a huge air entrainment process takes place. This hydraulic phenomenon is called hydraulic jump and it is a typical RVF condition. The flow expansion contributes to the transformation of the kinetic energy into potential energy and huge energy dissipation takes place in correspondence of the transition. Namely, the mechanical energy is partially dissipated because of the huge actions exerted by friction and turbulence, resulting in a slight increase of water temperature. Because of its characteristics, hydraulic jump is often adopted to dissipate the excessive flow energy. The upstream section of the hydraulic jump (Section 1) is termed jump toe, whereas Section 2 (downstream of the jump) is characterized by an almost hydrostatic distribution of pressure. The longitudinal distance between Sections 1 and 2 is called hydraulic jump length L_R, whereas the water depths at the sections 1 and 2 (y_1 and y_2, respectively) are termed conjugate depths. Figure 15 shows a diagram sketch of a hydraulic jump.

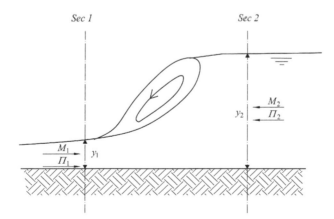

Fig. 15. Diagram sketch of a hydraulic jump.

Basic equations governing the hydraulic jump

To analyze the hydraulic jump phenomenon, the momentum equation is applied, considering the control volume included between the Sections 1 and 2 (i.e., where the conjugate depths occur), and the channel bed horizontal (Pagliara et al. 2008; Pagliara et al. 2009; Pagliara and Palermo 2015). The momentum equation is applied neglecting the friction forces due to walls roughness, that is, walls are assumed to be smooth. In addition, the distributed energy losses are neglected. Finally, the inertial forces are also considered negligible because of the steady flow conditions. Based on these hypotheses, the momentum equation assumes the following expression:

$$\Pi_1 + M_1 = \Pi_2 + M_2 \tag{76}$$

where Π_i is the hydrostatic force and M_i the momentum flux at section i (with i=1 and 2 for Sections 1 and 2, respectively). From a practical point of view, the hydraulic jump is essentially a concentrated energy loss occurring between two close sections; therefore the distributed losses are negligible. The total hydrodynamic force S_i, directed horizontally and acting on each vertical section of the control volume (Sections 1 and 2), can be expressed as follows:

$$S_i = \Pi_i + M_i = \gamma A_i y_{Gi} + \rho \frac{Q^2}{A_i} \tag{77}$$

where A_i is the cross-section area, y_{Gi} the distance of the center of gravity from the water surface, γ the specific weight of water and ρ is the water density. If Eq. (77) is specialized for Sections 1 and 2, then the following expression can be obtained:

$$A_1 y_{G1} + \frac{Q^2}{gA_1} = A_2 y_{G2} + \frac{Q^2}{gA_2} \tag{78}$$

Note that both A and y_G are functions of the water depth y, therefore $S=f(y)$, if Q is constant. Let's analyze the function $S=f(y)$. The water depth may vary between two extremes, that is, y=0 m and $y \rightarrow \infty$. Therefore, by doing limits of the function $S=f(y)$, we obtain:

$$\lim_{y \to 0} \left(\gamma A y_G + \rho \frac{Q^2}{A} \right) = \infty \tag{79}$$

$$\lim_{y \to \infty} \left(\gamma A y_G + \rho \frac{Q^2}{A} \right) = \infty \tag{80}$$

From the two previous equations, it is evident that S tends to infinity if y tends to the domain extremes. This means that S(y) should have a minimum in the y-domain. In order to calculate the value of y for which S is minimum, we can re-write Eq. (77) as follows:

$$S = \gamma \frac{1}{2} \int y^2 \, dB + \frac{\rho Q^2}{A} \tag{81}$$

By differentiating the previous equation, we obtain:

$$\frac{dS}{dy} = \gamma \int y \, dB - \frac{\rho Q^2}{A^2} \frac{dA}{dy} \tag{82}$$

But, it should be noted that:

$$\int y\,dB = A \tag{83}$$

Therefore:

$$\frac{dA}{dy} = B \tag{84}$$

By substituting Eq. (83) and Eq. (84) in Eq. (82), it is easy to derive the following expression:

$$\frac{dS}{dy} = \gamma\,A - \frac{\rho Q^2}{A^2}\,B \tag{85}$$

Therefore, S is minimum when:

$$\gamma A - \frac{\rho Q^2}{A^2}\,B = 0 \tag{86}$$

which can be written as follows:

$$1 - \frac{Q^2}{gA^3}\,B = 0 \tag{87}$$

Equation (87) coincides with Eq. (19), which identifies for critical flow conditions, that is, minimum energy. Therefore, the total hydrodynamic force is minimum for critical flow conditions. Figure 16 shows the plot of Eq. (77) along with the two conjugate water depth y_1 and y_2.

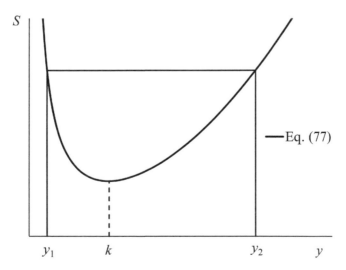

Fig. 16. Plot of Eq. (77).

Conclusions

The present chapter synthetized the basic principles of open-channel hydraulics. The aim was to furnish a synthetic but comprehensive analysis of the open channel flow behavior. Therefore, the main equations governing the flow characteristics were analytically derived and applied for the most common phenomena, such as the hydraulic jump.

Acknowledgements

The two authors equally contributed to the present chapter.

Keywords: Flow energy, hydraulic jump, open channels, uniform flow, water surface profiles.

References

Chow, V.T. 1959. Open-Channel Hydraulics. McGraw-Hill, New York, USA.
Citrini, D. and Noseda, G. 1987. Idraulica. Casa Editrice Ambrosiana, Milano, Italy [in Italian].
De Marchi, G. 1986. Idraulica. Hoepli, Milano, Italy [in Italian].
Henderson, F.M. 1966. Open Channel Flow. The Macmillan Company, New York, USA.
Pagliara, S., Lotti, I. and Palermo, M. 2008. Hydraulic jumps on rough bed of stream rehabilitation structures. Journal of Hydro-Environment Research 2(1): 29–38.
Pagliara, S., Palermo, M. and Carnacina, I. 2009. Scour and hydraulic jump downstream of block ramps in expanding stilling basins. Journal of Hydraulic Research 47(4): 503–511.
Pagliara, S. and Palermo, M. 2015. Hydraulic jumps on rough and smooth beds: aggregate approach for horizontal and averse-sloped beds. Journal of Hydraulic Research 53(2): 243–252.

CHAPTER 2

Hydraulics of Selected Hydraulic Structures

Hubert Chanson[1],* and *Stefan Felder*[2]

INTRODUCTION

Hydraulic structures are human-made systems interacting with surface runoff in urban and rural environments including structures to assist stormwater drainage, flood mitigation, coastal protection, or enhancing and controlling flows in rivers and other water bodies. The structures may be built across a natural stream to divert, control, store, and manage the water flow: for example, a weir across a waterway and its upstream reservoir controlling both upstream and downstream water levels (Figs. 1 and 2). Hydraulic structures can be designed pro-actively to control the water flow motion: for example, a series of drop structures along a mountain river course built to stabilize the river bed by dissipating the flow energy along the drops. The construction of weirs, dams, and hydraulic structures is possibly the oldest and most important civil engineering activity (Schnitter 1994; Levi 1995). Life on our planet is totally dependent upon water and only two species build hydraulic structures: humans and beavers. The latters are called *"the engineers of Nature"* (Dubois and Provencher 2012). Although the date and location of the most ancient hydraulic structures are unknown, some very famous heritage structure includes the Sumerian irrigation canals in Mesopotamia (BC 3,000), the Sadd-El-Kafara dam in Egypt (BC 2,500), the Marib dam and its irrigation canals in Yemen (BC 750), the Dujianyan irrigation system in China (BC 256), and the Vichansao canal and its diversion structure in the Moche Valley, South America (AD 200).

The two key technical challenges in hydraulic structure design are the conveyance of water and dissipation of kinetic energy. Conveyance implies the transport of water, for example, into the spillway of a dam. The conveyance of the structure is closely linked to the intake design, for example the spillway crest, and chute design (Fig. 3).

[1] The University of Queensland, School of Civil Engineering, Brisbane QLD 4072, Australia.
[2] The University of New South Wales, School of Civil and Environmental Engineering, Sydney NSW 2052, Australia.
 E-mail: s.felder@unsw.edu.au
* Corresponding author: h.chanson@uq.edu.au

Its estimate is based upon fundamental fluid dynamic calculations, with a range of proven solutions. Figure 1D illustrates a rounded spillway crest designed to increase the discharge capacity at design flow compared to a broad crest. The dissipation of energy occurs along the chute and at its downstream end. The available energy can be very significant and kinetic energy dissipation must take place safely before the water rejoins the natural river course. With many structures, a major challenge is the magnitude of the rate of energy dissipation at design and non-design flow conditions. The design of the energy dissipators relies upon some sound physical modelling combined with solid prototype experiences (Novak et al. 1996; Vischer and Hager 1998; Chanson 2015).

The rate of energy dissipation at hydraulic structures can be enormous and its design is far from trivial. For example, the Choufu weir (Fig. 1A) discharging 120 m³/s with a 2 m high drop would dissipate turbulent kinetic energy flux per unit time at a rate of 2.4 MW. Many engineers have never been exposed to the complexity of dissipator designs, to the hydrodynamic processes taking place, and to the structural challenges. Often hydraulic structures are tested for design discharges, but smaller discharges can result in instationary flow transitions leading to more complex flows and increased hydrodynamic forces on the structure. Numerous spillways, energy dissipators, and storm waterways failed because of poor engineering design (Hager 1992; Novak et al. 2001). A known issue has been a lack of understanding of basic turbulent dissipation processes and the intrinsic limitations of physical and numerical models (Novak and Cabelka 1981; Chanson 2015). Physical studies are conducted traditionally using a Froude similitude implying drastically smaller laboratory Reynolds numbers than in the corresponding prototype flows. Despite advances, the extrapolation of laboratory results to large size prototype structures must be carefully checked, including the implications in terms of numerical model validation and numerical data quality.

Hydraulics of Small Dams and Weirs

Presentation

Dams and weirs are built across a stream or river to facilitate water storage (Fig. 1). A weir is a structure designed to raise the upstream water level to increase water storage and irrigation capacity, and to enable navigation. During large floods, the water is allowed to pass over the top of the full length of the weir. There are several types of weirs defined by their weir shape and crest length. A particular type of structure is the minimum energy loss (MEL) weir (McKay 1975; Chanson 2003); a MEL weir is designed to minimize the total head loss of overflow, thus inducing zero afflux for the design flow in the case of the structure shown in Fig. 2A. A dam is defined as a large structure built across a valley to store water in the upstream reservoir for flood mitigation, hydroelectricity, or water supply. The upstream water elevation should not overtop the dam wall, because it would lead indeed to dam erosion and possibly destruction. During large rainfalls, large inflows enter the dam reservoir, and a spillway structure must be designed to spill the flood flow beside, below, through, or above the dams, under controlled conditions (USBR 1987; Novak et al. 2001). Most small dams are equipped with an overflow spillway system (Fig. 3). The overflow spillway consists typically of three sections: the crest, the steep chute, and the stilling structure

Fig. 1. Photographs of hydraulic structures and weirs along river courses: (A) Choufu weir on Tama River (Japan); (B) Shih Kang Dam (Taichung, Taiwan); (C) Bucca weir (Bundaberg QLD, Australia); (D) Jordan weir (Gatton QLD, Australia); (E) Goulburn weir (VIC, Australia); (F) Chenchung weir (Pingtung, Taiwan).

at the downstream end. The crest and chute are designed to carry safely the flood flow, while the stilling basin is designed to break down the kinetic energy of the flow before reaching the downstream river channel.

During the design stage, the engineers select the optimum spillway shape for the design flow conditions. Then the safe operation of the spillway must be checked for a range of operating flow conditions ($Q<Q_{des}$) and for emergency situations ($Q>Q_{des}$), where Q is the discharge and Q_{des} is the design discharge.

Fig. 2. Unusual weir designs in Australia: (A) Minimum energy loss weir on the Condamine River (Chinchilla, QLD); (B) Timber crib weir on MacIntyre Brook (Whetstone weir, QLD).

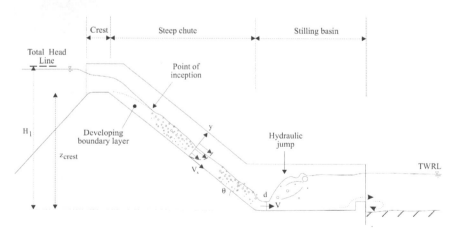

Fig. 3. Cross-sectional sketch of small dam spillway.

Conveyance and Crest Design

For an overflow spillway design, the crest of the spillway is basically designed to maximize the discharge capacity of the structure. In open channels and for a given specific energy, maximum flow rate is achieved for critical flow conditions (Bélanger 1828; Bakhmeteff 1912), and critical flow conditions are observed at the weir crest (Chanson 2006), unless the crest is drowned. For a rectangular spillway crest, the discharge per unit width q may be expressed as:

$$q = C_D \sqrt{g} \left(\frac{2}{3} (H_1 - z_{crest}) \right)^{3/2} \text{ with critical flow conditions} \quad (1)$$

where C_D is the dimensionless discharge coefficient, g is the gravity acceleration, H_1 is the upstream total head and z_{crest} is the spillway crest height (Fig. 3).

When the crest is broad enough ($L_{crest}/(H_1-z_{crest})$>1.5 to 3), the streamlines are parallel to the crest invert and the pressure distribution above the crest is hydrostatic (Hager and Schwalt 1994; Felder and Chanson 2012). Despite some scatter, experimental data for broad-crested weir with rounded corner yielded a discharge coefficient C_D of about unity:

$$C_D = 0.934 + 0.143 \frac{H_1 - z_{crest}}{L_{crest}} \text{ broad-crested weir for } H_1/L_{crest}>0.1 \quad (2)$$

where L_{crest} is the broad-crest length. Equation (2) is compared with data in Fig. 4A. For H_1/L_{crest}<0.1, the flow is near-critical along the crest and some stable undulations of the free-surface are present (Fig. 4A).

Rounded crest designs can achieve a larger discharge for the same head above crest than a broad-crested weir (C_D>1). A simple rounded design is the circular crest of radius R. The analysis of experimental data with partially-developed inflow conditions yielded (Chanson and Montes 1998):

$$C_D = 1.13 \left(\frac{H_1 - z_{crest}}{R} \right)^{0.18} \text{ circular crested weir } (0.5<(H_1-z_{crest})/R<3.5) \quad (3)$$

An efficient design is the ogee crest, for which the pressures at the face of the crest invert are atmospheric at design flow conditions. The profile is basically the trajectory of the underside of a free-falling jet downstream of a two-dimensional sharp crested weir for the design discharge Q_{des} and upstream head H_{des} (Creager 1917; Montes 1992). Figure 4B illustrates typical ogee crest profiles in dimensionless terms with coordinates $X/(H_{des}-z_{crest})$ and $Y/(H_{des}-z_{crest})$ representing the dimensionless locations in x- and y-directions respectively. For the design flow, the discharge coefficient $(C_D)_{des}$ is primarily a function of the crest shape (USBR 1987; Chanson 2004). For an ogee with vertical upstream wall, Fig. 4B presents typical values of $(C_D)_{des}$ as a function of the relative design head above crest $(H_{des}-z_{crest})/z_{crest}$. When the discharge differs from the design flow, the relative discharge coefficient $C_D/(C_D)_{des}$ becomes a function of the relative upstream total head $(H_1-z_{crest})/(H_{des}-z_{crest})$ (Fig. 4B). For H_1<H_{des}, Q<Q_{des} and the pressure on the crest invert is larger than atmospheric. For very low flows (i.e.,

$H_1 \ll H_{des}$), the discharge coefficient tends to unity: that is, the broad-crested weir case. When $H_1 > H_{des}$, the pressures on the crest are less than atmospheric and the discharge coefficient C_D is larger than the design discharge coefficient $(C_D)_{des}$. Such flow conditions are exceptional and cavitation might occur on the crest profile.

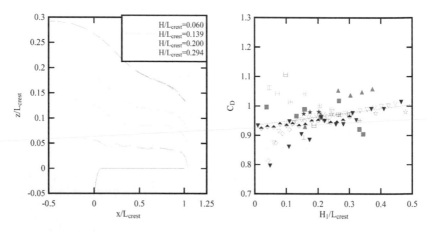

(A) Broad crested weir with upstream rounded corner: dimensionless free-surface profiles (Data: Felder and Chanson 2012) and dimensionless discharge coefficient data (Data: Vierhout 1973; Bazin 1896; Gonzalez and Chanson 2007; Felder and Chanson 2012; Zhang and Chanson 2015; Felder 2015).

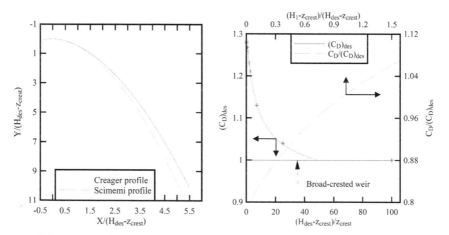

(B) Ogee crest with vertical upstream wall: typical crest profiles and discharge coefficient.

Fig. 4. Spillway crest overflow performances.

Downstream of the crest, the fluid is accelerated by gravity along the steep chute (Fig. 3). A turbulent boundary layer is generated by bottom friction at the upstream end and the boundary layer thickness increases with increasing distance along the chute. When the outer edge of the boundary layer reaches the free-surface, the flow becomes fully-developed and free-surface aeration takes place over a very small distance in rapidly varied flows (Fig. 3). Figures 1D and 7A illustrate the inception point of free-surface aeration, clearly visible with the apparition of white waters. Downstream of the inception point in the fully-developed flow region, the flow is gradually-varied until it reaches equilibrium. The gradually-varied flow properties may be calculated based upon the backwater equation, while the uniform equilibrium flow conditions (normal flow conditions) may be estimated based upon momentum considerations (Henderson 1966; Chanson 2004).

On a steep chute, both the flow acceleration and boundary layer development affect the flow properties. The complete flow calculations can be tedious. Combining physical observations with developing and uniform equilibrium flow calculations, a general trend may be derived for preliminary designs (Bradley and Peterka 1957; Henderson 1966; Chanson 1999, 2004). In absence of losses, the maximum ideal velocity at the chute's downstream chute would be:

$$V_{max} = \sqrt{2g(H_1 - d\cos\theta)} \tag{4}$$

where d is the downstream flow depth ($d = q/V_{max}$), q is the discharge per unit width and θ is the chute slope (Fig. 3). In reality the flow velocity V is smaller than the theoretical velocity V_{max} because of energy losses down the chute. Figure 5 summarizes the dimensionless flow velocity at the end of the steep chute V/V_{max} as a function of the dimensionless upstream head H_1/d_c where d_c is the critical depth ($d_c = (q^2/g)^{1/3}$). Both developing and uniform equilibrium flow calculations are presented for smooth and stepped chutes. Prototype smooth chute data and laboratory stepped chute data are included for comparison using an average Darcy friction factor of f=0.03 for the smooth chutes and of f=0.2 for the stepped chute. The results are valid for smooth and stepped spillways (concrete chutes) with slopes ranging from 18° to 55°.

Comparing stepped and smooth chutes, larger energy dissipation rates are systematically observed along a stepped spillway because of a predominance of form drag compared to friction drag on the smooth invert. Hence the residual energy at the

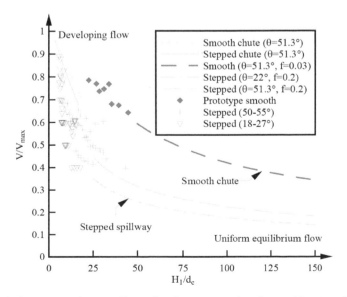

Fig. 5. Velocity at a steep chute toe. Comparison between smooth and stepped invert configurations.

downstream end of a stepped chute will be smaller and the size of the downstream stilling basin can be reduced (e.g., Felder and Chanson 2013). The form drag results in larger mean bottom shear stress, implying larger hydrodynamic loads on the steps than on a smooth invert.

Energy dissipation

At the downstream end of the spillway system, the excess in kinetic energy must be dissipated before the flow re-joins the natural stream. Energy dissipation may be achieved by a hydraulic jump stilling basin downstream of the steep chute, a high velocity water jet taking off from a flip bucket and impinging into a downstream plunge pool, and a plunging jet pool in which the spillway flow impinges and the kinetic energy is dissipated in turbulent recirculation (Figs. 1 and 2) (USBR 1965; Novak et al. 2001; Chanson 2015). The design of a stepped chute may assist also in energy dissipation (Figs. 1C and 2B).

A hydraulic jump stilling basin is the common type of dissipators for small dams and weirs. Most kinetic energy is dissipated in the hydraulic jump, sometimes assisted by appurtenances (step, baffles) to increase the turbulence. A hydraulic jump is the rapid and sudden transition from a supercritical to subcritical flow. The jump is an extremely turbulent process, and the large-scale turbulence region is typically called the roller. The downstream flow depth d_{conj} (conjugate depth) and energy loss DH in the jump may be deduced from the momentum principle. For a flat horizontal rectangular channel:

$$\frac{d_{conj}}{d} = \frac{1}{2}\left(\sqrt{1 + 8Fr^2} - 1\right) \tag{5}$$

$$\frac{\Delta H}{d} = \frac{\left(\sqrt{1 + 8Fr^2} - 3\right)^3}{16\left(\sqrt{1 + 8Fr^2} - 1\right)} \tag{6}$$

where $Fr = V/(gd)^{1/2}$ is the inflow Froude number, and d and V are the inflow depth and velocity (Fig. 3).

Hydraulic jump flows may exhibit different patterns depending upon the upstream Froude number spanning from undular jumps for Froude numbers close to and slightly above unity, to steady and strong jumps for large Froude numbers. In practice, it is recommended to design energy dissipators for $4.5 < Fr < 9$–10, although some standard stilling basins may be devised for lower inflow Froude number conditions. An important design parameter is the roller length L_r, which may be estimated from experimental observations:

$$\frac{L_r}{d} = 6 \,(Fr - 1) \tag{7}$$

Equation (7) is valid for rectangular horizontal channels and $2 < Fr < 10$.

The hydraulic design of the stilling basin must ensure the safe dissipation of the kinetic energy of the flow and minimize the size of the stilling structure. In practice, energy dissipation by the hydraulic jump may be assisted by elements (e.g., baffle blocks, sill) placed on the stilling basin apron (Fig. 6). In all cases, the length of a hydraulic

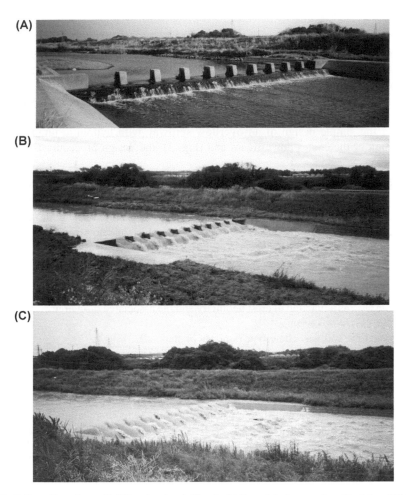

Fig. 6. Operation of a small stilling structure in Toyohashi (Japan) during very-low (Top), low (Middle), and medium flows (Bottom).

jump stilling basin must be greater than the roller length for all flow conditions. Basic aperture of stilling basins may include drop, sill, baffle blocks, sudden expansion (Hager 1992). A number of standard stilling basin designs were developed in the 1940s to 1960s (Chanson and Carvalho 2015). Each basin design was tested in models and prototypes over a considerable range of flow conditions. Their performances are well known, and they can be selected without further model studies within their design specifications. In practice, design engineers must ensure that the energy dissipation takes place in the spillway system and that the stilling basin can operate safely for a wide range of flow conditions. Damage (scouring, abrasion, cavitation) to the basin and to the downstream natural bed may occur for a number of reasons as discussed by USBR (1965), Chanson (1999), and Novak et al. (2001), including insufficient length of and dissipation structures in the basin or the design for an insufficient range of transitional flow rates as well as flows exceeding the design discharge.

Design procedure

The construction of a small dam and weir across a river has impacts on both the upstream and downstream flow conditions. The weir crest elevation must be selected accurately to provide the required water storage and upstream water level rise, while the spillway system must operate safely for a wide range of flow conditions including tailwater levels and non-design flow rates.

For the simple design of an overflow spillway with hydraulic jump energy dissipation at the chutes downstream end (Fig. 3), the design procedure is (USBR 1965; Chanson 1999):

a) Select the spillway crest elevation z_{crest} based upon reservoir bathymetry, topography, and required storage level.

b) Choose the spillway crest width B based upon site geometry and hydrology.

c) Determine the design discharge Q_{des} from risk analysis and flood routing. The peak spillway discharge is deduced from the combined analysis of storage capacity, and inflow and outflow hydrographs. Potential increases in rainfall over the life of the structure should be taken into account.

d) Calculate the upstream head above spillway crest $(H_{des} - z_{crest})$ for the design flow rate Q_{des}, as a function of crest geometry and the associated discharge coefficients (Eq. (1)).

e) Select the spillway chute type (smooth or stepped).

f) Choose the chute toe elevation. The stilling basin level may differ from the natural bed level (i.e., tailwater bed level).

g) For the design flow conditions, calculate the flow properties d and V at the end of the chute toe, the conjugate depth for the hydraulic jump, and the roller length L_r. Note that the apron length must be greater than the jump length.

h) Compare the jump height rating level (JHRL) to the natural downstream water level (i.e., tailwater rating level TWRL). If the JHRL does not match the TWRL, the apron elevation, spillway width, design discharge, chute type, and crest elevation must be altered, and the process becomes iterative.

The tailwater rating level TWRL is the natural free-surface elevation in the downstream flood plain (Fig. 3). The downstream channel typically operates in a subcritical flow regime controlled by the downstream/tailwater flow conditions. The jump height rating level JHRL is the free-surface elevation downstream of the hydraulic jump stilling structure. For a horizontal apron, the JHRL is deduced simply from the apron elevation and the conjugate depth. If the apron has an end sill or drop, the JHRL is deduced from the Bernoulli equation, assuming that the hydraulic jump takes place upstream of the sill/drop, and that no energy loss takes place at the sill/drop.

During design stages, engineers are required to compute both the JHRL and the TWRL for all flow rates. The location of the hydraulic jump is determined by the conjugate flow conditions. The upstream depth is the supercritical depth at the steep chute toe and the downstream depth is deduced from the tailwater conditions. The upstream and downstream depths must also satisfy the momentum equation, e.g., Eq. (5) for a horizontal stilling basin in absence of baffles. The results must be compared with the variations in tailwater level (TWRL) for the whole range of discharges.

For a hydraulic jump stilling basin, it is extremely important to consider the following points. The stilling basin is designed for the design flow conditions. For discharges larger than the design discharge, it may be acceptable to tolerate some erosion and damage, but the safety of the dam must be ensured. For discharges less than the design discharge, the energy dissipation must be controlled completely, it must occur in the stilling basin, and there must be no maintenance issue. Figures 6 and 7 show prototype weirs in operation for different flow rates, highlighting a range of tailwater effects on the hydraulic jump. In Fig. 7C, the weir became drowned for very large discharges.

(A)

(B)

(C)

Fig. 7. Operation of Mount Crosby weir on the Brisbane River (QLD, Australia) during low flow (Top), medium flow (Middle), and large flows (Bottom).

Culvert Hydraulics

A culvert is a covered channel designed to pass flood waters, drainage flows, and natural streams through earthfill and rockfill structures (e.g., roadway, railroad). The design can vary from a simple geometry (standard culvert) to a hydraulically-smooth shape (MEL culvert) (Figs. 8 and 9). A culvert consists of three sections: the intake or inlet, the barrel or throat, and the diffuser or outlet. The cross-sectional shape of the barrel may be circular (pipe) or rectangular (box and multi-cell box); a culvert may be designed as a single cell or multiple cell structure.

The hydraulic characteristics of a culvert are the design discharge, the upstream total head and the maximum acceptable head loss from inlet to outlet. The design discharge and upstream flood level are deduced from the hydrological investigation of the site in relation to the purpose of the culvert. Head losses must be minimized to reduce upstream flooding. From a hydraulic engineering perspective, a dominant feature is whether the barrel runs full or not.

Fig. 8. Box culverts. (A) Muscat, Oman on 31 Oct. 2010 afternoon; (B) Culvert outlet below Ridgewood St, Algester (QLD, Australia) in August 1999; (C) Culvert along Gin House Creek, Carrara, Gold Coast (QLD, Australia) on 5 Dec. 2007.

(A)

(B)

Fig. 9. Minimum energy loss culvert along Norman Creek beneath Ridge Street, Brisbane (Australia). (A) Inlet operation on 7 Nov. 2004 for $Q \approx 80$ to 100 m³/s; (B) Outlet on 13 May 2002.

The hydraulic design of a culvert is basically an optimum compromise between discharge capacity, head loss, and construction costs. While the key objective is to keep the cost of the culvert to a minimum, some consideration must be taken to avoid upstream afflux and flooding by keeping the head loss small and to avoid scour downstream of the culvert outlet if a hydraulic jump might take place by placement of some scour protection. The minimization of head losses and increase in flow performances can be assisted through streamlining of the culvert inlet and outlet including streamlined wings and fans (Figs. 8 and 9). Most culverts are designed to operate as open channel systems, with critical flow conditions occurring in the barrel in order to maximize the discharge per unit width and to reduce the barrel cross-section. Figure 9A shows a typical operation, for a discharge less than the design discharge.

Hydraulics of box culverts

For standard box culverts, the culvert flow may exhibit various flow patterns depending upon the discharge, the upstream head above the inlet invert (H_1-z_{inlet}), the tailwater depth d_{tw}, the bed slope S_o, and the barrel's internal height D (Hee 1969; Chanson 1999). Free-surface inlet flows take place typically for (H_1-z_{inlet})/D<1.2, and submerged inlet operation occurs for (H_1-z_{inlet})/D>1.2. In each case, different flow patterns may occur depending upon the hydraulic control location: that is, inlet control or outlet control (Hee 1969). The transition from free-surface to submerged inlet and the transition from free-surface flows along the barrel to fully drowned flow conditions can be associated with three-dimensional flows at the inlet and along the barrel and instationarities linked to changes from free-surface to pressurized flow conditions. The discharge capacity of the barrel is primarily related to the flow pattern: free-surface inlet flow, submerged entrance, or drowned barrel. When a free-surface flow occurs in the barrel, the discharge is set only by the entry conditions. When the entrance is submerged, the discharge is determined as an orifice flow using experimentally determined discharge coefficients. With fully drowned culverts, the discharge is determined by the culvert's flow resistance. Nomographs are also commonly used to estimate the discharge characteristics (USBR 1987; Concrete Pipe Association of Australasia 1991; Chanson 2004).

For standard culverts, the design procedure can be divided into two parts (Herr and Bossy 1965). First a system analysis must be carried out to ascertain the culvert purposes, design data, constraints. This first stage leads to the estimate of the design flow Q_{des} and the design upstream total head H_{des}. During the second stage, the barrel size is selected by an iterative procedure, in which both inlet control and outlet control calculations are conducted. At the end, the optimum size is the smallest barrel size allowing for inlet control operation. The construction cost may be optimized using a multi-cell culvert of precast circular or rectangular box elements.

Hydraulics of minimum energy loss (MEL) culverts

A minimum energy loss (MEL) culvert is a structure designed with the concept of minimum head loss. In the approach channel, the flow is smoothly contracted through a streamlined inlet into the barrel and then it is expanded in a streamlined outlet before being released into the downstream natural flood plain (Figs. 9 and 10B) (Apelt 1983, 1994; Chanson 1999). The inlet and outlet must both be streamlined to avoid major form losses. The barrel invert is sometimes lowered to increase the discharge capacity since:

$$\frac{Q}{B_{min}} = \sqrt{g}\left(\frac{2}{3}(H_1 - z_{inlet} - \Delta z)\right)^{3/2} \tag{8}$$

where Dz is the barrel invert elevation below the natural ground level (Fig. 10B). The inlet and outlet must both be streamlined to avoid major form losses. The barrel invert is sometimes lowervides less energy loss of the same discharge Q and barrel width B_{min}. An alternative design includes a narrower barrel width for the same discharge and head loss. Successful prototype experiences showed that there are a wide range of design options (McKay 1970, 1978; Cottman and McKay 1990; Chanson 2003, 2007).

The basic design concepts of MEL culverts are: (a) streamlining and (b) critical flow conditions from the inlet lip to the outlet lip at design flow conditions (Apelt 1983). The intake must be designed with a smooth contraction into the barrel and the outlet must be shaped as a smooth expansion back to the natural channel: that is, the flow streamlines must follow very smooth curves and no separation be observed. Minimum energy loss culverts are designed to achieve critical flow conditions in the entire waterway: that is, in the inlet, at the barrel and in the outlet (Fig. 10B).

Professor C.J. Apelt devised a simple design method to estimate the basic characteristics of a MEL culvert (Apelt 1983).

1) Decide the design discharge Q_{des} and the associated total head line (THL) based upon the flow conditions upstream of the culvert.

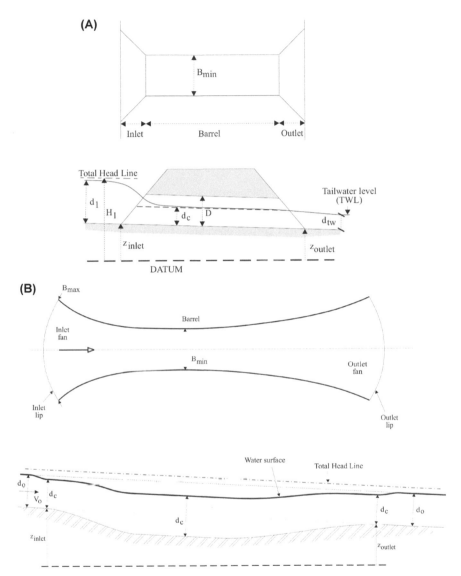

Fig. 10. Definition sketch of culvert hydraulics. (A) Box culvert; (B) Minimum energy loss (MEL) culvert.

2) First neglect energy losses
   2.1) Calculate the waterway characteristics in the barrel for critical flow conditions.
   2.2) Calculate the inlet and outlet lip width B_{max} assuming critical flow conditions and natural bed elevation. The lip width is an equipotential and must be measured along a smooth line normal to the streamlines (Fig. 10B).
3) Decide the shapes of the inlet and outlet.

4) Calculate the geometry of the inlet and outlet to satisfy critical flow conditions everywhere. The contour lines of the inlet and outlet are defined each by their bed elevation to satisfy critical flow conditions, the corresponding width B measured along a smooth line normal to the streamlines, and the longitudinal distance from the lip.
5) Then include the energy losses: namely friction losses along the culvert since the form losses are minimized.
    5.1) Adjust the bed profile of the culvert system to take into account the energy losses. For a long barrel, the barrel slope is selected to be the critical slope.
6) Check the performances of the MEL culvert for non-design conditions for $Q<Q_{des}$ and $Q>Q_{des}$.

The above method gives a preliminary design and full calculations are required later to predict accurately the free-surface profile, complemented by physical modelling (Apelt 1983; Chanson 1999). A correct operation of MEL waterways and culverts requires a proper design. Separation of flow in the inlet and in the outlet must be avoided and head losses must be accurately predicted. Since MEL culverts are designed for critical flow along the entire culvert structure from inlet lip to outlet lip, free-surface undulations may occur, typically in the culvert barrel, and the sidewall of the culvert must be sufficiently high.

Altogether the MEL design technique allows a drastic reduction in the upstream flooding associated with lower costs. The successful operation of MEL culverts for more than 40 years demonstrates the design soundness, while highlighting the importance of the hydraulic expertise of the design engineers (Apelt 1983; Chanson 2007).

Dam Failure and Dam Break Wave

Failures of dams, weirs, and reservoirs during the 19th and 20th centuries led to research into dam break waves. During the second half of the 20th century, two major failures were the Malpasset dam (France) break in 1959 and the overtopping of the Vajont dam (Italy) in 1963 (Fig. 11). Figure 11A shows the remains of the Malpasset dam; on 2 December 1959, the dam collapsed completely and more than 300 people died in the catastrophe.

The propagation of a dam break wave can be predicted analytically for a number of well-defined boundary conditions. For an ideal dam break wave over a dry rectangular channel, the method of characteristics yields an analytical solution first proposed by Ritter (1892):

$$U = 2\sqrt{g\,d_o} \tag{9}$$

$$\frac{x}{t} = 2\sqrt{g\,d_o} - 3\sqrt{g\,d} \tag{10}$$

$$V = \frac{2}{3}\sqrt{g\,d_o}\left(1 + \frac{x}{t\sqrt{g\,d_o}}\right) \tag{11}$$

where U is the dam break wave celerity, d_o is the initial reservoir height, d and V are the water depth and velocity at a location x at time t (Fig. 12).

(A)

(B)

Fig. 11. Dam failures. (A) Malpasset dam, Fréjus, France in September 2004 (Courtesy of Sylvia Briechle); (B) Farm dam failure near Biggenden (QLD, Australia) on 5 March 2013.

The dam break flow with friction may be analyzed as an ideal-fluid flow region led by a friction-dominated tip zone (Fig. 12). Whitham (1955) introduced this conceptual approach, and a complete solution was presented by Chanson (2009). In the tip region ($x_1 < x < x_s$), the flow velocity is about the wave front celerity U. When the friction is dominant and the acceleration and inertial terms are small, the shape of the wave front ($x_1 < x < x_s$) is given by:

$$d = \sqrt{\frac{f}{4} \frac{U^2}{g} (x_s - x)} \text{ for } x_1 < x < x_s \tag{12}$$

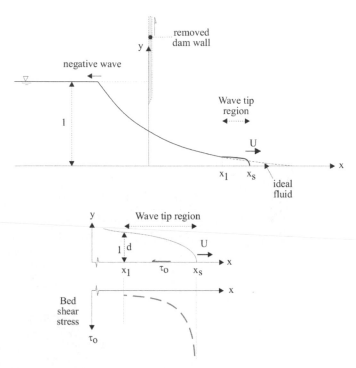

Fig. 12. Definition sketch of a dam break wave and details of the wave tip region.

where the Darcy-Weisbach friction factor f is assumed constant. For x<x_1, the free-surface profile follows Eq. (10). At the location x=x_1, the transition between ideal fluid and tip regions, the depth and velocity (d_1, V_1) must be continuous and satisfy:

$$d_1 = \frac{d_o}{9}\left(2 - \frac{x_1}{t\sqrt{gd_o}}\right)^2 = \sqrt{\frac{f}{4}\frac{U^2}{g}(x_s - x_1)}$$ (13)

$$V_1 = U = \frac{2}{3}\left(\sqrt{gd_o} + \frac{x_1}{t}\right)$$ (14)

Further the conservation of mass yields an exact solution in terms of the wave front celerity

$$t = \frac{8}{3f}\frac{\left(1 - \frac{U}{2\sqrt{gd_o}}\right)^3}{\frac{U^2}{\sqrt{gd_o}^3}}$$ (15)

The wave front location equals

$$x_s = \left(\frac{3}{2}\frac{U}{\sqrt{gd_o}} - 1\right)t\sqrt{gd_o} + \frac{4gd_o^2}{fU^2}\left(1 - \frac{U}{2\sqrt{gd_o}}\right)^4$$ (16)

Equation (15) gives the wave front celerity U at a time t. Equation (16) expresses the wave front location x_s as a function of the wave front celerity U. Equations (10) and (12) give the entire free-surface profile.

The theoretical solution is based upon a few key assumptions. Comparisons between the present solutions and experimental results were successful for a fairly wide range of experimental conditions obtained independently; such comparisons constitute a solid validation of the proposed theory (Chanson 2009). It is acknowledged that the present solution is limited to semi-infinite reservoir, rectangular channel, and quasi-instantaneous dam break. The latter approximation is often reasonable for concrete dam failure (Fig. 11A) but it is not applicable to many other applications, including an embankment breach (Fig. 11B).

Conclusion

For millennia, hydraulic structures have enabled the establishment of human settlements and the development of societies. Man-made hydraulic structures play an important role in mitigating, controlling, storing, and diverting waters in water supply, irrigation, drainage, and stormwater systems, as well as in and along rivers, natural streams, and artificial channels. The design of hydraulic structures must be based upon a sound knowledge of the hydrological and topographical conditions of the catchment to provide design flow estimates and upon advanced engineering expertise in the hydrodynamics of flow processes upstream, downstream, and along the hydraulic structure. For a range of flow rates smaller than or equal to the design flow rate, the structure must allow for conveyance and energy dissipation performances without any damage to the structure and its surroundings. For flow rates in excess of the design flow rate, the structural integrity must be guaranteed.

This chapter provides a brief overview of the engineering design of several hydraulic structures including weirs, small dams, and culverts. For a typical small dam, a key challenge is the design of a structure capable to pass a range of flow conditions without compromising the safety of the structure and the surrounding environment. A similar approach is essential for the design of culvert structures, including box and MEL culverts. Key considerations further encompass the safe passage of flood waters through embankments and associated energy dissipation, as well as a small afflux and minimum costs. Sound engineering design is typically accompanied by thorough physical modelling and engineering design expertise. Despite some advance in numerical modelling, a sound approach relies upon the expert knowledge and solid understanding of the flow physics by the design engineers.

The last section deals with an extreme scenario: the dam break. A simple analytical solution for an instantaneous dam break wave is derived from the method of characteristics. The method provides a simple explicit solution to the dam break wave problem that is easily understood by students, researchers, and professionals, and may be used in real-time by emergency services to estimate the dam break wave with bed friction.

Finally the design of hydraulic structures relies upon high level technical expertise and first-hand experience in hydrodynamics and hydraulic engineering, particularly in terms of conveyance and energy dissipation. The operational challenges are also numerous and they require a broad and solid technical experience and expertise. Such technical challenges are not always well understood and are often understated.

Acknowledgements

This research is supported by the Australian Research Council (Grants DP0878922 & DP120100481).

Keywords: Hydraulic engineering, weirs, small dams, culverts, dam break wave, conveyance, energy dissipation, minimum energy loss, hydraulic structures.

References

Apelt, C.J. 1983. Hydraulics of minimum energy culverts and bridge waterways. Australian Civil Engineering Transactions, I.E. Aust., Vol. CE25(2): 89–95.
Apelt, C.J. 1994. The Minimum Energy Loss Culvert. Videocassette VHS colour, Dept. of Civil Eng., University of Queensland, Australia, 18 minutes.
Bakhmeteff, B.A. 1912. O Neravnomernom Dwijenii Jidkosti v Otkrytom Rusle. (Varied Flow in Open Channel.) St Petersburg, Russia, 232 pages (in Russian).
Bazin, H. 1896. Expériences Nouvelles sur l'Ecoulement par Déversoir. (Recent Experiments on the Flow of Water over Weirs.) Mémoires et Documents, Annales des Ponts et Chaussées (Paris, France) Sér. 7, 12, 2nd Sem., pp. 645–731 (in French).
Bélanger, J.B. 1828. Essai sur la Solution Numérique de quelques Problèmes Relatifs au Mouvement Permanent des Eaux Courantes. (Essay on the Numerical Solution of Some Problems relative to Steady Flow of Water.) Carilian-Goeury, Paris, France, 38 pages & 5 tables (in French).
Bradley, J.N. and Peterka, A.J. 1957. Hydraulic design of stilling basins: short stilling basin for canal structures, small outlet works and small spillways (Basin III). Journal of Hydraulic Division, Proceedings ASCE, Vol. 83, No. HY05, Paper 1403, 22 pages.
Chanson, H. and Montes, J.S. 1998. Overflow characteristics of circular weirs: effect of inflow conditions. Journal of Irrigation and Drainage Engineering, ASCE 124(3): 152–162.
Chanson, H. 1999. The Hydraulics of Open Channel Flow: An Introduction. Butterworth-Heinemann, 1st edition, London, UK, 512 pp.
Chanson, H. 2000. Introducing originality and innovation in engineering teaching: the hydraulic design of culverts. European Journal of Engineering Education 25(4): 377–391.
Chanson, H. 2003. Minimum Energy Loss Structures in Australia: Historical Development and Experience. Proceedings of 12th National Engineering Heritage Conference, I.E. Aust., Toowoomba Qld, Australia, N. Sheridan Editor, pp. 22–28.
Chanson, H. 2004. The Hydraulics of Open Channel Flow: An Introduction. Butterworth-Heinemann, 2nd edition, Oxford, UK, 630 pp.
Chanson, H. 2007. Hydraulic performances of minimum energy loss culverts in Australia. Journal of Performances of Constructed Facilities, ASCE 21(4): 264–272 (doi:10.1061/(ASCE)0887–3828(2007)21: 4(264)).
Chanson, H. 2009. Application of the method of characteristics to the dam break wave problem. Journal of Hydraulic Research, IAHR 47(1): 41–49 (doi:10.3826/jhr.2009.2865).
Chanson, H. 2015. Energy Dissipation in Hydraulic Structures. IAHR Monograph, CRC Press, Taylor & Francis Group, Leiden, The Netherlands, 168 pp.
Chanson, H. and Carvalho, R. 2015. Hydraulic jumps and stilling basins. pp. 65–104. In: Chanson, H. (ed.). Energy Dissipation in Hydraulic Structures. IAHR Monograph, CRC Press, Taylor & Francis Group, Leiden, The Netherlands.
Concrete Pipe Association of Australasia. 1991. Hydraulics of Precast Concrete Conduits. Jenkin Buxton Printers, Australia, 3rd edition, 72 pp.
Cottman, N.H. and McKay, G.R. 1990. Bridges and Culverts Reduced in Size and Cost by Use of Critical Flow Conditions. Proceedings of the Institution of Civil Engineers, London, Part 1, 88: 421–437. Discussion: 1992, 90: 643–645.
Creager, W.P. 1917. Engineering of Masonry Dams. John Wiley & Sons, New York, USA.
Dubois, J.M.M. and Provencher, L. 2012. Beaver dams and ponds. pp. 110–113. In: Bengtsson, L., Herschy, R.W. and Fairbridge, R.W. (eds.). Encyclopedia of Lakes and Reservoirs. Springer, Encyclopedia of Earth Sciences Series.

Felder, S. and Chanson, H. 2012. Free-surface profiles, velocity and pressure distributions on a broad-crested weir: a physical study. Journal of Irrigation and Drainage Engineering, ASCE 138(12): 1068–1074 (doi:10.1061/(ASCE)IR.1943-4774.0000515).

Felder, S. and Chanson, H. 2013. Aeration, flow instabilities, and residual energy on pooled stepped spillways of embankment dams. Journal of Irrigation and Drainage Engineering 139(10): 880–887, http://dx.doi.org/10.1061/(ASCE)IR.1943-4774.0000627.

Felder, S. 2015. Pressure, velocity and boundary layer distributions on an embankment weir. Proc. 36th IAHR World Congress, The Hague, The Netherlands, 11 pp.

Gonzalez, C.A. and Chanson, H. 2007. Experimental measurements of velocity and pressure distribution on a large broad-crested weir. Flow Measurement and Instrumentation 18(3-4): 107–113 (doi:10.1016/j.flowmeasinst.2007.05.005).

Hager, W.H. 1992. Energy Dissipators and Hydraulic Jump. Kluwer Academic Publ., Water Science and Technology Library, Vol. 8, Dordrecht, The Netherlands, 288 pp.

Hager, W.H. and Schwalt, M. 1994. Broad-crested weir. Journal of Irrigation and Drainage Engineering, ASCE 120(1): 13–26. Discussion. 12(2): 222–226.

Hee, M. 1969. Hydraulics of culvert design including constant energy concept. Proc. 20th Conf. of Local Authority Engineers, Dept. of Local Govt, Queensland, Australia, paper 9, pp. 1–27.

Henderson, F.M. 1966. Open Channel Flow. MacMillan Company, New York, USA.

Herr, L.A. and Bossy, H.G. 1965. Capacity Charts for the Hydraulic Design of Highway Culverts. Hydraulic Engineering Circular, US Dept. of Transportation, Federal Highway Admin., HEC No. 10, March.

Levi, E. 1995. The Science of Water. The Foundation of Modern Hydraulics. ASCE Press, New York, USA, 649 pp.

McKay, G.R. 1970. Pavement drainage. Proc. 5th Aust. Road Res. Board Conf., Vol. 5, Part 4, pp. 305–326.

McKay, G.R. 1975. Appraisal of constant energy structures. Queensland Division Technical Papers, IEAust. 16(6): 2–15.

McKay, G.R. 1978. Design principles of Minimum Energy Waterways. pp. 1–39. In: Porter, K.F. (ed.). Proc. Workshop on Minimum Energy Design of Culvert and Bridge Waterways, Australian Road Research Board, Melbourne, Australia, Session 1.

Montes, J.S. 1992. Curvature Analysis of Spillway Profiles. Proc. 11th Australasian Fluid Mechanics Conference AFMC, Vol. 2, Paper 7E-7, Hobart, Australia, pp. 941–944.

Novak, P. and Cabelka, J. 1981. Models in Hydraulic Engineering. Physical Principles and Design Applications. Pitman Publ., London, UK.

Novak, P., Moffat, A.I.B., Nalluri, C. and Narayanan, R. 1996. Hydraulic Structures. E & FN Spon, London, UK, 2nd edition, 599 pp.

Novak, P., Moffat, A.I.B., Nalluri, C. and Narayanan, R. 2001. Hydraulic Structures. Spon Press, London, UK, 3rd edition, 666 pp.

Ritter, A. 1892. Die Fortpflanzung von Wasserwellen. Zeitschrift Verein Deutscher Ingenieure, Vol. 36, No. 2, 33, 13 Aug., 947–954 (in German).

Schnitter, N.J. 1994. A History of Dams: The Useful Pyramids. Balkema Publ., Rotterdam, The Netherlands.

USBR. 1965. Design of Small Dams. Bureau of Reclamation, US Department of the Interior, Denver CO, USA, 1st edition, 3rd printing.

USBR. 1987. Design of Small Dams. Bureau of Reclamation, US Department of the Interior, Denver CO, USA, 3rd edition.

Vierhout, M.M. 1973. On the boundary layer development in rounded broad-crested weirs with a rectangular control section. Report No. 3, Laboratory of Hydraulics and Catchment Hydrology, Agricultural University, Wageningen, The Netherlands, 74 pages.

Vischer, D. and Hager, W.H. 1998. Dam Hydraulics. John Wiley, Chichester, UK, 316 pp.

Whitham, G.B. 1955. The effects of hydraulic resistance in the dam-break problem. Proceedings of Royal Society of London, Series A 227: 399–407.

Zhang, G. and Chanson, H. 2015. Hydraulics of the developing flow region of stepped cascades: an experimental investigation. Hydraulic Model Report No. CH97/15, School of Civil Engineering, The University of Queensland, Brisbane, Australia, 76 pp.

Bibliography

Chanson, H. 2015. Introduction: energy dissipators in hydraulic structures. pp. 1–9. *In*: Chanson, H. (ed.). Energy Dissipation in Hydraulic Structures. IAHR Monograph, CRC Press, Taylor & Francis Group, Leiden, The Netherlands.

Chanson, H. 2015. Energy dissipation: concluding remarks. pp. 159–165. *In*: Chanson, H. (ed.). Energy Dissipation in Hydraulic Structures. IAHR Monograph, CRC Press, Taylor & Francis Group, Leiden, The Netherlands.

Chanson, H., Bung, D. and Matos, J. 2015. Stepped spillways and cascades. pp. 45–64. *In*: Chanson, H. (ed.). Energy Dissipation in Hydraulic Structures, IAHR Monograph, CRC Press, Taylor & Francis Group, Leiden, The Netherlands.

Dressler, R.F. 1952. Hydraulic resistance effect upon the dam break functions. Journal of Research of the National Bureau of Standards 49(3): 217–225.

Dressler, R. 1954. Comparison of theories and experiments for the hydraulic dam-break wave. Proc. International Association Scientific Hydrology Assemblée Générale, Rome, Italy 3(38): 319–328.

Hee, M. 1978. Selected Case Histories. Proc. Workshop on Minimum Energy Design of Culvert and Bridge Waterways, Australian Road Research Board, Melbourne, Australia, Session 4, Paper 1, pp. 1–11.

McKay, G.R. 1971. Design of minimum energy culverts. Research Report, Dept. of Civil Eng., Univ. of Queensland, Brisbane, Australia, 29 pages & 7 plates.

Montes, J.S. 1992. A Potential Flow Solution for the Free Overfall. Proceedings of Institution of Civil Engineers Water Maritime & Energy, Vol. 96, Dec., pp. 259–266. Discussion: 1995, Vol. 112, Mar., pp. 81–87.

Montes, J.S. 1998. Hydraulics of Open Channel Flow. ASCE Press, New-York, USA, 697 pp.

Tullis, B.P. 2013. Current and future hydraulic structure research and training needs. pp. 3–9. *In*: Bung, D. and Pagliara, S. (eds.). Proceedings of International Workshop on Hydraulic Design of Low-Head Structures, IAHR, 20–22 Feb., Aachen, Germany, Bundesanstalt für Wasserbau, Germany.

CHAPTER 3

Weir Classifications

Blake P. Tullis

INTRODUCTION

Traditional weirs are impermeable, wall-like, overflow structures. They have a fixed crest elevation and are typically used as flow control and/or flow measurement structures in rivers, channels, and reservoirs (see Fig. 1). The relationship between the upstream energy (or driving head) and the discharge over the weir is referred to as a head-discharge relationship. Various expressions for weir head-discharge relationships can be found in the literature. Two common empirical formulas are presented in Eqs. (1) and (2).

$$Q = \frac{2}{3} C_d L \sqrt{2g} H^{3/2} \tag{1}$$

$$Q = CLH^{3/2} \tag{2}$$

In Eqs. (1) and (2), Q is the volumetric discharge, L is the weir length, g is the gravitational acceleration constant, and H is the driving upstream head measured relative to the weir crest elevation. As shown by Eqs. (1) and (2), Q is proportional to the $H^{3/2}$ and an empirical discharge coefficient is required that varies with the weir geometry, H, and other factors that influence discharge characteristics. Equation (1) is dimensionless; the corresponding discharge coefficient (C_d) is likewise dimensionless and independent of the system of units (e.g., SI or ES). The derivation of Eq. (1) can be found in Henderson (1966). Equation (2) is non-dimensionless and is a simplification of Eq. (1) where all of the constants (i.e., 2/3, C_d, $\sqrt{2g}$) have been combined into a single non-dimensionless discharge coefficient (C). C has units of length$^{1/2}$/time and its value will vary with the system of units and the relevant units of volume (m^3, ft^3, liters, gallons, etc.) and time (seconds, minutes, etc.). The simplicity of Eq. (2) has led to its widespread use in practice; for example, the US Bureau of Reclamation's *Design of Small Dams* (USBR 1987) uses Eq. (2) exclusively to characterize weir head-discharge relationships. It is also used in Brater and King's *Handbook of Hydraulics* (1982).

Professor, Utah Water Research Laboratory, Civil & Environmental Engineering Dept., Utah State University.
E-mail: blake.tullis@usu.edu

Note that the piezometric head (h) or the total head (H_t) can be used to characterize H in Eqs. (1) and (2). The piezometric head is the vertical distance from the weir crest to the upstream water surface (see Fig. 2), measured at a distance equal to ~4–6 h upstream from the weir (USBR 1997). The total head is the sum of the piezometric head and the velocity head [$h+V^2/(2g)$] (see Fig. 2). When using a head-discharge relationship, it is important to select discharge coefficient values that are consistent with the empirical formula and its definition of upstream driving head (h or H_t). If Q is the dependent variable and H_t is required, an iterative solution technique may be needed if the velocity head component of H_t is unknown. For convenience, many weir problems historically were typically solved using the piezometric head to facilitate an explicit calculation of the discharge. However, nowadays H_t is more commonly used in an effort to better account for the effect of approach flow condition variations on the discharge coefficient.

Weir Classification

Common weir geometries can vary significantly and are typically classified by their shape, crest geometry, and the self-aerating nature of the nappe. Some typical weir geometries are presented in Fig. 1.

The jet of water passing over the weir, as shown in Fig. 1, is referred to as the nappe. If the weir length is narrower than the channel width, the weir is "contracted" and the air space between the downstream weir wall and the nappe is open to the atmosphere and, therefore, self-venting (Fig. 1a). When the length of a rectangular weir and the channel width are equal (Fig. 1e), then the weir is referred to as "suppressed" and the air space between the weir and nappe typically cannot self-vent without adding ventilation pipes. Figure 1(f) presents a weir whose geometry mimics the nappe bottom profile boundary, referred to as an ogee crest (a.k.a. WES, Creager) weir. Hager and Schleiss (2009) and USBR (1987) provide details related to the ogee crest geometry and head-discharge characteristics. Additional common weir classifications by crest shape include: sharp-crested, flattop, quarter-round, half-round, and ogee crest (see Fig. 3). Brater et al. (1996) presents discharge coefficient data for many different weir geometries.

Broad-crested weirs represent another general weir type. Broad-crested typically feature a stream-wise crest dimension that is large, relative to the flow depth over the weir. The stream-wise flow profile over a broad-crested weir is categorized as a gradually varied flow that transitions from subcritical to supercritical flow. Figure 4 shows examples of broad-crested weirs. Gonzalez and Chanson (2007), Brater et al. (1996), and Hager and Schwalt (1994) present broad-crested weir discharge coefficient data. Johnson (2000) found that flattop weirs (Fig. 4a) operate hydraulically as broad-crested weirs at low discharges (low upstream head). But if the vertically converging approach flow has sufficient momentum, the nappe will separate at the leading edge of the weir, creating a sharp-crested weir flow pattern. Figures 4(b) and 4(c) show a broad-crested weir (trapezoidal cross section in the stream-wise direction), referred to as an embankment weir, incorporated into an irrigation channel flow measurement flume. Fritz and Hager (1998) discuss the hydraulic behavior of embankment weirs. Broad-crested weirs are hydraulically less efficient generally than other weir designs.

Fig. 1. Examples of various linear weir types: (a) contracted rectangular (sharp-crested) weir (photo courtesy of Blake Tullis), (b) rectangular (sharp-crested) weir (photo courtesy of Blake Tullis), (c) Cipoletti (flattop) weir (photo courtesy of Steven Barfuss), (d) V-notch (sharp-crested) weir (Photo courtesy of Bret Dixon), (e) suppressed rectangular (flattop) weir (photo courtesy of Blake Tullis), (f) ogee crest weir (photo courtesy of Blake Tullis).

Nonlinear Weirs

Weirs are commonly used as the control structure for regulating spillway discharge as a function of reservoir elevation (head-discharge). According to Eqs. (1) and (2), the weir discharge is proportional to the weir length. One economic solution for increasing the weir length and spillway capacity for a given spillway chute width is to use a nonlinear weir. Common nonlinear weir types include labyrinth weirs, piano key weirs, and arced weirs (see Fig. 5). At a common upstream head, a labyrinth and piano key weirs can produce 3-to-4 times more discharge than a linear weir of the same channel width. Crookston and Tullis (2013a, 2013b, 2012), Falvey (2003), and Tullis et al. (1995) address labyrinth weir design and hydraulic performance. Piano key weir design details are found in Anderson and Tullis (2013, 2012a, 2012b), Machiels et al. (2013), and Leite Ribeiro et al. (2012). Non-linear weirs are popular design solutions for new dams and particularly for existing dams where the original spillway is found to have inadequate discharge capacity.

Reservoir Storage, Flood Routing, and Weir Characteristics

The primary function of most reservoirs is storing water (e.g., irrigation, municipal, industrial uses), flood control, and/or power generation. Reservoirs utilize spillways, which typically feature weirs, to regulate the reservoir water level and release surplus water in a controlled and predictable manner. Adequate spillway capacity is essential for dam safety, particularly for embankment dams. The design of spillways includes: the selection of an inflow design flood, a reservoir storage relationship, the spillway head-discharge relationship, flood routing, and downstream impacts. The weir discharge characteristics represent an influential component of a complex flood routing system. For example, the weir efficiency and crest elevation have a direct impact on the outflow hydrograph of a given storm event.

As an alternative to using nonlinear weirs to increase the weir discharge capacity (discharge efficiency), gates can be added on top of linear weirs (see Fig. 6). With the gate closed, water storage levels can exceed the weir crest elevation. Gate operation can be optimized for flood routing. The gated weir crest elevation is typically lower than the elevation of the non-gated or uncontrolled weir alternative, which increases the gated spillway maximum discharge capacity (larger driving head) relative to the non-gated spillway. When a gate is open but the bottom of the gates remains below the water surface, the gated weir structure loses discharge efficiency because it operates more like an orifice ($Q \propto H^{1/2}$) than a weir ($Q \propto H^{3/2}$). Gate operation typically requires the operator to be on site. This can be problematic when trying to manage flood flows through reservoirs in remote locations, as the operator may not be able to reach the site in the time required. Gate operating mechanisms must also be maintained and in good working condition to insure their operability. Furthermore, floating debris can also generate maintenance issues for gates.

Other Weir Applications

In addition to flow control, weirs and weir-like structures are used for a variety of other applications, including grade-control and flow diversion structures. In steep rivers and channels, weirs are often used in series to function as grade-control structures. The weirs increase the local flow depth, which decrease the local upstream velocity and reduces the potential for channel erosion. In a similar fashion, they are used in river lock-and-dam structures to maintain adequate flow depths for boat travel. To improve the efficiency of diverting water from a canal to a pipe intake, weirs are used to increase the upstream flow depth and reduce the approach flow velocity and momentum. This helps to facilitate the change in flow direction required for flow diversion. Figure 7 shows an example of a flow diversion structure with a weir located just downstream of a pipe intake (hidden from view by the trash rack assembly).

Parallel Flow Control Structures

For some applications, it may be beneficial to employ parallel flow control structures with differing characteristics. This can be as simple as installing a segmental linear weir with varied crest elevations (see Fig. 8a). The "staged" weir can restrict base-flows or more frequent flood flows to the lowest weir stage. At higher discharges, all weir stages or segments are engaged, increasing the discharge capacity. Weir segments can also be designed to modify the outflow hydrograph to generally match a range of outflow design requirements. This approach was studied by Dabling et al. (2013), whose work included the hydraulic behavior of staged labyrinth weirs.

Parallel control structures also allow the combination of "active" and "passive" flow control. Active control structures include gated weirs, which facilitate water discharge on-demand by opening the gate. Non-gated (traditional) weirs are passive flow control structures, which only discharge when the reservoir level exceeds the crest elevation. The head-discharge characteristics of a passive control structure are fixed by the weir geometry and the reservoir level. The head-discharge characteristics of an active control structure can be varied with gate opening. Figures 8B and 8C show combinations of parallel passive and active control structures. With such combinations, small floods and the rising leg of a larger flood hydrograph can pass over the non-gated weir without dam operator involvement. For remote dams, the ability to pass the early stage of the flood event over the non-gated weir allows dam operators more time to reach the dam before gate opening is required. Parallel flow control structures also give engineers options for modifying the reservoir outflow hydrograph, when limiting downstream flooding potential for higher return-period flood events is a priority.

Environmentally Sensitive Weirs Designs

Weirs with "tall" vertical weir walls can become barriers to aquatic animal species movement, which can significantly impact environmental processes. In an effort to reduce such impacts for grade-control structures, more environmentally friendly, non-traditional weir designs, such as block ramps, for example, have been implemented in river systems. Block ramps utilize natural materials, such as rock, gravel, etc. to create a ramped structure with velocities and flow depths that are conducive to fish passage and avoid sharp vertical discontinuities created by traditional weirs. Block ramps differ from traditional weirs in that they are typically porous (rock and gravel fill) and they do not have a uniform, horizontal weir crest elevation. Consequently, block ramps or rock weirs are typically not used as flow measurement structures per Eqs. (1) and (2). Figure 9 shows examples of single and multiple (series) environmentally friendly grade-control structures constructed using block ramps. Some recent block ramp studies include:

Other Weir Considerations

Some additional factors that can influence weir head-discharge characteristics included weir submergence caused by high tailwater levels (Dabling and Tullis 2013; Tullis 2011; Tullis and Nielsen 2008; Tullis et al. 2007; USBR 1987; Villemonte 1947), sedimentation, floating and/or submerged debris collecting at the weir (Pfister et al. 2013), approach flow angles, flow separation caused by weir abutment, and pier geometries (Anderson and Tullis 2012b; USBR 1987). Non-level weir crests, upstream staff gauges inaccurately referenced to the crest elevation and non-uniform approach flow patterns can also affect the flow measurement and/or head-discharge relationship accuracies of weirs.

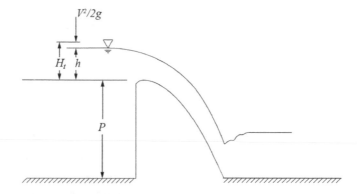

Fig. 2. Illustration of weir upstream piezometric and total head.

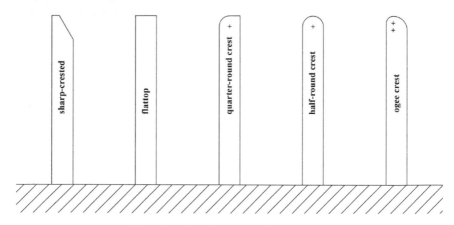

Fig. 3. Examples of common weir crest shapes.

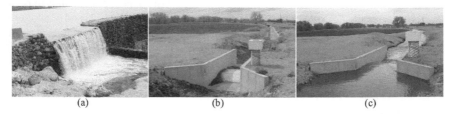

Fig. 4. Examples of broad-crested weirs: (a) flattop broad-crested weir spillway (photo courtesy of Bret Dixon); (b) dry overview of a broad-crested weir incorporated into a flow measurement flume; (c) flow measurement flume in operation (photos b and c courtesy of Stuart Styles and Walter Winder.).

Fig. 5. Nonlinear weir examples: (a) labyrinth weir (Brazo dam, TX, USA, photo courtesy of Freese and Nichols Inc.); (b) piano key weir (St-Marc dam, France, photo courtesy of Blake Tullis); (c) arced ogee crest weir (Iron Works dam, PA, USA photo courtesy of Aqua Pennsylvania); (d) arced labyrinth weir (scaled physical model of Isabella dam, CA, USA, photo courtesy of Donna Barry).

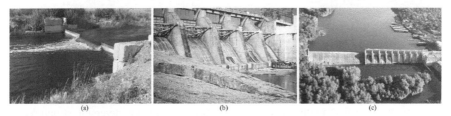

Fig. 6. Example of parallel flow control structures: (a) staged linear weir (photo courtesy of Steven Barfuss); (b) ogee crest weir with radial gates (Wirtz dam, TX, USA) (photo courtesy of Blake Tullis); (c) parallel gated and non-gated ogee crest spillways (Tom Miller Dam, TX, USA, photo courtesy of Freese and Nichols Inc.).

Fig. 7. Example of a flow diversion structure (photo courtesy of Blake Tullis).

Fig. 8. Examples of parallel flow control structures: (a) multi-staged linear weir (photo courtesy of Steven Barfuss); (b) St-Marc dam (France) piano key weir and gated spillways in parallel (photo courtesy of Blake Tullis); (c) Tom Miller Dam (Texas, USA) parallel gated and ungated ogee crest weirs (photo courtesy of Lower Colorado River Authority).

(a)   (b)

Fig. 9. Block or rock ramp grade control structures in river applications: (a) single rock ramp structure (photo courtesy of Blake Tullis); (b) two rock ramps in series (photo courtesy of Artur Radecki-Pawlik).

Keywords: Weir, open channel flow, head-discharge relationship, submergence, flood routing, linear weirs, nonlinear weirs, grade-control structure, diversion structure.

References

Anderson, R.M. and Tullis, B.P. 2012a. Comparison of Piano Key and rectangular labyrinth weir hydraulics. Journal of Irrigation and Drainage Engineering 138(6): 358–361.

Anderson, R.M. and Tullis, B.P. 2012b. Piano Key weirs: reservoir versus channel applications. Journal of Irrigation and Drainage Engineering 138(8): 773–776.

Anderson, R.M. and Tullis, B.P. 2013. Piano Key hydraulics and labyrinth weir comparisons. Journal of Irrigation and Drainage Engineering 139(3): 246–253.

Brater, E.F., King, H.W., Lindell, J.E. and Wei, C.Y. 1996. Handbook of Hydraulics, 7th Edition. McGraw Hill, Columbus, OH, USA.

Crookston, B.M. and Tullis, B.P. 2012. Arced labyrinth weirs. Journal of Hydraulic Engineering 138(6): 555–562.

Crookston, B.M. and Tullis, B.P. 2013a. Hydraulic design and analysis of labyrinth weirs. 1: Discharge relationships. Journal of Irrigation and Drainage Engineering 139(5): 363–370.

Crookston, B.M. and Tullis, B.P. 2013b. Hydraulic design and analysis of labyrinth weirs. 2: Nappe aeration, instability, and vibration. Journal of Irrigation and Drainage Engineering 139(5): 371–377.

Dabling, M.R. and Tullis, B.P. 2012. Piano Key weir submergence in channel applications. Journal of Hydraulic Engineering 138(7): 661–666.

Dabling, M.R., Tullis, B.P. and Crookston, B.M. 2013. Staged labyrinth weir hydraulics. Journal of Irrigation and Drainage Engineering 139(11): 955–960.

Falvey, H.T. 2003. Hydraulic Design of Labyrinth Weirs. ASCE Press, Reston, VA, USA.

Fritz, H.M. and Hager, W.H. 1998. Hydraulics of embankment weirs. Journal of Hydraulic Engineering 124(9): 963–971.

Gonzalez, C.A. and Chanson, H. 2007. Experimental measurements of velocity and pressure distribution on a large broad-crested weir. Flow Measurement and Instrumentation 18(2007): 107–113.

Hager, W.H. and Schwalt, M. 1994. Broad-crested weir. Journal of Irrigation and Drainage Engineering 120(1): 13–16.

Hager, W.H. and Schleiss, A.J. 2009. Constructions hydraulicques, Ecoulements stationnaires. Traite de Genie Civil, Vol. 15, Presses Polytechniques et Universitaires Romandes, Lausanne, Switzerland.

Henderson, F.M. 1966. Open Channel Flow. MacMillan Publishing Co., New York, New York, pp. 175.

Johnson, M.J. 2000. Discharge coefficient analysis for flat-topped and sharp-crested weirs. Irrig Sci 19: 133–137.

Leite Ribeiro, Pfister, M. and Boillat, J.-L. 2012. Hydraulic design of A-type Piano Key weirs. Journal of Hydraulic Research 50(4): 400–408.

Machiels, O., Pirotton, M., Archambeau, P., Dewals, B. and Erpicum, S. 2014. Experimental parametric study and design of Piano Key weirs. Journal of Hydraulic Research 52(3): 326–335.

Pfister, M., Capobianco, D., Tullis, B.P. and Schleiss, A. 2013. Debris blocking sensitivity of Piano Key weirs under reservoir type Approach flow. Journal of Hydraulic Engineering 139(11): 1145–1141.

Tullis, B.P., Young, J.C. and Chandler, M.A. 2007. Head-discharge relationships for submerged labyrinth weirs. Journal of Hydraulic Engineering 133(3): 248–254.

Tullis, B.P. and Neilson, J. 2008. Performance of submerged ogee crest head-discharge relationships. Journal of Hydraulic Engineering 134(2): 263–266.

Tullis, B.P. 2011. Behavior of submerged ogee crest weirs. Journal of Irrigation and Drainage Engineering 137(10): 677–681.

Tullis, J.P., Nosratollah, A. and Waldron, D. 1995. Design of labyrinth spillways. Journal of Hydraulic Engineering 121(3): 247–255.

USBR 1987. Design of Small Dams. US Bureau of Reclamation, Washington DC.

USBR 1997. Water Measurement Manual. US Bureau of Reclamation, Washington DC.

Villemonte, J.R. 1947. Submerged weir discharge studies. Eng News-Rec 139(26): 54–56.

CHAPTER 4

Hydrodynamics of River Structures Constructed with Natural Materials

Kohji Michioku

INTRODUCTION

Many rivers have been anthropogenically regulated to reduce flood and manage water resources, which resulted in significant degradation of landscape, biodiversity, and water quality of rivers in the past few decades. The paradigm of river management had already shifted towards the direction of near-nature river work half a century ago and river structures are expected to have multiple-functions not only in flow control, but also in being a part of the natural landscape and ecosystem. One of the promising technologies that meets both engineering and ecological demands is river restoration with the use of nature-friendly structures. Rational and wide use of natural materials could bring about a breakthrough in river restoration. By adapting this scheme of river restoration, groynes, spurs, shore protection works, ramps, weirs, and so on will tend to be constructed with natural materials such as stones, boulders, logs, timbers, live trees, branches, rootwads, and so on. Although most of such structures originated from traditional technologies developed before the modern ages, they are also state-of-the-arts technologies even in the present day.

Despite significant research activity, there is still much work to be done in understanding the complex interrelating systems between hydrodynamics, fluvial processes, and the ecosystem depicted in Fig. 1. The flow is dominated not only by artificial structures, but also by natural structures like vegetation and geomorphology. During a severe flood flow, interaction between flow and sediment transport causes significant changes in geomorphology and causes irreversible damage of structures and vegetation. This, in turn, causes marked impacts on hydraulic and fluvial processes. Since vegetation is a part of flora as well as a source of habitat for various aquatic and

Department of Civil and Environmental Engineering, Hosei University, 2-33 Ichigaya Tamachi, Shinjuku, Tokyo 162-0843, Japan.
E-mail: kohji.michioku.47@hosei.ac.jp

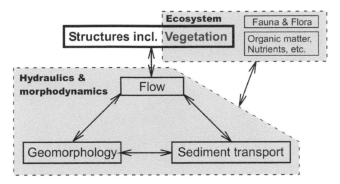

Fig. 1. Interrelating river system.

riparian species, a predominant flood event may even alter the whole structure of an ecosystem. The key issue is to properly understand such an interrelating mechanism in river systems in order to establish a comprehensive design scheme for nature-oriented structures.

The present article is to highlight recent works focusing on hydrodynamics of natural vegetation or rock-arranged structures. It is noted, however, that little discussion is provided on sediment transport in this chapter, since, unlike open channel hydraulics, there is still a gap between fluvial hydraulics and hydrodynamics for this type of structure. Sedimentation is such a critical issue in the management scheme of river structures that enhancement of tighter collaboration between these two disciplines may be achieved in the ensuing years.

Examples of Structures Constructed with Natural Materials

Flood protection forest

Photo 1 is a typical example of a flood protection forest in Japan "Manriki Forest (meaning 'Mighty Forest')", where pine trees were planted several hundred years ago at the apex of the alluvial fan to protect towns and farmlands downstream. Rubble mound dykes were distributed in the forest to divert and diffuse inundation flows. Many of this type of flood protection forest were constructed nation-wide by local feudal lords in the Middle Ages when modern infrastructures for flood disaster prevention had not yet been developed. The flood protection forest is now being reconsidered as a useful flood mitigation facility since there have been frequent occurrences of extreme flood events exceeding the designed level in recent years.

Rubble mound weir, Gabion weir

Photo 2 shows a rubble mound weir which is still active as an irrigation facility in this region. Although this type of weir was more commonly used hundreds years ago, most of the weirs have been retired and replaced by modern structures with rigid and impermeable bodies such as concrete, metal, and rubber. However, a few of the classical weirs are voluntarily still preserved by local communities expecting their performance as a nature-friendly river system. The rock elements are sometimes tighten with wires

Photo 1. Flood protection forest "Manriki Forest (Mighty Forest)".
(a) A view from floodplain, (b) Rubble mound dyke in the forest

Photo 2. Rubble mound weir in Jobaru River. (Kyushu Regional Development Bureau, Ministry of
Land, Infrastructure, Transport and Tourism, MLIT, Japan.)

or caged in wire baskets in order to stabilize the structure. The latter type of weir is given the term, "gabion weir".

Groynes and spurs

Groynes and spurs have recently been constructed with rocks, boulders, wood, gabions, and natural plants. In addition to the function of regulating flow and sedimentation, groynes and spurs constructed with natural materials provide various ecosystem services such as providing habitat for aquatic life, self-purification of water, re-aeration associated with turbulence generation, creation of desirable landscape, etc. The wooden crib spur in Photo 3 and Fig. 2 was a popular method to control floods in the Middle Ages. The wooden crib spurs are still used in shore protection works, especially in steep slope rivers. However, most of the crib spurs have been recently constructed with reinforced concrete pillars instead of timber, since building this traditional structure requires special techniques of skilled craftsmen. Other examples of groynes are shown in Photo 4.

Photo 3. Crib spur fighting against flood flow in 2011. (Photo taken by Kofu River and National Highway Office, Kanto Regional Development Bureau, MLIT.)

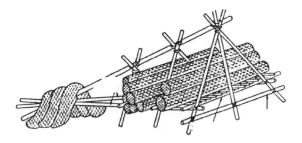

Fig. 2. Illustration of traditional crib spur.

a

b

c

Photo 4. Groynes built by natural materials.
(a) gabion, (b) wooden piles, (c) boulders

Photo 5. Rough ramp in Maruyama River, Japan.

Rough ramp

To bypass steep slopes in riverbeds, rough ramps built by natural rocks are used as a solution to re-naturalize creeks and rivers worldwide. The rough ramps serve the function not only as flow energy dissipaters, but also as fish ladders or fish ways. The block ramp maintains the morphological continuity of the longitudinal profile and ensures biological exchanges between the water and the river bed. Photo 5 shows a rough ramp installed in a river reach which Japanese giant salamanders, *Andrias japonicas*, designated as a special natural treasure inhabit. The boulder elements are interlocked and tightened together with steel rods to stabilize the structure.

Engineering issues of river structures constructed with natural materials

Common hydrodynamic and fluvial features found in this type of structure are as follows:

  i) The structure has a permeable body and the flow is in multi-phases consisting of water, solid substances, and air.
 ii) Velocity and turbulence intensity are extremely different between the inside and outside of the structure and thus, the structure is bounded by a shear-dominant mixing layer.
iii) Mass and momentum are exchanged between the inside and outside of the structure.
 iv) Sediment deposition and scouring around the structure, which dominates local geomorphology and sediment transport, are significant.

Hydrodynamics of Vegetation as a Natural Structure

Hydrodynamics, fluvial process, and geomorphology in rivers are profoundly influenced by riparian vegetation (Jang and Shimizu 2007; Yang et al. 2007; Corenblit et al. 2009). Riparian vegetation functions to enhance stream banks as roots are able to bind bank sediment. Large woods defend embankment from erosion and are frequently used as shore protection works and riparian buffers. On the contrary, vegetation decreases the channel's conveyance capacity due to flow retention and blockage. During floods, much driftwood is yielded from vegetated floodplains and sticks to weirs and bridges downstream as shown in Photo 6, which often blocks flow and causes overtopping and failure of levees. A proper management of riparian vegetation is a critical issue not only for landscape and ecology, but also for shore and bank protection technology. A better understanding of hydrodynamics and ecology of vegetation is incredibly necessary in order to establish a comprehensive design guideline of vegetation management. The following paragraphs are brief summaries on the most important aspects of vegetated flow dynamics that provide information necessary in analyzing vegetated flow.

Photo 6. Driftwoods sticking to a bridge after flood in 2004 (Sugihara River, Japan).

Impacts of riparian vegetation on river morphology

Figure 3 schematically illustrates the bathymetric evolution of a vegetated river channel after a flood. During a flood, the main channel or thalweg is entrenched in a region in which the velocity exceeds the critical value for sediment movement, while fine sediments are trapped and deposited in the vegetated floodplain to thicken the substrate layer. In this manner, the main channel becomes narrower, deeper, and the specific elevation difference between the floodplain and main channel increases after every flood by the fluvial process. This is how the geomorphology changes during a flood event and the vegetation grows and expands during the low-stage period. An example of the historical development of river morphology in the last fifty years in Kizu River, Japan, is shown in Fig. 4 (Takebayashi 2014). Vegetation overgrew on the floodplain due to less frequent inundation of the floodplain, which decreases flow conveyance capacity and eventually promotes sediment deposition. From an environmental perspective, vegetation overgrowth reduces biodiversity and produces a poor landscape, by which attractiveness and accessibility of the waterfront are spoiled and damaged.

The proper use and management of riparian vegetation may enhance its multi-functions as a natural structure for preventing flood disaster, controlling stream, sedimentation, and geomorphology, providing various ecosystem services, etc. Extensive research on vegetated flow dynamics provided indispensable information for designing, planning, and managing river systems. These researches are partially introduced below.

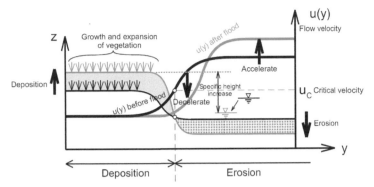

Fig. 3. Vegetation and geomorphological changes after a flood event in river channel with riparian vegetation.

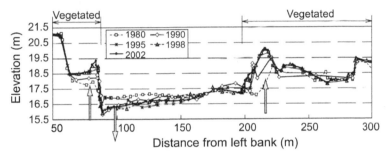

Fig. 4. Historical change in cross-sectional profiles in a vegetated reach, Kizu River (Takebayashi 2014).

Shear flow structure between vegetated and unvegetated areas

A typical hydraulic feature observed in a vegetated river is the retention effect of flow conveyance on a vegetated floodplain. This produces mixing layers associated with organized vortex streets both in horizontal and vertical directions between vegetated and free water domains. The internal shear stress in the mixing layer plays a role as an additional flow resistance component.

Horizontal shear layer

A horizontal shear layer often develops between vegetated and unvegetated areas during a flood as shown in Photo 7 (Fukuoka 2005). Turbulent flow structure in the shear layer has been investigated experimentally and theoretically by many researchers such as White and Nepf (2008). In a quasi-two-dimensional analysis of depth-averaged flow field, Rameshwaran and Shiono (2007), proposed a depth-averaged eddy viscosity v_t as:

$$v_t = \alpha u_* H + \beta \delta \Delta U \tag{1}$$

where $\alpha = \kappa/6 = 0.0683$ is an eddy viscosity constant, $\kappa = 0.041$ is the von Karman's constant, H, is the total water depth $u_* = \sqrt{f/8U}$ is the friction velocity defined by the Darcy-Weisbach's friction factor f and U is the depth-averaged velocity. In

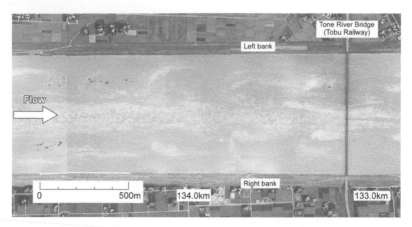

Photo 7. Aerial view of mixing layer and organized vortexes developed between the vegetated floodplain and main channel in Tone River during a flood in 1981. (Photo taken by Tonegawa Joryu River Office, Kanto Regional Development Bureau, MLIT, Fukuoka 2005.)

Eq. (1), ν_t has an extra term originated from the horizontal shear, $\beta\delta\Delta U$, in addition to the original diffusivity, $\alpha u_* H$. Here, the proportional constant β=0.008 is suggested. $\Delta U = U_m - U_f$ is the velocity step between the main channel and floodplain and (U_m, U_f) are the depth-averaged velocities at the center of main channel and floodplain, respectively. The following formula was proposed for the shear layer thickness δ:

$$\delta = 2(y_{75} - y_{25}) \tag{2}$$

where y_{75} and y_{25} are the streamwise coordinates at which $U=U_{75}=U_f+0.75(U_m-U_f)$ and $U=U_{25}=U_f+0.25(U_m-U_f)$, respectively (Van Prooijen et al. 2005). The analytical solution for U is compared with the laboratory data in Fig. 5.

White and Nepf (2007, 2008) also proposed a shear layer thickness δ as follows:

$$\delta = \frac{0.5 \pm 0.1}{C_D a} \tag{3}$$

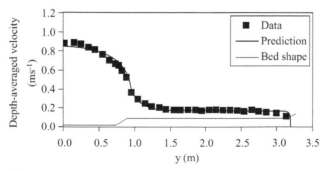

Fig. 5. A spanwise profile of the depth-averaged velocity compared between the quasi two-dimensional analysis and the laboratory data (Rameshwaran and Shiono 2007).

Here, C_D is a coefficient of vegetation drag as discussed later, $a=D/\Delta S^2=Dh_v/\Delta S^2 h_v$ is the vegetation's frontal area per unit volume, h_v, is the vegetation height and $(D, \Delta S)$ are stem diameter and spacing of vegetation, respectively.

Mixing layer at the top of vegetation canopy

When the vegetation is submerged, a mixing layer also develops at the top of vegetation canopy, which is another mechanism to reduce flow conveyance capacity. Field survey, laboratory experiments and flow modelling were extensively carried out regarding this topic (e.g., Sukhodolov and Sukhodolova 2010; Shimizu and Tsujimoto 1994). Characteristics of the vertical mixing layer and of the adjacent open flow are governed by various parameters such as vegetation density, morphology of plant community, relative submergence, flexibility and shape of plant, degree of foliation, and so on. The flow structure over a vegetation canopy is generally classified into three regimes, depending on the vegetation's roughness density, that is, (i) skimming flow, (ii) wake interference flow, and (iii) isolated roughness flow (Morris 1955). According to Nepf (2012), with denser vegetation as $\lambda=h_vD/\Delta S^2 \geq 0.23$, the velocity profile includes an inflection point near the top of canopy and a well-defined mixing layer develops at the canopy top. Here, $\lambda=ah_v$ is a dimensionless parameter representing the vegetation's frontal area per bed area, termed "roughness density". In this regime of dense vegetation, the bed is shielded by the canopy-scale turbulence and the flow resistance is more predominantly influenced by the canopy drag than by the bed shear stress. On the contrary, with sparser vegetation as $\lambda \ll 0.1$, the vertical velocity profile is better interpreted by the boundary layer scheme since the bed stress plays a more predominant role in turbulence generation (Luhar et al. 2008). From a viewpoint of flood control, more attention should be paid to denser vegetation bounded by the well-defined mixing layer rather than on sparser vegetation, since reduction of flow conveyance is more significant in the former vegetation regime.

In dense flexible vegetation, the mixing layer dynamics profoundly alters mass and momentum exchange across the shear layer interface. The key parameters are the mixing layer thickness, t_{ml}, and the frequency of vegetation waving f_{KH}. According to Ghisalberti and Nepf (2002), t_{ml} is well correlated with the momentum thickness θ as:

$$\frac{t_{ml}}{\theta} = 7.1 \pm 0.4 \tag{4}$$

Referring to Fig. 6, θ is defined as:

$$\theta = \int_{-\infty}^{\infty} \left[\frac{1}{4} - \left(\frac{U - \overline{U}}{\Delta U} \right)^2 \right] dz \tag{5}$$

where $\overline{U} = \dfrac{(U_1 + U_2)}{2}$. The vegetation waving (termed 'monami' by Ackerman and Okubo 1993) is driven by K-H billows generated in the mixing layer and its frequency, f_{KH}, is estimated by:

$$f_{KH} = 0.032 \left(\frac{U}{\theta} \right) \tag{6}$$

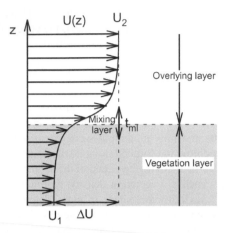

Fig. 6. Definition of vertical mixing layer.

Carollo et al. 2005 and Termini 2015, experimentally examined how the mixing layer parameters depend on roughness density of vegetation λ and on flow submergence H/h_v.

Flow conveyance and resistance in vegetated channels

Bulk drag coefficient of virtual vegetation: rigid cylinders

Flow resistance is exerted primarily by drag force of vegetation canopy in addition to the channel bed friction and the internal shear stress generated at the outer boundaries of vegetation canopy. In considering the vegetation drag, a rigid vertical cylinder is the simplest model to simulate the plant element and provides a good approximation of vegetation dynamics. The form drag of vegetation F is usually represented by a quadratic law in terms of the bulk drag coefficient $\overline{C_D}$ as:

$$F = \frac{1}{2}\rho\overline{C_D}aU_m^2 \tag{7}$$

where ρ is fluid density and U_m is a cross sectional average velocity. In general, $\overline{C_D}$ depends on Reynolds number under influence of wake structure. Discussions on $\overline{C_D}$ were reviewed in literature such as Nikora (2010), Nepf (1999), Nepf (2012), Luhar and Nepf (2013). Experimental results on $\overline{C_D}$ obtained from several studies on rigid emergent vegetation are summarized as a function of the normalized vegetation density or dimensionless population density aD as in Fig. 7. The solid and broken lines are analytical values for random and staggered arrays predicted by the wake interference model (Nepf 1999).

Although $\overline{C_D}$ is a convenient parameter for describing vegetation drag, information on flow conveyance, U_m/u_*, or Manning's roughness coefficient is necessary for practical use in river management. It is noted that the dimensionless form of Manning's roughness coefficient, $N_c(\sqrt{g}R^{1/6})$, is equivalent to the inverse of flow conveyance $(U_m/u_*)^{-1}$, where N_c is a composite roughness coefficient reflecting both vegetation drag and wall friction on unvegetated perimeters, g is gravity acceleration, and R is hydraulic radius.

Assuming a two layer flow consisting of overflow and vegetated layers, Luhar and Nepf (2013) found a functional dependency of $N_c(\sqrt{g}R^{1/6})$ on the flow submergence H/h_v. Michioku et al. (2015), measured N_c of vertical rigid cylinders being uniformly arranged in a test flume and also carried out a numerical analysis on N_c by applying their two-dimensional two-layer model to the uniformly vegetated flow system. Figure 8 shows solutions of rating curves (H, Q) and Manning's roughness coefficient, N_c, with the experimental data, which covers from the emergent ($H/h_v < 1.0$) to submerged vegetation ($H/h_v \geq 1.0$). The roughness coefficient, N_c, monotonically increases with increasing Q for the emergent vegetation. In contrast, it becomes less dependent on Q and is more dominated by vegetation properties, that is, D, h_v, and a, after submergence. Their solution of N_c represents a qualitatively similar tendency to the result suggested by Luhar and Nepf (2013).

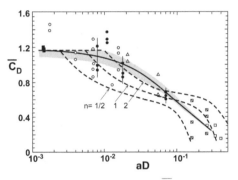

Fig. 7. Functional dependency of the bulk drag coefficient $\overline{C}_D$ on normalized vegetation density aD (Nepf 1999).

Case	D(cm)	h_V (cm)	a (cm⁻¹)	Cal.	Exp.
1	0.6	6.0	0.125	– –	□
2	0.6	6.0	0.005	——	◎
3	1.2	6.0	0.01	——	▲

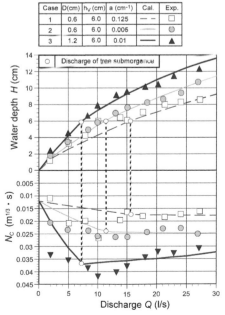

Fig. 8. Rating curves, (H, Q) and Manning's roughness coefficient N_c (Michioku et al. 2015).

Effects of flexibility, foliage and shape

Although vegetation's hydrodynamics are approximated with the rigid cylinder analogy to some extent, important vegetation properties such as plant flexibility, shape, and degree of foliation are missing in the modelling discussed above. The vegetation properties should be taken into consideration for analyzing dynamics of foliated natural vegetation. Based on de Langre et al. (2012), for example, Eq. (7) should be modified into a quadratic law for flexible plants as:

$$F \propto U_{\mathrm{m}}^{2+b} \tag{8}$$

The exponent b is determined by vegetation's stiffness and flexibility. A linear relationship between F and U_{m}, i.e., $b = -1.0$, was suggested for fully flexible plants. In contrast, $b = 0.0$ is consistent with the drag force law for rigid plants, Eq. (7). Comparison between Eqs. (7) and (8) implies that $\overline{C_{\mathrm{D}}}a \propto U_{\mathrm{m}}^{b}$, in other words, $\overline{C_{\mathrm{D}}}a$ has a monotonic decreasing tendency with flow velocity U_{m} in a case of partially flexible vegetation, that is, $-1.0 < b < 0$. This relationship demonstrates the dynamics of flexible vegetation well and also that both of the drag coefficient $\overline{C_{\mathrm{D}}}$ and the vertical projected area a are reduced by the bending over and streamlining of plants with increasing U_{m}. According to measurements from fully flexible to rigid natural plants, de Langre et al. (2012), reported that b ranged between $-1.2 < b < -0.2$.

Foliage is an additional component of vegetation drag. Equation (8) is modified so as to describe a functional dependency of F on degree of foliation or the leaf area index, LAI, as follows (Aberle and Järvelä 2013; Jalonen et al. 2013):

$$F \propto \mathrm{LAI} \times U_{\mathrm{m}}^{2+\chi} \tag{9}$$

where the parameter χ takes a negative value of $o(-10^{-1})$ and is uniquely determined for a particular species. Equation (9) implies that the foliage drag force gradually decreases with U_{m} due to the bending over and streamlining of leafs. The dependency of the drag coefficient on stem shape and foliage were experimentally determined for various species, for instance, by Wilson et al. (2008), and Christopher et al. (2008). Doncker et al. (2009), formulated Manning's roughness coefficient as a function of biomass.

Based on a dimensional argument, flow drag, $\overline{C_{\mathrm{D}}}$, or flow conveyance, U_{m}/u_*, is formulated for foliated flexible plants as functions of the relating dimensionless parameters such as:

$$(\overline{C_{\mathrm{D}}} \text{ or } \frac{U_{\mathrm{m}}}{u_*}) = Func .(\lambda, \frac{H}{h_{\mathrm{S}}}, \frac{u_* h_{\mathrm{S}}}{\nu}, \frac{h_{\mathrm{V}}}{h_{\mathrm{S}}}, \text{species - specific parameters}, \cdots) \tag{10}$$

where ν is kinematic viscosity, h_{S} is bent vegetation height and h_{V} is vegetation height in the absence of flow (see Fig. 9). Shucksmith et al. (2011), Nikora et al. (2008), Carollo et al. (2005), Termini (2015), Kouwen and Unny (1973), and Kouwen (1992), are among those who examined the functional relationship of Eq. (10) in a case of submerged flexible vegetation. According to their experimental results, U_{m}/u_* had an increasing tendency with H/h_{S}, while it had a decreasing trend with increasing $h_{\mathrm{V}}/h_{\mathrm{S}}$ and $u_* h_{\mathrm{S}}/\nu$.

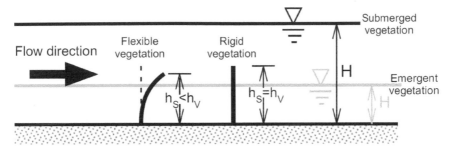

Fig. 9. Vegetation regimes classified by stem's flexibility and flow submergence.

Turbulence in a vegetated channel

Vegetation has a predominant impact on turbulent transport of mass, momentum, sediment, and other substances. Extensive research works have been performed to find out how the turbulence fields in and around vegetation canopies depend on vegetation properties such as roughness density, flow submergence, flexibility, shape, and degree of foliation. They are discussed in the literature such as Nepf (1999), Wilson et al. (2003, 2008), Velasco et al. (2008), and Yang and Choi (2009), and so on.

Two-layer flow modelling of submerged vegetation

During floods, most vegetation is submerged and bounded by the mixing layer at the canopy top. From a viewpoint of flood control, river managers pay more attention to denser submerged vegetation rather than to sparser emergent vegetation. In this case, the velocity profile no longer follows the logarithmic profile and a two-layer model is expected to give a reasonable approximation of vegetation dynamics. Yang and Choi (2010), proposed a two-layer model in which the velocity is assumed to be uniformly distributed in the vegetation layer and logarithmically distributed in the overlying layer, respectively. Applying a momentum balance to each layer, analytical solutions were successfully obtained for vertical profiles of velocity as well as of suspended sediment concentration. In addition, Manning's roughness coefficient was examined by taking a depth-average of velocity solution. Huai et al. (2013), proposed a two-layer model for predicting a velocity profile in an open channel with submerged flexible vegetation. A cantilever theory was introduced to describe the plant deformation in order to evaluate vegetation drag by considering a balance between the drag force and the restoring force associated with plant body stiffness. A polynomial velocity distribution was assumed so that a zero velocity gradient was expressed at the water surface, which was in good agreement with their measurements.

The two-layer approach was applied to a horizontally two-dimensional domain with submerged vegetation as shown in Fig. 10 by Uotani et al. (2014). In both of the unvegetated and vegetated domains, the system was vertically segregated into the vegetation and overlying open water layers by an interface that encompasses the vegetation canopy. Mass and momentum conservation were formulated in layer-averaged forms for analytical simplicity. The internal shear stresses in the mixing layer were represented in terms of entrainment velocity, q_i, and velocity difference between the two layers ($\Delta u, \Delta v$) as:

$$(\tau_{xi}, \tau_{yi}) = \rho q_i (\Delta u, \Delta v) \qquad (11)$$

These terms were added as an additional resistance component in the momentum equations for both of the two layers. Here, "entrainment" means a fluid motion perpendicularly crossing the two-layer interface. Figure 11 shows a numerical solution of stream-wise velocity in a vegetated reach compared with the field data collected by the H-ADCP (Michioku et al. 2012). Abscissas in the figures indicate a time coordinate and ordinate is a cross-wise distance from the right-side bank, respectively. Note that, according to the H-ADCP's specification, the reliable data is the signals surrounded by the broken line in the figures. The flow is accelerated with increasing discharge, especially in the main channel, while acceleration is not so significant on the floodplain because of vegetation drag.

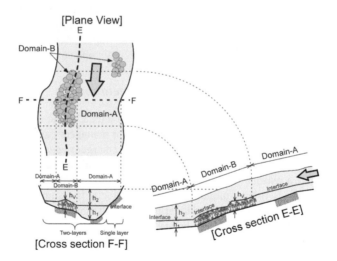

Fig. 10. Two-dimensional two-layer model.

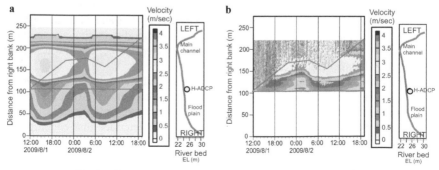

Fig. 11. Time-dependent velocity profile at the 23.6 km cross section.
(a) Numerical solution by, (b) H-ADCP observation

Hydrodynamics of Rock-Arranged Structures

River facilities such as dykes, embankments, weirs, spurs, groynes, ripraps, ramps, fish ladders, and so on, have increasingly been constructed with natural stones, rubbles, and boulders in recent years, which is a common solution for creating nature-oriented conditions. Another advantage is that materials are the original resources that can be collected *in situ*. Pores, concavities, and convexities on the granular surface allow aquatic animals to crawl and swim up the structure and allow the streamwise migration of aquatic life. Physical and chemical substances such as sediments and suspended organic matter can pass downstream through the permeable body, which eventually minimizes sedimentation and eutrophication in a pool formed behind the structures. Micro-organisms inhabiting the granular surface may decompose organic matter which contributes to the purification of river water as it flows through the permeable body. It is also expected that turbulence generated in the granular media will promote aeration, helping aerobic decomposition of organic matter. Besides biochemical and ecological functions of these structures, attention is paid more to the hydraulic perspective in order to provide fundamental knowledge necessary for the hydraulic design of structures.

Hydraulic conductivity and permeability of rock-fill structures

Conductivity and permeability are the key parameters in designing rock-fill structures. The through-flow is not like a Darcian porous media flow, since the structure is installed in running water and inertia force and turbulence are dominant in the porous body. Commonly used non-Darcian resistance laws in the literature are grouped under two broad forms, a quadratic law and a power relationship, as:

$$I = A_1 V + A_2 V^2 \tag{12}$$

$$I = A_3 V^W \tag{13}$$

where V is macroscopic or apparent velocity, $I = -(dp/dx)/\rho g$ is the hydraulic gradient, (dp/dx) is the pressure gradient in the flow direction, and ρ is the fluid density. The exponent, W, and the proportional coefficients, (A_1, A_2, A_3), are empirically or theoretically determined constants. Arbhabhirama and Dinoy (1973), Hansen et al. (1994), Mustafa and Rafindadi (1989), and Ward (1964), are among those who investigated the functional forms of Eqs. (12) and (13). A review of the non-Darcian resistance law was provided by Trussell and Chang (1999). George and Hansen (1992), proposed a conversion method between the two equations (Eqs. (12) and (13)). Although both formulae are originally based on experimental results, they are theoretically accepted after analytical verification by many researchers. In Eq. (12), the first term specifically coincides with Darcy's law, while the second term corresponds to the resistance law for a fully turbulent flow in a porous body. Therefore, Eq. (12) should be valid for a wide range of Reynolds number from laminar to turbulent flows.

Mccorquodale et al. (1978), examined a functional dependency of Eq. (12) for crushed rock and river gravel. A formula was proposed by considering effects of particle size, distribution and shape, surface roughness, porosity, and wall effects and verified for a wide range of pore Reynolds number. One of the popularly used formulae for Eq. (12) was proposed by Ward 1964 as:

$$i = \frac{v}{gK}V + \frac{c}{g\sqrt{K}}V^2 \tag{14}$$

where K is the permeability with a dimension of length squared and $c=0.55$ is a dimensionless coefficient proposed by Ward (1964). Arbhabhirama and Dinoy (1973), argued that c was dependent on particle mean diameter, d_m, and porosity, n, as:

$$c = f\left(\frac{d_m}{\sqrt{K/n}}\right)^{-3/2} \tag{15}$$

where f is a non-dimensional coefficient. Shimizu et al. (1990), assumed a linear functional relationship between the permeability, K, and the particle diameter, d_m, as:

$$\sqrt{K} = e \cdot d_m \tag{16}$$

Several parameters in the hydraulic conductivity formula must be experimentally determined, since shapes, sizes, and diameter distribution of particles and pores of materials are heterogeneously and irregularly distributed in the structure.

Li et al. 1998, provided theoretical basis for Eq. (12) by applying an analogy of flow in a pipe network and determined model parameters in Eq. (12) for a wide range of Reynolds number from experimental datasets collected by Stephenson (1979), and other experiments. A formula was proposed by Li et al. (1998), after a minor modification of Stephenson's proposal as follows:

$$i = \frac{1300v}{d^2 gn}V + \frac{3.84}{dgn^2}V^2 \tag{17}$$

where n is porosity and d is diameter of a rock particle. Other formulae of flow resistance were additionally summarized in Table 1.

Table 1. Flow resistance law summarized by Li et al. (1998).

Author	Equation	Remarks
Ergun (1952)	$i = 150\dfrac{(1-n)^2 v}{gn^3 d^2}V + 1.75\dfrac{(1-n)}{dgn^3}V^2$	Rearranged
McCorquodale et al. (1978)	$i = \dfrac{70v}{gnR^2}V + \dfrac{0.81}{gn^{1/2}R}V^2$	For crushed rockfill (Hansen 1992)
Wilkins (1956)	$i = \dfrac{0.0465V^{1.85}}{R^{0.925}n^{1.85}}$	$R=d/10$ (Parkin 1991)
Martins (1990, 1991)	$i = \dfrac{(1-n)V^2}{0.56^2 n^3 2gd}$	Rearranged for uniform rockfill material, $C_w=1$
Gent (1991)	$i = 1207.06\dfrac{(1-n)^2 v}{gn^3 d^2}V + 1.209\dfrac{(1-n)}{dgn^3}V^2$	Transformed for rockfill
Stephenson (1979)	$i = \dfrac{K_t V^2}{gdn^2}$	$K_t=4$ for rockfill; $K_t=2$ for semi-rounded stone; $K_t=1$ for smooth polished marbles

Rubble mound weir

The rock-fill's conductivity law can be applied to hydraulic analysis for rock-arranged structures. The first example is the rubble mound weir in Photo 2. Depending on water surface level and weir height, three different flow systems, (a) submerged weir, (b) transition and (c) emergent weir, appear as depicted in Fig. 12.

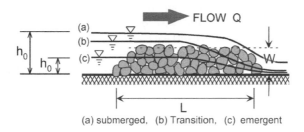

(a) submerged, (b) Transition, (c) emergent

Fig. 12. Flow around a rubble mound weir.

Emergent weir (Fig. 12(c))

Michioku et al. (2005a), obtained an analytical solution for the stage-discharge rating curve for an emergent weir in Fig. 13. The system was longitudinally partitioned into the three reaches; (I) the cross section at the leading edge $x=0$, where flow suddenly converges from the open channel to the porous body, (II) the reach between $x=0 \sim L$ in which the subsurface flow is gradually varied in the porous body, and (III) the cross section at the downstream end of the weir $x=L$, where flow rapidly diverges from the porous body to the open channel.

  I) Leading edge at x=0:
    Assuming a suddenly converging flow around $x=0$ from an open channel pouring into the weir of porosity n, a momentum principle was applied at $x=0$ to obtain a relationship between h_0 and h_1.

  II) Reach between $x=0 \sim L$:
    A solution of water surface profile in the reach of $x=0 \sim L$ was given by using the non-uniform flow equation with the quadratic law of flow resistance as:

$$\frac{d}{dx}\left(\frac{U^2}{2g}\right) + \frac{dh}{dx} - S_0 + \frac{\nu}{gK}U_s + \frac{c}{g\sqrt{K}}U_s^2 = 0 \tag{18}$$

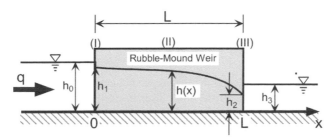

Fig. 13. Model system and definition of variables for an emergent weir.

Where, $U=q/nh$ and $U_s=q/h$ are the pore and macroscopic apparent velocities, respectively, q is the unit width discharge, and S_0 is the channel slope.

III) Downstream end at $x=L$:

The water depths h_2 and h_3 were correlated by applying a momentum principle to the suddenly diverging flow from the porous body to the open channel at $x=L$. For h_3 smaller than the critical water depth, a control section appears at $x=L$ and the flow rate, q, is determined only by the critical flow condition independently of h_3 (termed "C-Flow"). Otherwise, q depends on h_3 (termed "S-Flow").

By connecting the water surface profiles in the three reaches, the solution for the stage-discharge rating curve was obtained. In Fig. 14, the normalized discharge $F_0=q/(gh_0^3)^{1/2}$ is plotted against h_0/L and $\Delta h/h_0$ ($=(h_3-h_0)/h_0$) for a specific grain diameter $d_m/h_0=0.2$, where a substantial agreement between the theory and the laboratory data is recognized.

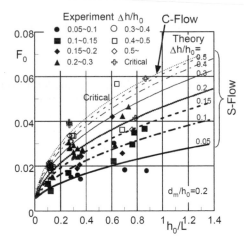

Fig. 14. Functional relationship between F_0 and h_0/L plotted for varied water level difference $\Delta h/h_0$ ($d_m/h_0=0.2$).

Submerged weir

Hydrodynamics of the submerged weir were examined assuming a two-layer flow structure consisting of a free surface flow over the weir and a turbulent seepage flow in the weir as depicted in Fig. 15 (Michioku and Maeno 2004). The continuity equations for the two layers are:

$$\frac{d}{dx}(U_U h) = q_i = -\frac{d}{dx}(U_L nW) = -\frac{d}{dx}(U_S W) \tag{19}$$

where, U_U is the average velocity in the overpassing layer, and U_L and $U_S=nU_L$ are the pore and macroscopic velocity in the lower layer, respectively, and W is the weir height. The variable, q_i, is the entrainment velocity as mentioned before. The total discharge per unit width q is given by:

$$q=q_U+q_S=U_U h+U_S W \tag{20}$$

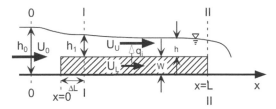

Fig. 15. Flow system in a submerged weir.

Momentum balances in the two layers are written as follows:

$$\frac{d}{dx}\left(\frac{U_U^2}{2g}\right)+\frac{dh}{dx}-i+\frac{\tau_w P}{\rho g A}+E\frac{q_i}{gh}(U_U-U_S)=0 \tag{21}$$

$$\frac{d}{dx}\left(\frac{U_L^2}{2g}\right)+\frac{dh}{dx}-i-E\frac{q_i}{gW}(U_U-U_S)+\frac{\nu}{gK}U_S+\frac{c}{g\sqrt{K}}U_S^2=0 \tag{22}$$

where, τ_w is wall friction on the top surface of weir, and (A, P) are the cross section area and wetted perimeter in the upper layer, respectively. A switching parameter, E, is included in the above equations for identifying the entrainment direction, which is defined as:

$$E=1 \text{ for } q_i>0 \text{ and } E=-1 \text{ for } q_i<0 \tag{23}$$

The last two terms in Eq. (22) are specifically equivalent to Eq. (14).

Similar to the emergent weir, a momentum principle was applied to rapidly varied flow around the leading edge at $x=0$. In addition, a control section appears with rapid drawing down at $x=L$ and then a singularity condition was applied to the differential equation system Eqs. (19–22) at $x=L$ to obtain solutions of discharge and critical water depth.

A solution of water surface profile is compared with the laboratory data in Fig. 16. Figure 17 demonstrates functional dependencies of dimensionless discharge $F_0=U_0/\sqrt{gh_0}=Q/B_0\sqrt{gh_0^3}$ on rubble's diameter d_m/h_0, weir length L/h_0 and weir height $w=W/h_0$. All the solutions from the two-layer model are well correlated with the laboratory data.

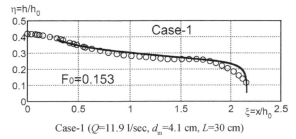

Case-1 (Q=11.9 l/sec, d_m=4.1 cm, L=30 cm)

Fig. 16. Analytical water surface profile (solid lines) with experimental data (open circles).

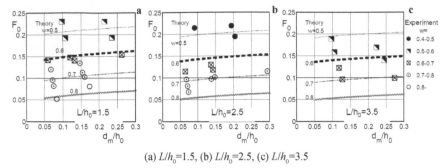

(a) $L/h_0=1.5$, (b) $L/h_0=2.5$, (c) $L/h_0=3.5$

Fig. 17. Dimensionless discharge F_0 versus normalized rubble's diameter $w=d_m/h_0$.

Rubble mound groyne

By extending the two-layer model for the rubble mound weir to a horizontal two-dimensional system, hydrodynamics of rubble mound groynes were examined (Michioku et al. 2005b, 2008). Configuration of the two-dimensional two-layer model is consistent with that for the vegetated flow system in Fig. 10. The only difference is that the vegetation canopy was replaced by the rubble mound body and a minor modification of flow resistance was given by replacing Eq. (7) by Eq. (22). Figure 18 indicates velocity vectors from the analysis and laboratory experiment, where the velocity was depth-averaged. Drag force acting on the rubbles can be evaluated by the last two terms in Eq. (22). A two-layer flow analysis and a field measurement were carried out to estimate drag force acting on the eleven units of groynes in Photo 8. The maximum drag force, $\tilde{F}_{P\,max}$, imposed on each groyne unit was evaluated for a flood which occurred in 2004 as in Fig. 19. The black-colored numbers in the figure denote the groynes destroyed during the flood. Estimation of $\tilde{F}_{P\,max}$ based on the two-layer model is in good agreement with the observed damage of groyne. Also note that groynes tend to be heavily damaged at the leading edge and downstream end.

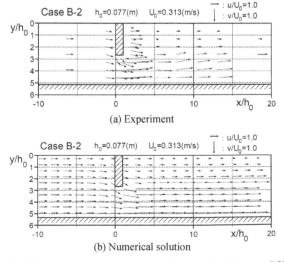

Fig. 18. Velocity vectors around submerged groyne (d_m=3.5cm, n=0.373).

Photo 8. Rubble mound groynes and Akashi River viewed from downstream.

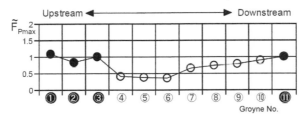

Fig. 19. Maximum drag force acting on 11 units of groyne, $\tilde{F}_{Pmax}$. The black colored numbers correspond to the groynes destructed during a flood event in 2004.

Rough ramps

Ramps and fish ways are the structures to provide continuities of hydraulic, geomorphological, and ecological structures along the streamwise direction. In recent years, natural stones and rocks are increasingly used to construct rough ramps. In some cases, rough ramps are expected to serve the additional function of fish ways. Based on Tamagni et al. (2014), rough ramps or block ramps are classified into (A) block-carpet type and (B) block-cluster type. The former is a tightly packed block carpet covering the entire river width. The latter is a ramp with regularly or randomly arranged block clusters. Ecohydraulic issues of rough ramps are flow retention or resistance, wall shear stress, drag force, stability and failure of the structure, sedimentation, turbulence and eddy structure, energy dissipation rate, aeration performance, functionality of fish migration, and so on.

Flow classification might be helpful in better understanding ecohydraulics and in making a more suitable design of rough ramps. The first classification is based on the relative submergence, H/h_B, where H is the water depth and h_B is the boulder's height. The regimes classified by H/h_B are (i) emergent boulder, $H/h_B<1$, (ii) overtopped boulder, $1<H/h_B<2$, and deeply submerged boulder, $2<<H/h_B$. The second concerns the obstruction ratio parameterized by the sum of boulder widths to channel width. A single solitary boulder rarely affects the outer flow regime, while densely arranged boulders bring massive influence on flow regime. This category primarily depends on arrangement and spacing of boulders. The third classification is concerned with subcritical, critical, moderate, and significantly-supercritical flows.

The flow in and around the rough ramps presents a complex three-dimensional free surface problem, and it is extremely difficult to find universal laws of hydraulics on a theoretical basis. This is why commonly accepted design guidelines are still limited for the rough ramps. In most cases, the performance of rough ramps are only empirically measured by relating parameters such as height and length of the ramp, shape and diameter of base materials and boulder, block placement density, maximum velocity, minimum or critical flow depth, relative submergence of boulders, target fishes or aquatic lives, and so on. A few examples of hydraulic characteristics of rough ramps are briefly demonstrated in the following paragraphs.

i) Energy Dissipation Rate:

Ahmad et al. 2009, proposed a formula for the energy dissipation on block ramps with submerged boulders based on a dimensional argument as:

$$\Delta E_{rB} = Func.(\frac{h_C}{H_r}, \frac{D_B}{h_C}, \Gamma) \tag{24}$$

Where, ΔE_{rB} is the energy dissipation rate, h_C is critical depth of flow, H_r is height of ramp, D_B is diameter of boulders and Γ is concentration of boulders, which is equal to the ratio of ramp surface covered with the boulders and the surface area of the ramp.

ii) Mean Flow Velocity:

Baki et al. (2014) examined mean flow velocity u_{avg} in staggered arrangement of boulders in a rock-ramp-type, nature-like fish way. The following formula for u_{avg} was proposed.

$$u_{avg} = \frac{2g}{C_D N A_p} \sqrt{R_v S} \tag{25}$$

Where C_D is drag coefficient of boulder element, N is number of boulders over a bed area, A_p is vertical projected area of boulder element, R_v is a volumetric hydraulic radius, and S is channel slope.

iii) Critical Discharge of Structural Failure:

Pagliara (2007) investigated failure mechanisms of rough ramps with regularly arranged boulders and he proposed a critical condition of failure as:

$$F_{dc} = 1.98 S^{0.18} \left(\frac{h}{D_{84}}\right)^{0.36} (1+\Gamma)^{a_1} \tag{26}$$

where, $F_{dc} = \dfrac{U_C}{\sqrt{(s-1)gD_{50}}}$ is a particle densimetric Froude number, U_C is a critical velocity for failure, $s = \rho_S/\rho_w$ is the relative particle gravity, D_{xx} is rock diameters for which $xx\%$ of the mixture is finer, and a_1 is an empirical exponent depending on the boulder's disposition such as rows, arc, random, and so on.

Concluding Summary

Extensive and wise use of nature-oriented river structures in restoration works might be a final solution for rejuvenating degraded river environments. Many varieties of complex mechanisms are involved in hydraulics, fluvial dynamics, geomorphology, and ecology in which multi-disciplinary approaches are indeed necessary to solve a wide range of engineering issues. In addition to state-of-the-art design schemes, a new paradigm of proper and adaptive river management is also required for creating sustainable river systems. Promoting the understanding of the interrelationship between hydrodynamics, geomorphology, and ecology and to integrate scientific knowledge from various disciplines should be tackled urgently. The first-step of this goal must be to establish a design and management guideline for the river structures built with natural materials based on a scientific platform.

Keywords: Rock-arranged structures, vegetation, drag force, conveyance capacity.

References

Aberle, J. and Järvelä, J. 2013. Flow resistance of emergent rigid and flexible floodplain vegetation. Journal of Hydraulic Research 51(1): 33–45.

Ackerman, J.D. and Okubo, A. 1993. Reduced mixing in a marine macrophyte canopy. Functional Ecology 7: 305–309.

Ahmad, Z., Petappa, N.M. and Westrich, B. 2009. Energy dissipation on block ramps with staggered boulders. Journal of Hydraulic Engineering 135-6: 522–526.

Arbhabhirama, A. and Dinoy, A.A. 1973. Friction factor and Reynolds number in porous media flow. Journal of Hydraulic Division ASCE 99-6: 901–911.

Baki, A.B.M., Zhu, D.Z. and Rajaratnam, N. 2014. Mean flow characteristics in a rock-ramp-type fish pass. Journal of Hydraulic Engineering 140-2: 156–168.

Carollo, F.G., Ferro, V. and Termini, D. 2005. Flow resistance law in channels with flexible submerged vegetation. Journal of Hydraulic Engineering, ASCE 131-7: 554–564.

Christopher, S.J., Uwe, K.G., Anthony, P. and Angelina, A.J. 2008. Influence of foliage on flow resistance of emergent vegetation. Journal of Hydraulic Research 46-4: 536–542.

Corenblit, D., Steiger, J., Gurnell, A.M., Tabacchi, E. and Roques, L. 2009. Control of sediment dynamics by vegetation as a key function driving biogeomorphic succession within fluvial corridors. Earth Surface Processes and Landforms 34: 1790–1810.

de Langre, E., Gutierrez, A. and Cossé, J. 2012. On the scaling of drag reduction by reconfiguration in plants. Comptes Rendus Mecanique 340(1-2): 35–40.

Doncker, L.D., Troch, P., Verhoeven, R., Bal, K., Meire, P. and Quintelier, J. 2009. Determination of the manning roughness coefficient influenced by vegetation in the river Aa and Biebrza river. Environmental Fluid Mechanics 9: 549–567.

Fukuoka, S. 2005. Flood Flow Dynamics and Design Method for River Channel. Morikita Publishing Co., Tokyo (in Japanese).

George, G.H. and Hansen, D. 1992. Conversion between quadratic and power law for non-Darcy flow. Journal of Hydraulic Engineering ASCE 118-5: 792–797.

Ghisalberti, M. and Nepf, H. 2002. Mixing layers and coherent structures in vegetated aquatic flow. Journal of Geophysical Research 107(C2): 3-1–3-11.

Hansen, D., Garga, V.K. and Townsend, D.R. 1994. Selection and application of a one-dimensional non-Darcy flow equation for two-dimensional flow through rockfill embankments. Canadian Geotechnical Journal 32(2): 223–232.

Huai, W., Wang, W. and Zeng, Y. 2013. Two-layer model for open channel flow with submerged flexible vegetation. Journal of Hydraulic Research 51(6): 708–718.

Jalonen, J., Järvelä, J. and Aberle, J. 2013. Leaf area index as vegetation density measure for hydraulic analyses. Journal of Hydraulic Engineering ASCE 139(5): 461–469.

Jang, C.L. and Shimizu, Y. 2007. Vegetation effects on the morphological behavior of alluvial channels. Journal of Hydraulic Research 45(6): 763–772.

Kouwen, N. and Unny, T.E. 1973. Flexible roughness in open channel. Journal of Irrigation and Drainage Engineering, ASCE 99(5): 713–728.

Kouwen, N. 1992. Modern approach to design of grassed channels. Journal of Irrigation and Drainage Engineering 118(5): 733–743.

Luhar, M., Rominger, J. and Nepf, H. 2008. Interaction between flow, transport and vegetation spatial structure. Environmental Fluid Mechanics 8(5-6): 423–439.

Luhar, M. and Nepf, H. 2013. From the blade scale to the reach scale: a characterization of aquatic vegetative drag. Advances in Water Resources 51: 305–316.

Mccorquodale, J.A., Hannoura, A.A. and Nasser, M.S. 1978. Hydraulic conductivity of rockfill. Journal of Hydraulic Research 16(2): 123–137.

Michioku, K. and Maeno, S. 2004. Study on flow structure and discharge over a permeable rubble mound weir. Proceeding of 4th International Symposium on Environmental Hydraulics and 14th APD-IAHR 2: 1801–1808.

Michioku, K., Maeno, S., Furusawa, T. and Haneda, M. 2005a. Discharge through a permeable rubble mound weir. Journal of Hydraulic Engineering ASCE 131(1): 1–10.

Michioku, K., Nanjo, M., Ishigaki, T. and Maeno, S. 2005b. Two-dimensional analysis on solid-liquid-phase flow in and open channel with a rubble mound groin. Proceeding of 31st IAHR Congress.

Michioku, K., Li, Z. and Kanda, K. 2008. Analysis on river flow in and around rubble mound groynes. Proceeding of River Flows 2008. 1: 843–850.

Michioku, K., Ohchi, Y., Aga, K., Miyamoto, H. and Kanda, K. 2012. Strategy for ecohydraulic management of vegetated river channel. E-Book: 9th International Symposium on Eco-hydraulics 2012 Proceedings.

Michioku, K., Kanda, K., Kometani, S., Irie, Y. and Sakamoto, C. 2015. Manning's coefficient of alternatively arranged sandbars with tree vegetation. Proceeding of 36th IAHR World Congress.

Morris, H.M. 1955. A new concept of flow in rough conduits. Transaction of ASCE 120: 373–398.

Mustafa, S. and Rafindadi, N.A. 1989. Nonlinear steady state seepage into drains. Journal of Irrigation and Drainage Engineering ASCE 115(3): 358–376.

Nepf, H. 1999. Drag, turbulence, and diffusion in flow through emergent vegetation. Water Resources Research 35: 479–489.

Nepf, H.M. 2012. Hydrodynamics of vegetated channels. Journal of Hydraulic Research 50(3): 262–279.

Nikora, V., Larned, S., Nikora, N., Debnath, K., Cooper, G. and Reid, M. 2008. Hydraulic resistance due to aquatic vegetation in small streams: field study. Journal of Hydraulic Engineering, ASCE 134(9): 1326–1332.

Nikora, V. 2010. Hydrodynamics of aquatic ecosystems: an interface between ecology, biomechanics and environmental fluid mechanics. River Research and Applications 26: 367–384.

Pagliara, S. 2007. Failure mechanisms of base and reinforced block ramps. Journal of Hydraulic Research 45(3): 407–420.

Rameshwaran, P. and Shiono, K. 2007. Quasi two-dimensional model for straight overbank flows through emergent vegetation on floodplains. Journal of Hydraulic Research 45(3): 302–315.

Shimizu, Y., Tsujimoto, T. and Nakagawa, H. 1990. Experiment and macroscopic modeling of flow in high permeable porous medium under free-surface flow. Journal of Hydrosci Hydraulic Engineering JSCE 8(1): 69–78 (in Japanese).

Shimizu, Y. and Tsujimoto, T. 1994. Numerical analysis of turbulent open-channel flow over a vegetation layer using a k-ε turbulence model. Journal of Hydrosci Hydraulic Engineering JSCE 11: 57–67 (in Japanese).

Shucksmith, J.D., Boxall, J.B. and Guymer, I. 2011. Bulk flow resistance in vegetated channels: analysis of momentum balance approaches based on data obtained in aging live vegetation. Journal of Hydraulic Engineering ASCE 137(12): 1624–1635.

Stephenson, D. 1979. Rockfill in Hydraulic Engineering. Elsevier Scientific Publishers, Amsterdam, The Netherlands, pp. 19–24.

Sukhodolov, A. and Sukhodolova, T. 2010. Case study: effect of submerged aquatic plants on turbulence structure in a lowland river. Journal of Hydraulic Engineering 136(7): 434–446.

Takebayashi, H. 2014. River Engineering. Corona Publishing Co. Ltd. (in Japanese).

Tamagni, S., Weitbrecht, V. and Boes, R.M. 2014. Experimental study on the flow characteristics of unstructured block ramps. Journal of Hydraulic Research 52(5): 600–613.

Termini, D. 2015. Flexible vegetation behavior and effects on flow conveyance: experimental observations. International Journal of River Basin Management 401–411.

Trussell, R.R. and Chang, M. 1999. Review of flow through porous media as applied to head loss in water filters. Journal of Environmental Engineering ASCE 125(11): 998–1006.

Uotani, T., Kanda, K. and Michioku, K. 2014. Experimental and numerical study on hydrodynamics of riparian vegetation. Journal of Hydrodynamics Ser B 26(5): 796–806.

van Prooijen, B.C., Battjes, J.A. and Uijttewaal, W.S.J. 2005. Momentum exchange in straight uniform compound channel flow. Journal of Hydraulic Engineering ASCE 131(3): 175–183.

Velasco, D., Bateman, A. and Medina, V. 2008. A new integrated, hydro-mechanical model applied to flexible vegetation in riverbeds. Journal of Hydraulic Research 46(5): 579–597.

Ward, J.C. 1964. Turbulent flows in porous media. Journal of Hydraulic Division ASCE 90(5): 1–12.

White, B. and Nepf, H. 2007. Shear instability and coherent structures in a flow adjacent to a porous layer. Journal of Fluid Mechanics 593: 1–32.

White, B. and Nepf, H. 2008. A vortex-based model of velocity and shear stress in a partially vegetated shallow channel. Water Resources Research 44(1): W01412.

Wilson, C.A.M.E., Stoesser, T., Bates, P.D. and Pinzen, A.B. 2003. Open channel flow through different forms of submerged flexible vegetation. Journal of Hydraulic Engineering ASCE 129(11): 847–853.

Wilson, C.A.M.E., Hoyt, J. and Schnauder, I. 2008. Impact of foliage on the drag force of vegetation in aquatic flows. Journal of Hydraulic Engineering ASCE 134(7): 885–891.

Yang, K., Cos, S. and Knight, D.W. 2007. Flow patterns in compound channels with vegetated floodplains. Journal of Hydraulic Engineering 133(2): 148–159.

Yang, W. and Choi, S. 2009. Impact of stem flexibility on mean flow and turbulence structure in depth-limited open channel flows with submerged vegetation. Journal of Hydraulic Research 47(4): 445–454.

Yang, W. and Choi, S. 2010. A two-layer approach for depth-limited open-channel flows with submerged vegetation. Journal of Hydraulic Research 48(4): 466–475.

CHAPTER 5

Block Ramps

A Field Example

Karol Plesinski[1,]* and *Artur Radecki-Pawlik*[1,2]

INTRODUCTION

The Water Framework Directive (Directive 2000/60/EC), introduced in many countries of the European Union, imposes on the Member States an obligation to bring their surface waters to at least good ecological status (Allan 2012; Lava et al. 2014). Therefore, it is more and more necessary to perceive the designed hydraulic construction not only as a technical object, but primarily as an artificial element that potentially can cause a significant damage to the river bed. Understanding of hydraulic processes occurring in hydraulic structures and within their range of impact on the river bed is therefore a crucial issue allowing for the determination of the type of impact on the water flow regime. If this impact is negative, then every effort should be made to have it reduced or even eliminated (Żelazo 1992; Żelazo and Popek 2002; Radecki-Pawlik et al. 2014).

However, in order to be able to do this, it is necessary to have very good understanding of the hydrodynamic conditions prevailing in a given type of construction, including rapid hydraulic structures, of which one is a subject of the presented study. These objects are increasingly often built in the Carpathian river beds. The principle of their operation is to dissipate the energy of flowing water on strongly inclined slope aprons, which constitute a main element of these objects or to a small extent on an energy dissipation pool, which is optional. The main advantages of using block ramps include (Oertel 2013; Oertel et al. 2011; Pagliara et al. 2008a,b,c,d, 2009a,b, 2011a,b, 2012, 2013; Radecki-Pawlik et al. 2013; Radecki-Pawlik et al. 2014):

- The use of natural elements for construction, which favorably interact and quickly fit into the landscape of a river valley (Hernik 2010; Radecki-Pawlik 2010a; Plesiński et al. 2013, 2013b; Abdel Sattar et al. 2013).
- They do not prevent the migration of aquatic organisms (Lusk 1980; Ślizowski 2002; Oertel and Schlenkhoff 2012).

[1] Department of Hydraulics Engineering and Geotechnic, University of Agriculture in Krakow, al. Mickiewicza 24-28, 30-059 Kraków, Poland.
[2] Institute of Structural Mechanics, Faculty of Civil Engineering, Cracow University of Technology, ul. Warszawska 24, 31-155 Kraków, Poland.
* Corresponding author: kplesinski@ar.krakow.pl

- They are cheaper to build and operate than other hydrotechnical structures (Ślizowski 1990).
- They increase the oxygenation of the river, which supports the process of its self-purification (Radecki-Pawlik 2010b).
- They reproduce or mimick gravel bars and/or riffle sequences therefore causing braiding of rivers (Radecki-Pawlik 2010b, 2013).
- They stabilize the river bed and prevent erosion, due to which they increasingly often replace the traditional dams.

However, despite a number of advantages of using block ramps in channels of the Polish rivers, no one so far has started their monitoring in the field. Until now, they have been studied and observed by the Austrians in the Alps (Schauberger 1957; Niel 1960; Scheuerlein 1968; Hartung and Scheurlein 1970), who are the forerunners in using this type of construction by Czechs and Slovaks in the Czech Sudeten (Jarabač and Vincent 1967; Kališ 1970; Jarabač 1973; Knauss 1980), and to a smaller extent, by Italians and Germans (Oertel 2013). Therefore, there is a need to carry out the study and analysis of the impact of this type of structure on the Carpathian river beds. For the future, it seems important also to analyze the water discharge through the gravel that forms the structure (Carling et al. 2006), but that is a subject for future study.

The aim of the present study is to determine the hydrodynamic parameters within the region of the interlocked-carpet block ramp and on the slope apron of this structure. This aim is an attempt to show the capability of monitoring the existing hydraulic structures of this type in river beds. Showing the hydrodynamic conditions will allow verification as to whether or not the hydraulic impact of the mentioned construction results in significant changes in the stream water flow regime.

One of the block ramps, situated in the bed of the Porębianka Stream in the Gorce Mountains, was selected for the study. The measurements of the velocity distribution upstream and downstream of the hydraulic constructions were made to investigate whether the hydraulic flow conditions are comparable, or do not cause excessive increase of dynamic forces on the river bed and a drastic change in the flow regime. Thus, the study analyzed the tangential velocities, average and dynamic velocities, occurring in the block ramp and within its impact zone. Additionally, basic characteristics associated with the water movement in the form of Froude's and Reynolds numbers, were analyzed.

Description of the Catchment and Research Facility

The catchment of the Porębianka Stream is located in the Western Carpathians, on the northern slopes of the Gorce Mountains in the Polish Carpathians (Kondracki 2000). Kudłoń (1276 m a.s.l.) is the highest point of the catchment, while the mouth of the River Porębianka to Mszanka in Mszana Dolna (400 m a.s.l.) is the lowest point. Geographically, the catchment is located in the Lesser Poland voivodeship, Limanowski poviat.

Porębianka stream is a 15.4 km-long watercourse (Fig. 1). Its spring is located on the slopes of Obidowiec (1000 m a.s.l.). Porebianka is a tributary of the Mszanka River. Porębianka stream catchment has an area of 71.8 km² and is located in the seepage spring—alimentation zone with the highest density of springs, that is, 7–12 springs · 1 km⁻². The density of the river network in this area is over 3 km · 1 km⁻² (Plesinski et al. 2013). The width of the Porębianka riverbed ranges from 1 to 140 m.

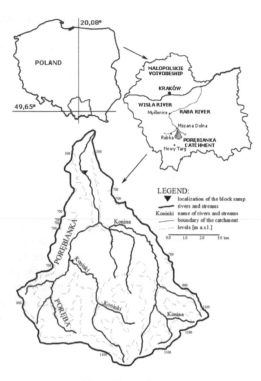

Fig. 1. Research region.

Its catchment is uncontrolled hydrologically; however in 1982–1991, there was a water gauge located in Niedźwiedź 5.2 km from the mouth. During that time water levels and flows were observed at the water gauge station of the Institute of Meteorology and Water Management (IMGW). The observed data indicate that the Porębianka is a typical mountain stream, characterized by high variability of water levels. The largest floods occur during summer rainfalls, while floods caused by the spring thaw are smaller, but long-term (Komędera 1993).

Kościelniak (2004) carried out geomorphological mapping according to the "Instructions for Riverbed Mapping" (Kamykowska et al. 1999) in which she recognized and separated the main riverbed sections:

- Erosion sections—cut into solid rock. Modeled by river bed erosion with large number of rock barrages. Those sections occur in the upper bank of the river, below the anti-rubble dams.
- Redeposition-erosion sections—cut into the sediments. Modeled by the deposition, redeposition, and side erosion. There are small sandbanks and occasional rock outcrops.
- Redeposition sections—cut into the sediments. Sandbanks are of large surfaces. These sections occur above the anti-rubble dams and at the sites of barrages and interlocked-carpet block ramps.

The formation of these sections in Porębianka stream bed is largely affected by the presence of check dams. They form a border between different types of beds. Erosion occurs below dams, while accumulation occurs above dams.

Along the 0+836–4+080 km section Porębianka stream watercourse was regulated using 25 interlocked-carpet block ramps. The purpose of the regulation was to stabilize the stream bed and banks of the river in order to protect the adjacent land and the asphalt road Mszana Dolna—Niedźwiedź together with neighboring buildings.

Photo 1. The block ramp and the water channel in the region influence it. 1A. The view of the block ramp from downstream. 1B. The block ramp. 1C. The downstream view from the block ramp. 1D. The upstream view.

The regulatory trapezoidal river bed is 28 m wide at the stream bed. The river bed inclination was reduced from 1.25% to 0.55%. In the vast majority of the sections regulated by interlocked-carpet block ramps, the river bed is of alluvial nature (redeposition section). There are rock outcrops in a few places. In contrast, above the regulated section (from 4+180 km), the river bed is of rocky and alluvial character (redeposition-erosion and erosion section). The block ramp No. 14 located in 2+890 km (Fig. 2) was selected for the analysis.

The construction consists of two rows of G-62 steel sheet pilings topped by a reinforced concrete pile cap. The space between sheet pilings was filled with rubble overhead with an average diameter of 0.90 m and layer thickness of 0.80 m. The stone overhead is placed on an even surface without any additional leveling. The inclination of the slope apron is 1:12 (length 12 m, slope 0.99 m).

The object has also a 3 m long and 1.2 m thick overhead in the stream bed above the upper sheet pilings. Below the lower sheet piling there is a 5 m long and 1.2 m thick energy dissipation pool (Fig. 2). In order to concentrate the stream during low water levels, to facilitate the fish migration, the over fall crest was lowered by 0.20 m over

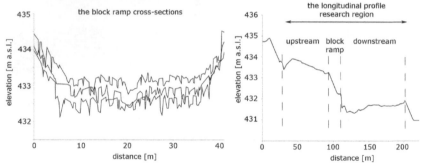

Fig. 2. The interlocked-carpet block ramp on Porębianka stream.

the length of 4 m of the central part of the upper and lower sheet piling. The previous block ramp No. 13 is situated at a distance of 98 m from the lower sheet piling of the studied block ramp (No. 14), while the block ramp No. 15 is at a distance of 68 m from the upper sheet piling.

Research Methodology

Field measurements were conducted at an interlocked-carpet block ramp located in the bed of Porębianka stream and within the region of hydrodynamic impact of this construction on the riverbed (i.e., 45 m above and 60 m below the object). The measurements were conducted in three series.

In the first series of measurements, the values of hydrodynamic parameters were determined based on 70 measuring points (Fig. 3) located within the stream bed, above and below the interlocked-carpet block ramp (points 1–23 and 44–70, respectively)

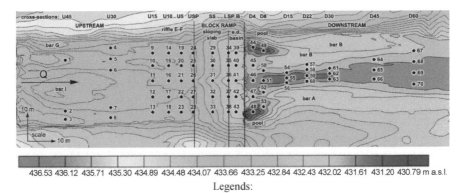

E.d. basin—energy dissipation basin

U45, U30, U15, etc.—the cross-section away from the upper sheet piling about 45, 30, 15, etc. meters

D45, D30, D15, etc.—the cross-section away from the lower sheet piling about 45, 30, 15, etc. meters

USP—the upper sheet piling cross-section

SS—the sloping slap cross-section

LSP—the lower sheet piling cross-section

B—the energy dissipation basin cross-section

Fig. 3. Location of the measuring points—first series of measurements.

as well as at the slope apron and energy dissipation pool of the construction (points 24–43). Their location was planned in the way to reflect the hydrodynamic situation in the river bed as precisely as possible. The hydrodynamic measurements were performed along the line of cross sections.

During the second series of measurements, the values of hydrodynamic parameters were determined based on 63 measuring points, located within the stream bed above and below the interlocked-carpet block ramp (points 1–25 and 40–63, respectively) as well as at the slope apron and the energy dissipation pool of the construction (points 26–39) (Fig. 4).

On the other hand, in the third series of measurements, the hydrodynamic parameters were measured in 70 points (points 1–23 above the construction, 44–70 below the construction, and 24–43 at the slope apron and energy dissipation pool of the object). Their location was planned in the way to reflect the hydrodynamic situation in the river bed as precisely as possible (Fig. 5).

In each series, hydrometric measurements of flowing water were conducted by an OTT Nautilus 2000 electromagnetic current meter. This device measures water flow velocity in the range from 0.001 m · s^{-1} to 10 m · s^{-1}. First, we measured a set of several instantaneous velocities measured just above the bed of the stream – V [m · s^{-1}], average velocity – V_{av} [m · s^{-1}] (Fig. 6) and maximum velocity – V_{max} [m · s^{-1}].

The dynamic velocity is obtained by plotting the regression line between the values of instantaneous velocities and the logarithmic values of the distance between the measurement from the river bed. If the line becomes straight, then we can calculate

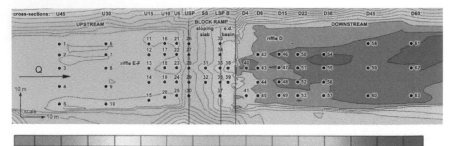

436.53 436.12 435.71 435.30 434.89 434.48 434.07 433.66 433.25 432.84 432.43 432.02 431.61 431.20 430.79 m a.s.l.

Legends:

E.d. basin—energy dissipation basin

U45, U30, U15, etc.—the cross-section away from the upper sheet piling about 45, 30, 15, etc. meters

D45, D30, D15, etc.—the cross-section away from the lower sheet piling about 45, 30, 15, etc. meters

USP—the upper sheet piling cross-section

SS—the sloping slap cross-section

LSP—the lower sheet piling cross-section

B—the energy dissipation basin cross-section

Fig. 4. Location of the measuring points—second series of measurements.

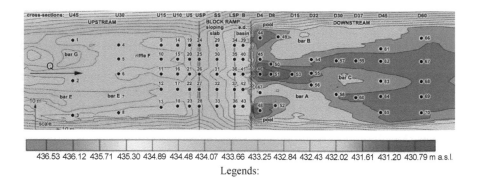

436.53 436.12 435.71 435.30 434.89 434.48 434.07 433.66 433.25 432.84 432.43 432.02 431.61 431.20 430.79 m a.s.l.

Legends:

E.d. basin—energy dissipation basin

U45, U30, U15, etc.—the cross-section away from the upper sheet piling about 45, 30, 15, etc. meters

D45, D30, D15, etc.—the cross-section away from the lower sheet piling about 45, 30, 15, etc. meters

USP—the upper sheet piling cross-section

SS—the sloping slap cross-section

LSP—the lower sheet piling cross-section

B—the energy dissipation basin cross-section

Fig. 5. Location of the measuring points—third series of measurements.

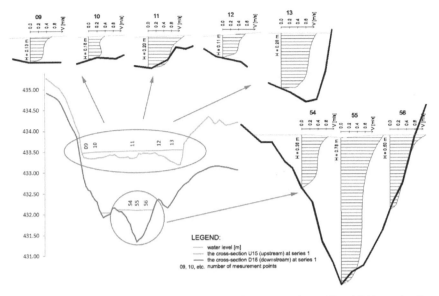

Fig. 6. The velocities profiles (chosen examples for the cross-sections U15 and D15 at series 1).

the dynamic velocity from the coefficient of its inclination to the abscissa axis (Gordon et al. 2007; Radecki-Pawlik 2011):

$$V_* = \frac{a}{5,75} [m \cdot s^{-1}]$$

Where:
a – coefficient of inclination of a straight $v=f(h)$ adopting the form of equation $y=ax+b$
(where: x – height above the river bed on which the velocity was measured; b – intercept of the equation)

The calculated value of the dynamic velocity was used to determine the forces acting on the stream bed, that is, shear stress, according to the formula (Gordon et al. 2007):

$$\tau = \rho \cdot (V_*)^2 [N \cdot m^{-2}]$$

Then the average and maximum Reynolds number was determined (Radecki-Pawlik 2011):

$$Re_{av} = \frac{v_{av} \cdot h}{\upsilon} [-]$$

$$Re_{max} = \frac{v_{max} \cdot h}{\upsilon} [-]$$

The average and maximum Froude number was also determined (Radecki-Pawlik 2011):

$$Fr_{av} = \frac{v_{av}}{\sqrt{gh}} \ [-]$$

$$Fr_{max} = \frac{v_{max}}{\sqrt{gh}} \ [-]$$

Results, Discussion and Conclusions

In total, we conducted three series of measurements and conducted analyses under different regime flow conditions. The first series of measurements was performed immediately after the flood wave, that reached the culmination flow of Q=55 m³ · s⁻¹, which corresponded to a probability of appearance of 1 x 5 years (p=21%). It was one of numerous components of the waves, which in the lower sections of the river valleys caused flooding on the Vistula River and its Carpathian tributaries in 2010. This caused the change in the streambed morphology. At that time, hydrometric measurements were performed and the bed material was collected from the river. The measurements were conducted during the flow Q=2.40 m³ · s⁻¹. During the mentioned flood, some ramp hydraulic structures located in the riverbed were damaged. Therefore, in November and December of 2010, the maintenance and repair works were conducted in the riverbed, which included the renovation of the damaged slope aprons of the interlocked-carpet block ramps. Additionally, the trims in the riverbed were present and gravel was mined. As a result, the bed of the stream was changed; therefore, the second measurement series was conducted in April 2011.

Hydrometric and granulometric measurements were conducted again at the low water-period flow of Q=1.15 m³ · s⁻¹. At the turn of July and August of 2011, a raised water level was observed with a culmination flow of Q=35 m³ · s⁻¹, corresponding to p=35%. As a result, the bed of the stream bed was changed again.

Other field measurements were performed in August 2011 (3rd series) at the flow of Q=3.50 m³ · s⁻¹. Also at that time, a series of measurements was conducted (hydrodynamics and granulometry).

Having obtained results of our observations (Fig. 7), Wilcoxon test was used (Figs. 9–12) for the output data. We gathered our observations in pairs—as it is required for the statistical analysis—for three series of observations and looked into numbers we obtained (Table 1).

The maximum values of flow velocity and shear stress were observed on the sloping slab of the block ramp (Fig. 7), where the values of velocities were up to V=2.86 m · s⁻¹ and shear stresses were up to their top values along the first measuring series. The values of water depths on the block ramp were low (H≤0.20 m) in comparison with upstream and downstream water depths. The maximum values of water depths were found downstream of the hydraulic structure, where the scour was formed-up to 1.5 m. The scours were formed by floods, when the length of hydraulic jump was longer than length of an energy dissipation pool.

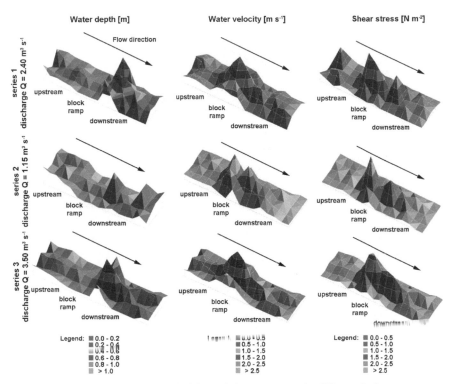

Fig. 7. Water depths, water velocities, and shear stresses under different discharges.

In Fig. 8, one can see the water depth to water velocity ratio values, which determines the types of flow movements. The turbulent—supercritical movement was observed only along the block ramp hydraulic structure. Upstream and downstream of the structure, flow was turbulent—we observed subcritical movement and, in the few cases, transition movement.

Table 1. Results of a Wilcoxon test for the significance of difference in the hydrodynamics parameters for three series of observations for the observed rapid hydraulic structure.

	Average Velocities	Average Shear stresses	Average Froude's numbers	Average Reynolds' numbers
Series 1	0.493	0.629	0.184	0.550
Series 2	0.157	0.265	0.043	0.148
Series 3	0.155	0.213	0.003	0.334

In the case of Series 1, 2, and 3, the comparison of water velocity between an upstream part of the river bed of the block ramp and downstream part are substantially comparable. The same occurred with shear stress values and Re number values. The only difference is in the case of the Froude number—for Series 2 and 3 of the upstream part, positions are different in comparison with the position of the downstream part. We can try to explain it with the fact that influence of the v shape of the notch in the

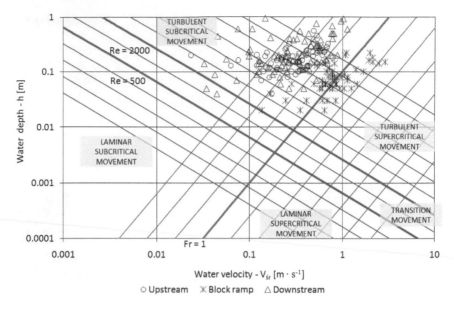

Fig. 8. Stream flow regime in the Porębianka stream during measurements.

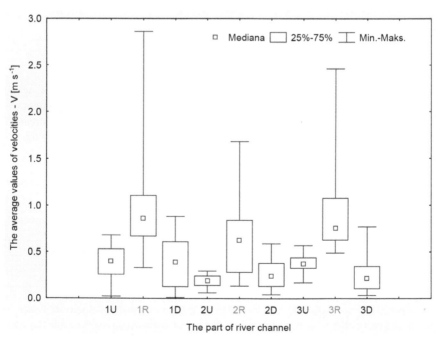

Fig. 9. Values of velocities measured in three series of measurements: 1, 2, 3—measurement series at axis x, U, R, D—the part of channel: U—upstream, R—block ramp, D—downstream.

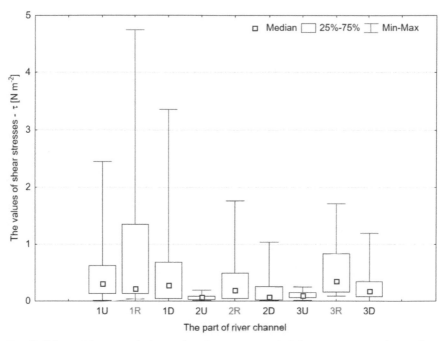

Fig. 10. Values of shear stress in three series of measurements: 1, 2, 3—measurement series at axis x, U, R, D—the part of channel: U—upstream, R—block ramp, D—downstream.

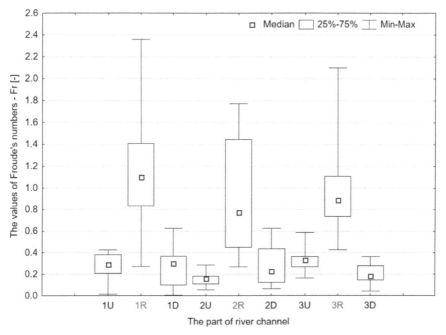

Fig. 11. Values of Fr numbers calculated in three series of measurements: 1, 2, 3—measurement series at axis x, U, R, D—the part of channel: U—upstream, R—block ramp, D—downstream.

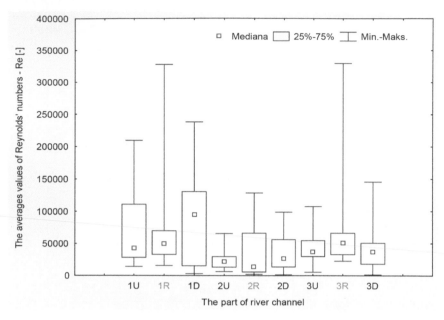

Fig. 12. Values of Re numbers calculated in three series of measurements: 1, 2, 3—measurement series at axis x, U, R, D—the part of channel: U—upstream, R—block ramp, D—downstream.

middle part of the ramp and the influence of the block ramp itself, which has got the same semi v shape along its slope apron. Definitely a future analysis here is necessary to better understand this phenomenon.

The following conclusions could be drawn from the conducted analyses of hydrodynamic parameters within the region of the ramp hydraulic structure:

1. Large variation was found in the hydrodynamic conditions in the river bed and in the construction slab within the region of impact of the interlocked-carpet block ramp (V_{av}=0.01–2.86 m · s^{-1}, τ=0.001–4.74 N · m^{-2}, Re=1 300–725 174 and Fr=0.003–2.36). The variation of hydrodynamic parameters was the greatest in the region of gravel bars (V_{av}=2.09 m · s^{-1}, τ=4.74 N · m^{-2}, Re=723 874 and Fr=0.628), which contributed to the formation of a variety of habitats for macroinvertebrate fauna living in the aquatic environment, which confirms past observations (Wyżga et al. 2012a,b, 2013, 2014; Radecki-Pawlik and Skalski 2008; Skalski et al. 2012).

2. Based on the observations, it can be concluded that the pools, formed below the block ramp, should not be buried, but should remain, because aquatic organisms wandering up the stream may rest there in order to gain strength to overcome the obstacle in the form of the construction. The values of flowing water velocity in the pools were very low, that is, ranged within V_{av}=0.01–0.35 m · s^{-1}.

3. The interlocked-carpet block ramps work properly, as the obtained values of hydrodynamic parameters above and below the constructions were comparable (V_{G}=0.02–1.90 m · s^{-1}, V_{D}=0.01–1.12 m · s^{-1}). This demonstrates the proper dispersion of flowing water energy on the slope apron of the block ramps (and on the energy dissipation pool). This is confirmed by the observations of Ślizowski et al. (2008). Additionally, properly designed interlocked-carpet block ramps

should be characterized by transverse symmetry of hydrodynamic parameter values (V_L=0.17–1.12 m · s^{-1} and V_p=0.04–1.42 m · s^{-1} and τ_L=0.01–1.97 N · m^{-2} and τ_p=0.01–1.57 N · m^{-2}).

4. The greatest values of hydrodynamic parameters were measured at the slope aprons of the constructions, and more precisely at the lowerings, which concentrated the flowing stream (block ramp in the stream bed of the Porębianka) (V_{max}=2.86 m · s^{-1}, τ_{max}=4.75 N · m^{-2}, Re$_{max}$=327 957, Fr$_{max}$=2.36). The values of hydrodynamic parameters on some boulders that form the slope apron of the construction were also high (V_{av}=0.52–1.42 m · s^{-1}), but were still much lower than at the site of the flow concentration.

5. The lowest values were measured in the crevices between the boulders (V_{av}=0.08–0.84 m · s^{-1}). Low velocity values of water flowing in the crevices of the slope apron of the block ramp should allow the migration of macrobenthos along the construction and prevent it from being washed out (Wyżga et al. 2012a,b, 2013, 2014; Radecki-Pawlik and Skalski 2008; Skalski et al. 2012).

6. Leveled and de-leveled river bed causes smaller dynamics and diversity of hydrodynamic parameters in the stream bed. This affects the deterioration of habitats for macroinvertebrate fauna, increases the flow of flood water, increases unit power of the flowing stream, causes the activation of pebbles with larger particle diameter at lower flow rate and thereby faster river bed erosion. Therefore, pebbles should not be taken out of the gravel streambed, but also the trims, basins, or pools should not be buried. Natural structure of the river bed, consisting of a sequence of rapids and pools together with bed forms must be preserved so that the river could remain in the hydrodynamic equilibrium.

Keywords: Block ramp, mountain stream, rapid hydrulic structure, shear stresses, water velocity.

References

Abdel Sattar, A.M., Radecki-Pawlik, A. and Ślizowski, R. 2013. Using genetic programming to predict scour downstream rapid hydraulic structures from experimental results. pp. 179–189. *In*: Bung, D.B. and Pagliara, S. (eds.). International Workshop on Hydraulic Design of Low-Head Structures. IWLHS. Bundesanstalt für Wasserbau, Aachen.

Allan, R. 2012. Water sustainability and the implementation of the Water Framework Directive—a European perspective. Ecohydrology & Hydrobiology 12(2): 171–178.

Carling, P., Whitcombe, L., Benson, I., Hankin, B. and Radecki-Pawlik, A. 2006. A new method to determine interstitial flow patterns in flume studies of sub-aqueous gravel bedforms such as fish nests. River Research and Applications 22(6): 691–701.

Dąbkowski, L., Skibiński, J. and Żbikowski, A. 1982. Hydrauliczne podstawy projektów wodnomelioracyjnych. Państwowe Wydawnictwo Rolnicze i Leśne, Warszawa.

Dyrektywa 2000/60/WE Parlamentu Europejskiego i Rady z dnia 23 października 2000r. ustanawiająca ramy wspólnotowego działania w dziedzinie polityki wodnej.

Gordon, N.D., McMahon, T.A., Finlayson, B.L., Gippel, C.J. and Nathan, R.J. 2007. Stream Hydrology. An Introduction for Ecologists. John Wiley & Sons, London.

Hartung, F. and Scheurlein, H. 1970. Design of Overflow Rockfill Dams. Beitrag no. 36 zu Q 36, Talsperren kongress in Montreal.

Hernik, J. 2010. Why do we protect cultural landscapes? *In*: Radecki-Pawlik, A. and Hernik, J. (eds.). Cultural Landscapes of River Valleys. Wydawnictwo UR w Krakowie, 11–13, Kraków.

Jarabač, M. and Vincent, J. 1967. Použiti zdrsněných skluzu v bystřinném korytě. Vodni hospodářstvi 11: 509–511.

Jarabač, M. 1973. Zkusenostise zavadenim balvanitych skluzu v Moravskoslezskych Beskydech. Sbornik DT-ČSVTS, Ostrava.

Kališ, J. 1970. Hydraulicky vyzkum balvanitych skluzu. Zaverecna sprava VVUVSH, Brno.

Kamykowska, M., Kaszowski, L. and Krzemień, K. 1999. River channel map ping instruction. key to the River Bed description. *In*: Krzemień, K. (ed.). River Channel. Pattern, Structure and Dynamics. Prace Geograficzne Instytutu Geografii UJ, 104.

Knauss, J. 1980. Drsneskluzy.Vodni Hospodarstvi, A C 1, Praha.

Komędera, M. 1993. Zmiany systemu korytowego Mszanki. Praca magisterska UJ. Maszynopis.

Kondracki, J. 2000. Geografia regionalna Polski. PWN. Warszawa.

Kościelniak, J. 2004. Zmiany funkcjonowania górskich systemów korytowych w wyniku przeprowadzonych regulacji hydrotechnicznych. Prace Geograficzne IGiPZ PAN, 200, Warszawa.

Lava, R., Majoros, L., Dosis, I. and Ricci, M. 2014. A practical example of the challenges of biota monitoring under the Water Framework Directive. Trends in Analytical Chemistry 59: 103–111.

Lusk, S. 1980. Balvanite skluzy a rybiosidleni toku. Vodni Hospodářstvi 9: 237–240.

Niel, A. 1960. Über die vernichtung kinetisch erenergie durchniederegef allsstufen. Öesterreichische Wasserwirtschaft, 4, 5, Wien.

Oertel, M., Peterseim, S. and Schlenkhoff, A. 2011. Drag coefficients of boulders on a block ramp due to interaction processes. Journal of Hydraulic Research 49(3): 372–377.

Oertel, M. and Schlenkhoff, A. 2012. Crossbar block ramps: flow regimes, energy dissipation, friction factors, and drag forces. Journal of Hydraulic Engineering 138(5): 440–448.

Oertel, M. 2013. *In situ* measurements on cross-bar block ramps. pp. 111–119. *In*: Bung, D.B. and Pagliara, S. (eds.). International Workshop on Hydraulic Design of Low-Head Structures, IWLHS. Bundesanstalt für Wasserbau, Aachen.

Pagliara, S. and Palermo, M. 2008a. Scour control downstream of block ramps. Journal of Hydraulic Engineering 134(9): 1376–1382.

Pagliara, S. and Palermo, M. 2008b. Scour control and surface sediment distribution downstream of block ramps. Journal of Hydraulic Research 46(3): 334–343.

Pagliara, S., Das, R. and Palermo, M. 2008c. Energy dissipation on submerged block ramps. Journal of Irrigation and Drainage Engineering 134(4): 527–532.

Pagliara, S., Lotti, L. and Palermo, M. 2008d. Hydraulic jump on rough bed of stream rehabilitation structures. Journal of Hydro-environment Research 2(1): 29–38.

Pagliara, S., Palermo, M. and Carnacina, I. 2009a. Scour and hydraulic jump downstream of block ramps in expanding stilling basins. Journal of Irrigation and Drainage Engineering 47(4): 503–511.

Pagliara, S., Palermo, M. and Lotti, I. 2009b. Sediment transport on block ramp: filling and energy recovery. KSCE Journal Civil Engineering 13(2): 129–136.

Pagliara, S. and Palermo, M. 2011a. Effect of stilling basin geometry on clear water scour morphology downstream of a block ramp. Journal of Irrigation and Drainage Engineering 137(9): 593–601.

Pagliara, S. and Palermo, M. 2011b. Block ramp failure mechanisms: critical discharge estimation. Proc Inst Civ Eng Water Manag 164(6): 303–309.

Pagliara, S. and Palermo, M. 2012. Effect of stilling basin geometry on the dissipative process in the presence of block ramps. Journal of Irrigation and Drainage Engineering 138(11): 1027–1031.

Pagliara, S. and Palermo, M. 2013. Rock grade control structures and stepped gabion weirs: scour analysis and flow features. Acta Geophysica 61(1): 126–150.

Plesiński, K., Janas, M. and Radecki-Pawlik, A. 2013. Analysis of hydraulic parameters in near rapid hydraulic structure (RHS) in Porębianka Stream in Gorce Mountains. Acta Scientiarum Polonorum: Formatio Circumiectus (Kształtowanie Środowiska) 12(1): 101–114 [in Polish].

Radecki–Pawlik, A. and Skalski, T. 2008. Bankfull discharge determination using the new invertebrate bankfull assessment method. Journal of Water and Land Development 12: 145–154.

Radecki-Pawlik, A. 2010a. Rapid hydraulic structures as close-to-nature models of mountain river and stream training. *In*: Radecki-Pawlik, A. and Hernik, J. (eds.). Cultural Landscapes of River Valleys. Wydawnictwo UR w Krakowie, 165–172, Kraków.

Radecki-Pawlik, A. 2010b. O niektórych bliskich naturze rozwiązaniach utrzymania koryt rzek i potoków górskich. Gospodarka Wodna 2: 78–85.

Radecki-Pawlik, A. 2011. Hydromorfologia rzek i potoków górskich. Działy wybrane. Wydawnictwo UR w Krakowie. Kraków.

Radecki-Pawlik, A. 2013. On using artificial rapid hydraulic structures (RHS) within mountain stream channels—some exploitation and hydraulic problems. pp. 101–115. *In*: Rowiński, P. (ed.).

Experimental and Computational Solutions of Hydraulic Problems, Series: GeoPlanet: Earth and Planetary Sciences, Monograph, Springer-Verlag Berlin Heidelberg.

Radecki-Pawlik, A., Plesiński, K. and Wyżga, B. 2013. Analysis of chosen hydraulic parameters of a Rapid Hydraulic Structure (RHS) in Porębianka Stream, Polish Carpathians. pp. 121–128. *In*: Bung, D.B. and Pagliara, S. (eds.). International Workshop on Hydraulic Design of Low-Head Structures, IWLHS. Bundesanstalt für Wasserbau, Aachen.

Radecki-Pawlik, A., Bucała, A. and Plesinski, K. 2014. Hydrodynamic and ecohydrological conditions in two catchments in the Gorce Mountains: Jaszcze and Jamne streams—Western Polish Carpathians. Ecohydrology & Hydrobiology 14(3): 229–242.

Schauberger, W. 1957. Naturgemäßer Wasserbau geschiebe führenden flusen. Wasser und Boden, 11, Wien.

Scheuerlein, H. 1968. Der Rauhgerinneabfluss. Bericht, 1, Versuchsanstalt fur Wasserbau der Technischer Uniwersitat, Munchen.

Skalski, T., Kędzior, R. and Radecki-Pawlik, A. 2012. Riverine ground beetles as indicators of inundation frequency of mountain stream: a case study of the Ochotnica Stream, Southern Poland. Baltic J Coleopterol 12(2): 117–126.

Ślizowski, R. 1990. Bystrza w świetle badań czechosłowackich. Zeszyty Naukowe AR w Krakowie 240(14): 19–35.

Ślizowski, R. 2002. Wpływ bystrzy o zwiększonej szorstkości na migrację ryb w potokach górskich. Zeszyty Naukowe AR w Krakowie 393(23): 331–336.

Ślizowski, R., Radecki-Pawlik, A. and Huta, K. 2008. Analysis of chosen hydrodynamics parameters along the rapid hydraulic structure with increased roughness—the Sanoczek Stream. Infrastruktura i Ekologia Terenów Wiejskich 2: 47–58 [in Polish].

Wyżga, B., Oglęcki, P., Radecki-Pawlik, A., Skalski, T. and Zawiejska, J. 2012a. Hydromorphological complexity as a driver of the diversity of benthic invertebrate communities in the Czarny Dunajec River, Polish Carpathians. Hydrobiologia 696(1): 29–46, Springer, Netherlands.

Wyżga, B., Zawiejska, J., Radecki-Pawlik, A. and Hajdukiewicz, H. 2012b. Environmental change, hydromorphological reference conditions and the restoration of Polish Carpathian rivers. Earth Surface Processes and Landforms 37(11): 1213–1226.

Wyżga, B., Oglęcki, P., Hajdukiewicz, H., Zawiejska, J., Radecki-Pawlik, A., Skalski, T. and Mikuś, P. 2013. Interpretation of the invertebrate-based BMWP-PL index in a gravel-bed river: in sight from the Polish Carpathians. Hydrobiologia 712(1): 71–88, Springer Netherlands.

Wyżga, B., Amirowicz, A., Oglęcki, P., Hajdukiewicz, H., Radecki-Pawlik, A., Zawiejska, J. and Mikuś, P. 2014. Different response of fish and benthic invertebrate communities to constrained channel conditions in a mountain river: case study of the Biała, Polish Carpathians. Limnologica 46: 58–69.

Żelazo, J. 1992. Badania prędkości i oporów przepływu w naturalnych korytach rzek nizinnych. Rozprawy Naukowe. Monografia SGGW. Warszawa.

CHAPTER 6

Hydraulic Calculations for Fish Passes

Mokwa Marian and Tymiński Tomasz*

INTRODUCTION—General Information on Fish Passes

As is generally known, all fish species migrate during the year over small or long distances. These migrations are related to the subsequent stages of their life cycle. For example, fish migrate to spawn or to feed. Migratory fish are faced with various obstacles. These include, particularly during upriver migrations, all kinds of damming structures.

In order to allow the fish to overpass abrupt level differences on rivers resulting from artificial damming by dams, weirs, drops, sills and other structures, and also natural barrages formed by waterfalls, hydraulic structures called fish passes (or fishways) are built.

There are several different designs of structures for fish migration.

Technical structures intended to facilitate fish migrations

These are usually chutes of arbitrary lengths, made of either concrete or stone (less often wood), with geometrically shaped walls, baffles, and spill openings. Their bottom is made of embedded rock material (stones) of various heights. Functionally, these structures are meant mainly for upstream migration (from tailwater to headwater).

The most common designs for technical structures of this kind are as follows:

- *Conventional pool passes*—in these designs, the fish pass consists of a concrete (or less often wooden) channel, either inclined or arranged in a series of steps/ pools separated by various cross walls with openings (orifices and weirs).

- *Slot pool passes*—these are a modification of the classical pool pass design: the openings are vertical slots which extend over the entire height of the cross wall.

Wrocław University of Environmental and Life Sciences, Institute of Environmental Engineering, pl. Grunwaldzki 24, 50-363 Wrocław, Poland.
* Corresponding author: tomasz.tyminski@up.wroc.pl

The slots can be located either at one-side or at both sides of cross walls. There are many modifications of slot pool passes.

• *Denil (baffle) fish passes*—built in the form of concrete chutes separated with resting pools. The water current is slowed down by means of densely arranged baffles of various shapes. The most common of all the baffle-type fish passes is the Denil fishway with U-shaped baffles.

• *Modular meander-type fish passes*—this type consists of a cascade of subsequent cylindrical chambers connected in such a way that the flow passing through the slots has a distinctly meandering character. These fishways are modular and can be arbitrarily extended.

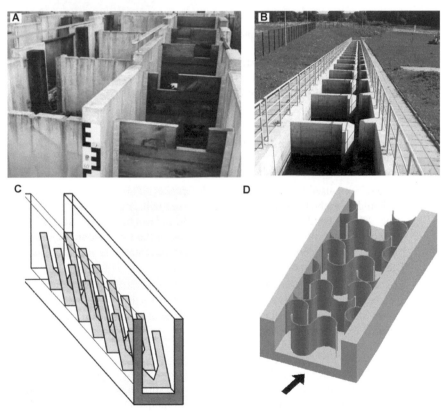

Fig. 1. Technical fish passes.

Semi-technical structures for fish migration

• *Cascaded pool fish pass with boulder weirs—step-pool rock ramp fishway*—in this design, in a channel made of either reinforced concrete or rock, weirs in the form of regular cascades are used instead of wooden or concrete cross walls. These weirs form pools which can be more or less pronounced. The cascades consist of embedded boulders or concrete pillars set upright. The bottom between the boulder bars is filled with substrate and stones of various heights.

Fig. 2. Semi-technical fish passes.

- *Fish pass*—this type of fishway is somewhere between the technical and the semi-natural designs. It resembles the layout of a fish pass with evenly arranged boulders or that with boulder bars, but dense plastic brush-type structures are used instead of boulders.

Structures which imitate natural conditions—"close to nature"

A. Built over the entire width of the drop or sill
- *Ramp riffles*—used for moderate level differences of up to 1.0 m, built of rocks and boulders embedded in the bottom, and fixed with smaller, evenly distributed stones. The embedded boulders are set upright and can be evenly distributed. The distances between boulders in rows and between the rows should be the same. However, the boulders must be positioned to form at least one migration channel. The slope should not exceed 1:10.
- *Cascaded riffles*—this structure consists of pools separated with cross walls made of embedded boulders. The design of cascaded riffles usually includes a concrete bottom slab placed over the entire width of the drop with embedded boulders arranged in the form of bars.

B. Extending only over a part of the width
- *Ramp riffles*—used for moderate level differences of up to 1.0 m, built of rocks and boulders embedded in the bottom and fixed with smaller stones. They are usually located at one of the banks and consist of two ramps. One of them is parallel to the direction of flow and has a slope of 1:20–1:30 and the other, adjacent to it, is directed towards the channel centre, with a much higher slope.
- *Cascaded riffles*—made in the form of concrete or stone channel. These are located at one of the banks or in the middle of the drop.

C. Natural—imitating a mountain stream or a lowland brook

A bypass is made with natural material (boulders, stones, gravel, tree trunks, wooden piles, fascines). The bottom is formed in the natural substrate. Such bypasses may include pools and overflow areas. Higher slopes sills or bars can be made in the bottom. These are built with loosely arranged boulders and trunks of trees.

Fig. 3. Fish passes close to nature.

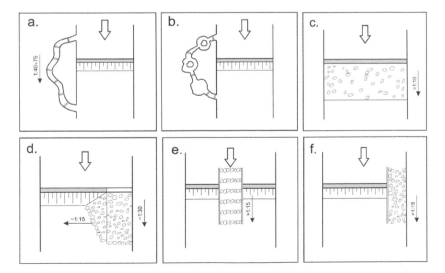

Fig. 4. Preferred modern fish pass designs (Mokwa and Wisniewolski 2008).

- *Cascaded bypasses*—these are channels which imitate a natural river, with a small, constant slope and considerable width at the bottom. They consist of evenly spaced rows of embedded boulders set upright and perpendicular to the direction of flow (boulder bars).
- *Bypass in the form of riffles*—in a channel imitating a natural river, with evenly distributed embedded boulders set upright. The distances between boulders in rows and between the rows must be the same. Instead of boulders, wooden piles can be used in the fish pass (which are either rammed or concreted into the bottom). These piles may be 10–30 cm in diameter and should be inclined in the downstream direction.

Special fish passes

- *Eel ladders*—these are usually small channels leading from tailwater to headwater. They can be made of concrete, steel, or plastics. The bottom should be covered with a material (e.g., nylon brushes) that allows eels to swim upstream. An eel ladder should be located just below the surface of water in the fish pass.
- *Fish locks, including normal locks for inland navigation*—the design of a fish lock is similar to that of a lock for boats and ships. Both consist of a lock chamber, the upper and lower culvert, and the gates. A navigation lock is not enough for the fish to migrate, as it is operated irregularly and does not generate any steady attraction current.
- *Fish lifts*—these are built when the head between tailwater and headwater is considerable. Due to the moveable components, power transmission, and control equipment, they require constant monitoring. Both their design and operation differ significantly from a typical fish pass.
- *Ropeways and railways*—for significant damming heights, mainly in reservoirs with considerable fluctuations of headwater level.

Other structures for fish migration

From this group of structures, the following are worth mentioning:

- The migration weirs for downstream migration,
- The fish passes which are also used for kayaking,
- The turbo fish pass from the Austrian company Connect—a bifunctional, double rotating Archimedes screw, which combines power generation with a fish lift.

The design of a fish pass is one of the most challenging processes in water engineering. This is because the designer is faced here with a problem that requires solving several interrelated tasks: the design work, the hydraulics of flow, and the behaviour of fish.

When designing a fish pass, it is important to combine good engineering practice, good biological assessment, careful laboratory tests, and numerical simulations.

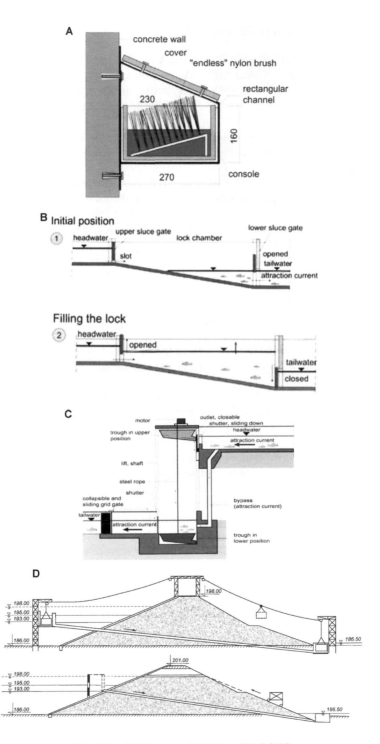

Fig. 5. Special fish passes (DVWK and FAO 2002).

General Requirements for a Fish Pass

1. The critical values of water flow rate (2.0 m/s – salmonids, 1.5 m/s – rheophilic cyprinids, 1.0 m/s – other fish) are only admissible on short distances such as inlets, narrows, slots and notches connecting subsequent parts of the fish pass. In all other parts of the fish pass, the flow rate must be reduced to ≤50% of the critical value.
2. The energy dissipation coefficient in pools must not exceed $E \leq 100$ or 200 W/m³, depending on the requirements of the fish species for which the fishway is made.
3. The limiting of the flow rate to the required critical values requires one to design the fishways with gentle slopes. Depending on the local conditions and the type of structure concerned, this can vary from 1:10 to 1:50 or more. Another factor related to this condition is that a fish pass may need to have sufficiently large pools.
4. The location of a fish pass must always be in the main course of the river and in the vicinity of the damming structure (hydroplant, spillway) with the highest discharge. If this condition cannot be met, special barriers must always be installed, so as to guide the migrating fish to the fish pass.
5. The outlet from a fish pass must be positioned at 30°–40° to the direction of the river current.
6. The velocity of water flowing out of a fish pass must exceed the flow rate in the river by about 0.20–0.30 m/s—only then will the attraction current be able to lure fish.
7. The velocity of water flowing out of a fish pass should not exceed 60–80% of the critical flow rate determined for a given fish species.
8. The bottom of the fish pass must gently connect with that of the river. This means that the bottom of the fish pass and that of the river below the damming structure must be at the same level, or should connect through a (cone-shaped) ramp with a 1:2 slope. This latest condition also applies to the upstream side of the structure.
9. In order to ensure suitable conditions for migration and safety for the fish in the fish pass, the minimum water depth should not be less than: 0.30 m in narrows and weirs and 0.6 m in pools. In the case of migratory fish and catfish these depths must be even greater: they must exceed 0.5 m in narrows and 0.80 m or even 1.30–2.00 m (sturgeon) in pools. It is always advantageous to design even greater depths. For most fish species which migrate through the fish pass, one must also appropriately adjust the dimensions of its pools.

Fundamental Notions and Formulae Used in Fish Pass Calculations

In terms of hydraulics, a fish pass is usually a pressure-free duct, in which the flowing liquid forms a free table. The motion of the liquid, caused by the (longitudinal) slope of the channel, is a gravity flow. This flow may be non-uniform (slow- and fast-changing) or uniform. Most of the phenomena related to the hydraulics of a fish pass are solved under the assumption of steady uniform motion, in which the stream with a constant discharge (Q=const.) has a constant cross section A, a uniform filling depth h and, consequently, the invert line, water surface line, and energy line are parallel:

$i = J_{zw} = J = \sin \alpha$

where: $i = \dfrac{z_1 - z_2}{L} = \sin \alpha$,

    i [–] – the invert slope

    J [–] – the hydraulic gradient (for the uniform motion equal to the invert slope)

$$J = \frac{h_s}{L}$$

    h_s – the head between cross sections I and II

    L – the distance between cross sections I and II

    J_w [–] – the slope of the water surface (in practice, equal to the hydraulic gradient)
    The value of slope or gradient is expressed in per mil [‰=m/km] or in percent [%]

By writing the Bernoulli equation for sections I and II of the channel (Fig. 1)

$$z_1 + \frac{p_1}{\gamma} + \frac{\alpha v_1^2}{2g} = z_2 + \frac{p_2}{\gamma} + \frac{\alpha v_2^2}{2g} + \lambda \frac{L}{4R} \cdot \frac{v^2}{2g}$$

and substituting: $p_1 = p_2$, $v_1 = v_2 = v$, and $z_1 - z_2 = i \times l = J \times L$,

one obtains a transformed form of the Darcy-Weisbach formula:

$$v = \frac{1}{\sqrt{\lambda}} \sqrt{8g \cdot R \cdot J}$$

Historically, the first formula for average velocity at a given channel cross section has been the Chézy formula:

$$v = C \sqrt{R \cdot J}$$

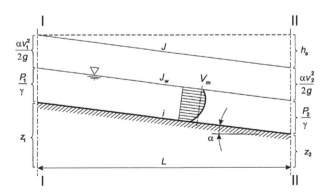

Fig. 6. The longitudinal profile of the fish pass channel.

Where: C—the coefficient (Chézy constant), which can be determined by means of many empirical relationships.

The formula of Chézy is not applicable in its original form, that is, with a constant C at the cross section. For the uniform motion, to calculate the flow rate in open channels, the empirical formula of Manning is most often used:

$$v = \frac{1}{n} R^{2/3} I^{1/2}$$

Where: n—the roughness coefficient for the bottom surface and the walls of the channel determined by Manning—an empirical parameter which depends on the nature of the bottom and walls.

The roughness coefficient n is a "non-physical" value, it is expressed in $[m^{-1/3} \cdot s]$. It contains information on those characteristics of the channel which influence the drag.

The roughness coefficient depends on the following parameters:

- The transverse, longitudinal, and horizontal shape of the channel.
- The cover (the granulometric composition of the material of which the channel is made, the vegetation which grows in it, its type and the area it covers).

This coefficient may vary from 0.009 to 0.03. The lower value is for very smooth enamel or glazed surfaces. The upper value is for channels which are particularly badly maintained, with breaches and slides of considerable size, overgrown with rushes, with big stones on the bottom, etc.

The coefficient n has the following values (Chow 1959):

n=0.009 for a glazed surface,
n=0.014 for a concrete surface,
n=0.025 for an earth channel,
n=0.050 for stones and boulders on the bottom.

If n is visibly variable over the perimeter, the value of this coefficient can be calculated as the weighted average of the roughness coefficients ni:

$$n = \frac{\Sigma n_i \cdot U_i}{U}$$

The relationship between C and n can be calculated, for example, from the Manning formula:

$$C = \frac{1}{n} R^{1/6}$$

$$R = \frac{A}{U}$$

Where: R [m] – the hydraulic radius of the channel cross section,
U [m] – the wetted perimeter of the channel cross section area,
A [m²] – the channel cross section area.

For a shallow cross section, when B≥10 h_{sr}, the hydraulic radius can be substituted with the average depth:

$R \approx h_{sr}$

The discharge is the product of the average velocity v and the flow cross section area A:

$$Q = v \cdot A \ [\text{m}^3/\text{s}]$$

or can be calculated from the Manning formula:

$$Q = v \cdot A = \frac{1}{n} A \cdot R^{2/3} \cdot I^{1/2}$$

or else from the Darcy-Weisbach formula:

$$Q = v \cdot A = \sqrt{\frac{8g}{\lambda}} \cdot A\sqrt{R \cdot I} = \frac{1}{\sqrt{\lambda}}\sqrt{8g \cdot R \cdot I}$$

Where the friction coefficient λ is determined from the Colebrook-White formula

$$\frac{1}{\sqrt{\lambda}} = -2\lg\frac{k}{3.71 \cdot 4R} = -2\lg\frac{k}{14.84R}$$

and the roughness coefficient k [mm] can have the following values for different channel surfaces:

    k=0.6 mm – smooth concrete surface,
    k=1 mm – sandy bottom,
    k=300 mm – stone bottom.

The relationship between the absolute roughness, k, and the Manning roughness coefficient n can be determined from the following approximation:

$$n = \frac{k^{1/6}}{21{,}1}$$

Where: k is expressed in [m].

Hydraulic Calculations for an Open Channel with a Compact Cross Section (calculations for a steady uniform flow)

In hydraulic calculations for a uniform flow through open channels, the following types of problems are most often solved:

To calculate the average velocity v (or the discharge Q) in the channel by substituting all the data to the formula of Manning.

To determine the invert slope J—the required slope value is calculated from an appropriately transformed Manning formula.

To determine an arbitrary linear dimension of the channel (usually the depth h)— even for the simplest channel cross section, such problems require one to determine the roots of a polynomial of a higher degree, this type of problems is solved by subsequent approximations or using engineering methods (using the rating curve).

Natural channel

As can be seen, the shape of the transversal velocity distribution is related to that of the channel cross section.

- The cross section area of a natural channel:

$$A = \frac{1}{2}\Sigma(h_i + h_{i+1})B_i$$

- The wetted perimeter:

$$U = \sum_{i=1}^{N}\sqrt{(h_{i+1} - h_i)^2 + B_i^2}$$

- The hydraulic radius:

$$R = \frac{A}{U}$$

- The discharge calculated from based on the formula of Manning:

$$Q = v \cdot A = \frac{1}{n}R^{2/3} \cdot I^{1/2} \cdot A$$

Where: substitute roughness $n = \dfrac{\Sigma n_i \cdot U_i}{U}$

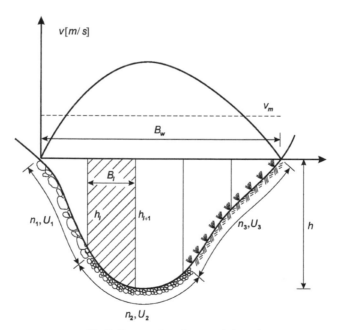

Fig. 7. Cross section of a natural channel.

Artificial rectangular channel

- The discharge is:

$$Q = \frac{1}{n} I^{1/2} \left(\frac{Bh}{2h+B} \right)^{2/3} Bh$$

- The formula for iterative calculation of channel filling is:

$$h = \left(\frac{Qn}{I^{1/2}B} \right)^{0.6} \left(2\frac{h}{B}+1 \right)^{0.4}$$

- The formula for iterative calculation of channel width is:

$$B = \left(\frac{Qn}{I^{1/2}h} \right)^{0,6} \left(\frac{B}{h}+2 \right)^{0,4}$$

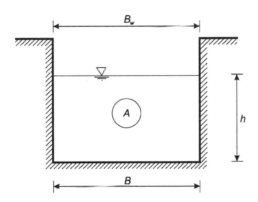

Fig. 8. Computational diagram for a rectangular channel.

Artificial trapezoidal channel

For: $m_1 \neq m_2$ and assuming that $m=0,5(m_1+m_2)$—the cross section area is $A=Bh+mh^2$ and the width at the surface level is $B_w=B+2mh$

- The discharge can be expressed as:

$$Q = \frac{1}{n} I^{1/2} \left(\frac{Bh+mh^2}{B+h(()\sqrt{1+m_1^2}+\sqrt{1+m_2^2})} \right)^{2/3} \left(mh^2+Bh \right)$$

- The formula for iterative calculation of channel filling is:

$$h = \frac{B}{2m} \left\{ \sqrt{1+\left(\frac{Qn}{I^{1/2}B} \right)^{0,6} \left[1+\frac{h}{B}\left(\sqrt{1+m_1^2}+\sqrt{1+m_2^2} \right) \right]^{0,4} \frac{4m}{B}} -1 \right\}$$

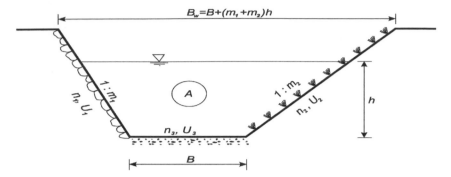

Fig. 9. Hydraulic diagram for a trapezoidal channel.

- The formula for iterative calculation of channel width is:

$$B = \left(\frac{Qn}{I^{1/2}h}\right)^{0,6} \left(\frac{B}{h} + \sqrt{1+m_1^2} + \sqrt{1+m_2^2}\right)^{0,4} - mh$$

For: $m_1 = m_2 = m$—the cross section area is $F = Bh + mh^2$ and the width at the surface level is $B_w = B + 2mh$

- The discharge can be expressed as:

$$Q = \frac{1}{n}I^{1/2}\left(\frac{Bh + mh^2}{B + 2h(()\sqrt{1+m_1^2}}\right)^{2/3} (mh^2 + Bh)$$

- The formula for iterative calculation of channel filling is:

$$h = \frac{B}{2m}\left\{\sqrt{1 + \left(\frac{Qn}{I^{1/2}B}\right)^{0,6}\left(1 + 2\frac{h}{B\sqrt{1+m^2}}\right)^{0,4}\frac{4m}{B}} - 1\right\}$$

- The formula for iterative calculation of channel width is:

$$B = \left(\frac{Qn}{I^{1/2}h}\right)^{0,6}\left(\frac{B}{h} + 2\sqrt{1+m^2} + \sqrt{1+m_2^2}\right)^{0,4} - mh$$

Rating curve (discharge versus stage curve)

This is a curve which represents the relationship between the stage in the channel (H) and the discharge (Q). This curve is generated by plotting, in a Cartesian coordinate system, the points obtained from the calculation or measurement of discharge for various water levels at a given cross section.

When designing a fish pass, the rating curve is mainly needed for determining the filling of the fish pass channel as a function of discharge and filling at the tailwater cross section (fish pass inlet) for a given discharge in the watercourse channel. In order to plot an analytic rating curve, one must determine the following values:

H – the channel filling height measured from the bottom,
I – the average slope of the water surface (or bottom),
n – the roughness coefficient for the bottom.

1. To determine the rating curve using the multipartite channel method:
   • Divide the channel into sections.
   • Calculate the cross sectional area for individual sections for various fillings.
   • Calculate the average velocity for individual sections for various fillings.
   • Calculate the discharge for various fillings.

2. To determine the rating curve using the averaged roughness coefficient method:
   • Calculate the cross sectional area for various fillings.
   • Calculate the average velocity for various fillings.
   • Calculate the discharge for various fillings.

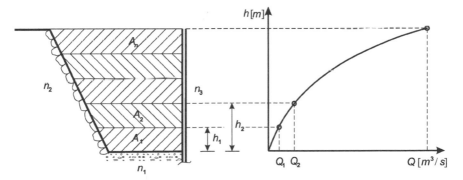

Fig. 10. The discharge curve for a trapezoidal channel determined using the averaged roughness coefficient method.

Hydraulic Calculations for Fish Passes

In order to reduce the velocity of flow, the fish pass channels are equipped with all kinds of cross walls and other elements intended to increase drag. The type of such structures determines the way the hydraulic calculations are carried out. The flow through barriers (wall with orifices or slots, the flow over or under a baffle, etc.) is calculated using classical equations for weirs, spillways, and openings. The drag for the case of roughness increased by various elements embedded in the channel (e.g., stones, boulders, and vegetation) is calculated based on the values of absolute roughness k_s from the formula of Colebrook-White. The irregularities in the distribution of embedded elements (e.g., boulders, piles, etc.) can be calculated under the assumption of the same number of elements with averaged geometrical parameters distributed in a regular way.

Calculation of the discharge in an open channel built of tiny grains (e.g., a fish pass in the form of bypass, Fig. 3a)

Calculation procedure

Value	Symbol	Unit	Formula
Depth	h	m	
Surface area of the channel cross section	A	m²	
Wetted perimeter	U	m	
Hydraulic radius	R	m	$R = \dfrac{A}{U}$
Bottom drag coefficient where: ($k_s \approx 0.25$) [m]	λ_d	-	$\lambda_d = \left[-2.0\log\left(\dfrac{k_s}{14.84R}\right)\right]^2$
Average flow velocity at the channel cross section	v_m	m/s	$v_m = \sqrt{\dfrac{8gRJ}{\lambda}}$
Longitudinal channel slope	J	m/km	
Discharge	Q	m³/s	$Q = A \cdot v_m$

In fish passes that are "close to nature", such as bypass channels with bottoms filled with material resembling that of the river floor, it is important to be able to forecast its stability, particularly under the conditions of a flood wave. Krüger (2008) gives a dozen or so empirical formulae developed by various authors for the specific discharge for the beginning of bed load movement. One of these equations is the formula of Whittaker and Jäggi from 1986:

$$q_{kr} = 0.257\sqrt{g\frac{\rho_s - \rho_w}{\rho_w}}\, I^{-7/6}\, d_{65}^{3/2}$$

Where: $\rho_s = 2600$ kg·m⁻³

The hydraulic conditions for the flow of water through a fish pass must also account for the bed load movement. This is because of the possible accumulation of transported material and the erosion of the bottom.

Calculation of the discharge in an open channel with equally distributed crushed stones (e.g., fish passes in the form of bypasses, riffles...Fig. 3a)

Calculation procedure

Value	Symbol	Unit	Formula
Depth	h	m	
Average channel width	b	m	
Surface area of the channel cross section	A	m²	

Wetted perimeter	U	m	
Hydraulic radius	R	m	$R = \dfrac{A}{U}$
Average boulder diameter	d_s	m	
Distance between boulders in the direction of flow	a_x	m	
Distance between boulders transversely to the flow	a_y	m	
Bottom drag coefficient where: $(k_s \approx 0,25)$ [m]	λ_d	-	$\lambda_d = \left[-2,0\log\left(\dfrac{k_s}{14,84R}\right) \right]^2$
Drag coefficient for boulders where: $(\cos\alpha \approx 1)$	λ_p	-	$\lambda_p = C_{WR}\dfrac{4d_s h\cos\alpha}{a_x a_y}$
Drag coefficient for the flow around a cluster of boulders (for C_W=1.1)	C_{WR}	-	$C_{WR} = \left(1,1+2,3\dfrac{d_s}{a_y}\right)\left(\dfrac{v_1}{v}\right)2 + \left(\dfrac{1}{1-\dfrac{d_s}{a_y}} - 1\right)$
Relationship, valid for: 0.05 ds/a_y<0.3 and 0.2<a_x/a_y<2	$\left(\dfrac{v_i}{v}\right)^2$	-	$\left(\dfrac{v_i}{v}\right)^2 = 0,6+0,5\log\left(\dfrac{a_x}{a_y}\right)$
Drag coefficient for boulders and the channel surface	λ	-	$\lambda = \lambda_d + \lambda_p$
Average flow velocity at the channel cross section	v_m	m/s	$v_m = \sqrt{\dfrac{8gRJ}{\lambda}}$
Maximum flow velocity	v_{max}	m/s	$v_{max.} = 1-\dfrac{v_m}{1-\dfrac{\Sigma A_s}{A}}$
Longitudinal channel slope	J	m/km	
Discharge	Q	m³/s	$Q = A \cdot v_m$
Froude number at a cross section with no stones	Fr	-	$Fr^2 = \dfrac{v_m^2 B}{gA}$
Froude number at the narrowest cross section	Fr	-	$Fr^2 = \dfrac{v_{max}^2 \cdot B_s}{gA_s}$

Froude number (Fr):

- Fr number <1   –   quiet flow,
- Fr number >1   –   rapid flow,
- Fr number >1.7 –   with a hydraulic jump.

Where: A_s – cross section surface area with no boulders [m²]
    B_s – cross section width with no boulders [m].

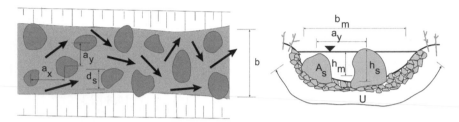

Fig. 11. Computational diagram for a riffle fish pass (DVWK 2002).

Hydraulic calculations for fish passes with vegetation
(either natural or artificial Fig. 2b) [Tymiński and Kałuża 2008, 2013]

The concept of using brushes in a fish pass has been investigated at the University of Kassel, Germany, since 2000. The creator and the main advocate of this concept is Dr. Hassinger (Hassinger and Reinhard 2002). The analysis is based on the assumption that brush, while splitting the volume of flow, absorb a considerable part of energy of the stream and generate a strong drag. In this case, the resistance of the fish pass walls and bottom (usually in the form of rough rock material) is practically irrelevant. The number and the location of brush must be chosen so as to ensure suitable flow conditions, appropriate for the requirements of the fish species living in the river stretch concerned. On the website of the Versuchsanstalt und Prüfstelle für Umwelttechnik und Wasserbau at the University of Kassel there is a ready-to-use, purpose-made downloadable EXCEL file helpful in the initial design of such a fish pass (Kałuża and Hämmerling 2015).

    The most important values which influence the flow conditions in a vegetated channel include the channel slope and the parameters that describe the vegetated area, such as the species and age of plants, the season of the year, the shape and density of the cluster as well as the way in which individual elements (plant stems or brush) are arranged in it (Tymiński and Kałuża 2012).

    The calculation procedure relies on the fact that the drag of the shape, F_R, for cylinders in the flow (stems, brushes) translates to shear stress on the so-called specific washed surface (for an individual stem, this amounts to the product $a_x a_y$).

$$F_R = \frac{1}{2} A_R c_w \rho\, v^2$$

In this particular case:

$$a_x = a_y = \sqrt{\frac{A_y}{n}}$$

A_y – the surface area of the vegetated part of the channel ($B \times L$)
n – the number of stems (brushes)

$$\lambda_R = \frac{4h \cdot d_p}{a_x \cdot a_y} c_{WR}$$

An analysis of equations (1–3) reveals that the problem of determining the drag coefficients, λ_R, can be reduced to the correct calculation of the shape drag coefficient c_{WR}—of a single element in a group of plants and the determining of the distance between them in the direction of flow a_x and perpendicularly to it a_y (Fig. 1). Lindner has given a relationship which can be used for the calculation of c_{WR}. In practice, it is convenient to use it in its slightly modified form of Rickert (1992):

$$c_{WR} = c_{W\infty}\left(1 + 1.9\,c_{W\infty}\frac{d_p}{a_y}\left(\frac{v_i}{v_R}\right)^2 + 2\left(\frac{1}{1-\dfrac{d_p}{a_y}} - 1\right)\right)$$

Where: $c_{w\infty}$—the drag coefficient for a single vegetative element (Fig. 3).

$\left(\dfrac{v_i}{v_R}\right)$—the relative velocity of inflow to the vegetative element, which depends on

the length (a_{NL}) and width (a_{NB}) of the Kármán vortex street behind the stem (see equations, Fig. 5).

During the flow around vegetative elements the so-called inertial standing waves are formed, which additionally increase the drag related to shape. In equation (4), this fact is taken into account in a simplified way with the following component:

$$\Delta c_W = 2\left(\frac{1}{1-\dfrac{d_p}{a_y}} - 1\right)$$

The characterization of drag in a vegetated fish pass λ_R should be complemented with a formula which includes the friction resistance on the bottom and banks of the channel λ_D. Moreover, a modification of the Colebrook-White equation, which is recommended, for example, by the German Association for Water, Wastewater and Waste (DVWK 1991; also Pasche 1984) ensures accuracy sufficient for the purpose of engineering designs:

$$\frac{1}{\sqrt{\lambda_D}} = 2log\left(\frac{14,84R_{h,D}}{k_{s,D}}\right)$$

In order to estimate the substitute sand roughness for the bottom of channel banks $k_{s,D}$, Table 6.2 from Kubrak (1998) may turn out useful. One may also use the knowledge of the roughness coefficient n (here n_D) in the Manning formula. According to Garbrecht (see Kubrak 1998), the following relationship is also helpful:

$$k_{St} = \frac{1}{n_D} = \frac{26}{k_{s,D}^{1/6}}$$

Which yields:

$$k_{s,D} = (26n_D)^6$$

At this point special attention should be given to another form of formula (6) preferred by Rickert (1992, 1994) for the calculation of natural river channels with complex cross sections:

$$\frac{1}{\sqrt{\lambda_D}} = 2\log\left(\frac{12,2R_{h,D}}{k_{s,D}}\right)$$

The answer to the question of which method is more appropriate for the dimensioning of "artificial structures"—the man-built fish passes—may be provided by the verification of formulae, for example, by means of laboratory measurements.

The overall drag coefficient for vegetation (brushes) λ_R and the channel bottom and banks λ_D, results from the Einstein/Banks superposition:

$$\lambda = \lambda_D + \lambda_R$$

In order to calculate the average velocity and the fish pass discharge one can use the general Darcy-Weisbach formula for flow:

$$v_{śr} = \frac{1}{\sqrt{\lambda}}\sqrt{8\,g\,R_h\,I_E}$$

The self-evident analogy between the cylinders used in the experiments of Lindner and Pasche and the plastic brush or natural vegetative elements used for example, in the construction of eel ladders was the starting point for investigating the possibilities of applying the calculation procedure of Lindner-Pasche to the hydraulic dimensioning of vegetated fish passes.

Thanks to the introduction of computational formulae (the above equations and the diagram in Fig. 12) to the computer programs, one can obtain a convenient designing tool ensuring results of high quality, which fully meet the requirements of hydraulic calculations for vegetated channels (brush-type fish passes).

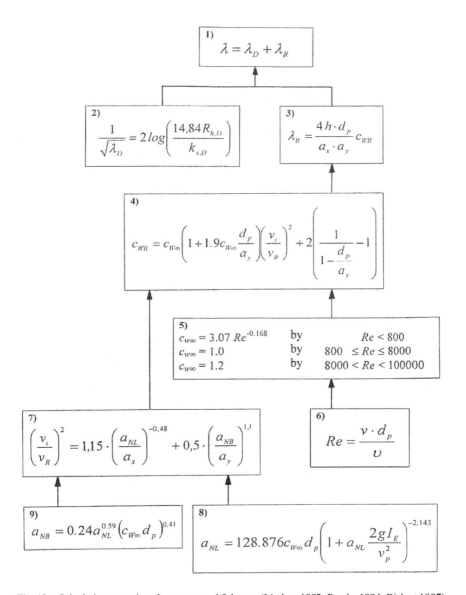

1)
$$\lambda = \lambda_D + \lambda_R$$

2)
$$\frac{1}{\sqrt{\lambda_D}} = 2\log\left(\frac{14,84 R_{h,D}}{k_{s,D}}\right)$$

3)
$$\lambda_R = \frac{4\,h \cdot d_p}{a_x \cdot a_y} c_{WR}$$

4)
$$c_{WR} = c_{W\infty}\left(1 + 1.9 c_{W\infty}\frac{d_p}{a_y}\right)\left(\frac{v_i}{v_R}\right)^2 + 2\left(\frac{1}{1 - \dfrac{d_p}{a_y}} - 1\right)$$

5)

$c_{W\infty} = 3.07\, Re^{-0.168}$	by	$Re < 800$
$c_{W\infty} = 1.0$	by	$800 \le Re \le 8000$
$c_{W\infty} = 1.2$	by	$8000 < Re < 100000$

7)
$$\left(\frac{v_i}{v_R}\right)^2 = 1,15 \cdot \left(\frac{a_{NL}}{a_x}\right)^{-0.48} + 0,5 \cdot \left(\frac{a_{NB}}{a_y}\right)^{1,1}$$

6)
$$Re = \frac{v \cdot d_p}{\upsilon}$$

9)
$$a_{NB} = 0.24 a_{NL}^{0.59}\left(c_{W\infty} d_p\right)^{0.41}$$

8)
$$a_{NL} = 128.876 c_{W\infty} d_p\left(1 + a_{NL}\frac{2 g I_E}{v_p^2}\right)^{-2.143}$$

Fig. 12a. Calculation procedure for a vegetated fish pass (Lindner 1982; Pasche 1984; Rickert 1987).

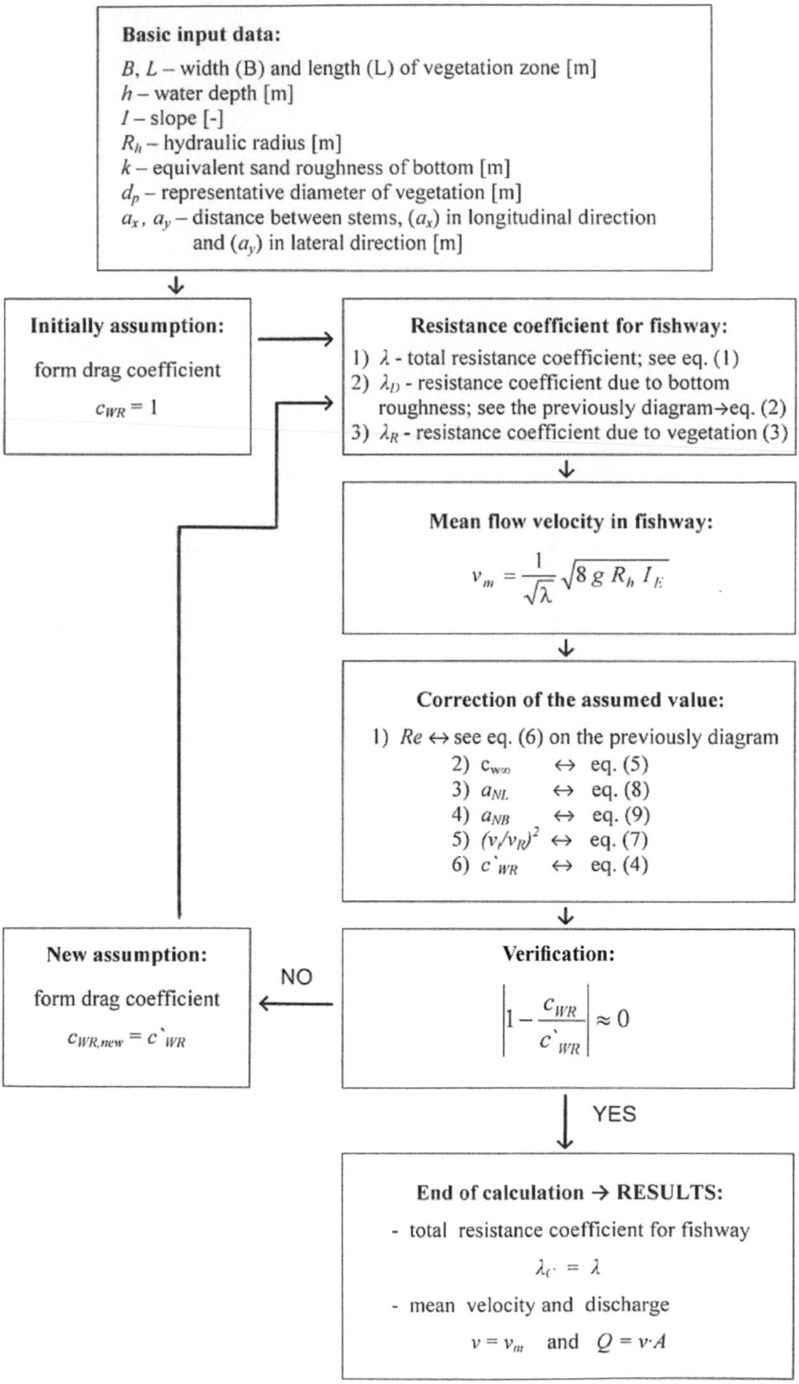

Basic input data:

B, L – width (B) and length (L) of vegetation zone [m]
h – water depth [m]
I – slope [-]
R_h – hydraulic radius [m]
k – equivalent sand roughness of bottom [m]
d_p – representative diameter of vegetation [m]
a_x, a_y – distance between stems, (a_x) in longitudinal direction
        and (a_y) in lateral direction [m]

Initially assumption:

form drag coefficient

$c_{WR} = 1$

Resistance coefficient for fishway:

1) λ - total resistance coefficient; see eq. (1)
2) λ_D - resistance coefficient due to bottom roughness; see the previously diagram→eq. (2)
3) λ_R - resistance coefficient due to vegetation (3)

Mean flow velocity in fishway:

$$v_m = \frac{1}{\sqrt{\lambda}} \sqrt{8\,g\,R_h\,I_E}$$

Correction of the assumed value:

1) $Re \leftrightarrow$ see eq. (6) on the previously diagram
2) $c_{w\infty}$ $\leftrightarrow$ eq. (5)
3) a_{NL} $\leftrightarrow$ eq. (8)
4) a_{NB} $\leftrightarrow$ eq. (9)
5) $(v/v_R)^2 \leftrightarrow$ eq. (7)
6) c'_{WR} $\leftrightarrow$ eq. (4)

New assumption:

form drag coefficient

$c_{WR,new} = c'_{WR}$

NO

Verification:

$$\left|1 - \frac{c_{WR}}{c'_{WR}}\right| \approx 0$$

YES

End of calculation → RESULTS:

- total resistance coefficient for fishway

$$\lambda_C = \lambda$$

- mean velocity and discharge

$$v = v_m \quad \text{and} \quad Q = v \cdot A$$

Fig. 12b. Calculation procedure for a vegetated fish pass (Lindner 1982; Pasche 1984; Lehmann 2005; Tymiński and Kałuża 2008).

Calculation of the discharge in an open channel with cross walls (bars)—cascaded fish pass (Fig. 3b)

Step-pool rock ramp fishway—a ramp-type fish pass with boulder steps which raise water.

The bars usually form pools of equal length. Long fish passes are divided into sections connected with pools of a more considerable size, the so called resting pools. In the bars, between the piles (made of stone, wood, etc.) there are gaps, of which at least one should be appropriate in width for the size of the largest fish. In the subsequent bars, the main gap should be located alternately, once on the left and once on the right, so as to have a meandering current. The channel can be natural (bypass, millrace, etc.) or profiled with stone/concrete/wood. The flow may occur though bars which are submerged either completely or only partially.

The hydraulic calculations for natural and artificial channels are similar. Each time, one must determine the correct drag coefficient. If the bars are formed of concrete or wooden piles and entered into a channel with regular shapes, the calculations are more accurate, as the elements are repetitive. For varied cross sections (various depths and widths) and when there are ponds, the discharge Q, must be determined stretch after stretch. In order to determine the discharge coefficient of a weir, one can use the values of coefficients for weirs having a sharp-edge, broad crown, and practical shapes. An equal damming height for drops made of boulders is difficult to achieve. Level Δh should be corrected during usage or the bars should be made so as to allow some adjustment of the slot width.

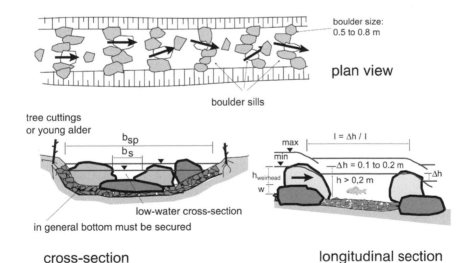

Fig. 13. Computational diagram for a step-pool rock ramp fishway (DVWK 2002).

Calculation procedure:

Value	Symbol	Unit	Formula
Head	H	m	
Water depth in pool	h	m	
Average pool width	b	m	
Width of the slot between boulders	b_s	m	
Pool cross section surface area	A	m	
Net distance between the bars	l	m	
Water level difference between the pools	Δh	m	$\Delta h = h_g - h$
Pool filling upstream of the bar	h_g	m	
Water flow velocity in narrows	v	m/s	$v_s = \sqrt{2g\Delta h} < v_{dpo.}$
Number of pools	n	pcs	$n = \dfrac{H}{\Delta h} - 1$
Fish pass length (net)	L	m	$L = n \cdot l$
Weir discharge coefficient	μ	-	$\mu = 0{,}5{-}0{,}6$ sharp stone $\mu = 0{,}6{-}0{,}8$ crushed stone, field stone
Weir submersion coefficient	σ	-	$\sigma = f\left(\dfrac{h}{h + \Delta h}\right)$ Fig. 14.
Discharge	Q	m³/s	$Q = \dfrac{2}{3}\mu\sigma\Sigma b_s \sqrt{2g}h_g^{3/2}$
Sum of the widths of a hydraulically active flow	Σb_s	m	$\Sigma b_s = \dfrac{Q}{\dfrac{2}{3}\mu\sigma\sqrt{2g}h_g^{3/2}}$
Energy dissipation	E	W/m³	$E = \dfrac{\rho Q\Delta h}{A \cdot l} < E_{dop}$

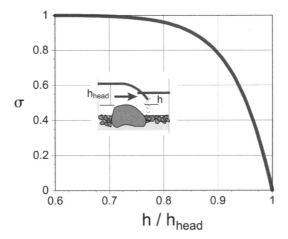

Fig. 14. Weir submersion coefficient σ (DVWK 2002).

Conventional pool fish pass (Fig. 1a)

Their principal feature is that the water flows through subsequent pools arranged in a cascade, which connects tailwaters with headwaters. The dimensions of pools and openings of the fish pass depend on the species of migrating fish, the level difference between the inlet and outlet of the fish pass, and the type of river.

A pool fish pass is a fish pass in the form of a series of pools separated with cross walls with weirs and (not necessarily) narrow bottom openings, arranged as a series of small dams.

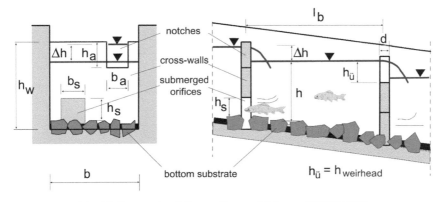

Fig. 15. Computational diagram for a pool fish pass (DVWK 2002).

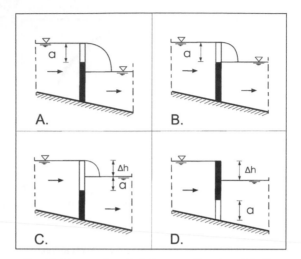

Fig. 16. Diagrams for calculating water flow through baffles.

Calculation procedure:

Value	Symbol	Unit	Formula
Head	H	m	
Height of the weir/ opening	a	m	
Width of the pool (weir/opening)	b	m	
Pool cross section surface area	A	m	
Pool length (net)	l	m	
Water level difference between the pools	Δh	m	$\Delta h = h_g - h$
Pool filling downstream of the baffle	h	m	
Pool filling upstream of the baffle	h_g	m	
Water flow velocity in narrows	v	m/s	$v_s = \sqrt{2g\Delta h}$
Number of pools	n	pcs	$n = \dfrac{H}{\Delta h} - 1$
Fish pass length (net)	L	m	$L = n \cdot l$
Discharge for cases "A" and "B"	$Q_{A,B}$	m³/s	$Q_{A,B} = \dfrac{2}{3}\mu\, b\sqrt{2g}\, a^{3/2}$ where: $\mu = 0.60\text{–}0.85$

Discharge for case "C" (partially submerged weir "a")	Q_C	m³/s	$Q_c = \dfrac{2}{3}\mu_1 b\Delta h\sqrt{2g\Delta h} + \mu_2 ba\sqrt{2g\Delta h}$ where: $\mu_1=0.85$ $\mu_2=0.65$
Discharge for case "D" (discharge from a submerged b x a opening)	Q_D	m³/s	$Q_D = \mu ba\sqrt{2g\Delta h}$ where $\mu=0.65$
Energy dissipation	E	W/m³	$E = \dfrac{\rho Q\Delta h}{A\cdot l} < E_{dop}$

Vertical slot fishway (Fig. 1b)

Vertical slot fishway → pool fish pass with vertical slots over the entire height of its pool walls. [sw]

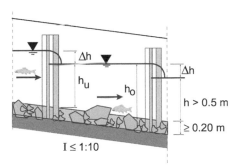

Fig. 17. Computational diagram for a slot fish pass (DVWK 2002).

   This type of fish pass was developed in the second half of the 20th century in North America. Its use proved to be successful for efficient migration of sea trout and salmon as well as the cyprinids. This fish pass is a modification of the classical pool design: the passage openings and weirs are replaced with slots located either on one side or on both sides of the pool. The slots extend over the entire height of the pool. Such a modification makes the fish pass more effective for larger water level differences and the slots do not clog up so often as the classical openings. The width of slots depends on the size of migratory fish for which the fish pass is designed and the number of slots (either one or two) depends on the amount of water flowing uninterruptedly through the fishway.

Calculation procedure:

Value	Symbol	Unit	Formula
Head (the difference between the headwater and tailwater level)	h	m	
Pool length (net)	l	m	
Pool width	b	m	

Pool cross section surface area	A	m	
Passage slot width	s	m	
Pool filling downstream of the baffle	h_{min}	m	
Pool filling upstream of the baffle	h_g	m	$h_g = h_{min} + \Delta h$
Water level difference between the pools	Δh	m	$\Delta h = h_g - h_{min}$
Water flow velocity in the slot	v_s	m/s	$v_s = \sqrt{2g\Delta h}$
Number of pools in the fish pass	n	pcs	$n = \dfrac{h}{\Delta h} - 1$
Fish pass length (net)	L	m	$L = n \cdot l$
Discharge	Q	m³/s	$Q = \dfrac{2}{3}\mu \cdot s\sqrt{2g}\,h_g^{3/2}$ where: $\mu = f(h_{min}/h_g)$
Average flow velocity at the pool cross section	v_k	m/s	$v_k = \dfrac{Q}{A}$
Energy dissipation	E	W/m³	$E = \dfrac{\rho \cdot g \cdot \Delta h \cdot Q}{b \cdot h_g \cdot l} < E_{dop}$

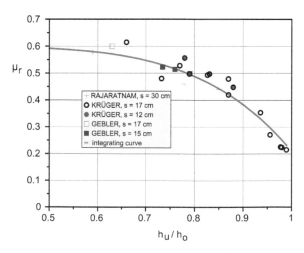

Fig. 18. Slot discharge coefficient (DVWK 2002).

Denil fishway (Fig. 1c)

A Denil fishway is a fish pass in the form of a series of inclined chutes with baffles mounted at regular intervals to slow down the flow and form the resting pools. Its slope and length are adjusted to the possibilities of the prevailing fish species and must ensure an appropriate attracting current. [sw]

The fish passes of this type are concrete chutes measuring a maximum of 10 [m] in length, with a slope up to 45°. They are separated with resting pools. The main elements are the embedded baffles of various shapes, which slow down the flow of water.

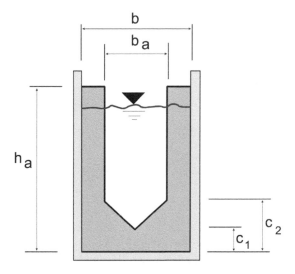

Fig. 19. Baffle in Denil fish pass (DVWK 2002 after Lonnebjcrg 1980).

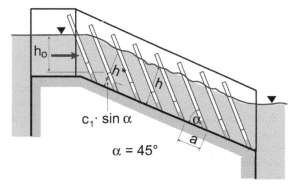

Fig. 20. Calculation diagram—longitudinal profile (DVWK 2002, after Larinier 1992).

Calculation procedure

Value	Symbol	Unit	Formula
Channel width	b	m	- salmonids b=0.8–1.2 m - brown trout and cyprinids b=0.6–0.9 m
Distance between the baffles	a	m	a=0.66b
Slot width	b_a	m	
Slot height from the water surface level	h*	m	h*=1.5b_a
Pool depth	h	m	
Resting pool length	l	m	
Channel slope	J	%	20%(1:5)–10%(1:10)
Discharge (Krüger 1994)	Q	m³/s	$$Q = 1{,}35 b_a^{2,5} \sqrt{gJ} \left(\frac{h^*}{b_a} \right)^{1,584}$$
Average flow velocity at the pool cross section	v_k	m/s	$$v_k = \frac{Q}{A} = \frac{Q}{b_a \cdot h^*}$$
Energy dissipation in the stilling pool	E	W/m³	$$E = \frac{\frac{\rho}{2} \cdot Q \cdot v_k^2}{b \cdot h \cdot l} < E_{dop} \approx 25\text{–}30 \text{ W/m}^3$$

Hydraulic Calculations for Fish Passes Using Mathematical Models

Computer simulations have been used as a support tool for technical design for many years. The modelling of flow through technical structures can be particularly useful in eliminating various design problems in fish passes, moreover, it also increases the effectiveness of the structure in transferring fish to the other side. A mathematical model is a finite set of symbols and mathematical relations as well as strict rules for using them. The symbols and relations contained in a model can be interpreted with respect to specific elements of the reality that is being modelled.

There are numerous computer programs which can be adapted to simulate the flow of water through a fish pass. Some of these advanced programs are developed by specialized companies; others are purpose-made by individual researchers to solve specific problems.

When constructing a model, some basic rules must be adhered to. A correctly developed model should lead to unequivocal solutions. These solutions should also be stable in terms of initial conditions and parameters.

An overview of models and mathematical methods used for describing the flow of water through a fish pass begins with the preparation of an EXCEL spreadsheet, which are helpful in the development of the initial concepts. During later stages of the designing process, it is recommended to use more advanced population models described with differential equations. Examples of such models include: HEC-RAS, FLOW 3d, CCHE_MESH 3.0, CCHE_GUI 3.290, etc.

It is recommended that the computer analysis of the water flow and fish behaviour in fish passes consist of three modules:

1. A computational model for fluid dynamics, which should describe the assumed discharge field and should allow one to understand the motion.
2. A model of particle motion, interpolating the hydraulic information at fixed nodes of the computational grid, in places important for fish migration.
3. A model which simulates the cognitive and conservative functions of individual fish in connection with the computational model of fluid dynamics.

Together, these three models should create a virtual reality in which virtual fish would behave realistically when passing through the fishway.

After the calibration and validation of computer simulation results, the designer can choose an optimal variant from among many competitive solutions.

Hydraulic calculations for fish passes using the HEC-RAS software

HEC-RAS is a computer program for one-dimensional hydraulic calculations in networks of natural and artificial channels. The software is developed by the Hydrologic Engineering Center, which is part of the United States Army Corps of Engineers. Since 1995 the program has been a freeware. Currently, HEC-RAS consists of four modules:

The critical flow in an open channel is a flow with a depth sufficient to ensure that the specific energy of the stream is minimal.

1. Calculation of water table profiles,
2. Simulations of unsteady flows,
3. Bed load transport,
4. Water quality analysis.

The most fundamental advantage of the system is that all the four components make use of the same type of representation of geometric data and of the same hydraulic calculation procedures.

The calculations consist of solving a one dimensional stream energy equation. The energy loss is expressed by friction (Manning formula) and contraction (an appropriate contraction coefficient multiplied by the change in velocity). In places with a fast changing motion, the calculation is performed using the equation of conservation of momentum. This equation is used for mixed flow regimes (a hydraulic jump is formed), within the clear span of bridges and in calculations of profiles at water nodes (connection or splitting of flow). The work on a new version of HEC-RAS has been carried out for two years and the source code is generally available. Currently, a beta version of HEC-RAS 5.0 is available. In the new version, it is possible to achieve a spatial distribution of the linear drag coefficient for 2D areas and some computational mechanisms for systems of channel networks have been improved. Another valuable novelty is the possibility of obtaining the distributions of velocities, both in one-dimensional and in two-dimensional areas of the model. The delay in publication of the full 5.0 version is due to the work intended to adjust the tool to the full 2D functionality in all the aspects of modelling of flow, both in networks of open channels and in structures such as culverts, sills, pumping stations, fish passes, riffles, etc. Both the stable version and the beta 5.0 version allow one to perform hydraulic calculations for fish passes of various designs and to calculate variants (changes in design) for various structures. Irrespective of the software version, the key to success is in the correct hydraulic analysis of the structure and the possible linearisation of some of its elements to facilitate and accelerate the calculations during the initial phase of design work.

Hydraulic calculations for fish passes using the CCHE2D software

CCHE2D is an integrated package allowing two-dimensional simulation and analysis of water flow, bed load transport, and various morphological processes. The package allows one to perform calculations for both the steady and unsteady flow (Zhang 2006). It enables modelling of transport of both the suspended load and the bed load, which results in a transformation of the original channel shape of the watercourse. The application accounts for the changes in the bottom roughness and the channel geometry. It is particularly important for simulations of flow over long periods of time and under the assumption of intensive erosion.

The CCHE2D package includes a program for making numerical models, a mesh generator (CCHE2D Mesh Generator) and a graphical user interface (CCHE2D-GUI) (Hasan et al. 2007) (Fig. 21).

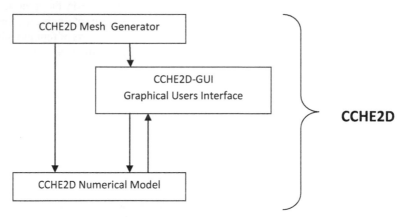

Fig. 21. Diagram of CCHE2D package (Hasan et al. 2007).

The application CCHE2D has been developed by the National Centre for Computational Hydroscience and Engineering at the University of Mississippi, USA. Calculations in the CCHE2D model are carried out based on the two-dimensional grid using the finite element method. This is a typical approach increasingly common in solving various engineering problems. In CCHE2D, the computational nodes are on the vertexes of quadrilaterals formed by the curvilinear grid. The flow parameters are directly determined (calculated) for these nodes. The values of parameters inside an element or on its sides are calculated using linear interpolation functions. The model makes use of the vertically averaged Navier-Stokes equations, the solving of which yields data on the vertical average velocity and the ordinate of water level (Zhang 2006):

$$\frac{\partial Z}{\partial t} + \frac{\partial(hu)}{\partial x} + \frac{\partial(hv)}{\partial y} = 0$$

$$\frac{\partial u}{\partial t} + u\frac{\partial u}{\partial x} + v\frac{\partial u}{\partial y} = -g\frac{\partial Z}{\partial x} + \frac{1}{h}\left[\frac{\partial(h\tau_{xx})}{\partial x} + \frac{\partial(h\tau_{xy})}{\partial y}\right] - \frac{\tau_{bx}}{\rho h} + f_{Cor}v$$

$$\frac{\partial v}{\partial t} + u\frac{\partial v}{\partial x} + v\frac{\partial v}{\partial y} = -g\frac{\partial Z}{\partial y} + \frac{1}{h}\left[\frac{\partial(h\tau_{yx})}{\partial x} + \frac{\partial(h\tau_{yy})}{\partial y}\right] - \frac{\tau_{by}}{\rho h} + f_{Cor}u$$

Where:

u, v – averaged velocities in the direction of x and y,

g – the force of gravity,

Z – water level,

ρ – the density of water,

h – local water depth,

f_{Cor} – the parameter of Coriolis,

$\tau_{xx}, \tau_{xy}, \tau_{yx}, \tau_{yy}$ – Reynolds stress averaged over the depth,

τ_{bx}, τ_{by} – shear stress on the surface of the bottom.

The process of model development using CCHE2D can be divided into five key stages:

- Generating of the mesh/grid.
- Determining the boundary conditions.
- Introducing the parameters of the model.
- Launching the simulation.
- Visualisation and interpretation of results.

It should be stressed that all of the elements listed above have a significant impact on the reliability of the results obtained.

In the example below, the simulation was carried out for a steady flow of Q=0.3 m³/s. The modelling accounted for obstacles in the form of boulders of approx. 1.0–1.5 m in diameter and logs of approx. 0.3–0.4 m in height, as well as for dense elastic vegetation modelled both by the change in channel geometry and by the use of coefficient n=0.150. The bottom of the watercourse is assumed to be made of stone, while its banks are vegetated with high grassy vegetation. The elements in the channel (such as plants, boulders, tree limbs, etc.) are modelled by changing the terrain geometry so that it resembles the obstacles being modelled. In practice, this has led to a reduction in the cross section by its partial shielding. A graphical visualisation of the water current under analysis and the result of modelling are shown in Fig. 22, Fig. 23 and Fig. 24.

Keywords: Fish pass, hydraulic calculations, flow conditions, ecological river continuity.

Fig. 22. Visualisation of natural elements in channel.

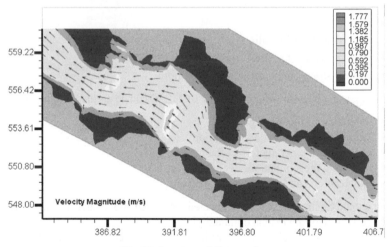

Fig. 23. Sample modeling results.

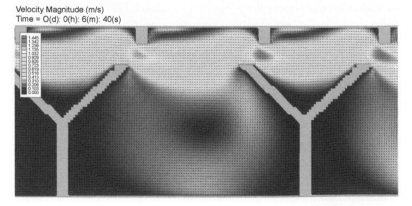

Fig. 24. Sample modeling results for a slot-type fish pass.

References

References

Adam, B. et al. 2005. Fischschutz- und Fischabstiegsanlagen - Bemessung, Gestaltung, Funktionskontrolle (in German), DWA, 2. Auflage, Hennef, 256 pp.

Chow, V.T. 1959. Open-Channel Hydraulics, McGraw-Hill Book Company, New York&London, 680 pp.

DVWK. 1991. Hydraulische Berechnung von Fließgewässern (in German), DVWK-Merkblatt, Nr. 220, Bonn, 70 pp.

DVWK. 1996. Fischaufstieganlagen: Bemessung, Gestaltung, Funktionskontrolle, Merkblätter zur Wasserwirtschaft (in German), Nr. 232, Bonn, 113 pp.

DWA. 2009. Fischaufstieganlagen und fischpassierbare Bauwerke: Gestaltung, Bemessung, Qualitätssicherung (in German), DWA-Regelwerk, Merkblatt DWA-M509/2009, Gelbdruck, 302 pp.

DWA. 2010. Durchgängigkeit und Habitatmodellierung von Fließgewässern (in German), Verlag der Bauhaus-Universität Weimar, 274 pp.

DVWK & FAO (Food and Agriculture Organization of the United Nations in arrangement with Deutscher Verband für Wasserwirtschaft und Kulturbaue.V.), 2002, Fish Passes - Design, Dimensions and Monitoring, Rome, 118 pp.

Hasan, Z.A. and Zakarina, N.A. 2007. Application of 2-D Modelling for Muda River Using CCHE2D, presented at the International Conference on Managing Rivers in the 21st Century: Solution Towards Sustanable River Basins, Riverside Kuching, Sarawak, Malaysia.

Hassinger, R. 2002. Der Borstenfischpass -Fischaufstieg und Bootsabfahrt in einer Rinne (in German), Wasserwirtschaft, 4-5, 38 ff.

Hassinger, R. and Kraetz, D. 2006. The canoe-fishway—a combination of fish migration facility and canoe passage in the same channel. EIFAC, 24th EIFAC Symposium on Hydropower. Flood Control and Water Abstractions: Implications for Fish and Fisheries, Mondesee.

Hassinger, R. 2009. Borsten-Fischpasse und Fisch-Kanu-Pass Beschreibung der Technik, Kassel (http://www.uni-kassel.de/fb14/vpuw/Download/FKP/Stand_der_Technik Borstenkonzept o309.pdf).

Kałuża, T. and Hämmerling, M. (eds.). 2015. Design and Using Problems of Fish Passes (in Polish), Monografia, Bogucki-Wydawnictwo-Naukowe, Poznań, 139 pp.

Krüger, F. 1994. Denil-Fischpässe (in German), Wasserwirtschaft/Wassertechnik, 3/94, 24-32 pp.

Krüger, F. 2008. Anforderungen an Fischaufstiegsanlagen, Beispiele aus der Praxis. Konferenzmaterialien: Vortrag zum Wasserbaulichen Kolloquium "Oekologische Durchgängigkeit von Fließgewässern" (in German), Universität Hannover, 15 pp.

Kubrak, J. 1998. Hydraulics (in Polish), Wydawnictwo SGGW, Warszawa, 371 pp.

Larinier, M. 1992. Passes a bassins successifs, pre barrages et rivieres artificielles (in French), Bull. Fr. Peche Piscic, No 326/327, 45-72 pp.

Larinier, M. 2000. Dams and Fish Migrations. World Commission on Dams, Environmental Issues, Final Draft.

Lehmann, B. 2005. Empfehlungen zur naturnahen Gewässerentwicklung im urbanen Raum (in German), PhD thesis, Mitteilungen des Instituts für Wasser und Gewässerentwicklung, Universität Karlsruhe, Heft 230, 250 pp.

Lindner, K. 1982. Der Strömunswiderstand von Pflanzenbeständen (in German), Mitteilungen des Leichtweiss-Instituts für Wasserbau, TU Braunschweig, Heft 75.

Mokwa, M. and Wiśniewolski, W. (eds.). 2008. Ichthyofauna Protection by Hydrotechnical Structures (in Polish), Monograph, Dolnośląskie Wydawnictwo Edukacyjne, Wrocław, 201 pp.

Mokwa, M. 2010. Obliczenia hydrauliczne przepławek dla ryb. Hydraulic calculations for fishways. Acta Scientiarum Polonorum. Formatio Circumiectus.

Mokwa, M. and Tymiński, T. 2015. Hydraulic Model Research on Meander Fish Pass (in Polish), In: Kałuża, T. and Hämmerling, M. (eds.). Design and Using Problems of Fish Passes, Monograph, Bogucki-Wydawnictwo-Naukowe, Poznań, 109–118 pp.

Mumot, J. and Tymiński, T. 2016. Hydraulic Research of Sediment Transport in the Vertical Slot Fish Passes. Journal of Ecological Engineering 17(1): 143–148 pp.

Pasche, E. 1984. Turbulenzmechanismen in naturnahen Fließgewässern und die Möglichkeit ihrer mathematischen Erfassung (in German), Mitteilungen des Instituts für Wasserbau und Wasserwirtschaft, RWTH Aachen, Band 52.

Rickert, K. 1992. Naturnahe Regelung von Fließgewässern, Kurs PW 06 (in German), Universität Hannover, Weiterbildendesstudium Bauingenieurwesen, Wasserwirtschaft, WS 1992/93.

Teppel, A. and Tymiński, T. 2013. Hydraulic Research for Successful Fish Migration Improvement—"nature-like" Fishways. Civil and Environmental Engineering Reports, No. 10, 125–137 pp.

Tymiński, T. 2007. Analysis of Impact of Flexible Vegetation on Hydraulic Conditions of Flow in Vegetated Channels (in Polish), Monograph, Wydawnictwo Uniwersytetu Przyrodniczego we Wrocławiu, Wrocław, 82 pp.

Tymiński, T. and Kałuża, T. 2008. Hydraulic calculations for fish passes with biotechnical build-up (in Polish). pp. 102–111. *In*: Mokwa, M. and Wiśniewolski, W. (eds.). Ichthyofauna Protection by Hydrotechnical Structures (in Polish), Monograph, Dolnośląskie Wydawnictwo Edukacyjne, Wrocław, 201 pp.

Tymiński, T. and Kałuża, T. 2012. Investigation of Mechanical Properties and Flow Resistance of Flexible Riverbank Vegetation. Polish Journal of Environmental Studies 21(1): 201–207.

Tyminski, T. and Kałuża, T. 2013. Effect of Vegetation on Flow Conditions in the "nature-like" Fishways, Annual Set The Environment Protection, Vol. 15/2013, 348–360 pp.

Tyminski, T. and Mumot, J. 2015. Model Tests of Hydraulic Flow Conditions in the Vegetation Build-up Fishway (in Polish), Inżynieria Ekologiczna. Journal of Ecological Engineering, Vol. 44/2015, 227–234 pp.

Vannote, R.L., Minshall, G.W., Cummins, K.W., Sedell, J.R. and Cushing, C.E. 1980. The River Continuum Concept. Canadian Journal of Fisheries and Aquatic Sciences 37: 130–137.

WFD/EC. 2000. Water Framework Directive 2000/60/EC of 23 October 2000.

Whittaker, J. and Jäggi, M. 1986. Blockschwellen (in German), Mitteilungen, Versuchsanstalt für Wasserbau, Hydrologie und Glaziologie der ETH Zürich, 91.

Zhang, Y.: CCHE-GUI - Graphical user interface for the NCCHE model. User's manual - Version 3.0. NCCHE, School of Engineering, The University of Mississippi. MS 38677.

CHAPTER 7

River Embankments

Magdalena Borys,[1] *Eugeniusz Zawisza,*[2,]* *Andrzej Gruchot*[2] and
Krzysztof Chmielowski[3]

INTRODUCTION

Of late, due to rapid economic development, flood protection of areas threatened by flooding is becoming increasingly important. In general, there are a few basic methods for protection from flooding, namely: the construction of river embankments and storage reservoirs, retention adaptation of the catchment, construction of dry polders, river planning, flood control channels, and preventing the formation of blockages. The methods mentioned above are not mutually exclusive and may be used simultaneously.

River embankments are still the primary form of flood protection of river valleys, despite being condemned by environmentalists as having adverse impacts on the environment; they represent indispensable means of allowing the economic use of river valleys. Although it is possible to reduce some of the agriculturally used areas within valleys and to give up these areas to rivers, it is difficult to even consider the liquidation of rivers and housing estates (Mioduszewski and Dembek 2009).

River embankments have been built for hundreds of years. The first documented embankments in the current Polish territory were built in the second half of the 12th century in the delta of the Vistula, before the arrival of the Teutonic Order in this area. The fragments of the so-called "old dam" from that period have been preserved to this day. This dam was probably supposed to restrict the backwater of the main branches of the Vistula from the west and to protect the Żuławy Gdańskie and Gdańsk from flood (Makowski 1993).

[1] Institute of Technology and Life Sciences, Falenty, Al. Hrabska 3, 05-090 Raszyn.
  E-mail: m.borys@itp.edu.pl
[2] University of Agriculture in Kraków, Faculty of Environmental Engineering and Land Surveying, Department of Hydraulics Engineering and Geotechnics, al. Mickiewicza 24/28, 30-059 Kraków.
  E-mail: rmgrucho@cyf-kr.edu.pl
[3] University of Agriculture in Kraków, Faculty of Environmental Engineering and Land Surveying, Department of Sanitary Engineering and Water Management, al. Mickiewicza 24/28, 30-059 Kraków.
  E-mail: k.chmielowski@ur.krakow.pl
* Corresponding author: kiwig@ar.krakow.pl

In Poland, there is a total of nearly 9 thousand km of river embankments that protect an area of about 1.1 million hectacres (ha) located in river valleys and polders. It is estimated that the areas protected by the embankments are inhabited by nearly 1.5 million people (Borys and Rycharska 2007).

In recent years, virtually no new flood embankments have been built in Poland. The existing majority of embankments must, however, be modernized and kept in good technical condition, as safety and various types of human activity depend on it.

2. Functions and Types of Embankments

Flood embankments are the constructions that restrict the cross-sectional profile of the river, thereby reducing the extent of flooding. Due to their function, they are built in cases when:

- Flood plain, which is to be subjected to protection, is flat and expansive, and other ways of lowering the flood culmination proved to be ineffective.
- It is impossible to regulate the stream bed in order to allow the passage of flood waves in a safe manner.
- The increase of flood is large enough to exceed the natural shores of a river.
- There are objects of great value within the flood range (cities, settlements, industrial plants, transportation routes, etc.).

When making a decision to build embankments, all the "pros" and "cons" should be considered and the embankments should be used moderately and only where there is no possibility of using other means of flood protection.

The basic elements of flood embankments are shown in Fig. 1.

Fig. 1. Names of the elements of high water river channel and embankment. 1: High water river channel (inter-embankment), 2: Main river channel, 3: Inter-embankment (flood area), 4: Embankment body, 5: Crest, 6: Embankment base, 7: Upstream slope, 8: Downstream slope, 9: Bench, 10: Embankment ditch, 11: Strengthening, 12: Drainage, 13: Sealing element of the body, 14: Anti-filtration barrier in the substratum (Adamski et al. 1986).

Depending on the location of the stream, the flood embankments may be divided as follows:

- Main, located along the main rivers;
- Transverse (partition), constructed so that in the case of interruption of the main embankments it is possible to locate the area of flooding;
- Directing, constructed in order to protect in some places (e.g., river bend), the main embankment from the focused river current and flowing ice.

- Reverse (backwater), which aims to protect the areas from waters retracting from the main stream to the higher order streams;
- Annular, which surrounds and protects residential estates or other important facilities against the flood.

The typical embankment of a river section is shown in Fig. 2. The right side of the embankment is an open embankment and the left part is a closed embankment. Constructing open embankments is appropriate for rivers with large falls and with broad valleys. Rainwater and water coming from snowmelt, accumulated in the area fenced with open embankment, runs freely downwards. With closed embankments, the local runoff stops during the flood in the hollows on the landward side of the levee and in order to let this water through to the river, culverts are made in the embankments and discharge the water after flood drops in the inter-embankment (Bednarczyk et al. 2006).

3. Embankment Location

One of the basic parameters determined at the design stage of the river embankments is their spacing, that is, the distance between the embankments. Figure 3 shows how important this issue is, as it presents the relationship between the height of the flood embankments and their spacing.

In general, the embankment route is defined by the flow conditions in the flood risk area—in particular the location of the flow areas, results of the calculations and analyses of the high water river channel dimensions, as well as the topography and geology. Its detailed course and unavoidable local deviations from the routes determined by the mentioned factors can depend on the following conditions, which are usually not included in the analyses: Currents in the high water river channel and other flow conditions, economic conditions along the route, in the inter-embankment and on the landward side of the levee, housing estates and sometimes other considerations (e.g., transition bridges, pipelines, cables, water intakes, ports, harbors, piers, and others).

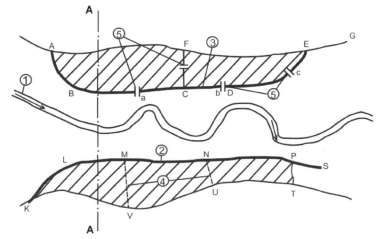

Fig. 2. Scheme of the river valley embankments. 1: River, 2: Open embankment, 3: Closed embankment, 4: Transverse embankments, 5: Culverts (Bednarczyk et al. 2006).

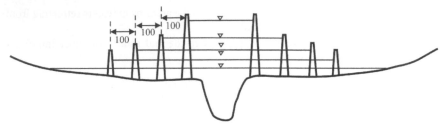

Fig. 3. Relationship between the embankment height and their spacing (Own study after [Rytel et al. 1973]).

While designing the embankment routes, the number and angles of currents' intersections in the main river channel and in the inter-embankment should be limited, because as assumed, they increase the flow resistance and may cause damage to the shores, buildings and sometimes embankments themselves. The system of currents in the high water river channel is significantly affected by many obstacles, such as single buildings or their groups, small clusters of very densely growing trees and shrubs, etc., because although they usually do not cause a significant rise in the water level, they still can cause watering down of the embankments. It may be difficult to assess the magnitude of the watering down events; therefore the aim is to (except when it is impossible due to the environmental protection requirements) eliminate these obstacles. It is also usually advisable to do this because of the threat of ice jam formation. Local irregular increase in spacing of embankments may be harmful in all cases, when it creates large areas of turbulence, and in such arrangements, the high water river channel with smaller embankment spacing may have a higher capacity than the wider river channel (Mosiej and Ciepielowski 1992).

The following rules should be followed at the design stage of the embankment route:

- The embankment route needs to be designed in the way that it follows the natural gradient (if possible) which will considerably reduce the cost of construction.
- Due to the geological, geotechnical, and hydrological conditions, crossing oxbow lakes, different types of excavations, wetlands, and ground where one can expect hydraulic failures, suffosion and seepage failures should be avoided.
- Due to the threat of scouring of the embankment toe, it is moved away from the river shore to a safe distance, which (when allowed by the overall width of the inter-embankment) should be at least 10 m, or preferably 20 to 50 m.
- As a result of scouring, the embankment should not be too close to the cavities and continuous channels or their discontinuous chains, where the flow concentration, and thus its high speed, may occur.
- As follows from the stability calculations, in order to avoid the threats to the stability of the body and base of the embankment, its toe should not be situated near excavations, exploited or operating gravel pits, and other mining sites of raw materials located in the inter-embankment or on the land side of the levee, and the distance should not be less than triple depth of the excavation or 30 m.
- Due to of the threat of digging dens by animals (e.g., beavers), the distance from the embankment toe of excavations or cavities in which the water table remains, should not be less than 15.0 m.

- Possible changes to the spacing between river embankments should be implemented gradually, without causing harmful interference to the high water flow.
- The embankment route should be adjusted to the direction of the valley and to the route of the low and medium water river channel.
- The embankment route should approach the building line on sags, while on summits the route should recede from it.

Embankments are associated with different types of constructions, such as: bridges, ferry crossings, ford crossings (natural or artificial), pipeline and cable passages, ports, overwintering areas, and marinas and waterfronts. The buildings existing on or by the river before the construction of embankments may cause the need to adapt the embankments to these facilities. These adaptations may include relatively small changes in the embankment routes, local strengthening of their bodies and the neighboring areas, etc. The greatest works are sometimes necessary in relation to the communication passages with small openings, but these are mainly regulatory works (increasing the cross-section of high water river channel and steering embankments). In some cases, building of embankments may require different types of reconstruction of port facilities and their access roads, as well as pipeline and cable passages (Mosiej and Ciepielowski 1992).

4. Basic Geometric Dimensions of Embankments

The cross-section of the embankment should be designed in a way that ensures its stability and that the phreatic surface, that is, the position of the percolating water table in the embankment body in the border location was at a distance less than 1 m from the surface of the downstream slope (Fig. 4).

The river embankment must meet the requirements for hydrotechnical buildings, including those stated in the Regulation of the Ministry of the Environment (2007). The minimum permissible dimensions of the embankment body depend on:

- Belonging of the embankment to the particular class of hydrotechnical buildings.
- Secure elevation of the embankment crest above the computed water table.
- Necessary minimum width of the embankment crest.
- Requirements for the embankment slope.

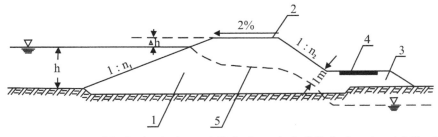

Fig. 4. Cross-section of the flood embankment. 1: Embankment body, 2: Embankment crest, 3: Flood bench, 4: Flood road, 5: Phreatic surface, $1:n_1$ – Inclination of the upstream slope, $1:n_2$ – Inclination of the downstream slope, h: Computed water table, Δh: Safe elevation of the embankment crest (Own study after [Borys 2006]).

According to the Regulation of the Ministry of the Environment (Rozporządzenie 2007), river embankments are classified as buildings intended for flood protection, to one of the four classes of hydrotechnical buildings, depending on the size of the area protected by the embankment (Table 1). The protected area is the area that can be flooded by flow of a probability, p=1%.

Table 1. Classification of flood embankments according to the regulation of the Ministry of Environment (Rozporządzenie 2007).

Description of the Index	Value of Index for Class				Remarks
	I	II	III	IV	
Protected area F, km²	$F>300$	$150<F<300$	$10<F\leq150$	$F\leq10$	The area that before being embanked was flooded by waters of a probability of $p=1\%$

The protected area is not a sufficient factor to determine the class of the construction, as the number of people living in the area that can be flooded also needs to be taken into account. For class I, it is more than 300 people, for II from 80 to 300 people, for III from 10 to 80 people, and for IV less than 10 people (Rozporządzenie 2007). Recall that this classification takes into consideration not only people permanently living in this area, but also temporary residents.

The class of hydrotechnical buildings is closely related to the concepts of base and control flow. The Regulation of the Ministry of the Environment (Rozporządzenie 2007) defines that the base flow is the one which hydrotechnical buildings are designed for and it is the calculated flow, Q_m, with probability of occurrence determined according to the Table 2, while the control flow, Q_k, is used for the assessment of the construction safety in a unique load system and it is a maximum calculated flow with a specific probability of occurrence.

The embankment crest should be located safely above the calculated water table (Fig. 3). The coordinate of the embankment crest results from the position of the calculated water table determined by base and control flows and the results of hydrological calculations for the assumed embankment spacing. Safe elevation of the embankment crest, according to the above quoted Regulation of the Ministry of Environment (Rozporządzenie 2007), should not be less than the one given in Table 3.

The size of the safe elevation of the embankment crest, determined in accordance with the presented rules should be considered as minimum. Increasing the embankment crest elevation should be considered, especially if:

• Hydrological calculations were based on very short strings of maximum flow or not very reliable material.
• Ice or frazil ice jams cannot be excluded.
• Inter-embankment may be overgrown or rubble may be deposited therein.
• Waving can be caused by wind or ship traffic.
• Vehicular traffic will take place on the embankment crest.

Crest of the embankment on which the flood action is to be conducted, needs to be additionally elevated (more than the calculated elevation) by 0.50 m, when the crest is passable and by 0.30 m, when it is impassable. In the places where passages, transhumances, pumping stations, and embankment culverts are located, the crest is elevated by approximately 0.20 m in relation to its position on the adjacent section (Żbikowski 1982).

The width of the embankment crest is generally comprised between 3–5 m. In order to drain the embankment crest, it is built up with the transverse inclination of 2% towards to the upstream slope.

The width of the entire embankment body results from the inclination of its slopes and the width of the crest. When it is expected that due to the impact of climatic factors, including wetting and drying, freezing and thawing, the soils in the embankment will crack or clod (e.g., as in the case of organic soil), the width of the embankment body should be increased on both sides by at least 0.5 m.

Table 2. Probability of occurrence (exceeding) of base and control flows for flood embankments according to the regulation of the Ministry of Environment (Rozporządzenie 2007).

Flow	Probability of Occurrence (p,%) for the Class			
	I	II	III	IV
Base Q_m	0.5	1.0	2.0	3.0
Control Q_k	0.1	0.3	0.5	1.0

Table 3. Safe elevation of the crests of solid hydrotechnical constructions according to the regulation of the Ministry of Environment (Rozporządzenie 2007).

Exploitation Conditions	Safe Elevation of the Crest of Hydrotechnical Constructions (in m) for Classes I-IV							
	Over-static Water Table				Over the Level Made by Waving			
	I	II	III	IV	I	II	III	IV
Maximum water level	2.0	1.5	1.0	0.7	0.7	0.5	0.5	0.5
Base flood flow	1.3	1.0	0.7	0.5	0.5	0.3	0.3	0.3
Extreme working conditions of the construction	0.3	0.3	0.3	0.3	Waving not considered			

Table 4. Minimum slope inclination of flood embankments (Żbikowski 1982).

Type of Soil in Embankment's Body	Slope Inclination		
	Upstream	Downstream	
		With Drainage	Without Drainage
Non-cohesive	1 : 2.5	1 : 2.0	1 : 2.25
Cohesive	1 : 2.0	1 : 2.0	1 : 2.0

The inclination of the embankment body slopes should be determined based on the calculations of stability, while the slope inclinations should not be steeper than those given in Table 4 (Żbikowski 1982).

A very important matter affecting the embankment security is the position of the phreatic surface. The embankment should be constructed so that the phreatic surface in the embankment body in its border position is at a distance less than 1 m from the surface of the downstream slope, as presented earlier in Fig. 4.

Such layout of the phreatic surface in the embankment body is possible under the conditions of unsteady filtration, that is, when the period of flooding is short. However, if the calculations of the course and layout of the phreatic surface for unsteady flow

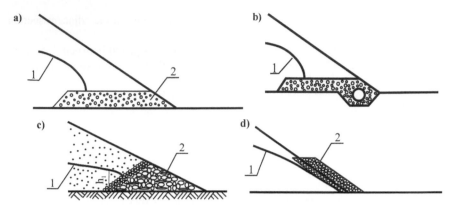

Fig. 5. Exemplary types of drainage. (a) Flat (band), (b) Combination of a flat and pipe drainage, (c) Triangular prism drainage, (d) Slope drainage; 1: Phreatic surface, 2: Drainage material (according to (Bednarczyk et al. 2006)).

show that the required distance from the downstream slope will not be maintained, the embankment body can be expanded or a bench on the downstream face of embankment can be constructed (Fig. 4), which is often very difficult due to the lack of available space. A safe distance of 1 m between the phreatic surface and the surface of the downstream slope may be also obtained by applying the drainage at the base of the downstream slope (Fig. 5).

When well permeable soil is spread in the substratum under the embankment, it serves as drainage, but if there is still a risk of water permeation through the embankment body, other means to secure the body are increasingly used, consisting in sealing the embankment (p. 8.2).

5. Scope and Methods for the Calculation of Stability, Filtration, and Settlement

In the case of river embankments, the stability and filtration analysis is performed for the needs of designing new embankments or their modernization as well as for periodic assessments of their technical condition.

Calculations of settlements should be performed only when designing new embankments or during their modernization consisting in the broadening of the embankment body located on low-bearing subsoils or bearing ones where there can occure significant differences in settlement, which can cause cracking of the embankment body.

5.1 Stability calculations

Calculation of the flood embankment stability is carried out for the building calculation case, when the embankment is not subjected to water damming and for the operating case, assuming that the damming level is as for as the base flood flow, adopting the position of the phreatic surface based on the filtration calculations.

Calculations for the building and operating case of embankments located on low-bearing soils should be performed in the total stress. In contrast, for the operating case on bearing soils it should be performed in effective stress.

In the case of cohesive soils occurring in the embankment body or directly in the soils under the hydrotechnical construction, the stability conditions of the construction should be checked for two calculation cases:

- With consideration of the drainage—including the existing pore water pressure and effective strength parameters in the calculations.
- Without consideration of the drainage—including in the calculations total stresses and strength parameters determined in the conditions without drainage.

The stability calculations are performed by applying specialized numerical programs, based either on the so-called accurate methods (methods of Morgenstern-Price, Spencer, GLE, FEM—fine element methods), described in literature in detail among others by Lechowicz and Szymański (2002), Dłużewski (1997), Popielski and Zaczek-Peplińska (2007), and Sroka et al. (2004).

For hydrotechnical constructions of class III and IV, it is allowed to perform stability calculations with simplified methods (e.g., Swedish or the wedge method).

In the case where there are interbeddings of soft soils or contact layers between weak layers and more durable ones either in the soils or in the embankment, or there are embedded elements (e.g., seals) in the embankment body that may cause slippage, the calculations of stability should be performed with methods that assume the slippage on inclined surfaces of arbitrary shape.

If there is an impermeable element (core or screen) in the embankment body from the upstream, the stability of the embankment body should be determined with consideration of this element in the conditions of normal operation, as well as rapid reduction in water level in the inter-embankment after a long period of flood. In the case of multi-layer screens, the slippage should be assumed not only on the screen-soil contact layer, but also on the junction between layers with the most unfavorable parameters, hence the lowest angle of internal friction (Borys 2007b).

The factors of safety (FS) should be adopted in accordance with the current technical conditions (Rozporządzenie 2007).

5.2 Filtration calculations

The main difference in relation to the earthfill dam situated by the water reservoir relates to the function of the river embankment which does not need to be as "tight" as the earthfill dam of comparable class and which should, even in principle, allow the natural water outflow towards the river after passing the flood wave. Flood embankment has contact with surface water only periodically, during the occurrence of high water levels. In the absence of active filtration for long periods, the embankment body may become relatively "dry". However, in general, the earth body of the embankment, particularly in the areas located deeper under the surface of the slope and crest, remains damp to varying extents, depending on the soil properties from which it was constructed, hydro-geological conditions of the soil under the embankment, and the remaining external factors. Partially saturated soil is essential for the course of filtration through the embankment during successive water accumulations, particularly if they occur shortly one after another.

The phenomenon of filtration in the river embankment and its substratum should be currently analyzed considering the water flow in completely and partially saturated soils, that is, the aeration zone and in the conditions of steady and unsteady flow in time. The water saturation degree of the soil is related to the effective pressure and current hydraulic conductivity, depending on the model adopted to describe the flow in the unsaturated zone. It is generally assumed that the degree of saturation has a value of 1.0 below the water level in the soil, that is, below the zero pressure line and drops to zero (or residual value) in the zone above the water table, the thickness of which depends on the model parameters. A good reproduction of both geometrical dimensions of the embankment and most frequently occurring non-uniform structure of the soil, as well as taking into consideration the relevant boundary conditions is also necessary. Meeting those requirements is possible only by means of numerical methods.

The analysis of filtration can be performed by many programs designed to conduct numerical modeling of groundwater movement, created using finite difference, finite elements, or boundary elements method, described in recent years in the literature among others by Wosiewicz (1996), Wosiewicz and Sroka (2000, 2001), Wosiewicz and Walczak (2002), Sroka (2001), and Sroka et al. (2004).

In the design practice, the evaluation of the filtration course during water accumulation on the river embankment often only comes down to the determination of time after which the permeation will occur in the downstream slope of the embankment in the unsteady flow conditions and the time needed to determine the filtration flow rate. On this basis, the following two boundary conditions are then verified:

- If the calculated time is longer than the time of water damming by the embankment, then the filtration does not pose any threat to the embankment;
- If the time is shorter than the time of water damming by the embankment, then further calculations should take into account the conditions of steady filtration.

The time of water damming by the embankment t_p is assumed as the period in which the level of accumulation is at the same level or higher than the base of the embankment from the side of the river.

The listed boundary conditions are universal and may be used for the assessment and comparison of the filtration times in the embankment both with and without the anti-filtration barriers, provided that the correct calculations are performed. This provision is of particular importance, especially with regard to the first condition, because the incorrect value of time after which the permeation will occur in the downstream slope may be the cause of an incorrect decision about the lack of need to seal the embankment. The problem results from the fact that in the design practice, it is currently uncommon to perform numerical calculations in the conditions of unsteady flow and simplified analytical methods are generally used (including the modified equations by Thiem or Erb) (Adamski et al. 1986; Borys and Mosiej 2003; Haurylkiewicz 2002; Żbikowski 1982), which were derived for idealized patterns, most frequently assuming homogeneous embankment on impermeable soils (the scope of applying these models was also broadened to the case of an embankment and soil of "similar" permeability after considering the amendments to the method of calculation, the hydraulic length of the embankment base, and the maximum water depth by the embankment).

The Thiem equation assumes a sudden rise in water level in the inter-embankment and allows one to determine the time in which the permeation will reach the toe of the downstream slope of the embankment body located on impermeable soil. The scope of the equation is limited to the embankments located on poorly permeable soils and, after assuming certain simplifications, to the case where the substratum contains the same soil as the embankment body. The Thiem equation is used mainly in order to clarify whether the filtration poses a threat to the embankment and whether it is necessary to perform further filtration calculations by using other methods.

The Erb equation allows one to determine the time to establish filtration flow in the body of the embankment located on an impermeable substratum and, after adoption of simplifications, also on a permeable substratum built from the same soil as the embankment body.

If the flood duration is shorter than the time calculated according to Thiem or Erb, we are dealing with unsteady movement and boundaries should be appointed, which will be reached by the water penetrating through the embankment body and substratum at a specified time. In the case where the duration of the flood is longer than the time to establish the filtration conditions, we are dealing with steady movement and then the following parameters should be determined:

- Phreatic surface in the body and, where necessary, a hydrodynamic grid.
- Water flow rate.
- Gradients in the body and substratum of the embankment and they should be compared with the values of permissible gradients.

For an objective comparison of possible embankment sealing variants and different construction solutions of the anti-filtration barrier, consisting in, for example, changing the length, material properties and/or the location of the barrier in the cross-section, it becomes necessary to implement reliable—preferably quantitative—comparison criteria, taking into account primarily technical aspects and secondly—economic benefits. It has been proposed to adopt the following comparative measures as the most reasonable ones (Borys et al. 2006):

- Time of occurrence of permeation through the embankment and substratum.
- Value of the maximum hydraulic gradient in the water outflow zone. i_{max} or equivalently—maximum water filtration speed in the outflow zone.
- Total filtration flow rate through the embankment and substratum.
- Coefficient of overall embankment stability.

The appropriate dimensionless comparative indices for the embankment before and after the modernization should be determined in relation to each of these measures. It is recommended to calculate all comparative measures in the conditions of unsteady filtration, accompanying the passage of the design flood wave.

Filtration pressure gradients occurring in the substratum of all hydrotechnical constructions, including river embankments, should meet the following relation (Rozporządzenie 2007):

$$v_i \cdot i \le i_{kr}$$

Where:
i – The filtration pressure gradient.
i_{kr} – Critical values of gradient for particular soil.
v_i – The confidence coefficient, which regardless of the construction type is: 1.5 for the basic load system and 1.3 for the exceptional load system.

The values of filtration pressure gradient should be determined for the conditions of steady and unsteady filtration, caused by fluctuations in water levels and consolidation processes in cohesive soils.

5.3 Settlement calculations

Settlement of the embankment substratum can be calculated based on the results of compressibility analyses specified for individual soil layers.

The substratum is divided into three calculation layers with thickness no greater than half the width of the embankment at the foundation level; but at the same time, it is calculated in a way that the boundaries of geotechnical layers separated in the substratum would coincide with the boundaries of processing layer. The stress in the individual layers needs to be determined based on equations or nomograms given in the literature (e.g., Czyżewski et al. (1973) and Wiłun (2001)).

Taking into consideration the actual stress distribution in the substratum, the construction is fairly laborious and for this reason it is often neglected in practice in the calculations conducted for low embankments, leading to considerable simplifications.

6. Soils for the Construction of Embankments

Embankments can be constructed with the soils meeting the requirements specified in the "Technical conditions..." (Wolski et al. 1994) and the standard (PN-B-12095:1997, PN–EN 1997-1:2008, PN–EN 1997-2:2009). These are:

- Mineral soils.
- Organic soils.
- Anthropogenic soils.

The following should not be used without special treatments:

- Swellable soils and soils soluble in water.
- Silts and clays with liquid limit value more than 65%.
- Soils that do not have the required moisture content w_w, of:
  - In relation to cohesive soils – $w_w = (0.95 \div 1.15)\ w_{opt}$, where w_{opt} – Optimum moisture content,
  - In relation to non-cohesive soils – $w_w \geq 0.7\ w_{opt}$.

Organic soils (silts, peat, gyttjas) can be used in special cases, for example, for the construction of embankments on poor organic soils or on areas which lack local mineral soils, under the specialist geotechnical supervision (Borys 1993; Warunki 1994).

It is not recommended to construct embankments from the following organic soils:

- High moor and transition peats.
- Low fibrous peat (with a degree of decomposition less than 30%).

- Carbonate peat (with the content of calcium carbonate more than 5%).

Anthropogenic soils, including industrial materials (e.g., lightweight aggregates or waste materials, e.g., from coal mines or power plants) can be used after specialist analyzes on a particular object and meeting specific requirements (Borys and Filipowicz 2008; Filipowicz and Borys 2007; Gruchot 2014; Gruchot et al. 2014; Pisarczyk 2004; Skarżyńska 1997; Zawisza 2001; Zydroń and Gruchot 2014).

Soils in the body of the river embankment should be properly compacted. Dry density of soils in the embankments is assessed using the compaction index I_s or the index of density I_D, depending on the soil type.

In the case of newly constructed embankments, the required relative density I_{Dw} or the compaction index I_{Sw} of soil compaction in the body can be assumed based on the values given in Table 5 (PN-B-12095:1995, Żbikowski 1982; Wolski et al. 1994).

In the existing or modernized embankments (not containing sealing elements) the parameters of soil compaction should be at a minimum:

- Non-cohesive soils – $I_{Dw} \geq 0.50$,
- Poorly cohesive and cohesive soils – $I_{Sw} \geq 0.92$.

These requirements also apply to the embankments from organic soils, and from the mixtures of organic and mineral soils provided, the methods and scope of their analyses should be developed individually. In the case of these soils, it is possible to employ other requirements for compaction, determined based on separate research and expertise.

Table 5. Required values of the compaction index I_{Sw} or the relative density I_{Dw}.

Type of Soil	Content of Fraction >2 mm, %	Required Compaction in the Body of New Embankments	
		I, II Klasa I, II Class	III, IV Klasa III, IV Class
Cohesive	0–10	$I_{Sw} \geq 0.95$	$I_{Sw} \geq 0.92$
	10–50	$I_{Sw} \geq 0.92$	
Non-cohesive	Fine Sands	$I_{Dw} \geq 0.70$	$I_{Dw} \geq 0.55$
	Medium Sands		
	Coarse Sands and Coarse-grained Soils	$I_{Dw} \geq 0.65$	

7. Causes of Damage to River Embankments

River embankments are exposed to a number of factors that can contribute to their failure. Damage to the flood embankments usually occurs due to one or more simultaneously occurring following factors:

- Faulty construction of embankments.
- Use of improper soil for the construction of the embankment body.
- Damage to the soil in the embankment body.
- Unfavorable layout and condition of soils in the substratum.
- Too low height of embankments.
- Improper condition of near-embankment areas (inter-embankment and landward side of the levee).

The risk of failure of an embankment's stability most frequently results from faulty construction and improper soil used for the construction. As a result of insufficient knowledge about the properties of soils and economic reasons in the past, distance from the embankment route was the only criterion for the selection of soil for the embankment construction. The requirements for the distribution of soils in the embankment body and their proper compaction were not respected. The studies on the existing embankments indicate that in many cases, they are structures characterized by a large variety and randomness of the built-in soils (Borys 2000; Borys and Rycharska 2004). Random arrangement of soil layers and lenses differentiates the moisture content, shear strength, and filtration conditions in the embankment body. Extensive and reaching close to the upstream slope inclusions of non-cohesive soils in the embankment body built of cohesive soils are particularly dangerous. They pose a threat to the formation of favored filtration routes, resulting in the embankment's destruction. Filtration phenomena within the embankment body and substratum may be of different intensity and generally can be divided into:

- Leaks, that is, filtration of small amounts of soil-free water, which manifests itself in the form of pure water outflow.
- Suffosion, that is, lifting of fine particles of soil by filtering water. These particles may be removed beyond the soil, which manifests itself in the form of water outflow together with soil. This phenomenon indicates that the so-called critical velocities or gradients were exceeded. As a result of suffosion, cavities or channels may be formed in the soil and this phenomenon has the traits of hydraulic perforation.
- Hydraulic perforations, that is, the formation of a channel (conduit) filled with soil of disturbed structure, connecting sites with higher and lower water pressure in pores. On the surface, the hydraulic perforation is visible in the form of a spring with elevation of soil from the substratum.

Some soils, mostly organic ones, but also mineral cohesive soils, undergo significant structural transformations during their use. Soils in embankments, subjected to alternating effect of wetting and drying as well as freezing and thawing, tend to change their structure (Borys 1993). Networks of cracks and disintegration of cohesive material into lumps or chunks occur in the embankment body, which may increase their permeability several times (Borys and Rycharska 2004).

Insufficient or uneven compaction of soil in the embankment body may have similar effects. Loose, usually fine-grained non-cohesive soils embedded in the area of the toe of the downstream slope may be liquefied under pore water pressure and flow out from under the body in the form of characteristic tongues with almost horizontal surface. Deprived of support, after outflow of soil, the part of downstream slope and then the body loses its stability and slips down.

Slipping of the downstream slope on the cylindrical or other surface can occur due to improper evaluation of parameters of embedded soil (or substratum), non-compliance to correctly assumed parameters during the construction process, changes in the parameters in the course of use or higher than expected position of phreatic surface and increase in the pore pressure. Higher position of the phreatic surface is usually caused by adverse variety in the internal structure of the embankment body. Local deformations of the downstream slope are caused by too large gradients and velocities of filtering water and lead to the formation of small, local collapses of the slopes, followed by larger collapses and in some cases—to the suffosion-loosened zones in the embankment body, which may influence failure of stability.

Threats to the stability may also arise when the organic soil turf is not removed from the soil during the construction of a new embankment or, in the case of extension of the body, from the slopes and crest of the existing embankment. On the contact zones of the embankment with the substratum or old embankment—heaped embankment, the routes of facilitated and concentrated filtration may be formed, which favors the erosion and removing of fine particles of soil. Particularly dangerous situations for the embankment stability arise when, due to an unfavorable course of plant decomposition or for other reasons, these routes are closed in the lower part at the outlet. High pressure at the junction causes the loss in the stability of the slope.

Damage to the flood embankments are also often caused by excessive deformations of the embankment substratum, associated with adverse type or layout of soil layers. Filtration under the embankment in the flood period may be the cause of reducing its stability mainly when the soil is non-cohesive or low-cohesive. This phenomenon is visible usually in the form of numerous springs of various sizes, from which fine-grained non-cohesive soils can be elevated, evidencing internal erosion of soil.

Another very common cause of embankment damage is the very low height of the embankment. Virtually each water overflow through the embankment crest causes watering down and destruction of the entire embankment body. This leads to flooding of the whole protected area. Most frequently the existing embankments do not meet the current requirements for their height, since:

- They were built at a time when different provisions existed or the views on the necessary degree of flood protection have changed.
- Reliable and control flows have increased.
- Water level increased under conditions of unchanged flow rate.
- Over the years there has been an increase in the class of embankments, which requires a change in their height.
- There was local excessive embankment settlement, caused by poor soil in the substratum.

Risks occurring in the shore areas of oxbow lakes, which are crossed by the embankment route, can be also very significant. In the slope of the oxbow lake, soils of the embankment substratum, including sands or gravels underlying the upper low-permeable layer are being exposed. Water flowing through the body embedded in the oxbow lake may enter at this point to deeper permeable layers, increasing pressure therein and thus deteriorating the embankment stability conditions. The edge constitutes the border between two types of soils—relatively little compressible soils of the floodplain and alluvial soils deposited in the old river channel. The difference in settlement of these two types of soils may cause cracks in the embankment body on the border of the oxbow lake. Floodplains of many rivers are built in such way that the upper layer of the soil is composed of cohesive (alluvial) soils of small thickness, typically 1–2 m, under which permeable sands, gravel sand, or gravels are spread. Very often in such cases the covering, the thin cohesive layer is characterized by diverse structures (inserts, breaks), is overgrown by roots and its infiltration coefficient has a small (range of 1 m·day^{-1}). The underlying non-cohesive soils have outcrops on the river channel slope, so that the infiltrating water does not encounter difficulties. Small coefficients of permeability of the upper layer and large permeability coefficients of the lower permeable layer cause the outflow from the latter towards the top to be difficult, resulting in high pressure of water in the pores. These pressures result in hydraulic perforations of the upper layer in places where it is the thinnest and weakest. Large gradients and velocities of filtration under the perforation cause the removal of small particles of non-cohesive soils from the subsoil and the small caverns or channels. When they are large enough, the embankment body may be deformed.

The interruption of embankment is often caused by watering down of the soil, caused by the presence, at a depth of several or even dozens of meters, trunks of trees felled by the erosive effects of rivers many centuries ago.

The potential sites of watering down of the soil and the embankment body, particularly during high and long-term floods, may also include oxbow lakes filled with soil loosely poured into water without clearing from mud, trees, and shrubs, as well as what is left of undeveloped sections of the old river channel right by the embankment slope.

The condition of the embankment and the resulting threats to the stability may be significantly affected by animals. In recent years, the embankments have been observed being destroyed by beavers or foxes. It is particularly dangerous, as these animals dig very spacious dens and corridors, sometimes over the entire width of the embankment body. Attempts are being made to secure the embankments by building in wire mesh into the body, which greatly increases the cost of modernization. Channels dug by the animals or remaining after the decomposition of old plant roots (trees, shrubs), can cause the occurrence of hydraulic perforations and concentrated exudates, as well as the formation of open channels intersecting the embankment and erosion of the soil. This may lead to the collapse of the embankment body.

Inadequately protected passages of different kinds of cables, for example, gas or telecommunication through embankments, as well as poorly compacted soils in the vicinity of culverts and floodgates, are also unsafe places.

Damage to embankments can be caused by the condition of adjacent areas (inter-embankment and land side of the levee). These damages most frequently occur due to the rise of the local water. Such rise can be caused by trees and shrubs intensively growing in the inter-embankment, which impede the water flow, resulting in the overflow by the embankment crest. Upstream face slope can be also watered down due to changes in the direction of flood water current. This may result, among others, from obstacles in the high water river channel in the form of trees and shrubs or malfunctioning regulatory constructions in the river channel. Often the intense current of water may cause damage and destruction to embankment buildings. The embankments can be also watered down by water incoming from the downstream side. This usually occurs in the absence of water outflow on the land side of the levee, causing accumulation or, in the case of water entering the land side of the levee, through the break in the embankment created at a higher section. Usually, this results in watering down the downstream slope and sometimes even overflowing of water through the crest towards the inter-embankment.

A very important cause of damage to the embankments includes trees, growing close to their body. Roots of those trees create the filtration routes in the soil and sometimes even in the embankment body. Therefore, efforts should be made to thoroughly remove trees (including their root system) in a zone at least 3 m wide on both sides of the embankment.

Improper use of the inter-embankment can also cause the damage to the embankments. Such areas are often agriculturally used. Cultivation measures that cause loosening of topsoil facilitate the filtration processes during the flood. The topsoil layer of compost (humus) and turf on permeable soil only seals the soil to a small degree, which is insignificant to the filtration processes.

Other factors affecting the safety of the embankments include the destruction of the crest, slopes, turf, or its irregular moving, allowing overgrow on the embankment body by shrubs and even trees and locating pits or water reservoirs too close to the embankment route.

Flood embankments should be protected by people, in a manner similar to the facilities and equipment that exist on private property, since the condition of these constructions affects the safety of the inhabitants of the protected areas. Therefore, destroying or devastating the embankments is prohibited by law (Water Law Act (Ustawa 2001)), in particular:

- Passing through the embankments or along the embankment crest with vehicles or horses as well as rushing animals, with the exception of places designed for these purposes.
- Soil cultivation, planting trees or shrubs on the embankments and at a distance less than 3 m from the embankment toe on the downstream face of the embankment.
- Digging the embankments, nailing poles, or setting road signs by unauthorized persons.
- Constructing buildings, digging wells, ponds, pits, and trenches at a distance less than 50 m from the embankment toe on the downstream face.
- Damaging the turf or other slopes' protection systems and embankment crest.

Accurate diagnosis of the reasons and extent of the embankment damage, or in the case of intact embankments, their technical conditions, affects their modernization.

8. Modernization of Flood Embankments

8.1 Factors affecting the selection of flood embankment modernization method

In recent years, new flood embankments have very rarely been constructed in Poland. However, it is necessary to conduct the modernization of the old and/or damaged ones and those that do not meet the requirements. While the scope and method of modernization depends on many factors, including the following (Borys 2006; Borys and Mosiej 2006a):

- Flow and level of both design and control water, and their relation to the ordinate of the embankment crest.
- Size of the protected area and the way of its use.
- Class of the embankment.
- Condition of the river channel and the conditions of high water flow.
- Condition and structure of the modernized embankment body, including the soil type and its condition, as well as the conditions of filtration and stability.
- Condition and structure of the ground, including soils, soil type and its condition, as well as the conditions of filtration and stability, the possible need for building up the oxbow lakes.
- The possibility of acquiring land for reconstruction of the embankment body, the type and size of land.

- Communication conditions along the embankments.
- Conditions for drainage of filtering water, rainwater, and water from the catchment on the land side of the levee.
- The impact of the embankment modernization on the environment;
- Restrictions in acquiring land for modernization.
- Cost of modernization.

It needs to be emphasized that each case of modernization should be individually examined and the relevance of the presented factors may vary in each case, as well as other factors, not included here, may occur.

8.2 Methods of flood embankments' modernization

Methods of modernization may be different. Several years ago the modernization of embankments consisted mostly of a significant expansion of their body. These procedures required large earthworks, associated with embedding large amounts of soil (typically from about 20,000 m^3 to 100,000 m^3 per length of 1 km). Generally, conducting such work involves large interference in the environment (removal of shrubs and tree plantings, acquiring significant amounts of soil, using a large number of heavy construction machinery). This resulted in the need to seek new solutions, towards a significant reduction in the expansion of the embankment body volume and replacing it with other, more modern procedures. The most commonly used solutions of modernization include:

- Increasing the embankment body by adding more soil or the construction of special barriers raised for the period of flood wave passing.
- Sealing the upstream face by a screen (e.g., from a geomembrane or bentomat).
- Compaction of the embankment body (currently used methods allow soil compaction for large depth of up to several dozen meters).
- Sealing the body with vertical anti-filtration barrier, made within the embankment body.
- Horizontal sealing of the soil from the upstream slope (apron).
- Footings preloading the soil and increasing the route of filtration, placed, depending on the permeability of the soil, on the land side of the levee or in the inter-embankment.
- Lowering water level on the land side of the levee by making drainages, ditches, pore water pressure reducing devices, for example, relief wells.

To improve flood protection, the construction of levee roads, in sections where there are none, should be included in the embankment modernization. Such roads should ensure access of heavy equipment, delivery of materials, and significantly improve flood protection action (Borys 2006).

The selection of a modernization method of a given object is based on the analysis of the local conditions. Due to the prevalent lack of space for expansion of the embankment body or high costs of purchase of adjacent lands, sealing of the embankment body and soil is mainly used in engineering practices.

The analysis of soil and water conditions as well as the results of water filtration and stability calculations is the basis for deciding whether or not to seal the embankment and its soil. It is significant that the flood embankment is only temporarily used for water damming. What differentiates it from the water reservoir dam, therefore, is that it does not have to be completely leak-proof and should even allow the natural drainage of water into the river.

The embankment body should be sealed when (Borys 2008; Borys et al. 2006):

- Time of permeation occurrence on the downstream slope of the embankment in the conditions of unsteady flow is shorter than the water damming by the embankment.
- Hydraulic gradient or velocity of filtration through the embankment body in the outflow zone, after considering the damage consequences coefficient, are greater than or similar to the limit values.
- General stability of the embankment body and soil in the conditions of steady filtration are smaller than the values required for a given embankment class.
- Water filtering through the embankment body contributes to the inundation of the protected area (it is estimated that such risk exists when the flow rate through the embankment body exceeds 1–2 $m^3/d/rm$).

The ground should be sealed when:

- Time of unsteady filtration through the soil is shorter than the water damming.
- Gradient or velocity of filtration through the soil in the outflow zone, after considering the damage consequences coefficient, are greater than or similar to the limit values.
- There is a danger of hydraulic perforation of poorly permeable soil layer spread throughout the area; at the same time it is assumed that if the poorly permeable layer has a thickness in the range from 1/3 to $1/2H_1$, then the hydraulic perforation should not occur.
- Water filtering through the soil contributes to inundation of the protected area.

General recommendations for the decision-making about the need for sealing of only the embankment body or the embankment soil were discussed based on the examples of two recently created and most frequently applied technical solutions, shown schematically in Fig. 6.

In the case where the upper layer of the soil is poorly permeable and its thickness is 1/3–1/2 of the height of water lifting, H_1, it can be assumed that the hydraulic perforation should not occur, as the filtration route extends beneath this layer and, therefore, there is no need to seal the soil. The decision to seal the soil needs to be based on the calculations showing how much the steady filtration value and its gradient will be reduced and how much the time of unsteady filtration through the soil will be increased.

The vertical barrier in the embankment soil should not reach the top of the impermeable layer. In order to allow the natural outflow of groundwater into the river, the unsealed zone should be left under the barrier having a minimum thickness L_1 equal to the height of water lifting H_1.

The anti-filtration barrier should be sunk into the soil to a depth equal to 2–3 water lifting heights, H_1, whereas the depth of the barrier equal to 3 H_1 and should be used if the embankment route passes through the oxbow lakes or if historic buildings or large housing estates are located on the land side of the levee.

If the selected method for the embankment body sealing is to use the geomembrane screens or bentomat, the stability of the upstream slope should be assessed, assuming the slippage of the soil layer covering the sealing is over its surface. The angle of the upstream slope inclination should be smaller than the angle of inclination of the sealing.

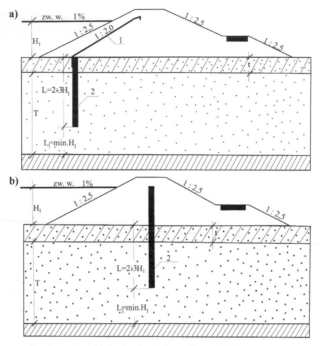

Fig. 6. A scheme of sealing: (a) With the barrier in the soil and with bentomate or geomembrane in the embankment's body: 1: Sealing screen of bentomate or geomembrane, 2: Anti-filtration barrier, H_m and H_k base and control water flow, respectively, H_1: Height of water lifting on the embankment, L: Depth of the anti-filtration barrier in the soil, L_1: Thickness of unsealed zone under the barrier, T: Thickness of permeable soil, t: Thickness of less permeable layer (Own study after [Borys 2005]).

Sealing of the embankment body and soil by a vertical anti-filtration barrier (Fig. 6b) should be used when such need is indicated by the calculations of the filtration and stability, and it is impossible to conduct modernization works on the upstream face of the embankment. Such methods have more disadvantages than the previous ones, due to the following factors:

- The general stability of the embankment after hydrating the upstream face part during damming is smaller.
- If there are leakages in the barrier, they may cause watering down of the downstream part of the embankment, which is relatively narrow after placing the barrier in its axis.

The advantage of such a sealing method is conducting work on the embankment crest, if it is wide enough. Often, however, the embankment crest should be broadened to 4–5 m by lowering it for the time of the barrier construction. The scope of calculations and analyses in the case of vertical barriers in the embankment axis should also include the calculations of the stability of the upstream slope while being loaded by flood water and after its rapid subsidence.

In recent years, the use of vertical anti-filtration barriers from hardening suspensions or waterproofing screens from geomembranes or bentonite mats are the most commonly used methods of sealing the embankments in practice.

8.2.1 Vertical anti-filtration barriers

The construction of anti-filtration barriers from hardening suspensions is one of the main methods of modernization of flood embankments, due to the possibility of improving the embankment tightness without major reconstructions of the body and the rapidity of realization.

Hardening suspension is the suspension that hardens over time. The anti-filtration barriers in flood embankments are constructed with suspensions containing cement, sodium bentonite, and fillers in the form of limestone dust, ground blast furnace slag, or fly ash. The suspensions are most recently prepared from ready mixtures containing all ingredients, or are prepared on-site from transported components, which requires the use of appropriate dispensers and is becoming less frequently used in practice. Dry components are mixed with water at the construction site in special mixers immediately before use. After preparation, the suspensions are in liquid state, but after hardening, they gain the characteristics of weak cement mortar.

Vertical anti-filtration barriers from hardening suspensions in flood embankments are constructed according to various methods described below.

Method of deep mixing, known under the names of DSM (deep soil mixing), DMM (deep mixing method), as well as MP (mixing in place), involves the construction of vertical overlapping columns, with a predetermined width and length, or blocks with rectangular cross section, formed by mechanical mixing of soil and the hardening suspension, pumped under pressure into the pipeline in the embankment body or in the substratum (Fig. 7). Forming of individual columns that create the barrier occurs alternately in the order "primary–secondary–primary". The admixing is carried out using the machines fitted with special mixing tips, wherein the mixing process needs to be repeated several times (minimum three times) in a vertical direction in order to obtain better homogeneity of the anti-filtration barrier. The stirrer speed and speed of its pulling up are adjusted according to the soil type. Deep mixing is performed without vibrations or shaking. The column diameters in the constructed anti-filtration barriers reach generally 0.6–0.8 m and result from the size of the mixing tip rotating in the soil. The spacing of the overlapping columns should be adapted in order to ensure the minimum width of the anti-filtration barrier of 0.30 m.

The basic drawbacks of this method should be considered while designing and construction, namely:

- Difficulty in obtaining homogeneous mixture of suspension soil in very cohesive soils.
- The need for very precise delineation of following columns and maintaining the mixer in a vertical position to ensure the barrier tightness.

Cut-off method involves drilling of a narrow slot in the soil, where the suspension is put, which after hardening, forms the anti-filtration barrier with thickness dependent on the tool used for digging—usually in the range of 0.30–0.60 m. The anti-filtration barrier may be constructed by the following methods:

- *Continuous dug slot,* which can be made with single-bucket or bucket wheel excavator (Fig. 8). Recently, a continuous deep mixing method (CDMM) is commonly used with a Trencher. Specialist equipment—Trencher, that digs trenches with the above mentioned method, is composed of a crawler with a cutting

Fig. 7. Method of deep mixing: (a) machine fitted with a single stirrer during the construction of the barrier; (b) Machine with double stirrer; (c) The view of the exposed fragment of the barrier after hardening of the suspension (photo 7a and 7c after [Borys 2005], photo 7b after [Borys 2006]).

disc, which the moving cutting and mixing elements are fastened to, acting as a chainsaw. The device forms the anti-filtration barrier continuously for a minimum width of 0.30 m. The extracted soil is mixed with the suspension pumped through a system of pipes connected to the mixer.

- *Section dug slot,* also called the method of successive sections or crotch, which involves digging subsequent fragments of trench separated by columns of soil, removed after hardening of the suspension in the separating sections (Fig. 9). The width of the barrier constructed with the dug slot method is generally 0.40–0.60 m and depends on the type of the excavator. This method is very rarely used in the case of embankments.

Drawbacks of the slot method include:

- Possibility of significant sedimentation of the suspension in the excavation.
- Possibility of uncontrolled tearing of the walls of narrow spatial excavation.
- Inability to fully control the works, particularly the accuracy of soil extraction by the excavator.
- High consumption of the suspension needed to fill the trench, which increases the cost of the barrier construction.

Fig. 8. Trench method of continuous crevice: (a) Single-bucket excavator while constructing the barrier, to the right visible pipe supplying the suspension to the trench; (b) Trencher; (c) The view of the exposed fragment of the barrier after hardening of the suspension (photo 8a and 8b after [Borys 2006], photo 8c by M. Borys).

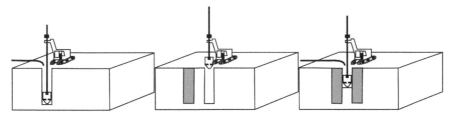

Fig. 9. The scheme of the barrier construction by the dug crevice (subsequent sections) (Own study).

Vibration method, called WIPS (vibrational—injected diaphragm), involves pushing apart the sealed soil while drilling in the steel hone with double-T (Fig. 10a) or butterfly section (Fig. 10b) under the action of a socket vibrator. The hardening suspension, which flows out through the outlet nozzle placed near the blade of the element drilled into the soil, is pumped both during plunging and pulling out of the device. During plunging, the suspension acts as a rinse, facilitates plunging the device into the soil, and stabilizes the wall of the slot. While pulling the device out, the entire space of the slot is filled with the hardening suspension and after hardening,

Fig. 10. WIPS method: (a) Double-T vibrator; (b) Butterfly vibrator; (c) Scheme of constructing the barrier with a double-T vibrator and the view from the top of the barrier; 1: Vibrator, 2: Profile, 3: Course of works, 4: Initial ditch, 5: Shape of double-T, 6: Shape of the barrier, 7: compaction area (photo 10a and 10b after [Borys 2006], photo 10c by M. Borys)..

forms a thin anti-filtration barrier. The thickness of the barrier constructed by this technology depends on the width of the element plunged into the soil, but also, to a large extent, on the soil type.

Obtaining tight connection of individual working elements of the barrier requires the construction of the barrier according to the rule "fresh into fresh", that is, one after another (Fig. 10c).

The following drawbacks of the vibration method should be considered while designing and construction:

- Small thickness of the barrier, which also largely depends on the type and condition of soil.
- Frequent hazard to the tightness of the barrier and the possibility of filtration windows, related to uncontrolled tightening of a thin barrier in unfavorable soil and water conditions.
- The need for very precise delineation of further places for plunging the device and holding it in a vertical position in order to maintain the tightness of the barrier.
- The impact of vibrations on buildings located nearby the construction site.
- Thickening of the lower parts of the soil, preventing the achievement of the expected depth of the anti-filtration barrier, particularly in soils containing gravel and cobble fractions.

Method of high pressure grouting (Jet grouting) involves local destruction of the existing soil structure by means of high pressure liquid jet while stirring and partial replacement into the binding agent, that is, the injectate (Fig. 11). The scope of impact of the cutting jet depends on the soil type and the applied technology variant and reaches up to 2.5 m. It is possible to form any geometrical solid in the soil.

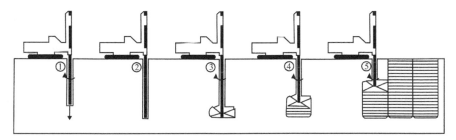

Fig. 11. Scheme of the barrier construction by high pressure grouting: 1: Drilling, 2: Finished drilling, 3: Starting injection, 4: Pulling the drill with simultaneous injection, 5: Repeating the process in the neighboring hole (own study).

The process of forming the barrier consists of several stages, involving:

- Drilling, in which the drilling rod with the monitor (that is, a device mounted on the end of the rod in order to allow the liquid stream inflow to the soil) and the drill bit are dug into the soil to the required level.
- Loosening the structure of the subsoil, using a very strong stream starting from the deepest point of the borehole.
- Cementation of soil, during which the injectate is pressurized simultaneously to the soil loosening, and is optimally mixed by turbulence with the remaining soil particles.

- Forming the barrier, when the single injected solids become combined. The solids can be combined, both in the fresh state—the order "fresh into fresh" when the injection elements are formed successively without waiting to bound the injectate, as well as after hardening—the order "primary–secondary" when the construction of overlapping injection elements cannot be combined before the certain time of hardening or before reaching the intended durability of adjacent, pre-shaped elements.

In the case of flood embankments, the method of jet grouting is mainly used when it is necessary to repair local leaks or sealing pipeline passages through the embankment.

Further information on the use of anti-filtration barriers, hardening suspensions for the modernization of embankments can be found, among others, in publications on the methods of embankment modernization and the performance guidelines (Borys 2006, 2008; Borys and Mosiej 2006b; Borys et al. 2006; Evans 1993; Evans and Dawson 1999; Evans et al. 2002; Fratalocchi and Pasqualini 2004; Jeffreis 1981) as well as in publications on the effect of the moisture content and temperature conditions on the barriers in flood embankments during their operation (Borys and Rycharska 2008).

8.2.2 Sealing screens

Bentonite mats are factory-made geogrid composites with very low permeability, in the form of soil bentonite inserted between permeable geotextiles (e.g., nonwoven geotextile fabric). The thickness of beontonite mats ranges between 4 and 7 mm. Bentonite is clay with unique physical properties. Generally, it is a sedimentary rock formed as a result of "*in situ*" weathering of volcanic ash—mostly Cretaceous and Tertiary—in a subsea environment. Its name comes from the first exploited deposit in Fort Benton, Wyoming (USA). The uppnusual swelling capacity of bentonite make the bentonite mats virtually impermeable and very homogeneous liners. The phenomenon of bentonite swelling also ensures that no perforations will occur in bentonite mats. Bentonite mats are much easier to be installed than poorly permeable soil layers, which are difficult to compact. Hence, the bentonite mats are so widely used in recent years for sealing the flood embankment bodies as part of their modernization. The success of such modernization depends mainly on careful construction.

Geomembranes are thin and elastic synthetic or bituminous products, impermeable for liquids and gases and come in the form of webs or sheets, composed of one or several layers.

Currently, geomembranes made of high density polyethylene (HDPE) are most widely used in civil engineering. They are characterized by very high chemical resistance, high mechanical resistance to loads and they are resistant to weather conditions, and they are not susceptible to microbiological degradation. They are practically impermeable, as the permeability coefficient is about $k=1 \times 10^{-14}$ cm·s^{-1}. Most commonly produced geomembranes have a thickness of 0.5–2.5 mm and width in the range 2–10 m. PVC geomembranes modified with rubbers are still used for sealing embankments. Due to the fact that normally they are not UV-stabilized, they need to be protected by a soil layer (in the period of a maximum six months after installation). Such protection eliminates the negative effect of sunlight or high and low temperatures. PVC geomembranes are mostly 6 m wide and their thickness for use in sealing flood embankments should reach a minimum of 1.0 mm.

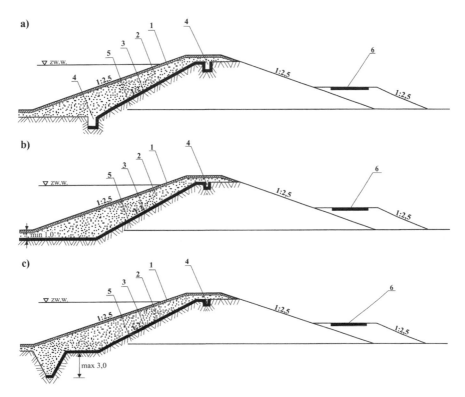

Fig. 12. Scheme of the screen sealing the embankment body made of geomembrane or bentonite mat: (a) Sealing screen anchored in the embankment crest and at the base of the slope, (b) Screen passing into a horizontal apron sealing the substratum, (c) Screen passing into Shallow substratum seal. 1: Humus layer, 2: Soil backfilling, 3: Geomembrane or bentonite mat, 4: Ditches anchoring the screen on the embankment crest and in the substratum, 5: Substratum, 6: Road on the bench (own study after [Borys 2009]).

If it is necessary to seal only the embankment body, which happens when the embankment body is located on impermeable or poorly permeable soil of considerable thickness, a sealing screen anchored in the embankment crest and at the base of the slope most commonly used, as schematically shown in Fig. 12a. However, flood embankments are most frequently situated on permeable soil and it is necessary to seal both the embankment body and the soil. It is then possible to apply the solution of sealing the embankment body with the screen which becomes a horizontal apron, sealing the soil below the embankment as shown in Fig. 12b. The use of a horizontal sealing apron requires performing very accurate calculations of the filtration in order to determine its optimum length, and at the same time using the area of considerable width for the construction of the apron, and hence this solution is rarely used in the case of flood embankments. A combination of screen sealing the embankment body with a vertical anti-filtration barrier is much more frequently used. If there is a need for sealing the soil of small thickness (up to a maximum of 3.0 m), a vertical seal may constitute the same bentonite mat which seals the body (Fig. 12c).

The most common solution is to combine the screen in the soil with a vertical anti-filtration barrier (as schematically shown earlier in Fig. 6a) made of different materials, for example, hardening suspensions.

Further information on anti-filtration screens can be found in publications on the methods of embankment modernization and performance guidelines (Borys 2006, 2008, 2009; Borys and Mosiej 2006b).

9. Inspections of Embankments

During their operation, flood embankments should be subjected to various inspections and assessments, including the inspections as defined by the Building Law (Ustawa 1994) and periodic assessments of the technical condition in accordance with the Water Law (Ustawa 2001).

According to the Building Law (Ustawa 1994) during their use, buildings should be subjected by the owner or the manager to the following inspections:

1. Periodically, at least once per year, the inspection of technical condition.
2. Periodically, at least once every five years, the inspection of technical condition and suitability for the use in the construction facility.
3. Inspection of safe use of the building, which should be carried out whenever there are external factors affecting the object, involving the actions of humans or forces of nature, which result in the damage of the object or imminent risk of such damage, that could endanger human life or health, property security, or environmental safety.

All water engineering facilities, used in flood protection, or the use of which is associated with floods, should be subjected to periodic inspections—seasonally or twice per year:

1. Spring inspection, after snowmelt floods, for the assessment of the technical condition before the summer floods.
2. Autumn inspection, after summer floods, for the assessment of the preparations for the winter and spring floods.

Seasonal inspections, performed twice per year, consist of:

- Technical inspection of the facility after passing of the flood wave.
- Assessment of damage caused to the facility.
- Assessment of the work necessary to be performed, or already completed to restore the efficiency of the facility and its equipment.
- Assessment of readiness of the facility, its buildings and equipment for the next flood wave.

Periodic inspections, according to the published guidelines (Borys and Mosiej 2008; Borys 2013a,b, 2014a,b), should include the following flood embankment elements:

1) **Embankment body,** including:
   a) Geometric dimensions:
      - Dimensional deviations from current requirements, project design, or from cross-sectional dimensions of the embankment representative for a particular section.

- Safe elevation of the embankment crest during the damming, the occurrence of local depressions, and excessive settlement.

b) Insufficient compaction of soil in the embankment body (in the case of embankments containing no sealing elements).

c) Filtration phenomena observed during the water damming, their location, severity and type, including:
  - Leakages (consisting in the filtration of small amounts of water free of soil particles, appearing as the outflow of pure water).
  - Suffosion (consisting in leaching away small soil particles by filtering water, while these particles can be carried beyond the soil, which appears as the outflow of water with soil).

d) Damages to the body, their location, severity and type, including:
  - Destruction of the crest and slopes as well as the embankment footing by the passing vehicles, equipment, or farm animals.
  - Local landslides.
  - Condition of slopes protection systems.
  - Wild growing shrubs and trees within the embankment body, as well as in the area directly adjacent to the embankment.
  - Dens dug by animals, particularly beavers and foxes; their density and range.
  - The condition of the embankment body in the passages of pipelines (water supply, sewerage, gas and others) and cables.
  - The presence of foreign structures in the embankment body, for example, cellars, bunkers, and other similar buildings.

e) Excessive settlement.

f) Other, not listed factors that may cause threat to the embankment stability.

2) **Soil** directly under the embankment and in the area adjacent to the embankment both on the upstream face and downstream from the face of embankment:
  a) Filtration phenomena, their location, severity and type, including leakages, suffosion, piping (the formation of a channel or duct filled with soil of disturbed structure, connecting sites with higher and lower pore water pressure; the piping is visible in the form of spring of water caring small particles from subsoil).
  b) Cavities.

3) **Embankment and associated buildings:** culverts, sluices, drainages, drainage facilities, drainage water discharge systems, embankment ditches, embankment passages, pipelines from the pumping stations, and pipes for liquids and gases passing through the embankment body, flood roads and access roads to embankments, measuring and control devices and other elements associated with a protective line formed by the embankment:
  a) Damage to buildings, lack of technical efficiency, in particular:
    - Condition of culvert seals (no seals, lack of tightness, inability to activate them).
    - Changes in the condition of concrete surface, appearance of deformations, cracks, scratches, dampness, and leakages in the concrete and reinforced concrete structures.
    - Poor condition of measuring and control devices (if they are present or should be present) including benchmarks or water gauges.
  b) Absence or poor condition of flood roads or embankment access roads.

4) **Inter-embankment, land side of the levee, and protected area** in the terrain adjacent to the embankment both on the upstream face and on the downstream face of embankment, especially in the area where the filtration phenomena occur most frequently during floods:

a) Local narrowing of the inter-embankment, which could cause water flow impediment.

b) Vegetation found in the inter-embankment, which could affect the level and direction of high water flow, the location, and type of vegetation.

c) Condition of the river channel and the river banks as well as regulatory structures, if the river channel is situated less than 50 m from the embankment, or more when the condition of the river channel and the shape of the shoreline indicate the potential threat to the embankment.

d) Intersection of the embankment route with oxbow lakes and the condition of oxbow lakes.

e) The presence of aggregate excavations, local land depressions, fish ponds, ponds, uncontrolled excavations, and landfills at a distance less than 50 m from the embankment.

f) Location of buildings and uncontrolled development directly by the embankment route.

g) Location of the embankment at the mining damage risk areas.

During the annual inspection, specialized studies or surveying are not usually conducted. These inspections can be performed based on:

- Analysis of archival materials, mainly documentation of previous inspections of particular buildings.
- Other available sources of information (e.g., obtained from municipality offices, residents of villages, and settlements located nearby the construction).
- Detailed site inspections.

Site inspection is a crucial element, which needs to be carried out with the utmost care and during which one can visually assess all key elements affecting the technical condition of the construction. The main source of information about the embankment should be its owner or manager, usually the Provincial Land Melioration and Water Units Board or Regional Water Management Council, that own the general design documentation, post-completion surveys, reports, and documentations of previous periodic inspections.

Periodic inspections involving the verification of the technical condition and suitability for use of the building, carried out every five years, should be conducted by teams of specialists provided with field and laboratory equipment adequate to conduct the analyses. Within the inspections conducted every five years, the evaluation of individual components of the construction should be based on, apart from visual observations, direct measurements and studies. For this purpose, the following steps should be completed:

- Land surveying.
- Geotechnical studies.

Land surveying is intended to:

- Control the condition of cross sections of the embankments.

- Control of the position of the embankment crest gradeline.
- Determine the location of soil penetration tests and boreholes sites.
- Control of settlement, deviations, and displacements of embankment structures.
- Determine the elevation changes in the inter-embankment.
- Locate other elements threatening the security (e.g., sites of the filtration phenomena, pipeline passages, and dens of beavers and foxes).
- Locate areas with shrubs and trees.
- Determine the altitude of traces of high waters after raised water stages as well as determine the water table drops on the background of the river channel and inter-embankment characteristics.

For earth structures, geotechnical investigations (soil penetration test and drilling combined with soil sampling for laboratory tests) should be performed in order to determine:

- Soil condition in the embankment body and subsoil (moisture content, content of organic particles, soil compaction parameters in the embankment body in the case of embankments containing no sealing elements).
- Basic parameters required for the calculations of the embankment stability and filtration flow.

For the flood embankments modernized by, among others, construction of anti-filtration elements (screens, vertical barriers), soil compaction parameters are not among the criteria that determine its technical condition, as it is in the case of embankments constructed from soil only (Borys 2014a,b). Geotechnical studies should, in this case, provide the parameters necessary to perform stability calculations and the results of these calculations should be among the criteria taken into account when assessing the embankment safety. Also, the condition of the soil protective layer arranged on the upstream face of the screen should be taken into consideration (e.g., cracks, local slippings, depressions, etc.).

Any abnormalities, found during the inspection, that could deteriorate the technical condition and safety of the embankment, should be documented (sketch, location, and photographs) in a way that allows the evaluation of changes of these abnormalities during subsequent inspections and surveys.

According to article 64 sect. 3 of the Water Law Act (Ustawa 2001), inspections of the technical condition and safety of hydraulic structures should be performed periodically and should assess in particular:

1. Condition of groundwater, its filtration through the construction, ground, and its surroundings.
2. Durability of the building and the subsoil.
3. Condition of outflow devices.
4. Changes in the upper and lower position of the building.

Evaluation of the technical condition and safety of the damming construction should be based on the assessment of individual elements of the structure and the associated elements. They consist of (Borys and Jędryka 2014, 2015):

- Substratum.
- Body.
- Anti-filtration and drainage equipment.

- Water passage devices.
- Control and measurement instruments.
- Upper and lower position of the building.

The evaluation of technical condition and safety of the structure should be developed based on: visual inspections, examinations, documentation of periodic inspections carried out in accordance with the provisions of the Building Law, analysis and interpretation of conducted and collected measurements, and:

- Site inspection.
- Calculations (of, e.g., stability, filtration flow).
- Analysis and interpretation of conducted and collected measurements, observations, and calculations.

The analysis should cover the entire lifetime of the structure (together with the pre-design and construction period). The assessment should include basic diagrams showing the diversity of the observed phenomena, their interpretation and conclusions about the technical condition of individual elements of the construction as well as the assessment of its safety and the global assessment of the construction safety, clearly indicating the factors affecting its condition. It should also contain proposals for necessary maintenance works, repairs, or modernization. The needs for the necessary construction works should be structured according to the urgency of their implementation.

Keywords: River embankments, flood embankments, survey of technical status, upstream and downstream slope, factor of safety, mineral and anthropogenic soils, modernization of embankments.

References

Adamski, W., Gortat, J., Leśniak, E. and Żbikowski, A. 1986. Małe budownictwo wodne dla wsi. (Small water construction in rural areas). Warszawa, Arkady, pp. 199.

Bednarczyk, S., Jarzębińska, T., Mackiewicz, S. and Wołoszyn, E. 2006. Vademecum ochrony przeciwpowodziowej. (Flood protection vademecum). Krajowy Zarząd Gospodarki Wodnej, Gdańsk.

Borys, M. 1993. Niskie nasypy z miejscowych gruntów organicznych dla potrzeb budownictwa wodno-melioracyjnego. (Low embankments of local organic soils for the needs of water and drainage construction). Rozprawa Habilitacyjna Falenty (DSc Thesis), Wydawnictwo IMUZ, pp. 159.

Borys, M. 2000. An analysis of the technical state of flood banks constructed from organic clay—the upper Vistula River example. Journal of Water Land Development 4: 97–111.

Borys, M. and Mosiej, K. 2003. Wytyczne wykonywania ocen stanu technicznego i bezpieczeństwa wałów przeciwpowodziowych. (Guidelines for the inspections of technical condition and safety of flood embankments). IMUZ Falenty, pp. 89.

Borys, M. and Rycharska, J. 2004. Ocena stanu technicznego obwałowań przeciwpowodziowych wykonanych z namułów organicznych na przykładzie Żuław Elbląskich. (Evaluation of technical condition of flood embankments made of organic silt based on the example of Żuławy Elbląskie). Woda-Środowisko-Obszary Wiejskie, 4, book 2b(12): 319–334.

Borys, M. (ed.). 2005. Podstawy techniczne modernizacji wałów przeciwpowodziowych i renaturyzacji małych rzek (Technical basis for the modernization of flood embankments and renaturization of small rivers). Woda-Środowisko_Obszary Wiejskie, Rozprawy naukowe i monografie, 15: 150.

Borys, M. 2006. Metody modernizacji obwałowań przeciwpowodziowych z zastosowaniem nowych technik i technologii. (Methods for modernization of flood embankments using new techniques and technologies). Falenty, Wydawnictwo IMUZ, pp. 126.

Borys, M. and Mosiej, K. 2006a. Podstawowe problemy przebudowy i modernizacji obwałowań przeciwpowodziowych. Cz. 1. Ogólne zalecenia, przykładowe rozwiązania. (The basic problems of reconstruction and modernization of flood embankments. Part 1. General recommendations, sample solutions). Gospodarka Wodna 4: 149–155.

Borys, M. and Mosiej, K. 2006b. Podstawowe problemy przebudowy i modernizacji obwałowań przeciwpowodziowych. Cz. 2. Uszczelnianie lub dogęszczanie korpusu i podłoża wału. (The basic problems of reconstruction and modernization of flood embankments. Part 2. Sealing and compacting the body and substratum of the embankment). Gospodarka Wodna 5: 188–193.

Borys, M., Mosiej, K. and Topolnicki, M. 2006. Projektowanie i wykonawstwo pionowych przegród przeciwfiltracyjnych z zawiesin twardniejących w korpusach i podłożu wałów przeciwpowodziowych. (Design and construction of vertical anti-filtration barriers from hardening suspensions in bodies and substratum of flood embankments). Falenty: Wydawnictwo IMUZ, pp. 64.

Borys, M. 2007a. Przepisy i wymogi oraz aktualny stan obwałowań przeciwpowodziowych w Polsce. (Regulations, requirements and the present status of flood embankments in Poland). Woda-Środowisko-Obszary Wiejskie, vol. 7 book. 2a (20). Wydawnictwo IMUZ: 25–44.

Borys, M. 2007b. Opory tarcia na styku geosyntetycznych ekranów przeciwfiltracyjnych w wałach przeciwpowodziowych. (Frictional resistance at the junction of geosynthetic anti-filtration screens in flood embankments). Woda-Środowisko-Obszary Wiejskie. Vol.7 book 2b (21). Wydawnictwo IMUZ: 21–31.

Borys, M. and Rycharska, J. 2007. Stan obwałowań przeciwpowodziowych w Polsce. (The condition of flood embankments in Poland). 12 Międzynarodowa Konferencja Technicznej Kontroli Zapór. Stare Jabłonki, 19–22 czerwca 2007. Warszawa: IMGW: 34–46.

Borys, M. 2008. Wytyczne wykonawstwa pionowych przegród przeciwfiltracyjnych z zawiesin twardniejących w korpusach i podłożu wałów przeciwpowodziowych. (Guidelines for the construction of vertical anti-filtration barriers from hardening suspensions in the bodies and substratum of flood embankments). Materiały instruktażowe 125/9. Procedury. Wydawnictwo IMUZ Falenty, pp. 20.

Borys, M. and Filipowicz, P. 2008. Właściwości geotechniczne odpadów powęglowych z lubelskiego zagłębia węglowego. (Geotechnical parameters of coal mining wastes from Lublin Coal Basin). Woda-Środowisko-Obszary Wiejskie, Rozprawy naukowe i monografie, 15, Wydawnictwo IMUZ Falenty, pp. 99.

Borys, M. and Mosiej, K. 2008. Oceny stanu technicznego obwałowań przeciwpowodziowych. (Assessments of the technical condition of flood embankments). Wydawnictwo IMUZ, pp. 82.

Borys, M. and Rycharska, J. 2008. Wpływ przemarzania i zmian wilgotności na przegrody przeciwfiltracyjne z zawiesin twardniejących w wałach przeciwpowodziowych. (The impact of freezing and moisture content changes on the anti-filtration barriers from hardening suspensions in flood embankments). Wydawnictwo IMUZ, pp. 73.

Borys, M. 2009. Projektowanie i wykonawstwo ekranów przeciwfiltracyjnych z geomembran i mat bentonitowych w wałach przeciwpowodziowych i obwałowaniach małych zbiorników wodnych. (Design and construction of anti-filtration screens of geomembranes and bentonite mats in flood embankments and embankments of small reservoirs). Wydawnictwo IMUZ Falenty, pp. 61.

Borys, M. 2013a. Wytyczne wykonywania kontroli corocznych stanu technicznego i doraźnych kontroli bezpiecznego użytkowania wałów przeciwpowodziowych. (Guidelines for conducting annual inspections and ad-hoc inspections of safe usage of flood embankments). Materiały instruktażowe. Procedury. 137/21. Falenty. Wydawnictwo ITP, pp. 20.

Borys, M. 2013b. Zasady wykonywania corocznych kontroli stanu technicznego wałów przeciwpowodziowych. (Principles of the flood embankments technical state control). Gospodarka Wodna 7: 260–266.

Borys, M. 2014a. Wytyczne wykonywania okresowych pięcioletnich kontroli stanu technicznego i przydatności do użytkowania wałów przeciwpowodziowych. (Guidelines for conducting periodic every five-year inspections of technical condition and suitability for use of flood embankments). Materiały instruktażowe. Procedury. 139/23. Falenty. Wydaw. ITP, pp. 32.

Borys, M. 2014b. Zasady wykonywania kontroli stanu technicznego wałów przeciwpowodziowych (raz na pięć lat). (Principles of the flood embankments technical state control (every five years). Gospodarka Wodna nr 4/2014. ISSN 0017-2448: 125–133.

Borys, M. and Jędryka, E. 2014. Warunki techniczne użytkowania budowli hydrotechnicznych istotnych dla rolnictwa. (Technical conditions for the use of hydraulic structures relevant to agriculture). Materiały instruktażowe. 140/23 Procedury. Wydawnictwo IMUZ Falenty.

Borys, M. and Jędryka, E. 2015. Utrzymywanie sprawności technicznej budowli hydrotechnicznych istotnych dla rolnictwa. (Preservation of technical efficiency of hydrotechnical structures having high importance for agriculture). Gospodarka Wodna 2: 41–46.

Czyżewski, K., Wolski, W., Wójcicki, S. and Żbikowski, A. 1973. Zapory ziemne. (Earth dams). Warszawa, Arkady, pp. 444.

Dłużewski, J. 1997. HYDRO-GEO. Program elementów skończonych dla geotechniki, hydrotechniki i inżynierii środowiska. (HYDRO-GEO. Finite element program for geotechnics, hydrotechnics and environmental engineering). Oficyna Wydawnicza PW, Warszawa.

Evans, J.C. 1993. Vertical cut-off walls. pp. 430–454. *In:* David E. Daniel (ed.). Geotechnical Practice for Waste Disposal. London: Chapman & Hall.

Evans, J.C. and Dawson, A.R. 1999. Slurry walls for control of contaminant migration—a comparison of UK and US practice. Proc. 3rd National Conference Geo-engineering for Underground Facilities, ASCE: Geotechnical Special Publication 90: 105–120.

Evans, J.C., Dawnson, A.R. and Opdyke, S. 2002. Slurry walls for groundwater control: a comparison of UK and US practice. ASCE/PENNDOT Centr. PA Geotechnical Conference.

Filipowicz, P. and Borys, M. 2007. Comparative analysis of the geotechnical properties of coal mining wastes from Lublin Coal Basin and from other basin. Journal Water Land Development, 11: 117–130.

Fratalocchi, E. and Pasqualini, E. 2004. Chemical compatability of cut-off walls to isolate polluted lands: a case study. Geoinżynieria Środowiskowa. Seminarium EU GeoEnvNet. Warszawa, Wydawnictwo SGGW, 15–20.

Gruchot, A. 2014. Wykorzystanie kompozytów z odpadów powęglowych i popiołu lotnego do budowy wałów przeciwpowodziowych. (Embankments from colliery spoils and fly ash composites). Przegląd Górniczy 7: 158–164.

Gruchot, A., Madej, A. and Koś, K. 2014. Usability of fly ash for filtration barriers. Annals of Warsaw University of Life Sciences – SGGW. Land Reclamation 46(2): 115–123.

Hauryłkiewicz, J. 2002. Czas przesiąkania wody powodziowej przez wał. (The time of permeation of the flood water through the embankment). Gospodarka Wodna 5: 194–195.

Jeffreis, S.A. 1981. Bentonite-cement slurries for hydraulic cut-offs. Proceedings 10th International Conference Soil Mech. Foundation Eng. Stockholm.

Lechowicz, Z. and Szymański, A. 2002. Odkształcenia i stateczność nasypów na gruntach organicznych. Cz. 2. Metodyka obliczeń. (Deformations and stability of embankments on organic soils. Vol. 2. Methods of calculations.). Warszawa, Wydawnictwo SGGW, pp. 143.

Makowski, J. 1993. Wały przeciwpowodziowe Dolnej Wisły, historyczne kształtowanie, obecny stan i zachowanie w czasie znacznych wezbrań. (Flood levees of Lower Vistula, historical evolution, present status and performance during larger flows). Wydawnictwo IBW PAN, Gdańsk, pp. 355.

Mioduszewski, W. and Dembek, W. (eds.). 2009. Woda na obszarach wiejskich (Water in rural areas). Warszawa, Falenty. Wydawnictwo IMUZ.

Mosiej, K. and Ciepielowski, A. (eds.). 1992. Ochrona przed powodzią. (Flood protection). Falenty, Wydawnictwo IMUZ, pp. 262.

Pisarczyk, S. 2004. Grunty nasypowe. Właściwości geotechniczne i metody ich badania. (Made soils. Geotechnical properties and methods of testing). Warszawa, Oficyna Wydawnicza PW, pp. 236.

PN-B-12095:1997. Urządzenia wodno-melioracyjne. Nasypy. Wymagania i badania przy odbiorze. (Water reclamation devices. Banks. Requirements and measurements for the acceptance). Polski Komitet Normalizacyjny, Warszawa.

PN–EN 1997-1:2008. Eurokod 7. Projektowanie geotechniczne. Część 1. Zasady ogólne. Polski Komitet Normalizacyjny, Warszawa.

PN-EN 1997-2:2009. Eurokod 7. Projektowanie geotechniczne. Część 2: Rozpoznanie i badanie podłoża gruntowego. Polski Komitet Normalizacyjny, Warszawa.

Popielski, P. and Zaczek-Peplińska, J. 2007. Wykorzystanie modeli numerycznych w eksploatacji budowli piętrzących. (Use of numerical models in the exploitation of damming structures). 12 Międzynarodowa Konferencja Technicznej Kontroli Zapór. Stare Jabłonki, 19–22 czerwca 2007. Warszawa: IMGW: 241–255.

Rozporządzenie Ministra Środowiska z dnia 20 kwietnia 2007 r. w sprawie warunków technicznych, jakim powinny odpowiadać obiekty budowlane gospodarki wodnej i ich usytuowanie. Dz.U. 2007 nr 86 poz. 579. (Regulation of the Ministry of Environment dated 20 April 2007 on technical conditions to be met by water management structures and their location. Journal of Laws 2007 No. 86 item 579).

Rytel, Z., Serafin, B. and Skibiński, J. 1973. Budownictwo i melioracje. (Civil engineering and irrigation). Państwowe Wydawnictwa Szkolnictwa Zawodowego, Katowice, pp. 304.

Skarżyńska, K. 1997. Odpady powęglowe i ich zastosowanie w inżynierii lądowej i wodnej. (Mining wastes and their use in civil and water engineering). Kraków, Wydawnictwo AR, pp. 199.

Sroka, Z. 2001. Modelowanie numeryczne ustalonej filtracji przez groblę. (Numerical modeling of steady filtration through a dam). Zesz. Probl. Post. Nauk Rolniczych, z. 477: 169–175.

Sroka, Z., Walczak, Z. and Wosiewicz, B. 2004. Analiza ustalonych przepływów wód gruntowych metodą elementów skończonych. (Analysis of steady groundwater flow by finite element method). Poznań, Wydawnictwo AR, pp. 178.

Ustawa z dnia 7 lipca 1994 r. Prawo budowlane Dz. U. 1994 nr 89, poz 414. (Act of July 7th 1994 - Building Law. Journal of Laws 1994 No. 89, item 414) z późn. zm. Dz. U. 2010 nr 243, poz. 1623. (amended Journal of Laws 2010 No. 243, item 1623).

Ustawa z dnia 18 lipca 2001 r. – Prawo wodne z późn. zm. Dz. U. nr 115 poz. 1229. (Act of Jul 18th 2001 – Water Law, amended. Journal of Laws No. 115, item 1229).

Wiłun, Z. 2001. Zarys geotechniki. (The outline of geotechnics). Warszawa: WKiŁ, pp. 723.

Wolski, W., Mirecki, J. and Mosiej, K. (eds.). 1994. Warunki techniczne wykonania i odbioru. Roboty ziemne (Technical conditions for execution and acceptance. Earthworks). MOŚZNiL, Warszawa, pp. 71.

Wosiewicz, B. 1996. O modelowaniu i modelach numerycznych zjawisk hydraulicznych. (About modeling and numerical models of hydraulic phenomena). Gospodarka Wodna, 3. 74–81 and 85.

Wosiewicz, B. and Sroka, Z. 2000. Wyspecjalizowane oprogramowanie do analizy i prognozowania zjawisk filtracyjnych. (The specialized software for analysing and forecasting of filtration processes). Gospodarka Wodna, 5: 178–181- vol. I and Gospodarka Wodna 6: 222–224 - vol. II.

Wosiewicz, B. and Sroka, Z. 2001. Ocena wpływu zabezpieczeń filtracyjnych modernizowanych wałów przeciwpowodziowych. (Evaluation of the filtration protection solutions of the modernized river embankments). Gospodarka Wodna 3: 104–110.

Wosiewicz, B. and Walczak, Z. 2002. Analiza numeryczna filtracji pod budowlą piętrzącą z pionową przesłoną uszczelniającą. (Numerical analysis of filtration under damming structure with vertical sealing barrier). Rocznik AR Poznań 338, Melioracje Inżynieria Środowiska 22: 129–145.

Zawisza, E. 2001. Geotechniczne i środowiskowe aspekty uszczelniania grubookruchowych odpadów powęglowych popiołami lotnymi. Zeszyty Naukowe AR w Krakowie, 280, Rozprawy, pp. 178.

Żbikowski, A. (ed.). 1982. Wały przeciwpowodziowe - wytyczne instruktażowe projektowania. (Flood embankments - instructional design guidelines). Melioracje Rolne, 2-3, pp. 49.

Zydroń, T. and Gruchot, A. 2014. Wpływ wilgotności i zagęszczenia na wytrzymałość na ścinanie popioło-żużli i stateczność budowanych z nich nasypów. (Influence of moisture and compaction on shear strength and stability of embankments with ash-slag mixture). Annual Set The Environment Protection. Rocznik Ochrona Środowiska 16: 498–518.

CHAPTER 8

Basics of River Flow Modelling

Tomasz Dysarz

INTRODUCTION

This chapter is a short presentation of main theoretical concepts, which are the basis of popular flow solvers. The term "flow solver" means computer implementation of mathematical models describing river and floodplain flow. The ideas presented here are applied in commercial software, for example, MIKE11, Sobek, CCHE-1D, as well as in non-commercial packages, e.g., HEC-RAS, BASEMENT, SRH-1D, etc. The popularity of such methodologies has been growing mainly for two reasons: (1) development of fast and broadly available computers, and (2) great opportunities for application of mathematical modelling in water engineering and management problems. One example is the successful use of mathematical models in implementation of the EU Flood Directive (EP 2003, 2007; KZGW 2009).

Problems of river flow description, including its variability in time and space, are not new. The first and still used version of the equation system suited for this problem was derived by *Barre de Saint-Venant*, French engineer and scientists, in 1871 (e.g., Szymkiewicz 2000, 2010). The St. Venant equations, as they were called later, make a system of two partial differential equations. They are a mathematical description of two fundamental principles determining the conditions of river flow, that is, the principle of mass balance and that of momentum balance.

The basic equations used here differ from those used in similar texts. Instead of long derivations of equations from mass and momentum balance principles, a short interpretation of the equations is presented. The interpretation leads the reader to the basic principles, which should be sufficient to understand the whole idea. This chapter is focused on the choice of boundary conditions which may be more important for an ordinary user of a typical flow solver. This is the element, which has to be configured properly by the user before simulation and post-processing. Hence, the "passive" understanding of the equations and the "active" understanding of boundary conditions may be crucial for preparation of stable and suitable simulation as well as a proper interpretation of results. The derivations, omitted here, may be found in other books on hydraulic modelling, for example, Cunge et al. (1980), Vreugdenhil (1994), Sawicki (1998), Szymkiewicz (2000, 2010), Wu (2008), and Brunner (2010).

Poznan University of Life Sciences, Department of Hydraulic and Sanitary Engineering, ul. Piatkowska 94A, 60-649 Poznan, POLAND.
E-mail: dysarz@au.poznan.pl

Basic Equations

Discussion of the St. Venant equations has to begin with description of the object and the problem. The object is a single river channel, for example, such as presented in Fig. 1. The river is marked as a blue line in a fragment of a topographic map. The channel is the Warta river reach, in the inlet of the Jeziorsko reservoir. The upstream cross-section of the reach is located in the town of Biskupice. The outlet is linked with the Jeziorsko reservoir in the town of Warta. There are several cross-sections measured along the channel, marked in red in Fig. 1a. The inlet cross-section is marked as $x = 0$, where x means the distance along the channel. The outlet is marked as $x = L$, where L is the total length of the reach. Each intermediate cross-section is identified with its specific position along the channel x_j. The typical cross-section is shown in Fig. 1b. The structure of the channel is characterized by a significant irregularity. In the inlet cross-section there is some inflow of water, which may vary in time t. An example of a flow hydrograph is shown in Fig. 1c.

The problem is the description of discharge Q and depth h variability along the channel and their variability in time. The temporal variability of Q and h may be presented as hydrographs made for subsequent cross-sections, for example in Fig. 1c. The blue and red lines represent changes in time of discharges in the inlet ($x = 0$) and outlet ($x = L$) cross-sections. The basic method for representation of the spatial variability are longitudinal profiles of stage H and/or discharge Q. This type of presentation is shown in Fig. 1d. The profile gives the bottom, water surface level, and bank lines. Two bridges are presented, too. The cross-sections indicated in the map (Fig. 1a) are also marked.

The mass balance equation for one dimensional flow may be written as follows:

$$\frac{\partial A}{\partial t} + \frac{\partial Q}{\partial x} = 0 \tag{1}$$

In the above equation, A is the cross-section area and Q is the discharge. There is a unique relationship between flow area A and depth h or water surface elevation H.

Using the same denotations, the momentum balance equation is presented below

$$\frac{\partial Q}{\partial t} + \frac{\partial}{\partial x}\left(\beta \frac{Q^2}{A}\right) + gA\frac{\partial h}{\partial x} - gA(s_0 - S_f) = 0 \tag{2}$$

where h is the average depth in the cross-section, g is the acceleration of gravity, s_0 is the bottom slope and S_f is the so-called hydraulic slope. The term β is called the Boussinesq coefficient. It describes the variability in flow velocity in a cross-section. It plays the same role as the St. Venant coefficient in Bernoulli's equation, namely it enables the calculation of momentum fluxes on the basis of average flow velocity. Similarly, the values of this coefficient vary between 1.1–1.2 for flows in open and not-compound channels. The latter parameter describes the dispersion and loss of mechanical energy during the water flow. According to one of fundamental assumptions used for the construction of this model, the hydraulic slope is determined in the same way as for the steady flow conditions. This assumption enables application of empirical formulae in the general form

$$S_f = \frac{|Q|Q}{K^2} \tag{3}$$

where K is the conveyance (e.g., Brunner 2010). This parameter depends on local characteristics of flow, defined in a single cross-section, such as flow area A, wetted perimeter P, and bottom roughness n. According to the popular Manning's equation, the conveyance may be calculated as follows

$$K = \frac{A^{5/3}}{nP^{2/3}} \qquad (4)$$

Equations (1) and (2) with closer formula (3) make the set called today the St. Venant equations. The form presented here is one of the practically implemented versions.

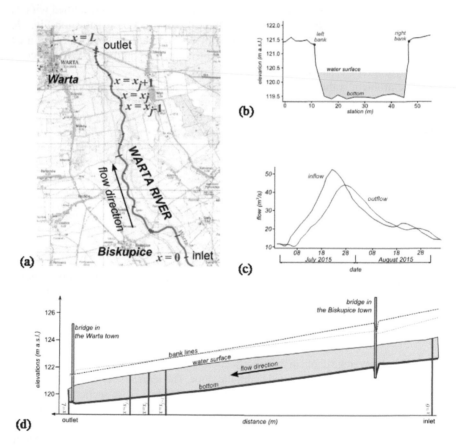

Fig. 1. Example—the Warta river reach: (a) river course in the inlet of the Jeziorsko reservoir, (b) cross-section in the end of the reach, (c) inflow hydrograph, and (d) longitudinal profile of the reach.

Interpretation of Equations

A comprehensive interpretation of the St. Venant equations is given below. Let's assume that the analyzed river section (Fig. 2) is bounded by cross-sections x_{j-1} and x_{j+1}. The cross-section x_j is located between them. Let's define a control volume assigned to the cross-section x_j closed with intermediate cross-sections $x_{j-1/2}$ and $x_{j+1/2}$. Let's assume that the two St. Venant equations are integrated in such a control volume.

$$\int_{x_{j-\frac{1}{2}}}^{x_{j+\frac{1}{2}}} \left(\frac{\partial A}{\partial t} + \frac{\partial Q}{\partial x} \right) dx = 0 \tag{5}$$

$$\int_{x_{j-\frac{1}{2}}}^{x_{j+\frac{1}{2}}} \left[\frac{\partial Q}{\partial t} + \frac{\partial}{\partial x} \left(\beta \frac{Q^2}{A} \right) + gA \frac{\partial h}{\partial x} - gA(s_0 - S_f) \right] dx = 0 \tag{6}$$

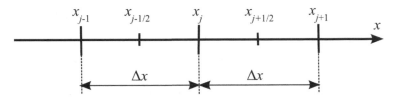

Fig. 2. Basic scheme of single channel reach used in the interpretation of St. Venant equations.

The first of the above equations may be easily simplified by defining the average cross-section area in the reach Δx as:

$$A_j = \frac{1}{\Delta x} \int_{x_{j-\frac{1}{2}}}^{x_{j+\frac{1}{2}}} A \, dx \tag{7}$$

It is important to notice that the product $A_j \cdot \Delta x$ is equal to the volume of water V_j stored in the control volume. Hence, Eq. (5) may be written as follows:

$$\frac{dV_j}{dt} + Q_{j+1/2} - Q_{j-1/2} = 0 \tag{8}$$

Assuming that $Q_{j-1/2}$ is inflow I_j into the control volume and $Q_{j+1/2}$ is outflow O_j from the control volume, we obtain a simple equation describing changes in the water volume V_j in a single reservoir

$$\frac{dV_j}{dt} = I_j - O_j \tag{9}$$

The Eq. (6) has to be changed before it is integrated. It may be proved, that the simplification proposed below is valid for rather regular channels

$$A \frac{\partial h}{\partial x} \approx \frac{\partial}{\partial x} \left(\frac{1}{2} Ah \right) \tag{10}$$

Equation (6) may be written as follows

$$\Delta x \frac{\partial Q_j}{\partial t} + \left(\frac{Q^2}{A} \right)_{j+1/2} - \left(\frac{Q^2}{A} \right)_{j-1/2} +$$
$$+ \frac{1}{2} g \left[(Ah)_{j+1/2} - (Ah)_{j-1/2} \right] - \int_{x_{j-1/2}}^{x_{j+1/2}} \left[gA(s_0 - S_f) \right] dx = 0 \tag{11}$$

where Q_j is the average discharge in the control volume. The subscripts $j-1/2$ and $j+1/2$ are used in the same way as previously. They denote values in inlet and outlet cross-sections respectively. The interpretation of the above equation is possible, if the whole formulae is multiplied by water density ρ and the relationship between the discharge Q and flow velocity u is used

$$Q = uA \tag{12}$$

Now it is possible to transform Eq. (11) into the form shown below

$$\frac{d}{dt}(\rho \Delta x A_j u_j) + (\rho u^2 A)_{j+1/2} - (\rho u^2 A)_{j-1/2} =$$
$$= \left(\frac{1}{2}\rho g A h\right)_{j-1/2} - \left(\frac{1}{2}\rho g A h\right)_{j+1/2} +$$
$$+ \int_{xj-1/2}^{xj+1/2}(\rho g A s_0)\, dx - \int_{xj-1/2}^{xj+1/2}(\rho g A s_f)\, dx \tag{13}$$

Because the product $\rho \cdot A_j \cdot \Delta x$ is a mass of water stored in the control volume, the terms in brackets on the left side are the momentum. The first term of the above equation describes a local change in the momentum in time. According to (12), the second and third terms on the left side are the momentum flux out of and into the control volume, respectively. The whole left side of the above equation represents the so-called substantial derivative of momentum (e.g., Puzyrewski and Sawicki 1998; Sawicki 1998; Durst 2008). Hence, Eq. (13) may be written as follows

$$\frac{D\tilde{P}_j}{Dt} = \left(\frac{1}{2}\rho g A h\right)_{j-1/2} - \left(\frac{1}{2}\rho g A h\right)_{j+1/2} +$$
$$+ \int_{xj-1/2}^{xj+1/2}(\rho g A s_0)\, dx - \int_{xj-1/2}^{xj+1/2}(\rho g A s_f)\, dx \tag{14}$$

In equation (14), the momentum is denoted as $\tilde{P}_j$.

The term $1/2 \cdot \rho \cdot g \cdot h$ is the area of a triangle representing the distribution of unit hydrostatic pressures along the depth. Because of that, the two first terms on the right side of (14) represent the hydrostatic pressure forces in cross-sections $x_{j-1/2}$ and $x_{j+1/2}$. The third term is easily identified as a total gravity force written under the assumption of a small bottom slope. This assumption is characteristic of analyses of gradually-varied flow in open channels and implies that

$$\sin\theta \approx \tan\theta = s_0 \tag{15}$$

where θ is the angle between the bottom and the horizontal plane. In such case, the product of gravity acceleration g and bottom slope s_0 is the projection of gravitational force on the direction parallel to the bottom. The last term may be transformed applying the assumption that the friction losses are calculated in the same way as they are determined for the steady flow. According to this assumption, the bottom stress τ caused by friction is calculated as follows

$$\tau = \rho g \tfrac{A}{P} S_f \tag{16}$$

The above equation is a generalization of equilibrium relationship derived basically for a steady uniform flow. The relationship (16) describes the unequilibrium conditions, such that the longitudinal component of gravity is not balanced by the friction force. It is well seen that the product of density ρ, gravity acceleration g, and flow areas A may be replaced by the product of bottom stresses τ and wetted perimeter P. The elements in the last integral may be written as $\tau \cdot P \cdot dx$. The product $P \cdot dx$, where dx is the unit distance along the channel, is the elementary friction force. The last integral describes total friction force acting along the whole control volume.

The Eq. (14) may be written as follows

$$\frac{D\tilde{P}_j}{Dt} = \Sigma F \tag{17}$$

where the right side is the sum of forces taken into account in the description of water flow, including hydrostatic pressure forces at the inlet (F_{p1}) and the outlet (F_{p2}) cross-sections, longitudinal component of the gravity force (G_x) and friction force (T).

$$\Sigma F = F_{p1} - F_{p2} + G_x - T \tag{18}$$

The above Eqs. (17) and (18) are mathematical form of the Newton's second principle for moving fluid. This principle is equivalent to the momentum balance (conservation).

Other Forms of the St. Venant Equations

In strictly mathematical terms, there are two independent variables in Eq. (1) and (2). These are distance x and time t. In the basic approach, the dependent variables are discharge $Q(x,t)$ and water surface elevation $H(x,t)$ or water depth $h(x,t)$. Other terms in the St. Venant equations may be expressed as functions of water surface elevation $H(x,t)$, for example, flow area A, or a function of both dependent variables, elevation $H(x,t)$ and discharge $Q(x,t)$, for example, hydraulic slope S_f.

In channels with non-movable bed, the relationship between the water surface elevation and other hydraulic parameters, for example, flow area A, wetted perimeter P, surface width B, the depth h, etc. is unique. It enables easy inversion of this relationship, for example, we may consider, that water surface elevation H is a function of flow area A. In such a case the other parameters, P, B, etc., maybe also expressed as functions of A. Then the flow area A may replace the water surface elevation H as dependent variables. Such an approach may be useful in some particular cases.

Equations (1) and (2) are not the only forms of the St. Venant equations used in practice and theory. Although there are a number of the St. Venant equation forms presented in literature, all of them are equivalent. Another popular form may be obtained by introduction of the basic relationship between the increment of flow area and increment of water level or water depth. The relationship is presented in Fig. 3. It may be written as:

$$dA = BdH \tag{19}$$

where B is width of the water surface and H is the water surface elevation. Equation (19) is valid only for stationary bottom cases. In fluvial channels with significant movement of the bed, the changes in bottom elevations should be also included in (19).

Application of (19) enabled transformation of Eq. (1) into the following form

$$\frac{\partial H}{\partial t} + \frac{1}{B}\frac{\partial Q}{\partial x} = 0 \tag{20}$$

where the first term is the time variation in water surface elevation. The second is the gradient of discharge per unit channel width. The meaning of Eq. (20) is the same as that of Eq. (1). Because the increment of the water surface elevation is the same as the increment of the depth, Eq. (20) may be also written as follows

$$\frac{\partial h}{\partial t} + \frac{1}{B}\frac{\partial Q}{\partial x} = 0 \tag{21}$$

where the depth h replaces the elevation H.

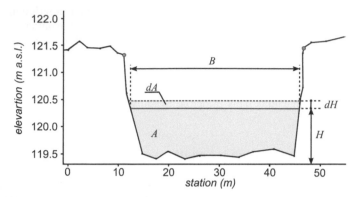

Fig. 3. Dependence between increments of water surface elevations and flow area.

The second change can be introduced if we notice, that the bottom slope is the gradient of bottom elevations with the opposite sign. It is presented in formulae (22). The second element of (22) is the obvious equivalence between water surface elevation and the sum of depth and bottom elevation.

$$s_0 = -\frac{\partial Z_b}{dx} \qquad\qquad H = h + z_b \qquad\qquad (22)$$

These two formulae are necessary to write Eq. (2) in the form presented below

$$\frac{\partial Q}{\partial t} + \frac{\partial}{\partial x}\left(\beta\frac{Q^2}{A}\right) + gA\left(\frac{\partial H}{\partial x} + S_f\right) = 0 \qquad\qquad (23)$$

In (23) the pressure and gravity terms are linked together. They are represented by the product $gA\dfrac{\partial H}{\partial x}$.

The Eqs. (1), (2), (20), (21), and (23) may be used in different combinations with the only constraint that the system of St. Venant equations consist of one mass balance equation and one momentum balance equation. For instance, Eq. (1) may be used with (23) and the dependent variables are A and Q. If Eq. (2) and Eq. (21) are used, the more common choice of dependent variables set is h and Q.

The equations presented above are the so-called conservative forms of the St. Venant equations. These forms are well suited for modelling of flows in regular and irregular channels without floodplains. In fact the equations similar to those are implemented in the majority solvers available today. However, these equations are not the best choice for theoretical analyses including features of wave propagation phenomena. Hence, the other forms the St. Venant equations can be found in scientific literature.

The so-called non-conservative form of St. Venant equations is more suitable for theoretical analyses. These St. Venant equations can be derived from (1) and (2), on imposing some additional assumptions on channel regularity and continuity of flow. To present the main idea, we have to consider the rectangular and not alluvial channel with constant width b. The relationship between flow area and the depth is presented below.

$$A = hB \qquad\qquad B = b = const \qquad\qquad (24)$$

The relationship (12) between discharge, flow area, and average velocity is also useful for this purpose. If Eqs. (12) and (24) are inserted into Eqs. (1–2) and then all differential terms are expanded, the resulting system is as follows

$$\frac{\partial h}{\partial t} + u\frac{\partial h}{\partial x} + h\frac{\partial u}{\partial x} = 0 \tag{25}$$

$$\frac{\partial u}{\partial t} + g\frac{\partial h}{\partial x} + u\frac{\partial u}{\partial x} = g(s_0 - S_f) \tag{26}$$

The first equation describes the mass balance equation, while the second equation —the momentum balance. In Eqs. (25–26), the dependent variables are the depth $h(x,t)$ and the velocity $u(x,t)$. The system (25)–(26) may be also written in matrix form

$$g\frac{\partial \varphi}{\partial t} + \mathcal{M}\frac{\partial \varphi}{\partial x} = \mathbf{f} \tag{27}$$

where φ is vector of dependent variables, $\mathcal{M}$ is jacobian matrix, and $\mathbf{f}$ is vector of right-hand sides. These terms are defined as follows

$$\varphi = \begin{bmatrix} h \\ u \end{bmatrix} \quad \mathcal{M} = \begin{bmatrix} u & h \\ g & u \end{bmatrix} \quad \mathbf{f} = \begin{bmatrix} 0 \\ g(s_0 - S_f) \end{bmatrix} \tag{28}$$

Characteristics and Initial-Boundary Conditions

The solutions of St. Venant equations in any of the presented forms are two functions. These may be either discharge Q and depth h in conservative form or velocity u and depth h in non-conservative case. The values of these functions vary with time t and distance measured along the channel (represented by the coordinate x). Hence, in any case, the domain of the solution is defined in the plane Oxt. In the simplest case of a single river reach, the domain is defined as follows:

$$0 \le x \le L \qquad t \ge 0 \tag{29}$$

where L is the channel length.

By definition, the characteristics or characteristic curves are the lines in the domain along which the partial differential equation (PDE) or the system of PDEs becomes ordinary differential equation (ODE) (e.g., LeVeque 2002). This sophisticated definition does not explain well why the idea of characteristics is so powerful. It has become the basis of one solution method, namely the method of characteristics (e.g., Cunge et al. 1980; LeVeque 2002), and it is used in other methods, for example, in the finite volume method with Riemann invariants (LeVeque 2002; Toro 2009). In simpler words, the characteristics are the lines in the domain, which determine the directions of any changes in propagation. For instance, if some amount of water is released to the stream in time t_0 and cross-section x_0, the effects of this action may be observed in other cross-sections x_j at times t_k ($t_k > t_0$) as an increase in water amount, but all (x_j, t_k) have to lie on the characteristics that have the common point of intersection (x_0, t_0).

The characteristics in domain (29) are defined by the following equation

$$\frac{dx}{dt} = \lambda \tag{30}$$

where λ is eigenvalue of the jacobian matrix. The eigenvalue of any matrix is a root of its characteristic equation, which is defined as follows

$$\det(\mathcal{M} - \lambda E) = 0 \tag{31}$$

Above, det() means determinant and E is unit matrix. The characteristic equation for the system (27) is written below

$$\begin{vmatrix} u-\lambda & h \\ g & u-\lambda \end{vmatrix} = 0 \qquad\qquad (u-\lambda)^2 - gh = 0 \tag{32}$$

There are two roots of this equation, which is typical of the so-called hyperbolic PDEs (e.g., Sawicki 1998; Szymkiewicz 2000, 2010). The characteristics are easily determined as the roots of quadratic equation

$$\left(\frac{dx}{dt}\right)_{1/2} = \lambda_{1/2} = u \mp \sqrt{gh} \tag{33}$$

It is clearly visible that the course of characteristics depends on the flow velocity u and the so-called celerity $\sqrt{gh}$, which is the velocity of wave propagation.

There are two cases possible: (1) the flow velocity is lower than celerity and (2) the flow velocity is greater than celerity. In the first case, one characteristic is negative and one is positive. In the second, both characteristics are positive. These two cases are well classified by the Froude number Fr

$$Fr = \frac{U}{\sqrt{gh}} \tag{34}$$

which is the ratio of velocity and celerity. If $Fr < 1$ then the first case occurs and the flow in the channel is subcritical. Otherwise ($Fr > 1$) the second case is realised and the flow in the channel is supercritical.

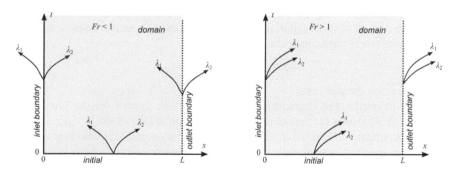

Fig. 4. Schematic view of boundary characteristics of St. Venant equations.

Equations (1) and (2) and the domain do not uniquely determine the solution. Because the problem solved is varying in space and time, it has to be completed with proper initial and boundary conditions. The number of additional conditions imposed on the domain boundaries must be equal to the number of characteristics incoming to the domain from this boundary (e.g., Cunge et al. 1980; Sawicki 1998; Szymkiewicz 2000, 2010). Hence, the initial-boundary conditions have to be imposed according to the rules illustrated in Fig. 4 and presented in Table 1. In the initial line ($0 \leq x \leq L, t = 0$), two conditions have to be imposed. This rule is independent of the flow characteristics, because two characteristics are incoming from this line to the domain in each regime. However, the number of conditions in the inlet ($x = 0, t \geq 0$) and outlet ($x = L, t \geq 0$) boundaries strongly depends on the flow regime. If the flow regime is subcritical ($Fr < 1$), one condition is imposed at the inlet and one at the outlet. If Froude number is higher than 1, two conditions are imposed at the inlet and none at the outlet.

Table 1. Rules for definition of initial and boundary conditions.

No.	Type of condition	Boundary		# of conditions	
		Space	Time	$Fr < 1$	$Fr > 1$
1.	Initial	$0 \leq x \leq L$	$t = 0$	2	2
2.	Inlet boundary	$x = 0$	$t \geq 0$	1	2
3.	Outlet boundary	$x = L$	$t \geq 0$	1	0

The basic form of initial condition is distribution of dependent variables along the channel reach. In the case of conservative equations, it can be as follows

$$\begin{cases} Q(x, t = 0) = Q_{in}(x) \\ H(x, t = 0) = H_{in}(x) \end{cases} \qquad 0 \leq x \leq L \qquad (35)$$

where Q_{in} and H_{in} are given functions of x. So detailed information may not be available in many cases. Fortunately, accurate reconstruction of the initial condition is not necessary. According to the idea of characteristics, the influence of initial condition on the solution may be neglected after some time of the simulation, as illustrated in Fig. 5. If the outermost characteristics leave the domain, the impact of this condition is nonexistent.

If the importance of the initial condition is neglected, the most common approach is to apply the steady flow solution to determine the distribution of the dependent variables in the channel reach. In fact, such initial condition is very popular and sometimes it is used as default option in the flow solvers, for example, HEC-RAS.

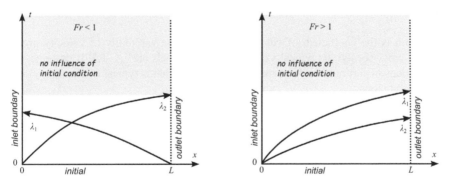

Fig. 5. Vanishing of initial condition influence.

As the boundary conditions the hydrographs of flow or stages may be imposed. The examples for the conservative form are presented below

$$Fr < 1 \quad \begin{cases} Q(x = 0, t) = Q_0(t) \\ H(x = L, t) = H_L(t) \end{cases} \qquad t \geq 0 \qquad (36)$$

$$Fr > 1 \quad \begin{cases} Q(x = 0, t) = Q_0(t) \\ H(x = 0, t) = H_0(t) \end{cases} \qquad t \geq 0 \qquad (37)$$

where Q_0, H_L, and H_0 are given functions of time. Of course, the order 'flow-stage' may be the opposite in the case of subcritical flow. It is also possible to impose the same type conditions, for example, stage for $x = 0$ and stage for $x = L$, in this regime.

It is important to notice that the flow or stage hydrograph imposed as outlet boundary condition in subcritical flow regime may be questionable in some cases. If the model is used for generation of flow or stage variability forecast in the outlet cross-section, the information (hydrograph) in this cross-section is the expected solution. Hence, these should not be the data provided to run the model. In such cases other forms of the boundary condition may be used for the outlet.

The most basic approach for generation of outlet boundary condition, not based on hydrograph, is to use some relationship between the flow and the stage. There are three very common relationships: (1) normal depth, (2) critical depth, and (3) rating curve.

The first is based on the equation for normal flow, which means that steady uniform flow is assumed at the outlet. Such an approach may be considered as outflow in the conditions unconstrained by any impacts. The basis of this method is the solution of the Manning equation in the following form:

$$Q = K \sqrt{s_0} \tag{38}$$

where K is defined by (4) and s_0 is the bottom slope as previously. Equation (38) implies the equality between hydraulic slope S_f, which should be present in (38), and bottom slope s_0, which is actually present. Comparison with (2) shows which simplifications are represented by (38). In fact, this condition means that inertia and pressure forces are neglected in the outlet cross-section.

The second condition, critical depth, may be written as Froude number equal one. But, in the case of irregular channels, it is better to present this condition as follows

$$\alpha \frac{Q}{g} = \frac{A^3}{B} \tag{39}$$

where α is the coefficient of velocity describing the variability of velocity profile in the cross-section. Its typical values in open channels are 1.1–1.2. It describes the conditions of fast outflow from the channel reach, what is typical of channels closed by abrupt slopes.

The third condition is generalization of the above two and it also reflects other cases. The form of the condition is as follows

$$Q = f(H) \tag{40}$$

where $f(H)$ is a given function of water stage or elevation. In the most cases the condition is provided in tabular form. It is the most typical condition of cross-sections with gauge stations.

Summary

The main purpose of this chapter is to present and explain the basic ideas used in modelling of river flow. The theory hidden behind the popular flow solvers is quite sophisticated. Hence, the information given in this chapter may be helpful for proper use of the software available.

Although, the number of flow solvers available is not small, there is one common concept driving their performance. This is the set of St. Venant equations. The equations express in mathematical form the mass and momentum balance principles written as partial differential equations. The presentation of this concept is focused on the needs of a potential user. The equations are explained in brief. Their interpretation is presented in the form enabling understanding of their importance. Another key issue of this chapter is analysis of the choice of boundary conditions. This aspect of the software configuration is explained. The reasons for specific form of the boundary conditions used in the case of St. Venant equations are discussed on the basis of characteristics.

The author of the chapter hopes that such a form of the presentation would enable quite an easy adaptation of the theory explained. It could help in application of the flow solvers in several ways. First of all, better understanding of the theory behind the software used should help in program configuration. This knowledge should also support the user in the process of results interpretation and application to the specific problem solved. The ideas presented here could also be useful in handling simulation errors and stability problems.

The user interested in deeper studying of the concepts presented here should look into the proper bibliography. Some useful books are quoted in this chapter and listed below. However, the bibliography presented here should not be considered as complete. There are also other useful books.

Keywords: River hydraulics, mathematical modeling, the St. Venant equations.

References

Brunner, G.W. 2010. HEC-RAS River Analysis System Hydraulic Reference Manual, US Army Corps of Engineers, Hydrologic Engineering Center (HEC), Report No. CPD-69.

Cunge, J.A., Holly, F.M., Jr. and Vervej, A. 1980. Practical Aspects of Computational River Hydraulics. Pitman Advanced Publishing Program. Boston, London, Melbourne.

Durst, F. 2008. Fluid Mechanics: An Introduction to the Theory of Fluid Flow. Springer-Verlag Berlin Heidelberg.

EP. 2003. Best practices on flood prevention, protection and mitigation. Meeting of Water Directors of the European Union (EU), Norway, Switzerland and Candidate Countries, Athens, June 2003.

EP. 2007. Directive 2007/60/EC of the European Parliament and of the Council of 23 October 2007 on the assessment and management of flood risks. European Parliament.

KZGW. 2009. Methodology of Flood Hazard and Risk Maps Elaboration. National Board on Water Management (pol. Krajowy Zarząd Gospodarki Wodnej), Warsaw (in Polish).

LeVeque, R.J. 2002. Finite Volume Methods for Hyperbolic Problems. Cambridge University Press.

Puzyrewski, R. and Sawicki, J. 1998. The Fundamentals of Fluid Mechanics and Hydraulics. Scientific Publishing PWN, Warsaw (in Polish).

Sawicki, J. 1998. Flows with Free Surface. Scientific Publishing PWN, Warsaw (in Polish).

Szymkiewicz, R. 2000. Mathematical Modelling of Flow in Open Channels. Scientific Publishing PWN, Warsaw (in Polish).

Szymkiewicz, R. 2010. Numerical Modeling in Open Channel Hydraulics. Water Science and Technology Library, Volume 83, Springer.

Toro, E.F. 2009. Riemann Solvers and Numerical Methods for Fluid Dynamics A Practical Introduction. Springer-Verlag Berlin Heidelberg.

Vreugdenhil, C.B. 1994. Numerical Methods for Shallow-Water Flow. Kluwer Academic Publishers, Dordrecht, Boston, London.

Wu, W. 2008. Computational River Dynamics. Taylor & Francis Group, London, UK.

CHAPTER 9

Bed Shear Stresses and Bed Shear Velocities
Ubiquitous Variables in River Hydraulics

Magdalena M. Mrokowska and *Paweł M. Rowiński**

INTRODUCTION

Wolfgang Pauli (1900–1958), one of the pioneers of quantum physics used to say "God made the bulk; the surface was invented by the devil". This reflects how complex is the situation when a solid surface shares its border with the external world. Hence, it refers also to the motion of water that causes friction with the bed and banks.

We realize that environmental hydraulics is mainly about various aspects of the physical, chemical, and biological attributes of flowing water. Thus, nothing seems more apropos for scientists and practitioners dealing with hydraulics than to understand, measure, describe, and model the roughness and resistance to flow. According to Knight et al. (2010), the roughness denotes a surface textural property, whereas resistance is the cumulative effect of the roughness resisting any hydraulic flow. A complete description of how roughness is treated in hydraulic research can easily fill several extensive volumes. When describing the roughness or any so-called roughness coefficients, one should remember about a variety of factors that are complicated enough when considered alone. Jain (2001) names eight such factors, that is, surface roughness characterized by the size and shape of the grains forming wetted perimeter; vegetation, channel irregularity, channel alignment, silting and scouring, obstruction, stage and discharge, and finally suspended and bed load.

In this chapter, we aim to present one concept pertaining to the notions of the bed shear stress and bed shear velocity which are a result of physical principles, namely the momentum conservation law. These notions have an important role to play in hydrodynamics, and this is the main reason why in our opinion greater and more specific attention must be focused on them. The bed shear stress and velocity are parameters present in the description of numerous aspects of the river flow and associated processes,

Institute of Geophysics, Polish Academy of Sciences, Ks. Janusza 64, 01-452 Warsaw, Poland.
  E-mail: m.mrokowska@igf.edu.pl
* Corresponding author: p.rowinski@igf.edu.pl

such as near-bed turbulence characteristics (Bagherimiyab and Lemmin 2013; Dey et al. 2011), normalization of various turbulence features (Campbell et al. 2005; Nikora and Smart 1997), sediment transport (Cooper and Tait 2010; De Sutter et al. 2001; Mrokowska et al. 2016), contaminants transport (Kalinowska et al. 2012), and eco-hydraulics problems (Graba et al. 2014). It is due to the fact that the bed shear stress generates resistance to flow which is the main factor affecting the pattern of water flow and behavior of diluted substances, microorganisms, and solids.

Theoretical Background

The shear stress (τ [N m^{-2}]) may be described in simple terms as a force acting on a surface tangential to the force and expresses flow resistance, that is, friction forces. Shear stresses in water are a consequence of water viscosity and flow turbulence and are generated by the transfer of momentum within the water volume. This transfer has a form of diffusion between layers of water in the laminar flow and is triggered by the mass exchange between turbulent eddies in the turbulent flow (Powell 2014; Yen 2002). Consequently, the total shear stress τ_{tot} is composed of the laminar shear stress τ_l and the turbulent shear stress τ_t:

$$\tau_{tot} = \tau_l + \tau_t. \tag{1}$$

Practically all river flows should be treated as turbulent since the criterion Reynolds number usually exceeds relevant threshold value. In other words, if a velocity probe of any kind were placed into the turbulent river flow, we might get a signal like that shown in Fig. 1. Therefore, in the river flow, turbulence is a dominant source of the shear stress. For detailed characteristic of the water flow, information about 3-dimensional velocity field is necessary, that is, temporal variability of velocity in horizontal u, vertical v, and lateral w direction. Each of these velocity components is a subject of fluctuations due to turbulence and may be decomposed into an average velocity denoted by $\bar{u}$, $\bar{v}$, $\bar{w}$ and fluctuations denoted by u', v', w' (Eqs. 2–4). This procedure is known as Reynolds decomposition. It is depicted in Fig. 1 for a sample of velocity data in horizontal direction.

$$u = \bar{u} + u' \tag{2}$$

$$v = \bar{v} + v' \tag{3}$$

$$w = \bar{w} + w' \tag{4}$$

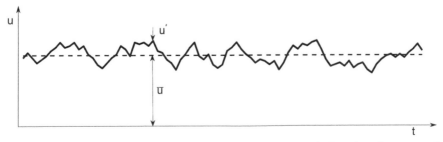

Fig. 1. Decomposition of horizontal component of instantaneous velocity u into time-averaged component $\bar{u}$ and fluctuations u'.

The turbulent shear stress (Reynolds shear stress) is defined as the time-averaged product of velocity fluctuations $\overline{u'v'}$ and is given by:

$$\tau_t = -\overline{\rho u'v'}. \tag{5}$$

Shear stresses occur within the whole volume of flowing water; however, a particular interest is given to the bed shear stress that represents the friction at the water-bed interface. As is shown in right-hand panel of Fig. 2, the bed shear stress is equivalent to the maximum total shear stress which occurs at the boundary between the water and the channel bed. The total shear stress decreases with distance above the channel bottom. The viscous shear stress reaches the maximum in the viscous sublayer; outside, this sublayer drops to zero, and the turbulent shear stress reaches its maximum instead. However, flow in rivers is hydraulically rough, and the viscous sublayer has no practical meaning.

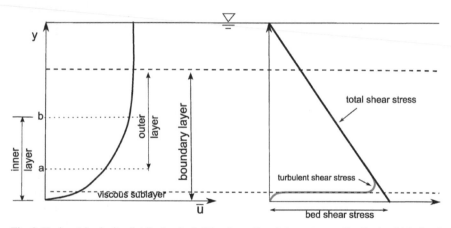

Fig. 2. Horizontal velocity distribution in (left hand panel) and shear stresses distribution (right hand panel) in open channel.

As mentioned by Powell (2014), the flow resistance is commonly discussed with reference to boundary resistance, that is, the mechanical drag (or friction) exerted on the flow by the rough channel perimeter in the form of bank shear and bed shear (see Fig. 3). In this figure, friction at the air-water interface is also shown but its value can be meaningful only in case of very strong wind countering the motion of water (Meyer 2011; Westenbroek 2006).

The explanation of the formation of shear stresses in the water flow in open channels is provided by the boundary layer theory. According to the theory, a region could be identified near the channel boundary in which the flow is influenced by the friction exerted by the water-bottom interface. This region is responsible of the variation of velocity and shear stress along the depth of a channel (Fig. 2). The boundary layer is divided into the inner region and outer region. The inner region is characteristic of viscous shear stresses near the bed (but only in hydraulically smooth flow) and small-scale turbulence in buffer layer (between a and b in Fig. 2). Large-scale turbulence occurs in the outer layer. The shear stress is tightly connected with the velocity, and for this reason, it is very often expressed in dimensions of velocity as the shear velocity (or friction velocity) u_* [m s^{-1}]:

$$u_* = \sqrt{\frac{\tau}{\rho}},\tag{6}$$

where ρ – water density [kg m^{-3}].

As an additional example, Fig. 4 presents the distributions of shear velocity and the local shear stresses at a measuring cross-section of a compound channel, that is, natural channel subject to seasonal floods consisting of a main channel and two overbank channels. Various roughness characteristics in the main channel (smooth) and on the floodplains (rough) have been taken into account (Rowiński et al. 2002). Understanding of these basic phenomena is necessary for effective measuring and next applying shear stresses and velocities to solve river flow problems.

Fig. 3. Boundary resistance in a river (photograph by M. Mrokowska).

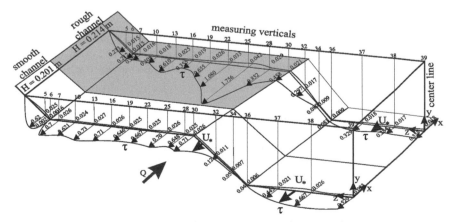

Fig. 4. Shear velocity and bed shear stress distribution in a compound channel (Rowiński et al. 2002).

As the velocity profile is connected with the flow resistance, there are a number of classical functional relationships involving the horizontal velocity profile and the shear velocity. One of the most widely used says that the horizontal velocity profile varies logarithmically with the distance above the bed according to the relation:

$$\bar{u} = \frac{u_*}{k} \log(y) + A, \tag{7}$$

where κ – von Karman constant = 0.41 (for clear water), y_0 – roughness height, and A – integration constant. The above expression forms the basic way of the evaluation of the shear velocity and is named as the law of the wall or simply the logarithmic profile method (Rowiński et al. 2005). This method obviously applies only to that part of the boundary layer that can be well described by logarithmic functions; that is, a near wall (near bed) region – inner region but outside the viscous sublayer.

Various forms of logarithmic profiles for velocity distributions are assumed in hydraulic research but one should remember that there are situations in rivers when those distributions do not hold at all. Examples may appear when flows through vegetated channels are considered or in many cases of unsteady flow conditions. Very often the velocity profiles in compound channels, particularly in the verticals above sloping banks of the deeper channel, do not collapse with log law. Then other methods, presented further in the chapter, are sought for the evaluation of the shear velocities.

An important part of our considerations deals with the problem of how certain or rather uncertain are the computations of bed shear stresses in real rivers. Nate Silver in "The Signal and the Noise: Why Most Predictions Fail—but Some Don't" writes "Both experts and laypeople mistake more confident predictions for more accurate ones. But overconfidence is often the reason for failure. If our appreciation of uncertainty improves, our predictions can get better too." And this is what we always should have in mind when measuring/computing the values or distributions of bed shear stresses in river channels. During field experiments, the bed roughness is not known and is generally assigned an arbitrary value. Therefore, bed shear stress estimates are uncertain. Taking into account that the knowledge of bed shear stresses is often a prerequisite for computations of other hydraulic or transport characteristics, uncertainty in its evaluation transmits to other values as well.

Examples of Problems Involving Bed Shear Stress

As previously mentioned, the bed shear stress or shear velocity are simple concepts traditionally used in hydraulic research to describe a variety of situations. Below, we present examples of the hydraulic problems in which those variables are extensively applied.

Characteristics of turbulent velocity field

In a turbulent flow, energy is spread over a range of frequencies. We may observe so-called energy cascade, that is, the turbulent kinetic energy is transferred from larger scale motions to smaller scale motions. Correlations occur between velocities at two points in space or at two points in time at a single space which is guaranteed by the existence of the viscosity and eddy-like motions in the turbulence. The turbulent kinetic energy, or its part when one velocity component is considered, represents the degree of velocity fluctuations. It may be represented in the form of root mean square value of velocity fluctuations, e.g., $(\overline{u'u'})^{0.5}$ in longitudinal direction.

To compare various turbulent flows, the intensity is usually normalized (divided) by either the shear velocity or mean velocity—each of those approaches has its advantages or disadvantages but the first method reveals the importance of the proper evaluation of the bed shear velocity. An example of such approach is the one proposed in the classical work of Nezu and Nakagawa (1993) in which universal formulae for normalized turbulence intensity in each direction were proposed. This formula in longitudinal direction reads:

$$\frac{\overline{(u'u')}^{0.5}}{u_*} = 2.3\exp\left(\frac{y}{h}\right), \tag{8}$$

where h – water depth [m], and y – vertical spatial coordinate [m]. Turbulence intensity represents the first statistical moment of the turbulent velocity field. Further moments in the form of skewness or flatness factors are normalized with use of bed shear velocity, as well.

In turbulence studies, a crucial value is also eddy viscosity ε_t which depends upon the characteristics of the flow under investigation. It is affected among others by the nature and shape of the solid boundaries and freestream turbulence intensity. It is often assumed that it is proportional to spatially averaged bed shear velocity $\overline{u}_*$ in the transverse direction, for example in the form proposed by Ikeda (1981):

$$\varepsilon_t = \frac{k}{6}\overline{u}_*h. \tag{9}$$

The above is just to show that friction velocity is essential to adequately represent turbulence structure in rivers.

Evaluation of bed load transport

Sediment transport is one of the most difficult processes to evaluate in open channels. Sediment is transported by water flow along the bed (i.e., bed load) and in suspension (i.e., suspended load). The form of transport depends on flow conditions as well as on the size of sediment particles (van Rijn 1993). It is a common practice to evaluate sediment transport by analytical formulae. Analytical methods have been developed mainly by laboratory and *in situ* research. Formulae are usually empirical (derived by regression methods); nonetheless, some have theoretical background (e.g., Bagnolds energy concept), and some were derived by dimensional analysis (Garcia 2007; van Rijn 1993; Yang and Wan 1991). The comprehensive summary of sediment transport formulae can be found in Dey (2015), Garcia (2007), Julien (2010), and van Rijn (1993), among others. A common feature of most of the formulae for the bed load transport is that among others they assume a priori knowledge of the values of the bed shear stress. As examples of these, let us briefly discuss the popular Meyer-Peter and Mueller formula used for the evaluation of the bed load and the Ackers-White formula used for the total load evaluation. Both formulae were derived for the steady flow conditions.

The Meyer-Peter and Muller relation is an empirical formula derived based on laboratory experiments on the transport of sand and gravel sediment (Garcia 2007; van Rijn 1993). The dimensionless bed load transport rate is given in the following way:

$$q^* = 8(\tau^* - \tau_c^*)^{1.5}, \tag{10}$$

where $q^* = q/(d_m\sqrt{grd_m})$ – dimensionless bed load transport rate [–], g – gravitational acceleration [m^2 s^{-1}], q – volumetric bed load transport rate [m^2 s^{-1}], $r = (\rho_s - \rho)/\rho$ – submerged specific gravity of sediment [–], $\tau^* = \tau/(\rho grd_m)$ – dimensionless Shield stress [–] d_m – mean particle diameter [m], ρ_s – sediment density [kg m^{-3}], and τ_c^* – dimensionless critical Shields stress for incipient motion [–], $\tau_c^* = 0.047$.

Ackers and White derived their formula based on Bagnold's power concept and expressed the total load rate in terms of dimensionless parameters presented below (Garcia 2007):

$$q = \rho_s c d_m U \left(\frac{U}{u_*}\right)^n \left(\frac{F_{gr}}{A_{aw}} - 1\right)^m ,$$

(11)

where F_{gr} – Ackers-White mobility number given by:

$$F_{gr} = \frac{u_*^n u_*'^{1-n}}{\sqrt{rdg_m}},$$

(12)

$$u_*' = \frac{U}{\sqrt{32}\,log(10\frac{r}{d_m})}.$$

(13)

The authors of the formula established the values of parameters n, m, c, and A_{aw} by fitting laboratory data as a function of the dimensionless grain size D_{gr}:

$$D_{gr} = (\sqrt{grd_m}\,d_m/v\,)^{2/3},$$

(14)

where v – kinematic viscosity [m^2 s^{-1}]. If $D_{gr} > 60$ then $n = 0$, $m = 1.5$; $A_{aw} = 0.17$; $c = 0.025$. In other cases, $1 - 0.56\,log(D_{gr})$; $m = \frac{9.66}{D_{gr}} + 1.34$; $A_{aw} = \frac{0.23}{\sqrt{D_{gr}}} + 0.14$; $log(c) = 2.86\,log(D_{gr}) - [log(D_{gr})]^2 - 3.53$.

Without going into details and following an excellent recent overview of Petit et al. (2015), we realize that shear stresses can explain the shapes of river beds, particularly meandering beds; the formation and destruction of pebble clusters, the stability of step-pool systems, and the evolution of other bedform patterns, such as bedrock patterns and cascades. We also remember from basic textbooks of hydraulics that the so-called Shields criterion uses critical shear stress, that is, the force required to move an element of a given size. Shields number is simply a non-dimensional number used to calculate the initiation of motion of sediment in a fluid flow. The Shields diagram has become the preferred basis for predicting eroding-flow conditions. It may also be used to describe the effects of grain sheltering and packing.

Contaminants transport

Bed shear stress influences all transport in rivers in two ways. First of all, as mentioned before, river flows are turbulent in nature. Therefore, transport of various constituents depends on the characteristics of flow which, in turn, are very often described with use of the friction velocity or bed shear stresses directly. Besides, diffusion and dispersion coefficients that are present in the governing equations for transport processes are often shown as dependent on friction velocity.

Dispersion coefficients controlling the rate of mixing are essential to solve the transport equations and, for example, in the simplest one dimensional case, an overall formula for its derivation yields:

$$D = ahu_*,$$ (15)

where D – longitudinal or transverse dispersion coefficient and a – dimensionless parameter that theoretically may assume values from a relatively large range. This general relation shows the strong dependence of transport processes on the bed shear velocity. Studies of the influence of the shear velocity upon dispersion coefficients can be found in Kalinowska and Rowiński (2012).

Bed Shear Stress Evaluation in Steady Flow

A bench of methods allowing researchers to evaluate bed shear stresses and velocities under steady flow conditions exists and recently critical overviews of those methods have been presented in numerous publications (Biron et al. 2004; Dey 2015; Rowinski et al. 2005; Powell et al. 2014).

The first and most frequently used method is based on the force balance approach and requires only morphological parameters of the reach of a channel: a hydraulic radius R [m] and the bed slope I [–] which is parallel to the water slope in the steady flow. The bed shear stress is evaluated from relation between friction forces and forces exerted by flowing water, which are in balance in steady flow, giving the following relationship:

$$\tau = \rho g R I,$$ (16)

The bed shear stress evaluated by this method is an average value and is used in studies where bulk variables are considered (MacVicar and Roy 2007).

The next method is based on the horizontal velocity profile (Nikora and Smart 1997; Smart 1999; Wilcock 1996). Velocity distribution is measured, for example, by ADVP (Acoustic Doppler Velocimetry Profiling) (Bagherimiyab and Lemmin 2013). Then the profile is fitted to theoretical velocity distribution, for example, logarithmic distribution (Eq. 7), and shear velocity is assessed. Next, bed shear stress may be calculated from Eq. (6). The bed shear stress evaluated by this method is a point value, as it is measured based on velocity profile measured above a certain point. Since this is one of the most popular methods, let us present it in more details for the convenience of the Reader.

To remove integration constant form Eq. (7), a boundary condition ($\overline{u}(y = y_0) = 0$) is imposed. y_0 denotes height above the bed $y = 0$. In a distance y_0 above the bed, flow velocity is 0 and flow is assumed to be laminar in a region between $y = 0$ and $y = y_0$. After introducing boundary condition, Eq. (7) takes the form:

$$0 = \frac{u_*}{k}\log(y_0) + A,$$ (17)

and A is evaluated as:

$$A = i\frac{u_*}{k}log(\mathrm{y}_0).$$ (18)

Incorporating Eq. (18) into Eq. (7) gives:

$$\overline{u} = \frac{u_*}{k}\log\left(\frac{y}{y_0}\right).$$ (19)

After rearranging Eq. (19), a formula for shear velocity is obtained:

$$u_* = \frac{\bar{u}k}{log\left(\dfrac{y}{y_0}\right)}.$$ (20)

Let us present an artificial sample data set to illustrate how to use measured velocity profile to calculate shear velocity. In this example, flow depth $h = 1.2$ m and velocity profile is logarithmic in the inner region of flow ($y/h < 0.2$), that is for $y < 0.24$ m. Figure 5 presents the vertical profile of time-averaged velocity. To evaluate shear velocity, the following steps must be taken:

1. Check if the velocity profile is logarithmic in the inner region of flow:
   When velocity profile data are plotted in semi-logarithmic graph (Fig. 6), data that obey the law of the wall fit a straight line. This condition is met herein for the sample data.

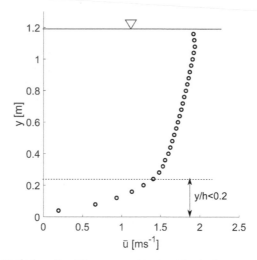

Fig. 5. Vertical profile of time-averaged horizontal velocity – sample data.

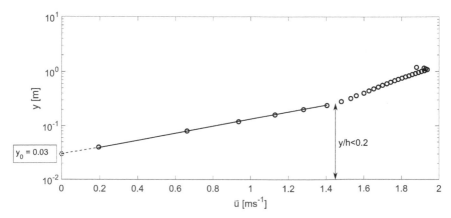

Fig. 6. Vertical profile of time-averaged horizontal velocity in a semi-logarithmic graph.

2. Evaluate the value of y_0:

   y_0 is evaluated as the vertical coordinate of a point where fitted straight line intersects the vertical axis ($u = 0$, y_0). In our case $y_0 = 0.03$ m.

3. Apply Eq. (11) to evaluate the shear velocity:

   Having all necessary data: $y_0 = 0.03$ m, $\kappa = 0.4$, $u(y)$, the shear velocity is calculated, and the result is:

   $u_* = 0.15$ m s^{-1}.

Afzalimehr and Anctil (2000) propsed boundary-layer characteristics method (BLCM) that uses velocity distribution:

$$u_* = \frac{(\delta_* - \theta)u_{max}}{\alpha\delta_*} \tag{21}$$

where α is a constant that has to be evaluated based on flow conditions, u_{max} is maximum velocity observed in a velocity profile, δ_* is displacement thickness, and θ is momentum thickness. These thicknesses are defined as (Schlichting and Geresten 2000):

$$\delta_* = \int_0^h \left(1 - \frac{u}{u_{max}}\right)dy , \tag{22}$$

$$\theta = \int_0^h \frac{u}{u_{max}}\left(1 - \frac{u}{u_{max}}\right)dy \tag{23}$$

A large group of bed shear stress evaluation methods takes advantage of the turbulence theory. The first method applies the definition of Reynolds shear stress (Eq. 5). In order to collect input variables, instantaneous velocities are measured near the channel bottom. They are usually measured by ADV (Acoustic Doppler Velocimetry) or ADVP (Acoustic Doppler Velocimetry Profiling) (Nikora and Goring 2000; Nikora and Smart 1997; Wilcock 1996).

Another method uses the kinetic energy (Bagherimiyab and Lemmin 2013; Pope et al. 2006) given by:

$$E_{TK} = 0.5\rho(u'^2 + v'^2 + w'^2) \tag{24}$$

The shear stress is proportional to the turbulent kinetic energy: $\tau = CE_{TK}$, where C is a coefficient usually taken as $C \approx 0.19$ or $C \approx 0.20$ (Pope et al. 2006).

We also have to keep in mind that in favorable conditions it is possible to directly measure the tangential force per unit area (with use of shear plates, or floating-element balances (Gmeiner et al. 2012; Kaczmarek and Ostrowski 1995)) and to indirectly determine the bed shear stress from pressure, heat, and mass-transfer measurements in the near bed region, from which the bed shear stress can be derived based on theoretical or empirical dependencies.

The above short review of methods does not exhaust the subject, and much more methods can be found in the literature. Some of them require advanced mathematical apparatus, pertaining for example to double averaging methodology with extensive roots in multi-phase hydrodynamics and porous media hydrodynamics (Pokrajac et al. 2006). The application of different methods of shear stress evaluation to data from laboratory experiments and measurements in natural settings may be found in the literature (Bagherimiyab and Lemmin 2013; Biron et al. 2004; Pope et al. 2006).

Bed Shear Stress and Shear Velocity Evaluation in Unsteady Flow

The evaluation of bed shear stress and shear velocity in the unsteady flow, which occurs during the flood wave propagation, is somewhat complicated due to the spatial and temporal variability of flow parameters. The majority of methods presented in the previous sections are inapplicable in the unsteady flow due to both technical and theoretical constraints. First of all, it would be difficult to perform detailed measurements of instantaneous velocities in transient flow. Secondly, the theory of boundary layer is applicable only to the steady flow conditions. To tackle this far more complicated problem of unsteady flow, the reader has to cast his/her mind back to the theory of unsteady varied flows which may be caused by the movement of natural flood waves but also the propagation of tides, the operation of the control gates, the failure of a dam, etc.

In unsteady flow, the relation for shear velocity is derived from basic flow equations, for example, Reynolds 2D model or 1D Saint-Venant model (Mrokowska et al. 2015; Rowinski et al. 2000) may be applied. The 1D flow is represented by the Saint-Venant model expressed by Eq. (25) and Eq. (26):

$$U\frac{\partial h}{\partial x} + h\frac{\partial U}{\partial x} + \frac{\partial h}{\partial t} = 0, \tag{25}$$

$$\frac{\partial U}{\partial t} + U\frac{\partial U}{\partial x} + g\frac{\partial h}{\partial x} - gI + \frac{u_*^2}{h} = 0, \tag{26}$$

where t – time [s], U – mean cross-sectional velocity [m s^{-1}], and x – horizontal spatial coordinate [m]. Equation (25) represents the law of flow continuity in a rectangular channel, and Eq. (22) the law of momentum conservation. The terms of Eq. (26) represent as follows: local acceleration, advective acceleration, hydrostatic pressure, bed slope, and friction slope expressed in terms of the shear velocity. The shear velocity derived from above set of equations is given by:

$$(u_*)_{SV} = \left[gR\left(I + \frac{U}{gh}\eta + \left(\frac{U^2}{gh} - 1\right)\vartheta - \frac{1}{g}\zeta \right) \right]^{\frac{1}{2}}, \tag{27}$$

where $\eta = \dfrac{\partial h}{\partial t}$, $\vartheta = \dfrac{\partial h}{\partial x}$, $\zeta = \dfrac{\partial U}{\partial t}$. The examples of evaluation of shear velocity in unsteady flow could be found in Ghimire and Deng (2011), Mrokowska et al. (2015), Mrokowska et al. (2016), Rowiński et al. (2000), and Shen and Diplas (2010).

The case of a small lowland river is presented for illustrative purposes (analyzed widely in Mrokowska et al. 2015). Figure 7 depicts the relation between flow (Q) and water level (H). It takes the form of a hysteresis which is characteristic of the majority of flood waves. Figure 8 presents the temporal variability of the mean velocity (U) and water depth (h) from which the shape of the flood wave with a rising and falling limb may be inferred. Figure 9 shows the results of shear velocity evaluated by Eq. (27) dedicated to unsteady flow and by Eq. (16), the formula for steady flow conditions. Both evaluations are presented with the uncertainty bounds of the results. It is evident from the Fig. 9 that Eq. (16) must not be used in the region of rising limb of a wave, as results lie outside uncertainty bounds of shear velocity evaluated for unsteady flow conditions. Hence, a simplified formula for shear velocity is not recommended in unsteady flow.

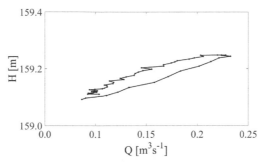

Fig. 7. Relation between flow rate (Q) and water level (H) during flood wave propagation.

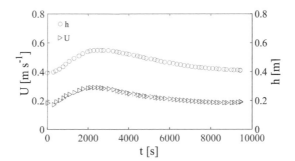

Fig. 8. Mean flow velocity (U) and water depth (h) during flood wave propagation.

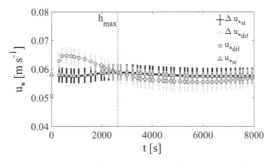

Fig. 9. Shear velocity for flood conditions in a small lowland river. Results of shear velocity evaluation by Eq. (27) and formula for steady flow Eq. (16) with uncertainty bounds.

The obtained uncertainty bound requires some commentary. Shear velocity evaluated from Eq. (23) depends on seven input variables, namely, mean flow velocity U, water depth h, bottom slope I, hydraulic radius R, derivatives: $\dfrac{\partial u}{\partial t}$, $\dfrac{\partial h}{\partial t}$, $\dfrac{\partial h}{\partial x}$. The reasonable method to evaluate the uncertainty of this kind of dependent variable is the law of propagation of uncertainty based on the total differential (Mrokowska et al. 2013; Fornasini 2008). This method is appropriate when input data originates from measurements in natural conditions which cannot be repeated. To evaluate the uncertainty of shear velocity, one needs to assess uncertainties of input variables, that is, ΔI, ΔR, Δu, Δh, $\Delta \frac{\partial u}{\partial t}$, $\Delta \frac{\partial h}{\partial t}$, and $\Delta \frac{\partial h}{\partial x}$. They are assessed based on knowledge of measurement techniques and experimental settings. Knowing the maximum uncertainties of input variables, the maximum deterministic uncertainty of shear velocity is calculated as follows:

$$\Delta(u_*)_{SV} = \left|\frac{\partial u_*}{\partial I}\right|\Delta I + \left|\frac{\partial u_*}{\partial R}\right|\Delta R + \left|\frac{\partial u_*}{\partial u}\right|\Delta U + \left|\frac{\partial u_*}{\partial h}\right|\Delta h + \left|\frac{\partial u_*}{\partial \vartheta}\right|\Delta \vartheta + \left|\frac{\partial u_*}{\partial \eta}\right|\Delta \eta + \left|\frac{\partial u_*}{\partial \varsigma}\right|\Delta \varsigma \quad (28)$$

The impact of each input variable on the shear velocity may be studied thanks to the sensitivity analysis (Gong et al. 2006; Hupet and Vanclooster 2001). The sensitivity of the shear velocity on an input variable may be evaluated by a relative sensitivity coefficient defined as:

$$WW = \left|\frac{\partial u_*}{\partial a_i} \frac{a_i}{u_*}\right|. \quad (29)$$

where a_i – i-th input variable. The sensitivity coefficient indicates to what extent the uncertainty of an input variable affects the uncertainty of shear velocity. Thanks to this coefficient one may identify which variables are most significant for the evaluation of shear velocity in a studied case and decide which should be measured more precisely in future studies.

Figure 10 presents the results of sensitivity analysis for shear velocity evaluated for the sample data. In the figure, only the variables of significant sensitivity coefficient values are presented: I, R, $\frac{\partial h}{\partial x}$; for the others WW is negligible. The results indicate that these three variables are the most significant, and special attention has to be paid during their measurement and evaluation. Bed slope I needs only measurements of channel geometry, hydraulic radius R – channel geometry and water depth, while spatial derivative of water depth requires simultaneous measurements of water depth in at least two cross-sections according to the definition of derivative and approximation methods (e.g., difference quotients).

The evaluation of the spatial derivative is one of the most problematic issue in hydrology and hydrodynamics, as simultaneous measurements in different locations are very often not feasible, especially in considered case of flood. Although simplified methods have been devised to evaluate the derivative based on measurements of water depth in one cross-section (kinematic wave approximation) (Song and Graf 1996; Ghimire and Deng 2011), they are not universal, and it is recommended to perform measurements in at least three cross-sections to obtain reliable results (Mrokowska et al. 2015).

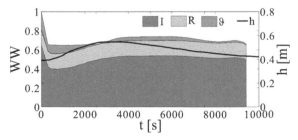

Fig. 10. Sensitivity analysis for shear velocity.

As presented above, the evaluation of bed shear stress and shear velocity in unsteady flow is much more complicated during flood wave propagation than in steady flow conditions due to unsteadiness and non-uniformity of flow. There are ongoing research on this subject, especially in the context of sediment transport processes (Ghimire and Deng 2013; Guney et al. 2013; Mrokowska et al. 2016; Tabarestani and Zarrati 2015).

Conclusions

In this chapter, a short introduction of the crucial notions of the bed shear velocity and bed shear stress in open channels has been provided. The main aim was to explain the physical interpretation of both values and to depict how problematic is the evaluation of them in natural water courses. There are no universal methods that can be applied in every situation and each evaluation method is burdened with potential large uncertainties. In-depth understanding of bed shear stresses requires advanced methods of recognition of river turbulence and therefore further reading is advised for those who want to apply those methods, especially when they aim to incorporate the bed shear velocity to other problems of hydraulics, sediment transport, or eco-hydraulics.

Keywords: Boundary layer, velocity distribution, shear stress, bed load transport, unsteady flow.

Acknowledgements

This work was supported by the Institute of Geophysics, Polish academy of Sciences within the statutory activities No. 3841/E-41/S/2016.

References

Afzalimehr, H. and Anctil, F. 2000. Accelerating shear velocity in gravel bed channels. Hydrological Sciences Journal 45(1): 113–124 doi:10.1080/02626660009492309.

Bagherimiyab, F. and Lemmin, U. 2013. Shear velocity estimates in rough-bed open-channel flow. Earth Surface Processes and Landforms 38: 1714–1724 doi:10.1002/esp.3421.

Biron, P.M., Robson, C., Lapointe, M.F. and Gaskin, S.J. 2004. Comparing different methods of bed shear stress estimates in simple and complex flow fields. Earth Surface Processes and Landforms 29: 1403–1415 doi:10.1002/esp.1111.

Buxton, T.H., Buffington, J.M., Yager, E.M., Hassan, M.A. and Fremier, A.K. 2015. The relative stability of salmon redds and unspawned streambeds. Water Resources Research 51: 6074–6092 doi:10.1002/2015wr016908.

Campbell, L., McEwan, I., Nikora, V., Pokrajac, D., Gallagher, M. and Manes, C. 2005. Bed-load effects on hydrodynamics of rough-bed open-channel flows. Journal of Hydraulic Engineering-ASCE 131: 576–585 doi:10.1061/(asce)0733-9429(2005)131:7(576).

Cooper, J.R. and Tait, S.J. 2010. Examining the physical components of boundary shear stress for water-worked gravel deposits. Earth Surface Processes and Landforms 35: 1240–1246 doi:10.1002/esp.2020.

De Sutter, R., Verhoeven, R. and Krein, A. 2001. Simulation of sediment transport during flood events: laboratory work and field experiments. Hydrological Sciences Journal-Journal Des Sciences Hydrologiques 46: 599–610 doi:10.1080/02626660109492853.

Dey, S., Sarkar, S. and Solari, L. 2011. Near-bed turbulence characteristics at the entrainment threshold of sediment beds. Journal of Hydraulic Engineering-ASCE 137: 945–958 doi:10.1061/(asce)hy.1943-7900.0000396.

Dey, S. 2014. Fluvial Hydrodynamics: Hydrodynamic and Sediment Transport Phenomena. Springer-Verlag Berlin Heidelberg, pp. 1–687.

Fornasini, P. 2008. The Uncertainty in Physical Measurements. An Introduction to Data Analysis in the Physics Laboratory. Springer, New York, USA.

Garcia, M. 2007. Sedimentation Engineering: Processes, Measurements, Modeling and Practice. American Society of Civil Engineers, Virginia, USA.

Ghimire, B. and Deng, Z.-Q. 2011. Event flow hydrograph-based method for shear velocity estimation. Journal of Hydraulic Research 49: 272–275 doi:10.1080/00221686.2011.552463.

Ghimire, B. and Deng, Z. 2013. Event flow hydrograph-based method for modeling sediment transport. Journal of Hydrologic Engineering 18: 919–928, doi:10.1061/(ASCE)HE.1943-5584.0000710.

Gmeiner, P., Liedermann, M., Tritthart, M. and Habersack, H. 2012. Development and testing of a device for direct bed shear stress measurement. *In*: Water – Infinitely Deformable but Still Limited – Proceedings of the 2nd IAHR Europe Congress, Technische Universität München, München, Germany, D23, 27–29 June 2012.

Gong, L., Xu, C., Chen, D., Halldin, S. and Chen, Y. 2006. Sensitivity of the Penman-Monteith reference evapotranspiration to key climatic variables in the Changjiang (Yangtze River) basin. Journal of Hydrology 329: 620–629 doi:10.1016/j.jhydrol.2006.03.027.

Graba, M., Sauvage, S., Majdi, N., Mialet, B., Moulin, F.Y., Urrea, G., Buffan-Dubau, E., Tackx M., Sabater, S. and Sanchez-Perez, J.-M. 2014. Modelling epilithic biofilms combining hydrodynamics, invertebrate grazing and algal traits. Freshwater Biology 59: 1213–1228 doi:10.1111/fwb.12341.

Guney, M., Bombar, G. and Aksoy, A. 2013. Experimental study of the coarse surface development effect on the bimodal bed-load transport under unsteady flow conditions. Journal of Hydraulic Engineering - ASCE 139: 12–21, doi:10.1061/(ASCE)HY.1943-7900.0000640.

Hupert, F. and Vanclooster, M. 2001. Effect of the sampling frequency of methodological variables on the estimation of the reference evapotranspiration. Journal of Hydrology 243: 192–204 doi:10.1016/S0022-1694(00)00413-3.

Ikeda, S. 1981. Self forced straight channels in sandy beds. Journal of the Hydraulics Division 107(4): 389–406.

Jain, S.C. 2001. Open-Channel Flow. Wiley & Sons, New York, pp. 328.

Julien, P.Y. 2010. Erosion and Sedimentation, 2nd Edn.. Cambridge University Press, Cambridge, UK

Kaczmarek, L.M. and Ostrowski, R. 1995. Modelling of bed shear stress under irregular waves. Archives of Hydro-Engineering and Environmental Mechanics 42: 29–51.

Kalinowska, M.B. and Rowiński, P.M. 2012. Uncertainty in computations of the spread of warm water in a river – lessons from Environmental Impact Assessment case study. Hydrology and Earth System Sciences 16: 4177–4190 doi:10.5194/hess-16-4177-2012.

Kalinowska, M.B., Rowinski, P.M., Kubrak, J. and Miroslaw-Swiatek, D. 2012. Scenarios of the spread of a waste heat discharge in a river—Vistula River case study. Acta Geophysica 60: 214–231 doi:10.2478/s11600-011-0045-x.

Knight, D.W., Mc Gahey, C., Lamb, R. and Samuels, P.G. 2010. Practical Channel Hydraulics: Roughness, Conveyance and Afflux. CRC Press/Balkema, London, UK, p. 354.

MacVicar, B.J. and Roy, A.G. 2007. Hydrodynamics of a forced riffle pool in a gravel bed river: 1. Mean velocity and turbulence intensity. Water Resources Research 43 doi:10.1029/2006wr005272.

Meyer, Z. 2011. Wind set-up of water level in a river. Acta Geophysica 59(2): 317–333, doi:10.2478/s11600-011-0005-5.

Mrokowska, M.M., Rowiński, P.M. and Kalinowska, M. 2013. The uncertainty of measurements in river hydraulics – evaluation of friction velocity based on an unrepeatable experiment. pp. 195–206. *In*: Rowiński, P.M. (ed.). Experimental and Computational Solutions of Hydraulic Problems: 32nd International School of Hydraulics. GeoPlanet: Earth and Planetary Sciences. Springer-Verlag, Berlin, Heidelberg, Germany, doi:0.1007/978-3-642-30209-1_13.

Mrokowska, M.M., Rowiński, P.M. and Kalinowska, M.B. 2015. A methodological approach of estimating resistance to flow under unsteady flow conditions. Hydrology and Earth System Sciences 19: 4041–4053 doi:10.5194/hess-19-4041-2015.

Mrokowska, M.M., Rowiński, P.M., Książek, L., Strużyński, A., Wyrębek, M. and Radecki-Pawlik, A. 2016. Flume experiments on gravel bed load transport in unsteady flow – preliminary results. pp. 221–233. *In*: Rowiński, P.M. and Marion, A. (eds.). Hydrodynamic and Mass Transport at Freshwater Aquatic Interfaces. Geoplanet: Earth and Planetary Sciences. Springer International Publishing, Switzerland, doi:10.1007/978-319-27750-9_18.

Nezu, I. and Nakagawa, H. 1993. Turbulence in Open-Channel Flows. IAHR Monograph, Balkema, Rotterdam.

Nikora, V.I. and Smart, G.M. 1997. Turbulence characteristics of New Zealand gravel-bed rivers. Journal of Hydraulic Engineering-ASCE 123: 764–773 doi:10.1061/(asce)0733-9429(1997)123:9(764).

Nikora, V. and Goring, D. 2000. Flow turbulence over fixed and weakly mobile gravel beds. Journal of Hydraulic Engineering-ASCE 126: 679–690 doi:10.1061/(asce)0733-9429(2000)126:9(679).

Ockelford, A.-M. and Haynes, H. 2013. The impact of stress history on bed structure. Earth Surface Processes and Landforms 38: 717–727 doi:10.1002/esp.3348.

Petit, F., Houbrechts, G., Peeters, A., Hallot, E., Van Campenhout, J. and Denis, A.-C. 2015. Dimensionless critical shear stress in gravel-bed rivers. Geomorphology 250: 308–320 doi:10.1016/j.geomorph.2015.09.008.

Pokrajac, D., Finnigan, J.J., Manes, C., McEwan, I. and Nikora, V. 2006. On the definition of the shear velocity in rough bed open channel flows. In: Ferreira, R., Alves, E., Leal, J. and Cardoso, A. (eds.). Proc. of the International Conference on Fluvial Hydraulics, Lisbon, Portugal, 6–8 September 2006, River Flow 1: 89–98.

Pope, N.D., Widdows, J. and Brinsley, M.D. 2006. Estimation of bed shear stress using the turbulent kinetic energy approach—a comparison of annular flume and field data. Continental Shelf Research 26: 959–970 doi:10.1016/j.csr.2006.02.010.

Powell, D.M. 2014. Flow resistance in gravel-bed rivers: progress in research. Earth-Science Reviews 136: 301–338 doi:10.1016/j.earscirev.2014.06.001.

Radice, A., Giorgetti, E., Brambilla, D., Longoni, L. and Papini, M. 2012. On integrated sediment transport modelling for flash events in mountain environments. Acta Geophysica 60: 191–213 doi:10.2478/s11600-011-0063-8.

Rowiński, P.M., Czernuszenko, W. and Pretre, J.M. 2000. Time-dependent shear velocities in channel routing. Hydrological Sciences Journal-Journal Des Sciences Hydrologiques 45: 881–895 doi:10.1080/02626660009492390.

Rowiński, P.M., Czernuszenko, W., Kozioł, A.P. and Kubrak, J. 2002. Properties of a streamwise turbulent flow field in an open two-stage channel. Archives of Hydro-Engineering and Environmental Mechanics 49(2): 37–57.

Rowiński, P.M., Aberle, J. and Mazurczyk, A. 2005. Shear velocity estimation in hydraulic research. Acta Geophysica Polonica 53(4): 567–583.

Schlichting, H. and Geresten, K. 2000. Boundary-Layer Theory. Springer Verlag, Berlin, Heidelberg.

Shen, Y. and Diplas, P. 2010. Modeling unsteady flow characteristics of hydropeaking operations and their implications on fish habitat. Journal of Hydraulic Engineering-ASCE 136: 1053–1066 doi:10.1061/(asce)hy.1943-7900.0000112.

Silver, N. 2012. The Signal and the Noise: Why So Many Predictions Fail—but Some Don't. The Penguin Press, US.

Smart, G.M. 1999. Turbulent velocity profiles and boundary shear in gravel bed rivers. Journal of Hydraulic Engineering-ASCE 125: 106–116 doi:10.1061/(asce)0733-9429(1999)125:2(106).

Song, T. and Graf, W. 1996. Velocity and turbulence distribution in unsteady open-channel flows. Journal of Hydraulic Engineering 122(3): 141–154.

Tabarestani, M.K. and Zarrati, A.R. 2015. Sediment transport during flood event: a review. International Journal of Environmental Science and Technology 12: 775–788 doi:10.1007/s13762-014-0689-6.

van Rijn, L. 1993. Principles of Sediment Transport in Rivers, Estuaries and Costal Seas. Aqua Publications, Amsterdam, The Netherlands.

Westenbroek, S.M. 2006. Estimates of shear stress and measurements of water levels in the Lower Fox River near Green Bay, Wisconsin, Scientific Investigations Report 2006–5226, USGS.

Wilcock, P.R. 1996. Estimating local bed shear stress from velocity observations. Water Resources Research 32: 3361–3366 doi:10.1029/96wr02277.

Yang, C.T. and Wan, S. 1991. Comparisons of selected bed-material load formulas. Journal of Hydraulic Engineering 117(8): 973–989 doi:10.1061/(ASCE)0733-9429(1991)117:8(973).

Yang, J.Q., Kerger, F. and Nepf, H.M. 2015. Estimation of the bed shear stress in vegetated and bare channels with smooth beds. Water Resources Research 51: 3647–3663 doi:10.1002/2014wr016042.

Yen, B.C. 2002. Open channel flow resistance. Journal of Hydraulic Engineering-ASCE 128: 20–39 doi:10.1061/(asce)0733-9429(2002)128:1(20).

CHAPTER 10

Gene Expression Programming in Open Channel Hydraulics

Ahmed M. Abdel Sattar[1,]* and *Bahram Gharabaghi*[2]

INTRODUCTION

While the complex relationship between variables can be represented by regression analysis (linear and non-linear), machine learning methods have proven to be more efficient in representing these complex relations. Among the machine learning methods, Gene Expression Programming (GEP) has recently proven to be efficient in many practical engineering hydrology and hydraulic applications, for example, Sattar (2014a,b), El Hakeem and Sattar (2015), Najafzadeh and Sattar (2015), Sattar and Gharabaghi (2015), Sabouri et al. (2016), Sattar et al. (2016), Thompson et al. (2016), Atieh et al. (2017), and Sattar and Beltagi (2017). The GEP was introduced in 1999 by mathematician Ferreira as an alternative that surpasses the genetic algorithms and genetic programming (Ferreira 2001) where simple linear elements encode complex non-linear relations between variables. This chapter introduces the GEP and shows the basics of its evolution principles and then introduces the successful applications of GEP on real river processes.

Gene Expression Programming: The Karva Language

Relationship between variables is expressed in form of expression tree (ET) as developed by Ferreira (2001). The ET represents the gene via simple linear rules, which are the basics for the unequivocal Karva language. This language encodes the gene as a sequence of codons: it starts with a codon, continues with amino codons, and ends with a termination one. The following example has been given by (Sattar 2014a) for a simple algebraic expression:

$$\frac{a \times b}{c} \tag{1}$$

---

[1] Irrigation & Hydraulics Department, Faculty of Engineering, Cairo University, Orman, 12316 Giza, Egypt (On leave to the German University in Cairo).
[2] Associate Professor, School of Engineering, University of Guelph, Guelph, Ontario, NIG 2W1, Canada.
* Corresponding author: ahmoudy77@yahoo.com

where a, b, and c are the set of the variables used in the problem; and $\times$ is the rule that determine the spatial organization of the terminals/variables. Using the ET representation, this algebraic expression can be expressed as shown in Fig. 1.

Using the Karva language, the algebraic expression in Eq. (1) is as follows:

$$
\begin{array}{ccccc}
0 & 1 & 2 & 3 & 4 \\
\div & \times & a & b & c
\end{array}
\tag{2}
$$

In this Karva expression, the algebraic expression is read from left to right and from top downward with the numbers representing the position of each element in the gene. The gene starts at position zero with the '$\div$' operator and ends at position four with the c variable. In the Karva language, the gene always consists of a head and a tail. The gene head contains the functions and variables as shown in Eq. (2), while the tail of the gene contains 'junk sequence' of variables that are extremely important for the evolution of the GEP. This gene tail allows the gene to evolve with any possible combinations for next generations with allowable genetic operators. Operators in the Karva language can range from basic simple operators ($+$, $-$, x, $/$), and complex trigonometric operators (e.g., sinh, tanh) to logical operators (e.g., more than, less than). The length of this junk sequence depends on the modeled variables and the gene head (variables + operators) such that the tail size = chosen head x (number of variables -1) $+1$. Applying this equation on the above algebraic expression, the head size is 5, number of variables is 3, therefore the tail consists of 11 junk sequences and the total length of the gene is the summation of both yields 16 elements. The gene is thus written as (with junk sequence starting from position 5 to 16):

$$
\begin{array}{cccccccccccccccc}
0 & 1 & 2 & 3 & 4 & 5 & 6 & 7 & 8 & 9 & 0 & 1 & 2 & 3 & 4 & 5 \\
\div & \times & a & b & c & c & b & c & b & b & a & a & c & b & a & a
\end{array}
\tag{3}
$$

This is a chromosome with one gene in the GEP Karva language. However, one gene chromosome can lead to poor evolution and thus the GEP depends on multi-genic chromosomes where evolution occurs as different genes share the inherited characteristics in each generation. A two gene chromosome can have the following form:

Gene 0:

$$
\begin{array}{cccccccccccccccc}
0 & 1 & 2 & 3 & 4 & 5 & 6 & 7 & 8 & 9 & 0 & 1 & 2 & 3 & 4 & 5 \\
\div & \times & a & b & c & c & b & c & b & b & a & a & c & b & a & a
\end{array}
$$

Gene 1: $\hspace{8cm}$ (4)

$$
\begin{array}{cccccccccccccccc}
0 & 1 & 2 & 3 & 4 & 5 & 6 & 7 & 8 & 9 & 0 & 1 & 2 & 3 & 4 & 5 \\
\div & \times & a & b & c & c & b & c & c & c & a & b & b & a & a & a
\end{array}
$$

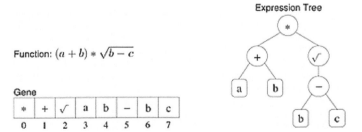

Fig. 1. Expression tree of a gene.

These two genes can evolve indecently and with one another to produce different evolved expression. This increases the effectiveness of the GEP compared to other available methods. However, it is to be noted that it largely depends on the problem under study and the increase in number of genes is not necessarily the best solution for all cases. The operators defined before are the set of functions that are needed to determine the relation between genes in the same chromosome and between different genes in a multi-genic chromosome. These set of functions are chosen according to the problem under study depending on the achieved performance. For certain type of problems, choosing the basic arithmetic functions seem to produce the best results, while in others, complex trigonometric functions are needed for better representation of the problem under study. However, simple models are always the best target equations where the relationship between various parameters is clear and logical.

The GEP needs the initial setup of the problem in terms of: number of variables, number of genes, and head and tail for each, gene after which the GEP forms a random distribution for the functions and variables in genes and creates the first expression. The first expression is called the 'parent' and afterwards, the evolutionary strategy yields 'offspring' from the parent. Evolution continues to produce offspring for various generations until one offspring is fit enough to be within a certain error of the correct value of the function. The fitness function specified can be any error measure, for example, the fitness function based upon the root relative squared error (RRSE) is as follows:

$$f_i = 1000 \cdot \frac{1}{1 + RRSE_i} \tag{5}$$

where $f_i =$ is the i-th offspring fitness, which ranges from 0 to 1000, with 1000 corresponding to a perfect fit. The root relative square error RRSE of the i-th offspring is defined by the following equation:

$$RRSE_i = \sqrt{\frac{\sum_{j=1}^{n}(P_{(ij)} - T_j)^2}{\sum_{j=1}^{n}(T_j - \overline{T})^2}} \tag{6}$$

where $P_{(ij)}$ is the value predicted by the program i for fitness case j, T_j is the target value for fitness case j, $\overline{T} = 1/n\sum_{j=1}^{n} T_j$, and n is the number of samples. For a perfect fit, $RRSE_i = 0$ thus, the RRSE ranges from 0 to infinity, with zero corresponding to the ideal.

Genetic mutations

Gene mutation is the strength of the GEP allowing genes and thus expressions to continuously evolve for better ones until the best and most fit offspring expression is reached. The mutation can occur in any part of the gene head and tail. Ferreira (2001) recommended using a mutation rate equivalent to two one-point mutations per chromosome (see Fig. 2). The mutation can change the elements in the previous multi-genic chromosome by changing the positions from gene to the other. Thus, the expression in Eq. (4) becomes:

Gene 0:

```
0  1  2  3  4  5  6  7  8  9  0  1  2  3  4  5
×  ÷  a  b  c  c  b  c  b  b  b  a  a  c  b  a  a
```

Gene 1: (7)

```
0 1 2 3 4 5 6 7 8 9 0 1 2 3 4 5
÷ × a a c c b c c c a b b a a a
```

Another important process involved in the GEP is the transposition. In this process, the GEP selects a sequence in a random manner and places it in any position in head or tail of the other gene. For the above example, the GEP randomly chooses to insert the sequence '*ab*' from gene 0 into position 3 of the first gene. The following expression is obtained:

Gene 0:

```
0 1 2 3 4 5 6 7 8 9 0 1 2 3
× ÷ c c b c b b a a c b a a
```

Gene 1: (8)

```
0 1 2 3 4 5 6 7 8 9 0 1 2 3 4 5 6 7
÷ × a a a b c c b c c c a b b a a a
```

Ferreira (2001) explained other genetic operations that play important roles in the evolution strategy of the GEP: insertion, IS element transposition, root transposition, one-point recombination, two-point recombination, and gene recombination. They

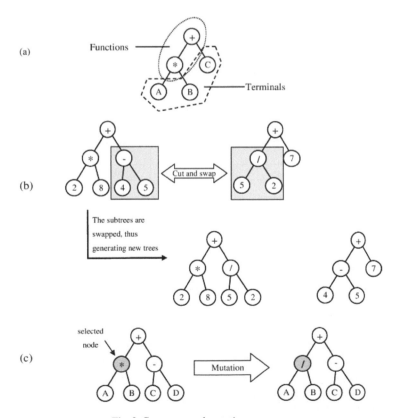

Fig. 2. Cross over and mutation processes.

can be assigned separately or based on the assigned value for the mutation rate; more details are given in Ferreira (2001). Therefore, the GEP needs the user choice of various parameters, which if correctly specified can lead to faster convergence and evolution to better expressions. These control parameters are shown in Table 1 with a recommended range from the author's experience that has proven to be the best for model evolution.

Many researchers have developed their in-house codes for GEP, for example, Radi and Poli (2002), or one can use the commercial nonlinear data-mining software GenXProTools (www.gepsoft.com).

Table 1. Optimum parameter settings for GEP models.

Parameter	Setting	Comments
Number of generations	5,00 to 20,000	Normally, generation number is set till convergence
Number of chromosomes	30 to 50	A 40 chromosomes is enough for all problems
Number of genes	2 to 4	2 is optimum, 3 and 4 are only used with small head size
Head size	3 to 5	Small head size for big number of genes and vice versa
Linking function	x, /,+, −	Multiplication is optimum
Fitness function error type	RRSE	Best fitness function for accurate models
Mutation rate	0.005–0.05	A value of 0.005 works fine with most models
Inversion rate	0.1	
One point recombination rate	0.3,0.5	
Two point recombination rate	0.3,0.5	
Gene recombination rate	0.1	
Gene transposition rate	0.1	
Function set	x, /, power	
Random numerical constants (RNC)	1,2,3,4	Four random constants are the best option
RNC mutation	0.005–0.05	Taken equal to mutation rate

Analysis procedure for GEP model development

For developing any GEP model, the following steps describe the process (more details are found in Fig. 3):

1. Choose a random set of control variables as terminals.
2. Define the chromosome architecture (number of genes, head size, functions) and mutation rates.
3. The GEP randomly arranges the variables and functions to form the 'parent' expression.
4. The GEP applies the genetic and other processes to produce mutant 'offspring'.
5. Using the specified fitness criteria, the GEP tests the offspring to see the score and offspring with high scores are kept for next mutation.
6. The GEP implements the genetic operators on the kept offspring to produce more second-generation offspring.

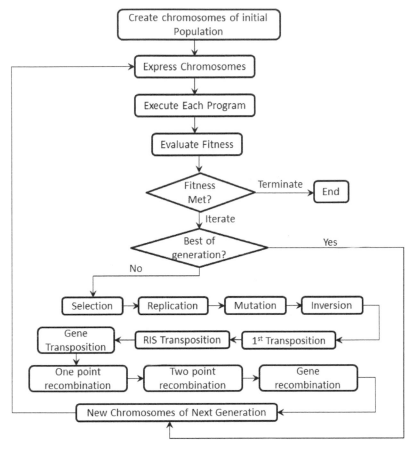

Fig. 3. The GEP flow chart.

7. The GEP evolution continues per steps 3, 4, and 5 till either the specified fitness is reached, or the specified number of generations is reached.

8. Following the choice of the GEP model, further testing is performed using a set of indicators suggested by Najafzadeh and Sattar (2015): the square of the Pearson product moment correlation coefficient (R^2), the relative absolute error (RAE), Coefficient of efficiency (E_{sn}), and Index of agreement (D).

The indicators are calculated by the following equations:

$$R_t^2 = \left(\frac{\frac{1}{n}\Sigma_{j=1}^n (T_j - \bar{T})(P_{(ij)} - \bar{P})}{\sqrt{\Sigma_{j=1}^n (T_j - \bar{T})^2/n}\sqrt{\Sigma_{j=1}^n (P_{(ij)} - \bar{P})^2/n}} \right)^2 \tag{9}$$

$$RAE_i = \frac{\Sigma_{j=1}^n |P_{(ij)} - T_j|}{\Sigma_{j=1}^n |T_j - \bar{T}|} \tag{10}$$

$$E_{sn} = 1 - \frac{\sum_{i=1}^{n}(T_i - P_i)^2}{\sum_{i=1}^{n}(T_i - \bar{T})^2} \tag{11}$$

$$D = 1 - \frac{\sum_{i=1}^{n}(T_i - P_i)^2}{\sum_{i=1}^{n}(|P_i - \bar{T}| + |T_i - \bar{T}|)^2} \tag{12}$$

where $\bar{P} = 1/n \sum_{i=1}^{n} P_j$.

Validation of Developed GEP Models

Further validation measures for developed models as developed by Tropsha et al. (2003) are adopted in Sattar and Gharabaghi (2015). These measures require that the developed model performs well is one or more of them at least. These measures are as follow:

- The gradient of the regression line between the predicted versus the observed value should be close to one

$$k = \sum_{i=1}^{n}(T_i \times P_i)/P_i^2 \ or \ k' = \sum_{i=1}^{n}(T_i \times P_i)/T_i^2 \tag{13}$$

- The coefficient of determination for the regression line should be less than 0.1

$$m = (R^2 - R_o^2)/R^2 \ \text{and} \ n = (R^2 - R_o'^2)/R^2 \tag{14}$$

- The cross validation should be more than 0.5

$$R_m = R^2 \times (1 - \sqrt{|R^2 - R_o^2|}) > 0.5 \tag{15}$$

- where R_o^2 and $R_o'^2$ are calculated as:

$$R_o^2 = 1 - \sum_{i=1}^{n} P_i^2(1-k)^2/\sum_{i=1}^{n}(P_i - \bar{P})^2 \ \& \tag{16}$$

$$R_o'^2 = 1 - \sum_{i=1}^{n} T_i^2(1-k')^2/\sum_{i=1}^{n}(T_i - \bar{T})^2$$

Uncertainty Analysis of GEP Models

In most studies, the GEP models are developed using available experimental data and there is always a degree of uncertainty of prediction outcomes; therefore it would be useful assess the quantitative effect of the stochastic nature of the GEP models. The deterministic GEP model is converted to a stochastic model using the Monte Carlo Simulation (MCS) method. The MCS is a numerical technique developed back in 1940 for simulating probabilistic events (Frey 2001). This technique is easy, robust, and can be applied in a straight-forward way for obtaining the output distribution based on joint input distributions (Vose 1996; Sattar et al. 2009; Sattar 2016b). Assuming that the input variables (uncertain model parameters) are uncorrelated and independent, samples can be drawn randomly from their probability distributions without affecting the output distributions (Burmaster 1994).

MSC procedure is applied as follows (Sattar 2016a):

1. Fit the input parameters to probability density functions (PDF).
2. Random parameters are selected from the PDF of the input variables and fed to the developed GEP model to solve for one deterministic output.

3. The MCS is initiated by selecting one random sample from the PDF of each parameter and input variable and the combination of parameters is entered into the GEP models to solve for deterministic output that is stored.
4. The previous step is repeated for many Monte Carlo (MC) realizations, thus we now have MC outputs.
5. The outputs are fitted to a distribution function from which properties can be determined.
6. The Mean Absolute Deviation (MAD) is calculated around the median of the output distribution as:

$$\text{MAD} = \frac{1}{250000} \sum_{i=1}^{250000} |P_i - Median(P)| \tag{17}$$

7. Once the MAD is calculated, the uncertainty of the model output can be given as;

$$Uncertainty \% = \frac{100 \times MAD}{Median(P)} \tag{18}$$

Verbeeck et al. (2006) reported accepted uncertainties in the range of 25–30% for good reliable prediction models. The MCS can be used together with multiple linear regression to determine the importance of various random parameters. This is known as the error budget and performed using the least square linearization (LSL) method (Sattar 2014b). The error budget can partition the contribution of the model input parameters in terms of output variance and rank these parameters according to their influence on the model output uncertainty. Considering the output variable y function of a number of independent variables, $v_1, v_2,..., v_n$. For small variations of these variables, the variation in output shall be calculated as:

$$\Delta y = \frac{\partial y}{\partial v_1} \Delta v_1 + \frac{\partial y}{\partial v_2} \Delta v_2 + \dots \dots \dots + \frac{\partial y}{\partial v_n} \Delta v_n \tag{19}$$

where Δv_i is the difference between v_i, the random chosen sample of parameter i and $\bar{v}_i$, the mean value of parameter i of all the random samples.

$$\Delta v_i = v_i - \bar{v}_i \tag{20}$$

When m Monte Carlo simulations are carried out, Δv_i for each parameter and the model output y are calculated for each simulation. Next, a multi-linear regression on the obtained dataset is performed. The Δv_i values are considered as independent variables and the output y is the dependent variable. This gives the following regression equation:

$$y \approx b + w_1 v_1 + w_2 v_2 + \dots \dots \dots + w_n v_n \tag{21}$$

The regression coefficients (w_i) are estimated by minimizing the sum of squared errors. Comparing this with Equation 3.38., it can been seen that the coefficients w_1, w_2,..., w_n are estimates of the partial derivatives of y with respect to v_i and b is an estimate of the value of y at default parameter values (i.e., when $\Delta v_i = 0$ for all i). If the uncertainties of the independent parameters are statistically independent, the overall variance of the model output Var(Δy) can be calculated as:

$$Var(\Delta y) \approx \sum_{i=1}^{n} w_i^2 . Var(\Delta v_i) \tag{22}$$

where: $Var(\Delta v_i)$ is the variance of the calculated difference Δv_i. Based on the regression coefficients and the variations of the parameter uncertainties, the sensitivity coefficient of each parameter i (S_{Vi}) can be approximated as:

$$Sv_i \frac{w_i^2 . Var(\Delta v_i)}{Var(\Delta y)} * 100\% \tag{23}$$

GEP and Dam Breaching

One of the most important tasks in dam failure analysis is the prediction of the breach outflow hydrograph, which is then routed through the downstream valley. The dam breach is an embankment failure due to erosion or structural reasons causing the release of flood water in an uncontrolled manner. This failure has the shape of a hole in the body of the dam embankment close to a trapezium. Sattar (2014b) utilized a database of more than 140 dam breach cases to develop a GEP model for the main breach key parameters: breach flow, time to failure, and average breach width. The following models (Sattar 2014b) were reported to have the highest fitness (more than 750) and lowest root square error of 0.1. The GEP model for the peak flow can be written as:

$$Q_p^* = \frac{C_1}{RS} c_2^{H_d^*} + \frac{C_3}{RSH_d^*} + \frac{C_4}{RS} c_5^{H_d^*} \tag{24}$$

where $c_1 = -0.15$, $c_2 = 0.13$, $c_3 = 0.069$, $c_4 = -0.16$, and $c_5 = 0.12$. While the GEP model for the average width can be written as:

$$B_{avg}^* = c_1 D_e D_f^4 RS^{c_2} \tag{25}$$

where $c_1 = 0.096$ and $c_2 = 0.85$. On the other hand, the GEP model for the failure time is written as:

$$t_f^* = sin(D_e^{c_1 RS}) + RS^{\frac{1}{2}} - D_e^{c_2} + sin^2(D_e^{RS}) \tag{26}$$

where $c_1 = 0.55$ and $c_2 = 0.48$.

The parameters in the above equations are defined as: $Q_p^* = Q_p/\sqrt{gV_w^{5/3}}$, $B_{avg}^* = B_{avg}/h_b$, $t_f^* = t_f/t_r$, $h_b^* = h_b/H_d$, $RS = V_w^{1/3}/H_w$, and $H_d^* = H_d/H_r$. H_r is set to 15 m and t_r is set to 1 hour. For D_t, 4 is given for HD, 3 for DC, 2 for ZD, and 1 for FD. For D_e, 3 is assigned for HE, 2 for ME and 1 for LE. And for D_f, 1.1 denotes piping failure, and 1.2 overtopping failure. And H_d = dam height, V_w = volume of water above breach invert, H_w = water depth above breach invert, h_b = breach height, B_{avg} = average width, D_f = failure mode, Q_p = peak outflow rate, t_f = failure time, D_t = dam type, and D_e = erodibility.

The developed GEP models (Sattar 2014b) predictions agreed with the physical behavior of the dam breach problem. Figure 4 shows the GEP model predictions versus some of the available prediction models. In general, the GEP model predictions agree well with those of Froehlich (1995a and 1995b). The breach peak flow shows an increase with the increase in H_w and V_w, which is due to the resulting high velocity through the breach that leads to widening the opening area Chinnarasri et al. (2004). And as the H_w, V_w decrease, the water depth above the breach is low leading to smaller velocities and smaller outflows. In case of average breach width prediction, Sattar

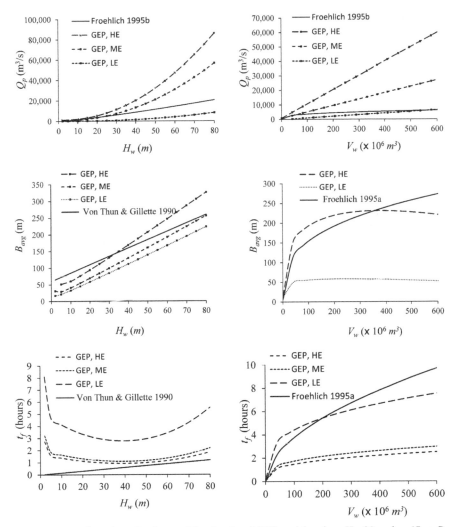

Fig. 4. Parametric analyses for Q_p, t_f and B_{avg} developed GEP models; where H_d= 20 m, h_b = 17 m, D_f = 1.1, and D_t= 4, 2, 1 for Q_p, t_f and B_{avg} respectively.

(2014b) reported discrepancy with Von Thun and Gillette (1990) regression relation and a good agreement with Xu and Zhang (2009) regression relation. For failure time predictions, GEP models deviated from Von Thun and Gillette (1990) relations. This has been attributed to the poor performance of available failure prediction methods with error reaching more than 40 orders of magnitude (Wahl 2004). Moreover, it is observed from the same figure that the embankment erodibility played an important role in peak flow, where it resulted in higher peak flow rates, and shorter failure times. This is due to the fact that the highly erodible embankment forces the breach to be formed due to progressive surface erosion resulting in wider breach opening, and thus, higher flow (Temple et al. 2005).

GEP and Sediment Erosion

In many alluvial rivers, the bed consists of cohesive sediments and especially in some parts of the River Nile in Egypt. These cohesive soils have many chemical characteristics affecting their behavior and interaction with river flow. The erosional stability of these soils have not been studied much and therefore, Sattar (2013) utilized the GEP on a dataset of 60 cohesive soil samples (Fig. 5) erosion to develop a prediction model for the cohesive soil critical shear stress as a function of their chemical composition. The following predictive equation has been developed (Sattar 2013):

$$\tau_{critical} = e^{\{Cl^3(Kt \times Q - Cl)\}^9} - 0.85819 \sqrt{Il} + 2M + 0.795014 + Kt^2 \times Q^2$$
$$- Kt - 137.684Q \times M(Il - Ca) - Ca \qquad (27)$$
$$+ H^3[(Kt - Q)^2 + 64.444 + 16.0557(Kt - Q)]$$

where all minerals are in percentages and $\tau_{critical}$ is in Pascal; Q = Quartz percentage in soil sample, Kt = Kaolinite percentage in soil sample, Cl = Chlorite percentage in soil sample, Il = Ilite percentage in soil sample, M = Montmorillonite percentage in soil sample, H = Halite percentage in soil sample, and Ca = Calcite percentage in soil sample. The developed GEP model was scored in various statistical and error measures as shown in Table 2.

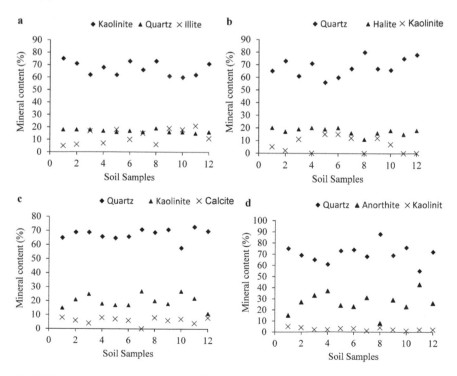

Fig. 5. Selected soil sample mineral compositions at (a) Aswan, (b) Shark El Tafreea, (c) Fayoum, and (d) Kafr ElSheikh, Egypt.

Table 2. Performance of produced functions for critical shear stress.

Performance Indicators	GEPI
Fitness	780
RSE	0.09131
R-Square	0.942

The developed GEP model shows that the critical shear stress for the onset of erosion depended on all mineral compositions of the soil sample with other soil parameters assumed constant (water content and density). No single mineral had a higher influence than other minerals on erosion. Sattar (2013) reported that clay samples from Shark El Tafreea had the least erodibility potential with the highest shear stress measured of 2Pa. This cohesive soil had a higher percentage of Quartz followed by Hallite and Kolinite. The highest erodible cohesive soil was that of Aswan with Kaolinite as a major constituent followed by Quartz and Illite. The developed GEP model can be used to provide estimates for the critical shear stress versus soil type in Egypt Nile River according to soil mineral composition and location of samples.

GEP and Sediment Entrainment

The movement of various bed grain sizes from the river sediment is very important in understanding river bed morphology. The entrainment rates for these various sediment sizes are also required for many ecologic and hydraulic engineering applications. In this regard, El Hakeem and Sattar (2015) developed GEP models for the prediction of the dimensionless rate of entrainment as a function of the dimensionless river bed shear stress. The GEP model describing the dimensionless rate of entrainment E_{Bj} for various grain size fractions d*j within different sediment mixtures is written as (El Hakeem and Sattar 2015):

$$E_{Bj} = \frac{E_j}{f_{Ej}\sqrt{(G_s - 1)gd_j}} = \left(0.024 + 0.021e^{-132(0.245 - c_u)^2}\right)d_{*j}^{-0.5}(\tau'_{*g} - \tau_{*cg})^2 \qquad (28)$$

where f_{Ej} is the j-th grain size fraction entrained from bed, G_s is the specific gravity of the sediment, g is the acceleration due to gravity, d_j is the mean grain size of the j-th fraction in the sediment mixture, c_u is the Kramer coefficient of uniformity defined as $c_u = \sum_0^{50} f_{Oj}d_j / \sum_0^{100} f_{Oj}d_j$, E_{Bj} dimensionless entrainment rate of the j-th fraction in the sediment mixture respectively, τ'_{*r} is the Shields stress of the sediment mixture representative size due to grain resistance only, τ_{*cr} is the critical Shields stress of the mixture representative size, and d_{*j} is the ratio of the mean grain size of the j-th fraction in the sediment mixture to the mixture representative size.

The external validation scoring of the GEP model is shown in Table 3. The developed GEP model statistically performed well in all measures on the testing data set and can be considered a good predictive model.

Table 3. External validation statistical measures for the prediction models.

Model	$(R > 0.8)$	$(0.85 < k < 1.15)$	$(0.85 < k' < 1.15)$	$(m < 0.1)$	$(m' < 0.1)$	$(R_m > 0.5)$
E_{Bj}	0.98	1.10	0.89	−0.02	−0.02	0.82

The developed GEP model was compared by El Hakeem and Sattar (2015) to three available models for entrainment of non-uniform sediment, McLean (1992), Tsujimoto (1999), and Wright and Parker (2004) and predictions are shown in Fig. 6. Close agreements are reported especially with predictions of McLean (1992). However, the agreement between the GEP and other two models predictions was for certain classes of sediment, fine sediment in case of Wright and Parker (2004), and coarse sediment in case of Tsujimoto (1999) model. For sediment mixtures M1 and M2, the GEP model provided better predictions for both coarse and fine sediment as compared to other models and had the advantage of being applicable to wider range of sediments, while the model of Wright and Parker is not appropriate for mixtures with coarse sediments and the model of Tsujimoto is not appropriate for mixtures with fine sediments.

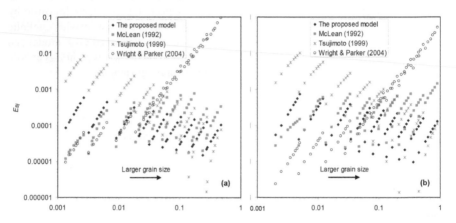

Fig. 6. Comparison between different models: (a) Mixture 1; (b) Mixture 2.

GEP and Scour Depth

In river training, sometimes hydraulic structures are constructed to regulate the river by changing its slope. While these structures are environmental friendly, downstream river bed erosion might be a main cause for failure and structure instability. Therefore, the prediction of the scour depth downstream rapid hydraulic structure is an important research issue. Sattar (2014a) utilized the GEP to develop a scour depth prediction model for mountain streams in Poland. The following GEP models have been presented:

$$h_{max} = 7.89 \frac{h}{s}(Q - d_{90}) + 2.973 \left(Q - \frac{d_{90}}{s}\right) - Q^2(s + 9.488) \tag{29}$$

and:

$$h_{max} = (7.38Q)h \frac{Q^{0.5}}{1.67^Q} h^{(\frac{d90}{2}+0.25)} \tag{30}$$

These GEP models have been compared with many of the existing models (Table 4) showing superiority with respect to representation of experimental results with the highest coefficient of determination and the lowest error. Other existing regression-based formulae showed large errors in predictions such that the R^2 could not be calculated. The error was very high reaching 60.99 for Mason and 27.1 for Veronese. Some formulae gave acceptable—though much less than GEP models—predictions such as those of Van Der Meuler and Yang.

Table 4. Performance of developed equations for scour downstream rapid hydraulic structures.

Scour prediction models	R-Square	RAE	D
GPI	0.938	0.0073	0.000079
GPIV	0.85	0.0112	0.000193
Veronese	0	0.233	27.1
Martin	0	0.1599	13.9
Chian Min Wu	0	0.108	6.3
Mason	0	0.333	60.99
Chividini	0	0.0626	2.66
Eggenberg	0.223	0.0706	3.2
Jaeger	0	0.0335	0.773
Yang	0.43	0.0609	2.35
Mavis	0	0.0246	0.563
Van Der Meuler	0.63	0.0221	0.389

This shows that the GEP model successfully predicted the scour depth formation downstream rapid hydraulic structure and surpassed existing regression-based models as a function of both the flow characteristics over the rapid hydraulic structure and the bed sediment size.

Conclusions

As has been seen, the GEP constitutes a strong alternative for traditional regression-based methods and for many of the machine learning methods, for example, artificial neural networks. They were employed in many problems in hydraulic engineering with great success and were capable of formulating complex relations between non-linear variables. They have the advantage of producing easy clear and simple equations linking variables together and having predictions that follow the physical behavior of the real process. The GEP is still in its infancy and more research is required to enhance its capabilities for handling large amounts of data and still produce simple equations. The issue of the curse of dimensionality still poses a challenge for all equation development models and has to be addressed in data prior to using any machine learning method.

Keywords: Gene Expression programming, evolutionary algorithms, longitudinal dispersion coefficient, sediment entrainment, critical shear stress.

References

Atieh, M., Taylor, G., Sattar, A.M. and Gharabaghi, B. 2017. Prediction of flow duration curves for ungauged basins. Journal of Hydrology. February 2017 545: 383–394.

Burmaster, D.E. 1994. Principles of good practice for the use of Monte Carlo techniques in human health and ecological risk assessments. Risk Anal 14(4): 477–481.

Chinnarasri, C., Jirakitlerd, S. and Wongwises, S. 2004. Embankment dam breach and its outflow characteristics. Civil Engineering and Environmental Systems 21(4): 247–264.

El Hakeem, M. and Sattar, A.M. 2015. An entrainment model for non-uniform sediment. Earth Surface Processes and Landforms. doi:10.1002/esp.3715.

Ferreira, C. 2001. Gene expression programming: a new adaptive algorithm for solving problems. Complex Systems 13(2): 87–129.

Frey, H.C. 2001. Quantification of variability and uncertainty in stationary natural gas-fuelled internal combustion engine NOx and total organic compounds emission factors. Proc Ann Meeting Air and Waste Management Assoc., Abstract No.695, A&AWMA, Pittsburgh, PA.

Froehlich, D.C. 1995a. Peak outflow from breached embankment dam. Journal of Water Resour Plan Manage 121(1): 90–97.

Froehlich, D.C. 1995b. Embankment dam breach parameters revisited. pp. 887–891. *In*: Proc 1995 ASCE Conf. on Water Resources Engineering, New York.

McLean, S.R. 1992. On the calculation of suspended load for non-cohesive sediments. Journal of Geophysical Research 97(C4): 1–14.

Najafzadeh, M. and Sattar, A.M. 2015. Neuro-fuzzy GMDH approach to predict longitudinal dispersion in water networks. Water Resources Management 29: 2205–2219, 10.1007/s11269-015-0936-8.

Radi, A. and Poli, K. 2002. Genetic programming discovers efficient learning rules for the hidden and output layers of feedforward neural networks. pp. 120–134. *In*: Poli, R. et al. (eds.): EuroGP'99. Volume 1598 of the series Lecture Notes in Computer Science. Springer-Verlag Berlin Heidelberg.

Sabouri, F., Gharabaghi, B., Sattar, A.M.A. and Thompson, A.M. 2016. Event-based stormwater management pond runoff temperature model. Journal of Hydrology, 540, 306-316, doi:10.1016/j.jhydrol.2016.06.017.

Sattar, A.M., Dickerson, J. and Chaudhry, M. 2009. A wavelet Galerkin solution to the transient flow equations. Journal of Hydraulic Engineering 135(4): 283–295.

Sattar, A.M. 2013. Using gene expression programming to determine the impact of minerals on erosion resistance of selected cohesive Egyptian soils. pp. 375–387. *In*: Pawel rowinski (ed.). Experimental and Computational Solutions of Hydraulic Problems. GeoPlanet: Earth and Planetary Sciences, Springer-Verlag, Berlin.

Sattar, A.M. 2014a. Gene Expression models for the prediction of longitudinal dispersion coefficients in transitional and turbulent pipe flow. J Pipeline Syst Eng Pract ASCE 5(1): 04013011.

Sattar, A.M. 2014b. Gene expression models for prediction of dam breach parameters. Journal of Hydroinformatics, IWA 16(3): 550–571.

Sattar, A.M. and Gharabaghi, B. 2015. Gene expression models for prediction of longitudinal dispersion coefficient in streams. Journal of Hydrology 524: 587–596.

Sattar, A.M. 2016a. Prediction of organic micropollutant removal in soil aquifer treatment system using GEP. Journal of Hydrologic Engineering. doi:10.1061/(ASCE)HE.1943-5584.0001372 (in press).

Sattar, A.M. 2016b. A probabilistic projection of the transient flow equations with random system parameters and internal boundary conditions. Journal of Hydraulic research. doi:10.1080/00221686.2016.1140682.

Sattar, A.M., Gharabaghi, B. and McBean, E. 2016. Predicting timing of watermain failure using gene expression models for infrastructure planning. Water Resources Management. doi:10.1007/s11269-016-1241-x (in press).

Sattar, A.M. and Beltagy, M. 2017. Stochastic solution to the water hammer equations using polynomial chaos expansion with random boundary and initial conditions. Journal of Hydraulic Engineering 143(2), doi:10.1061/(ASCE)HY.1943-7900.0001227.

Temple, D.M., Hanson, G.J., Nielsen, M.L. and Cook, K.R. 2005. Simplified breach analysis model for homogeneous embankments: Part I, Background and model components. Proc 25th Annual USSD Conference, Salt Lake City, Utah, pp. 151–161.

Thompson, J., Sattar, A.M., Gharabaghi, B. and Richard, W. 2016. Event based total suspended sediment particle size distribution model. Journal of Hydrology 536: 236–246.

Tropsha, A., Gramatica, P. and Gombar, V.K. 2003. The importance of being Earnest: validation is the absolute essential for successful application and interpretation of QSPR models. QSAR & Combinatorial Science 22(1): 69–77.

Tsujimoto, T. 1999. Sediment transport processes and channel incision: mixed size sediment transport, degradation and armoring. *In*: Darby, S.E. and Simon, A. (eds.). Incised River Channels. John Wiley & Sons.

Verbeeck, H., Samson, R., Verdonck, F. and Raoul, L. 2006. Parameter sensitivity and uncertainty of the forest carbon flux model FORUG: a Monte Carlo analysis. Tree Physiology 26(6): 807–817.

Von Thun, J.L. and Gillette, D.R. 1990. Guidance on Breach Parameters. Internal Memorandum, U.S. Dept. of the Interior, Bureau of Reclamation, Denver, CO.

Vose, D. 1996. Quantitative Risk Analysis: A Guide to Monte Carlo Simulation Modeling. John Wiley & Sons, New York.

Wahl, T.L. 2004. Uncertainty of predictions of embankment dam breach parameters. Journal of Hydraulic Engineering 130(5): 389–397.

Wright, S. and Parker, G. 2004. Flow resistance and suspended load in sand-bed rivers: simplified stratification model. J Hydraul Eng 130(8): 796–805.

Xu, Y and Zhang, L.M. 2009. Breaching parameters for earth and rockfill dams. Journal of Geotechnical and Geoenvironmental Engineering 135(12): 1957–1970.

CHAPTER 11

Basics of Hydrology for Streams and Rivers

Agnieszka Cupak and *Andrzej Wałęga**

NTRODUCTION

The field of hydrology is of fundamental importance to civil and environmental engineers, hydrogeologists, and other earth scientists. Hydrology is a multi-disciplinary subject that deals with the occurrence, circulation, storage, and distribution of surface and ground water on the earth (Bedient et al. 2013). As an earth science, hydrology is closely related to other natural sciences. Hydrology is concerned with three issues: water use, water control, and pollution control (Maidment 1993). The term water use refers to the withdrawal of water from lakes, rivers, and aquifers for city water supplies, industries and agriculture, the instream use of water for hydropower, and protection of wildlife. Hydrology is called upon to specify the inflows to the system for both normal and drought conditions. Water control refers to the manipulation of hydrological extremes, particularly floods, and the erosion and sediment transport which occur during floods. Pollution control is the prevention of the spread of pollutants or contaminants in natural water bodies, and the cleanup of existing pollution.

Hydrologist Curves

A number of problems in river training projects are solved by means of hydrological curves. They show the distribution of a number of hydrological values, such as water stages, flows, or the course of natural processes taking place in rivers and catchments vs. various physical-and-geographic and hydrological parameters over time (Ciepielowski 1985). They are useful in identifying, for instance, the nature of a watercourse (stage curve), and also for sizing hydroengineering works in design (stage frequency curve, stage duration curve).

University of Agriculture in Krakow, Faculty of Environmental Engineering and Land Surveying, Department of Sanitary Engineering and Water Management. Mickiewicza Av. 24-28, 30-059 Krakow
* Corresponding author: awalega@ar.krakow.pl

Stage-discharge curve

The values of daily flows are found from recorded data. Unlike water stage observations, which are conducted regularly during a hydrologic year, flow measurements are taken in respective water gauge profiles at irregular intervals, typically a dozen or so times a year. Therefore, hydrologists use a method in which the results of stage observation are used for calculating the values of flow from the relationship between the flow Q and the stages observed for the specific flow H (Byczkowski 1979). The quality of the stage-discharge relation, or rating curve determines the quality of computed streamflow data. Hydraulic theory helps in determining the general form of the rating curve (Maidment 1993). Streamflow or discharge measurements normally involve obtaining a continuous record of water levels, or stage above datum; establishing the relationship between water level and discharge and transforming the record of stage into a record of discharge (Maidment 1993).

The value of water flow in a river depends on its flow velocity v and the wetted field, F, or cross-sectional area (Lambor 1971):

$$Q = f(f, v) \ m^3 \cdot s^{-1} \tag{1}$$

If flows are compared with the water stages in a river, which determine the depth of the river bed, the higher the stage, the higher the value of F. Moreover, the larger the depth, the higher the mean velocity at the cross-section. Therefore, both the water velocity and the flow cross-sectional area are water stage functions (Lambor 1971):

$$F = f_1(H) \ v = f_2(H) \tag{2}$$

Hence: $Q = f(H)$ (3)

The shape of the flow curve for the cross-section of a natural river bed depends on the shape of the cross-section and the slope of the water surface. A flow curve for the whole range of water stage variation from zero flow to the peak observed stage flow is termed "complete curve". A curve plotted for a single zone or covering an incomplete range of stage variation is termed "range curve" (Byczkowski 1979).

In practice, flow curves are described by various types of equations, although the following is typically used (Byczkowski 1979):

n-degree parabola equation with its vertex at the origin of coordinates (4) – so-called Harlacher equation (Maidment 1993):

$$Q = C(h - a)^N \tag{4}$$

Where: Q – flow, C and N – constant, h – water stage, a – stage shown by the gauge, for which the flow is 0.

Natural channels are often approximately parabolic in cross section, thus a value of about 2 for the exponent N is appropriate where there is channel friction control. Where there is a series of natural controls for different ranges of stage, different values of C, and N may apply for each range of stage (Maidment 1993).

The value of the constant "a" is positive or negative, depending on the gauge zero level above or below a theoretical bottom (Materiały ... 1994). The constant "a" is found by various methods, typically the following (Byczkowski 1979):

• Determination of a from the cross-section,
• Determination of a from the river profile,

- Gluszkow method,
- Determination of a by plotting the dependence $v = f(H)$ and $t = f(H)$,
- Determination of a by the trial method from the curve graph in the logarithmic scale.

The parameters C and N in the equation can be found mathematically by finding the best-fitting curve to the measurement points, using the least-squares method. The flow curve has different profiles for the rising and the falling water stages. This is especially noticeable during flood wave passages. It is a typical hysteresis phenomenon, frequently observed in geophysical processes. Consequently, neither Q_{max} nor v_{max} correspond to the peak water stage H_{max}, and a flow relationships hysteresis loop is obtained (Fig. 1). The flow curve is plotted more or less centrally across the loop. The phenomenon is

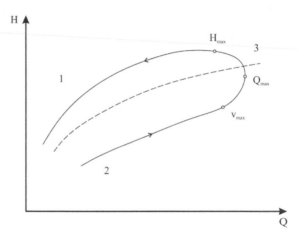

Fig. 1. Diagram of flow curve hysteresis: 1 – Falling stage phase; 2 – Rising stage phase; 3 – Flow curve (Lambor 1971).

caused by changes in the river bed which accompany the passage of flood waves: they are different in the rising and in the falling stages. The rising stages account for the river bed erosion and for increased depths and flows, in contrast to the falling stages (especially spectacular ones), when deposition of material is observed and depths are smaller than in the rising phase (Lambor 1971).

Variation in the slopes of surface table before and after culmination is another cause of the hysteresis. Before the culmination reaches the profile in question, the water surface slopes are smaller than after the passage of the culmination stage (Lambor 1971). The water surface slope variation, combined with a lack of explicit stage-discharge relationships, is mainly caused by the varying backwater or rapidly changing flows or both of the above (Szkutnicki et al. 2010). These two factors are capable of causing the phenomenon of flow curve hysteresis. The stronger the rising and the falling, the larger the loop opening. Therefore, the loop courses for different high water stages are also different, even though they always run in the same direction (Lambor 1971). If differences in the flow values for the same stages are small, then a middle curve is plotted. However, the relationship is often insufficient and other solutions are sought in many cases. Szkutnicki et al. (2010) proposed methods to find instantaneous flow values in conditions of varying water table falling below the gauge, when the flow curve based on relationship (3) is not applicable. Their studies were based on the causes of water surface slope variations that prevail in Poland (Szkutnicki et al. 2010):

1. Variation in the slope of the water table caused by hydroengineering works,
   1a. Channeled river with barrages maintaining a suitable depth for navigation purposes, water gauge station under influence of water lifting by the barrage,
   1b. Water gauge station under influence of backwater or discharges from a water reservoir,
2. Variation in the slope of the water surface caused by the main river swelling the tributary mouth,
3. Variation in the slope of the water surface caused by the tributary swelling the main river mouth,
4. Variation in the slope of the water surface caused by the sea swelling the river mouth,
5. Variation in the slope of the water surface due to rapid flow variations during high water stages.

Ice related phenomena and river bed overgrowing were not included in the study, although they are also known to contribute to water table slope variation. The authors propose two methods for flow determination in swelling conditions:

- Constant fall method,
- Flow module method.

Stage frequency curve (flow frequency curve)

In a number of hydroengineering issues, it is essential to know the typical range of water surface slope variation, the frequency of the occurrence of certain water stages in various seasons, and which stage has the highest frequency of occurrence or the longest duration. These are the river parameters which describe its nature and are factors of interest to design engineers in the dimensioning of hydroengineering works, e.g., the crest of hydroengineering structures, bottom of bridge structure, etc. (Lambor 1971). The frequency and duration of a given water stage (or flow) per annum are illustrated

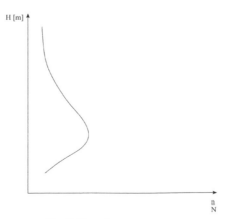

Fig. 2. Stage frequency curve.

by the stage (flow) frequency curve and the flow duration curve. It is also called water stage (flow) duration curve (Fig. 2) (Lambor 1971).

The frequency of events is the ratio of the number of events n in a given class time interval to the total number of events N ($\frac{n}{N}$). The result is expressed as an absolute value or a percentage. Its graphs are plotted for various lengths of time: months, years, multi-annual periods, and for special periods of time, such as navigating season, winter, summer half-year (Lambor 1971).

Stage (flow) duration curve

A flow duration curve is one of the most informative methods of displaying the complete range of river discharges from low flows to flood events. It is a relationship between any given discharge value and the percentage of time that this discharge is equalled or exceeded, or in other words—the relationship between magnitude and frequency of streamflow discharges (Smakhtin 2001).

A flow duration curve is constructed by reassembling the flowtime series values in decreasing order of magnitude, assigning flow values to class intervals and counting the number of occurrences (time steps) within each class interval (Smakhtin 2001). Every point of the stage (flow) duration curve shows the duration (in days) of a given stage (or flow) including its higher values within that interval (Fig. 3). Cumulated class frequencies are then calculated and expressed as a percentage of the total number of time steps in the record period. A flow duration curve may be constructed using different time resolutions of streamflow data: annual, monthly, or daily. A flow duration curves constructed on the basis of daily flowtime series provide the most detailed way of examining duration characteristics of a river. Curves may also be constructed using some other time intervals, for example, from day or month average flow time series (Institute of Hydrology 1980; McMahon and Mein 1986; Searcy 1959; Smakhtin 2001).

The nature of a river can be identified from its stage (flow) duration curve (Fig. 4). Curve 1 denotes a mountainous river with frequent high water stages and medium-to-low water stages of long duration. Curve 3 depicts a typical lowlands river with balanced discharges, flowing across lakes and wetlands. Curve 2 illustrates a middle type between the mountainous and lowland rivers. Curves 4 and 5 are theoretical types of rivers, not encountered in practice: curve 4 is a river having a balanced flow with no

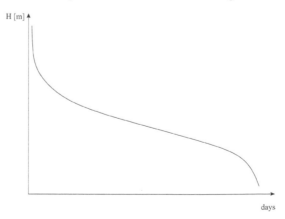

Fig. 3. Stage (flow) duration curve, including higher values.

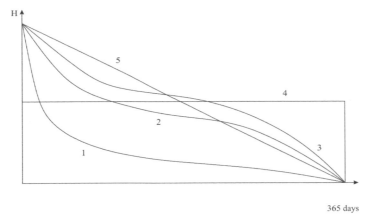

Fig. 4. Types of stage (flow) duration curves (Lambor 1971).

maximum or minimum, curve 5 is one for which the flow duration is the same number of days in each water stage zone (Lambor 1971).

A stage (flow) duration curve is useful in a number of operations in hydrology and hydroengineering, for instance, in identifying periodic stages and characteristic stages as well as in stage zoning, for example (Lambor 1971):

- A 9-month water or stage is a stage of which the duration (together with higher stages) is a total of 9 months a year (275 days); it is used for river training (in Poland, mainly for middle sections of navigable rivers). The stage is similar to mean low water;
- A 11-month stage is also used in river training;
- A 6-month stage is similar to mean annual stage;
- A 10-day stage is a stage of which the duration – together with higher stages – is merely 10 days a year and is essential for identifying the highest navigable waters.

Frequency Discharge

Flood frequency for gauge station

One of the most important hydrological analyses is the assessment of the frequency which discharges of a given magnitude are exceeded at a site and one of the most important data sets is the measured instantaneous flood peak discharge estimated from the stage records at a site (Shaw et al. 2011). The longer a record continues, homogeneous and with no missing peaks, the more its value is enhanced. Even so, it is very rare to have a satisfactory record long enough to match the expected life of many engineering works required to be designed or to assess flood risk for extremely low probability. Very often, flood discharge estimates are based on an extrapolation of the stage-discharge rating curve. Thus, where possible, the uncertainty in the flood peak estimates should be assessed as part of a flood frequency analysis.

Flood frequency analysis entails the estimation of the peak discharge, which is likely to be equalled or exceeded on average once in a specified period, T years. The estimation of flood magnitudes can be achieved by making use of two types of flood peak series, namely Annual Maximum (AM) series and Peaks over Threshold (PoT). The AM series consists of one value, the maximum peak flow, from each year of record, while the PoT series consists of all well defined peaks above a specified threshold value. Each model attempts to represent the flood peak aspects of the entire series of flow hydrographs by a simple series of flood peak values. PoT series are also denoted by some authors as Partial Duration Series (PDS) because the flood peaks can be considered as the maximum flow values during hydrograph periods of variable length. The periods divide the full flow record in subseries with partial durations (Rosbjerg et al. 1992). Frequency distributions of extreme events such as annual maximum discharges are normally positively skewed and can be described by functions such as Gumbel EVI, gamma (Pearson III type), lognormal, log Pearson III type, and Generalized extreme value distribution.

Gumbel (EVI) distribution

The equation of the Gumbel or EVI is given by:

$$F(X) = exp[-e^{-a(X-\mu)}] \tag{5}$$

Where F(X) is the probability of an annual maximum $Q_{max} \leq X$, and α and μ are parameters. The quantile Q_{maxp} is calculated from the formula:

$$Q_{maxp} = \mu - \frac{1}{\alpha} \ln(-lnp) \tag{6}$$

Where: p – exceedance probability

Gamma (Pearson III type) distribution

A gamma distribution with three parameters (Pearson III type) is defined by the following function of density:

$$F(X) = \frac{\alpha^{\lambda}}{\Gamma(\lambda)} (X - \epsilon)^{\lambda-1} e^{-a(X-\epsilon)}, X > \epsilon, \alpha, \lambda > 0 \tag{7}$$

Where $\Gamma(\lambda)$ is gamma Euler function. The function is available from tables or spreadsheets. Maximum quantiles of annual flows Q_{maxp} with a set exceedance probability p ($p = P(Q_{max} \geq Q_{maxp})$) are calculated from the formula (Kaczmarek 1970):

$$Q_{maxp} = \epsilon + \frac{t_p(\lambda)}{\alpha} \tag{8}$$

Where: ϵ – bottom limit of flows in $m^3 \cdot s^{-1}$: $Q_{max} \geq \epsilon$, α – parameter of scale in ($m^3 \cdot s^{-1}$), λ – parameter of shape, $t_p(\lambda)$ – standardized variable. The theoretical values x_p of maximum annual flows for the set values of exceedance probability "p" for the given values of ϵ, λ and α, can be calculated using available software, for instance, MS Excel spreadsheet GAMMA.DIST.INV:

$$Q_{maxp} = \epsilon + GAMMA.DIST.INV(1-p; \lambda; \frac{1}{\alpha}) \tag{9}$$

Lognormal distribution

If the values of hydrological events are logarithmized, then they are subject to normal distribution. Therefore, if a random variable has asymmetric distribution then it can be transformed into normal distribution by substituting the value of the variable X with the logarithms of that variable lgX. Thus, the variable X has a normal logarithmic distribution with the distribution function (Byczkowski 1972; Maidment 1993):

$$F(x) = \frac{1}{\sigma(x - \epsilon)\cdot\sqrt{2\pi}} \cdot \exp\left[-\frac{1}{2}\left(\frac{\ln(X)-\mu}{\sigma}\right)^2\right], x > \epsilon, \mu \text{ any value, } \sigma > 0 \tag{10}$$

Where: μ − Mean value of the sequence of values lgX, ϵ − Bottom limit of the sequence, σ − Standard deviation of the sequence of values lgX.

Maximum annual flow Q_{maxp} with a set exceedance probability p in normal logarithmic distribution of three parameters is calculated using the following equation (Metodyka ... 2009):

$$Q_{maxp} = \epsilon + \exp(\mu + \sigma \cdot \mu_p) \tag{11}$$

Where: ϵ − Bottom limit of flows, $Q_{max} \geq \epsilon$; value found in the graph, μ − Distribution parameter (mean value of variable $\ln(Q_{max}-\epsilon)$, u_p − Quantile of order p, where p is exceedance probability, in normal standardized distribution. In this case, the tables or the MS Excel spreadsheet function NORM.DIST.S.INV(1-p) will be helpful.

Log-Pearson III type distribution

Analogous to the normal-lognormal distributions, when the three-parameter gamma distribution is applied to the logs of the random variables, it is customarily called the log Pearson type III distribution. It plays an important role in hydrology because it has been recommended for application to flood flows by the (U.S.DIGS 1982) in the committee's Bulletin 17B method. The recommended technique for fitting a log Pearson III type distribution to observed annual peaks Q_{max} is to compute the base 10 logarithms of the discharge Q, at selected exceedance probability p, by equations:

$$logQ_{maxp} = \overline{logX} + SK \tag{12}$$

Where: $\overline{logX}$ - mean logarithm of annual peak flow, S − Standard deviation of logarithms, K is factor that is a function of the skew coefficient and selected exceedance probability.

Generalized Extreme Value (GEV) distribution

The Gumbel distribution is a special case of the generalized extreme value (GEV) distribution, which is described by destiny functions (Węglarczyk 2010):

$$F(X) = \begin{cases} \frac{1}{\alpha}\left(1 - \kappa\, \frac{X-\xi}{\alpha}\right)^{\frac{1}{\kappa}-1} exp\left[-\left(1 - \kappa\, \frac{X-\xi}{\alpha}\right)^{\frac{1}{\kappa}}\right], \kappa \neq 0 \\ \frac{1}{\alpha} - exp\left(-\frac{X-\xi}{\alpha}\right) exp\left[-\left(-\frac{X-\xi}{\alpha}\right)\right], \kappa = 0 \end{cases} \tag{13}$$

The quantile Q_{maxp} in the GEV distribution is calculated from the formula:

$$Q_{maxp} = \begin{cases} \xi + \dfrac{\alpha}{\kappa}\,[-1(-lnp)^{\kappa}], \; \kappa \neq 0 \\[2mm] \xi - aln(-lnp), \; \kappa = 0 \end{cases} \tag{14}$$

The second form with $\kappa = 0$ is the form of the Gumbel (EVI) distribution (Shaw et al. 2011). Whether a random variable is bounded from below ($\kappa > 0$), non-bounded ($\kappa = 0$), or bounded from above ($\kappa < 0$) depends on the value of the parameter κ. According to FLOODFREQ (2013), this type of distribution is most frequently recommended in many countries in Europe.

In many practical applications, it is necessary to find the probability distribution of a random variable X. More often than not, a random sample $\{x_1, x_2, \ldots x_n\}$ is the only information available and the procedure is typically as follows (Węglarczyk 2010):

1. Adoption of a probability distribution function for the variable X,
2. Estimation of the parameters of the adopted distribution function,
3. Validation of the hypothesis that the function is a good approximation of the real distribution function.

The estimation of the parameters of the adopted distribution is based on a simple random sample $\{x_1, x_2, \ldots x_n\}$, from which it is possible to obtain the empirical picture of a real distribution function. In practice, the estimation of distribution parameters is performed by the methods of moments, quantiles, or maximum likelihood. It is assumed that the estimators of distribution parameters provided by the maximum likelihood method have the following advantages: they are consistent, are at least asymptotically unbiased, they are asymptotically most efficient, and their mean squared errors are the lowest. Owing to all this, in practice, distribution parameters are most frequently estimated by the maximum likelihood method. The method to estimate the parameters of Pearson type III distribution using the maximum likelihood method is described below (Metodyka ….2009):

1. Sort the maximum annual flow time series $\{Q_{max,1}, Q_{max,2}, \ldots, Q_{max,N}\}$ in the decreasing order: $\{Q_{max,(1)} \geq Q_{max,(2)} \geq \ldots \geq Q_{max,(N)}\}$.
2. For each value $Q_{max,(i)}$, $i = 1, 2, \ldots, N$, sorted in a decreasing order in the series, calculate the empirical exceedance probability p_i using the formula:

$$p_i = \frac{i}{N+1}, \; i = 1, 2, \ldots, N \tag{15}$$

Where: i – number of i-th highest value, $Q_{max,(i)}$, in the data series.
3. Apply the resulting points (p_i, $Q_{max,(i)}$) onto Pearson's probability scale, smooth manually the bottom part of the curve up to the exceedance probability point $p = 100\%$, and find the lower bound value $\in$ in $m^3 \cdot s^{-1}$ for that probability.
4. For a known lower bound value $\in$, calculate the ancillary value A_λ:

$$A_\lambda = \ln\left(\frac{1}{N}\sum_{i=1}^{N} Q_{max,i} - \in\right) - \frac{1}{N}\sum_{i=1}^{N}\ln\left(Q_{max,i} - \in\right) \tag{16}$$

5. Estimate the parameter λ using the formula:

$$\lambda = \frac{1}{4A_\lambda}\left(1 + \sqrt{1 + \frac{4A_\lambda}{3}}\right) \tag{17}$$

6. Calculate the parameter α using the formula:

$$\alpha = \frac{\lambda}{\frac{1}{N}\sum_{i=1}^{N}Q_{max,i} - \in} \qquad (18)$$

The calculated values $\in$, λ, and α define unambiguously the distribution (8) of annual maximum flows Q_{max}.

Recently, the method of linear moments (L-moment) has been used more and more frequently for the estimation of distribution parameters. The main advantage of L-moments over conventional moments is that L-moments, being linear functions of the data, suffer less from the effect of sampling variability: L-moments are more robust than conventional moments to outliers in the data and enable more secure inferences to be made from small samples about an underlying probability distribution (Hosking 1990; Izinyon and Ehiorobo 2015). The L-moments methods are often used in regional frequency analysis. Bahmani et al. (2013) decided upon the impact of duration and temporal pattern of precipitation with different return periods on peak flow through integration of L-Moment theory and the HEC-HMS hydrologic model. Atiem and Harmancioglu (2006) performed a regional flood frequency analysis on the 14 stations annual maximum flood data in the Nile river basin by means of specific flood and L-moment method. Modarres (2008) used L-moment method for frequency analysis of wind speed in arid and semi-arid regions in Iran. The results of this study show that the generalized logistic distribution is the best regional three-parameter-distribution for average annual speed data.

Once the distribution parameters have been defined, the next step is to validate the hypothesis that the adopted distribution function is true. Validation is often performed by means of non-parametric tests: the Kolmogorow, Anderson-Darling, or χ^2 test (Węglarczyk 2010). According to Węglarczyk (1998), when testing the distributions of maximum values for consistency, somewhat better results are obtained by the Anderson-Darling in comparison with the Kolmogorow test; the former responds much more sharply to differences between empirical and theoretical values in the right-hand tail of distribution (low probability zone). The Kolmogorow test assigns the same weight to differences between theoretical and empirical distributions over the entire range of variation of the test variable. However, for the sake of simplicity, the Kolmogorow test is very often used in the engineering practice to check whether theoretical distribution correctly matches empirical data. On the other hand, it must be remembered that, in the case of estimated distribution parameters, the critical value of the Kolmogorow test is approximately 1/3 lower in comparison with the critical value established for distribution parameters which are known *a priori* (Węglarczyk 1998). If the Kolmogorow test is used for testing the consistency of theoretical and empirical distributions, the following procedure is used (Metodyka … 2009):

1. In the case of empirical data and theoretical quantiles, the value D_i is calculated for each value $Q_{max,(i)}$, $i = 1, 2, ..., N$, of a maximum annual flow data series, sorted in a decreasing order:

$$D_i = \max\left[\left|\frac{i}{N+1} - p_{teor}\left(Q_{max,(i)}; \in, \alpha, \lambda\right)\right|, \left|\frac{i+1}{N+1} - p_{teor}\left(Q_{max,(i)}; \in, \alpha, \lambda\right)\right|\right] \qquad (19)$$

Where: $p_{teor}(Q_{max,(i)})$ – Theoretical probability of exceeding the value $Q_{max,(i)}$:

2. A maximum value D_{max} for a series of differences D_i is defined:

$$D_{max} = \max_{i=1,\ldots,N} \{D_i\} \tag{20}$$

As an alternative, D_{max} is found in the graphs showing theoretical and empirical distributions.

3. The value λ_{Kol} of the Kolmogorow test statistics is calculated:

$$\lambda_{Kol} = \sqrt{N} \cdot D_{max} \tag{21}$$

4. Assuming that the level of significance of the test is $\alpha_{test} = 5\%$ of the value, λ_{Kol} is compared with the critical value of the test $\lambda_{kryt}(\alpha_{test} = 5\%) = 1.36$. If $\lambda_{Kol} < 1.36$, there are no grounds for the test distribution to be rejected. Otherwise, a different distribution, such as logarithmic normal or Weibull distribution ought to be sought.

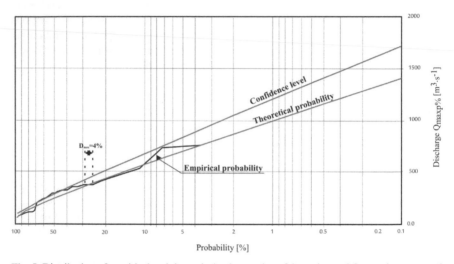

Fig. 5. Distribution of empirical and theoretical values and confidence interval for maximum annual flows in the San River, gauging station Lesko.

A representative empirical and theoretical curve of maximum annual flows in the river San, section Lesko, is shown in Fig. 5. The curve was obtained by means of Pearson type III distribution with estimation of parameters by the highest likelihood method. Also, shown in the figure is the value D_{max} used for calculating the Kolmogorow test.

Low discharge for gauge station

Hydrologists are not interested in the probability of the occurrence of a certain numerical feature of a variable of interest; rather, they seek to know the probability of failure for the value (in the case of minimum flows) to occur. This corresponds to the sum of probabilities together with higher (or lower) values of past phenomena (Byczkowski 1979). Probability of failure for a given limiting value x_p to occur is expressed by the following relationship (22):

$$P(X < x_p) = p\% \tag{22}$$

Where: X – any, incidental value of a variable, x_p – particular value of a variable with a specified percentage of probability of occurrence, with higher values, P – probability of the occurrence of the parenthesized event, and p – the value of probability of the occurrence of the event of interest, expressed as an absolute number or a percentage.

Probability distributions

The normal (N) or Gaussian distribution is certainly the most popular distribution in statistics. It is the basis of the lognormal (LN) and three-parameter lognormal distributions (LN3), which have seen many applications in hydrology (Maidment 1993).

In the case of minimum flows, the lognormal distribution (discussed earlier) and the Fisher-Tippet III distribution are used.

Fisher-Tippet III distribution

For the Fisher-Tippet III distribution, the distribution equation takes the following form (Byczkowski 1979):

$$P_{min(x)} = P(X_{min} \geq x) = e^{-(\frac{x-\varepsilon}{\theta-\varepsilon})^k} \tag{23}$$

Where: θ is the characteristic minimum flow.

The variable x has the lower limit: $\varepsilon < x < +\infty$.
The distribution function takes the form:

$$F_{min}(x) = 1 - e^{-(\frac{x-\varepsilon}{\theta-\varepsilon})^k} \tag{24}$$

For $X \geq \varepsilon \geq 0$.

Gumbel's method, in which distribution parameters are calculated by the method of moments, is one of the most popular methods for the estimation of the minimum probable flows. Calculations are made for two or three distribution parameters ($\xi=0$). Estimates of river low flow characteristics are needed for a range of purposes in water resources management and engineering. In water quality management they are used for issuing discharge permits and locating treatment plants, in water supply planning they are used to determine allowable water transfers and withdrawals (Tallaksen and van Lanen 2004), in hydropower development they are used for determining environmental flows (Gustard et al. 2004), and in an environmental context they are used for streamflow habitat assessment (Jungwirth et al. 2003). Application of the European Water Framework Directive to assess the state of water bodies requires low flow estimates. Several countries have developed national estimation procedures to provide a consistent methodology for determining low flows at gauged and ungauged sites (Laaha and Blöschl 2007).

The procedure for calculating the minimum probable flows in gauged catchments by Gumbel's method for three distribution parameters is as follows (Byczkowski 1979):

1. Sort an observation sequence composed of n terms in a decreasing order.
2. Apply the terms of the sequence to a normal scale of probability.
3. Define the following characteristics of the sequence:
    • Mean value:

$$\bar{x} = \frac{\Sigma x_i}{N} \tag{25}$$

Where: Σx_i sum of minimum flows, N – number of terms in the sequence.

- Standard deviation:

$$\sigma = \sqrt{\frac{N}{N-1}(\bar{x^2} - \bar{x}^2)} \tag{26}$$

- Minimum observed value – x_1,
- Calculate the test function $t(\lambda;N)$ and estimate its corresponding parameter λ,

$$t(\lambda, N) = \frac{\bar{x} - x_1}{s} \tag{27}$$

- Find the value of λ in the appendix using the calculated value of the function $t(\lambda,N)$ and estimate the parameter λ as follows:

$$k = \frac{1}{\lambda} \tag{28}$$

- Calculate the lower bound of the sequence ε (29) and the characteristic minimum flow θ (30):

$$\varepsilon = x_1 \frac{\bar{x} - x_1}{N^\lambda - 1} \tag{29}$$

$$\theta = \frac{\bar{x} - \varepsilon}{T(1+\lambda)} + \varepsilon \tag{30}$$

Where: $T(1+\lambda)$ – the value of the function.

4. For the assumed probability of failure for p' to occur, use Eq. (31) to calculate the minimum flow values with a given probability of failure to occur:
$$Q_{minp'} = \varepsilon + (\theta - \varepsilon) \cdot e^{\lambda yp'} \tag{31}$$
The results are applied to a normal scale of probability.
5. Then the empirical curve is checked for compliance with the assumed distribution type.

Flood Frequency for Ungauged Catchments

If the catchment of interest is ungauged, the necessary data in the form of values $Q_{maxp\%}$ are found by intermediate methods using hydrological analogy (interpolation, extrapolation, or differential catchment methods), empirical formulae (in the form of regional regression equations), or rainfall-runoff models. The rainfall-runoff models are recommended in (Handbook 2007) for defining both the maximum flows and the shapes of typical high water waves. If the available hydrological information is insufficient (a series of hydrometric parameters was observed for less than 30 years), the gauged series can be extended, for instance, by means of statistic methods or suitable models (Wałęga and Cupak 2011a).

Peak-flow transposition

Gaged flow data may be applied at design locations near, but not coincident with, the gauge location using peak flow transposition. Peak flow transposition is the process of adjusting the peak flow determined at the gauge to a downstream or upstream location. Peak flow transposition may also be accomplished if the design location is between two gauges through an interpolation process. The design location should be located on the same stream channel near the gauge with no major tributaries draining to the channel in the intervening reach. The definition of "near" depends on the method applied and the changes in the contributing catchment between the gauge and the design location (NHI 2002). Two methods of peak flow transposition have been commonly applied: the area-ratio method and the Sauer method. The area-ratio method is described as:

$$Q_{X\max} = Q_{W\max}\left(\frac{A_X}{A_W}\right)^n \tag{32}$$

Where: $Q_{X\max}$ = peak flow at the design location, $Q_{W\max}$ = peak flow at the gauge location, A_X = watershed area at the design location, A_W = catchment area at the gauge location, n = transposition exponent.

Equation (32) is limited to design locations with drainage areas within 25 percent of the gauge drainage area (according to NHI 2002) or 50 percent (according to Stachy and Fal 1986). The transposition exponent is frequently taken as the exponent for catchment area in an applicable peak flow regression equation for the site and is generally less than 1. The transposition exponent is frequently taken as the exponent for catchment area in an applicable peak flow regression equation for the site and is generally less than 1 (in Polish hydrology, n is equally 2/3, but Sygut et al. (2014) showed that the value should rather not be adopted for measured hydrometric data. In their paper, they specified the values of the parameter n for selected rivers in the upper Vistula basin, which varies with flow probabilities).

If the computational cross-section is located between gauged cross-sections on the same watercourse, then interpolation (a variant of the area-ratio method) is used. The maximum annual flow in the computational cross-section is defined from the formula:

$$Q_{X\max} = Q_{G\max} + \left(\frac{Q_{D\max} - Q_{G\max}}{A_D - A_G}\right)(A_X - A_G) \tag{33}$$

Where: $Q_{X\max}$ – flow in the computational cross-section, $Q_{G\max}$ – flow in the upper gauged cross-section, $Q_{D\max}$ – flow in the lower gauged cross-section, A_X – catchment area to the computational cross-section, A_G – catchment area to the upper gauged cross-section, A_D – catchment area to the lower gauged cross-section.

Sauer's method performed slightly better than the area-ratio method when tested on data from seven states for the 10- and 100-year events. Sauer's method is based first on computing a weighted discharge at the gauge from the log-Pearson Type III analysis of the gauge record and the regression equation estimate at the gauge location. Then, Sauer uses the gauge drainage area, the design location drainage area, the weighted gage discharge, and regression equation estimates at the gauge and design locations to determine the appropriate flow at the design location. More detailed descriptions of Sauer's method are found in Sauer (1973) and McCuen and Levy (2000).

Regional regression equations

Regional regression equations are commonly used for estimating peak flows at ungauged sites or sites with insufficient data. Regional regression equations relate either the peak flow or some other flood characteristic at a specified return period to the physiographic, hydrologic, and meteorologic characteristics of the watershed (NIH 2002). The typical multiple regression model utilized in regional flood studies uses the power model structure:

$$Y_t = aX_1^{b1} \cdot X_2^{b2}...X_n^{bn} \tag{34}$$

Where: Y_t = the dependent variable, X_1, X_2, ..., X_n = independent variables, a = the intercept coefficient, and b_1, b_2, ..., b_n = regression coefficients.

The dependent variable is usually the peak flow for a given return period T or some other property of the particular flood frequency, and the independent variables are selected to characterize the watershed and its meteorologic conditions. The parameters: a, b_1, b_2, ...,b_n are determined using a regression analysis. The most important watershed characteristic is usually the drainage area and almost all regression formulas include drainage area above the point of interest as an independent variable. The choice of the other watershed characteristics is much more varied and can include measurements of channel slope, length, and geometry, shape factors, watershed perimeter, aspect, elevation, basin fall, land use, and others. Meteorological characteristics that are often considered as independent variables include various rainfall parameters, snowmelt, evaporation, temperature, and wind.

For instance, in the whole territory of Poland, in non-urbanized ungauged catchments below 50 km² with the impervious surface areas below 5%, the rainfall formula ought to be applied for calculating the maximum annual flows with a set exceedance probability (Stachy and Fal 1986). The procedure for calculating the maximum annual flow with a set exceedance probability by means of the rainfall formula is as follows: find the borders of the catchment in a topographic map, calculate its surface area, calculate an averaged slope of the watershed and use appropriate tables to find the runoff coefficient for the maximum flows as well as the coefficients that influence runoffs in the catchment and in the river bed. The maximum daily rainfall, with exceedance probability $p = 1\%$, is an essential parameter affecting the wave culmination. Its value is available from the Institute of Meteorology and Water Management National Research Institute (IMWM-NRI) or it can be found using a map.

The following is the rainfall formula:

$$Q_{maxp} = f \cdot F_1 \cdot \varphi \cdot H_1 \cdot A \cdot \lambda_p \cdot \delta_j \tag{35}$$

Where: f – dimensionless wave shape coefficient, F_1 – maximum specific runoff module, φ – runoff coefficient for maximum flows, H_1 – maximum daily rainfall with probability 1%, A – surface area of catchment, λ_p – quantile of distribution of variable for a set probability p, δ_j – lake reduction factor.

Punzet formula (Punzet 1978) is one of the regional formulae. The equation for mountainous catchments in the range 50 km² $\leq A \leq$ 500 km² is used for the Carpathian region and for the upper parts of the Świętokrzyskie Mountains. The equations for upland catchments can be used for the sub-Carpathian region, those for flatland catchments of areas in the range 50 km² $\leq A \leq$ 600 km² can be used outside the Carpathian region, for the right-bank basin of the San river:

$$Q_{maxp\%} = \Phi_{p\%} \cdot Q_{max50\%} \tag{36}$$

Where: $Q_{maxp\%}$ – maximum flow with a defined exceedance probability, $\Phi_{p\%}$ – function of distribution probability reflecting the ratio $\dfrac{Q_{maxp\%}}{Q_{max50\%}}$, $Q_{max50\%}$ – ordinary high water.

The probability function for distribution $\Phi_{p\%}$ is estimated from the formula:

$$\Phi_{p\%} = \alpha_{p\%} \cdot C_v^{Wp\%} + 1 \tag{37}$$

Where: $\alpha_{p\%}$, $W_{p\%}$ – coefficients, expressed vs. distance in the probability scale t for a normal right of error, calculated from the following formulae, respectively:

$$\alpha_{p\%} = 0.944 \cdot t^{1.48} \tag{38}$$

$$\alpha_{p\%} = 0.144 \cdot t^{1.48} \tag{39}$$

C_v – Empirical variation coefficient, calculated from the formula:

$$C_v = 3.027 \cdot \Delta W^{0.173} \cdot A^{-0.102} \cdot L^{-0.066} \tag{40}$$

Where: ΔW – elevation difference between the highest sources of a watershed and the test profile, A – catchment area, and L – length of the watershed measured from the farthest source in the river basin of the test catchment.

For upland, flatland, and Carpathian catchments, the value $Q_{max50\%}$ can be found using the following formulae:

Flatland catchments:

$$Q_{max50\%} = 0.0138 \cdot A^{0.757} \cdot P^{0.372} \cdot N^{0.561} \cdot J^{0.302} \tag{41}$$

- Upland catchments:

$$Q_{max,50\%} = 0,00033 \cdot A^{0.872} \cdot P^{1.065} \cdot N^{0.07} \cdot I^{0.089} \tag{42}$$

- Carpathian catchments:

$$Q_{max50\%} = 0.00166 \cdot A^{0.747} \cdot P^{0.536} \cdot N^{0.603} \cdot J^{-0.075} \tag{43}$$

Where: P – annual normal rainfall, N – soil imperviousness index, and J – assumed slope indicator.

In the U.S., in most statewide flood-frequency reports, the analysts divided the state into separate hydrologic regions. Regions of homogeneous flood characteristics were generally determined by using major watershed boundaries and an analysis of the areal distribution of the regression residuals, which are the differences between regression and station (observed) T-ear estimates (NIH 2002). For example, in Maine, the regression equations for estimating peak discharges (QT) having recurrence intervals T that range from 2 to 100 years. The explanatory basin variables used in the equations are drainage area (A), in squaremiles; channel slope(s), in feet per mile; and storage (St), which is the area of lakes and ponds in the basin in percentage of total area (Morrill 1975):

$$Q_2 = 14 \cdot A^{0.962} \cdot s^{0.268} \cdot St^{-0.212} \tag{44}$$

$$Q_{100} = 50.9 \cdot A^{0.907} \cdot s^{0.358} \cdot St^{-0.282} \tag{45}$$

Rainfall-runoff models

Forecasting floods based on mathematical modelling allows experts to convert information on the past-to-present rainfall into a river flow forecast (discharge, stage, and inundated area for a future time horizon). It helps to reduce flood damage by permitting the public to act before the flood level increases to a critical level. Prognosis of the size of maximum discharge and discharge hydrographs can be created using rainfall-runoff model (Wałęga et al. 2011b). Values of model's parameters are estimated using registered episodes of rainfall-runoff. In many parts of the world, rainfall and runoff data are seldom adequate to determine a unit hydrograph of a basin or watershed. In the absence of rainfall-runoff data, unit hydrographs can be derivated by the synthetic means (Limantara 2009). A synthetic unit hydrograph is a unit hydrograph derived following an established formula, without the need for the rainfall-runoff data analysis (Ponce 1989). This includes Snyder's method, Soil Conservation Service (SCS) Method, Gray's Method, and Clark's Instantaneous Unit Hydrograph Method. The peak discharges of stream flow from rainfall can be obtained from the design storm hydrographs developed from unit hydrographs generated from established methods (Salami et al. 2009). Parameters of methods mentioned above are estimated on regional regression equations basis (Belete 2009; Straub et al. 2000).

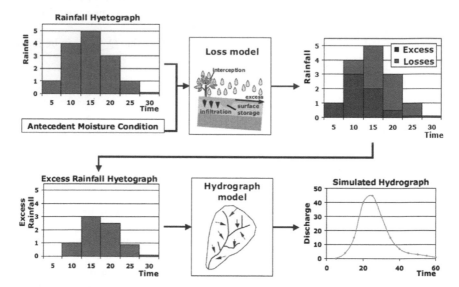

Fig. 6. Steps of the construction of the rainfall-runoff model (after: McEnroe and Gonzalez 2003).

The rainfall-runoff hydrological model is constructed in the following steps—Fig. 6:

1. A hyetograph is defined for total rainfall.
2. Rainfall losses (effective rainfall) are calculated.
3. Effective rainfall is transformed into direct runoff.
1. The form of the rainfall hyetograph is essential in the results of calculations of flows based on the mathematical model. The problems is of special importance in calculating the so-called hypothetical waves, as well as flows with a set exceedance probability, in which case a certain model for rainfall distribution vs. time must be adopted. Synthetic rainfall hyetographs are constructed by a number of methods

that are recommended for use in specific climatic conditions. Good examples include the synthetic hyetographs that were used in the UK (Butler and Davies 2011), the SCS curves that were recommended in the USA (Application ... 2010, Ponce 1989), or the DVWK curves (1985). In Poland, the latest methods for determination of synthetic hyetographs include the method developed by Kupczyk and Suligowski (1997) and, most recently, the proposed method to determine these characteristics using beta distribution. Detailed analyses of the effect of the shape of the rainfall hyetograph on the runoff volume are found in (Huff 1990; Kaczor and Wałęga 2011; Oliveira and Stolpa 2003; Wałęga et al. 2011c; Wałęga et al. 2012).

2. Runoff amount is commonly determined using NRCS CN method (formerly known as SCS CN), developed by the National Resources Conservation Services. This method represents an event-based lumped conceptual approach (Ven Te Chow et al. 1988). In the words of Ponce and Hawkins (1996), "The SCS-CN method is a conceptual model of hydrologic abstraction of storm rainfall, supported by empirical data. Its objective is to estimate direct runoff volume from storm rainfall depth, based on a curve number CN". The SCS-CN is a very simple method developed for predicting surface runoff from *hortonian overland flow* dominated watersheds. This method is straightforward and easy to apply. A primary reason for its wide applicability and acceptability is the fact that it accounts for major runoff generating watershed characteristics, namely soil type, land use/treatment, surface conditions, and antecedent moisture conditions (Banasik et al. 2014a and b; Mishra and Singh 2002; Váňová and Langhammer 2011). Currently, this method is included in widely used hydrological software, such as WinTR55, WinTR20, HEC-HMS, EPA-SWMM, SWAT, GLEAMS, EPIC, NLEAP, and AGNPS, and it is consequently applied in a large number of scientific studies (De Paola et al. 2013). Although several modifications to this method have been suggested and reported in the literature (Chauhan et al. 2013; Kowalik and Wałęga 2015; Mishra et al. 2005; Wałęga et al. 2015), further improvements are still needed. The greatest limitations of the original NRCS CN method are as follows: the three AMC levels used with this method permit unreasonable sudden jumps in CN and hence corresponding sudden jumps in the computed runoff are possible, there is a lack of clear guidance on how to vary antecedent moisture conditions, and there is no explicit dependency between the initial abstraction and the antecedent moisture (Sahu et al. 2012).

The SCS-CN method is based on the water balance equation and two fundamental hypotheses. The first hypothesis equates the ratio of actual amount of direct surface runoff Q to the total rainfall P (or maximum potential surface runoff) to the ratio of actual infiltration (F) to the amount of the potential maximum retention S. The second hypothesis relates the initial abstraction (Ia) to the potential maximum retention (S) (Deshmukh et al. 2013). The derivation of the basic equation for estimating the volume or depth of runoff is illustrated in Fig. 7.

No runoff occurs until rainfall equals an initial abstraction I_a. After allowing for I_a, the depth of runoff Q is the residual after subtracting F, the infiltration or water retained in the drainage basin (excluding I_a) from rainfall P. The potential retention S is the value that $(F+I_a)$ would reach in very long storm.

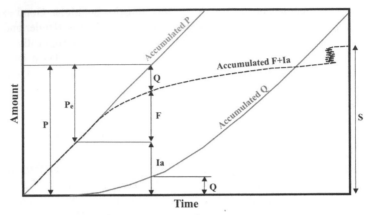

Fig. 7. Accumulated rainfall, and runoff during a uniform storm (after: Maidment 1993).

The popular form of the SCS-CN method is expressed as:

$$Q = \frac{(P - I_a)^2}{P - I_a + S} \qquad\qquad if\, P > I_a \qquad\qquad (46)$$

$Q = 0$ otherwise

$$I_a = \lambda S \qquad\qquad (47)$$

Where Q – direct runoff (mm), P – total precipitation (mm), I_a – initial abstraction (mm), S – potential maximum retention (mm), and λ – initial abstraction coefficient (dimensionless).

The parameter S of the NRCS-CN method depends on soil type, land use, hydrologic conditions, and antecedent moisture conditions (AMC). The parameter S is expressed as:

$$S = \frac{25400}{CN} - 254 \qquad\qquad (48)$$

Where S is in [mm] and CN is the curve number, which depends on the soil type, land cover and land use, hydrological conditions, and antecedent moisture condition (AMC). $CN = 100$ represents a condition of zero potential maximum retention ($S = 0$), that is, an impermeable watershed. Conversely, $CN = 0$ represents a theoretical upper bound to potential maximum retention ($S = 1$), that is an infinitely abstracting watershed. A theoretical value of CN parameter was determined based on an orthophoto, with reference to current land use and 1:25,000 scale soil maps. It was found to correspond to normal catchment moisture conditions AMCII.

3. Transformation of the effective rainfall into direct runoff is a final step of the computations. A unit hydrograph (UH) is defined as the direct runoff hydrograph resulting from a rainfall event that has uniform temporal and spatial distributions and the volume of direct runoff represented by the area under the unit hydrograph is equal to one unit of direct runoff from the drainage area—Fig. 8. A different unit hydrograph exists for each duration of rainfall. In all probability, the unit

hydrograph for a 1-hour storm will be quite different from the unit hydrograph for a 6-hour storm. The unit hydrograph is also affected by the temporal and spatial distributions of the actual rainfall excess. In other words, two rainfall events with different distributions over the drainage area may give different unit hydrographs, even if their respective durations are identical. Variations of the temporal and spatial distributions of rainfall contribute to variations in computed unit hydrographs for different storm events on the same watershed.

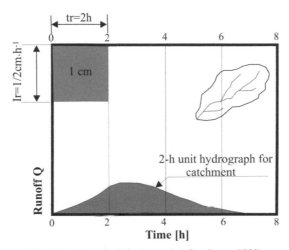

Fig. 8. Concept of unit hydrograph (after: Ponce 1989).

In the absence of rainfall, runoff data, unit hydrograph can be derived be synthetic means. A synthetic unit hydrograph (SUH) is a unit hydrograph derived following an established formula, without the need for rainfall-runoff data analysis (Ponce 1989). The development of SUH is based on the following principle: the volume under the hydrograph is known (volume is equal to catchment area multiplied by 1 unit of runoff depth), therefore, the peak discharge can be calculated by assuming a certain unit hydrograph shape. For instance, if a triangular shape as assumed—Fig. 9, the volume V and peak discharge Q_p is equal to:

$$V = \frac{Q_p \cdot T_{bt}}{2} = A \cdot (1) \tag{49}$$

$$Q_p = \frac{2A}{T_{bt}} \tag{50}$$

Where: T_{bt} – duration of base flow.

Synthetic Unit Hydrographs usually relate time base to catchment lag. In turn, catchment lag is related to the timing response characteristics of the catchment, including catchment shape, length and slope. Therefore, catchment lag is a fundamental parameter in SUH determinations. Two widely used methods, the Snyder and SCS methods are using for calculation of SUH.

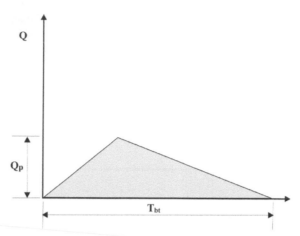

Fig. 9. Triangular unit hydrograph (after: Ponce 1989).

Snyder SUH

In this method, the unit hydrograph is described using three parameters, that is, peak discharge q_p, lag to peak T_1, and base time T_b, which are expressed, as:

$$q_p = \frac{2.78 \cdot C_p \cdot A}{T_l} \qquad (51)$$

$$T_l = C_t (L \cdot L_c)^{0.3} \qquad (52)$$

$$T_b = 72 + 3T_l \qquad (53)$$

Where: A is the area of the catchment, C_p and C_t are non-dimensional coefficients; L is the length of the main stream from the outlet to the catchment boundary, L_c is the distance from the outlet to a point on the stream nearest to the centroid of the catchment. Note that Eq. (52) is for unit rainfall-excess duration. For other duration, the lag time is adjusted as

$$T_{IR} = T_l + \frac{t_R - t_r}{2} \qquad (54)$$

Where, T_{IR} is adjusted lag time corresponding to a duration t_r. Snyder gave values of C_t varying in range 1.35 to 1.65 and C_p in range 0.56 to 0.69. The lower the value of C_p (the lower peak flow), the greater the value of C_t.

Snyder UH is recommended for use in catchments ranging from 250 to 5000 km².

Since the parameters C_t and C_p have been defined for specific local conditions, which in many times prevail in other climatic zones and are different from those observed in Poland, attempts are being made to optimize them for specific local conditions. For instance, Wałęga (2012) proposed the following equations that enable C_t and C_p to be defined for the Upper Vistula basin:

$$T_{L1} = 55.124 - 0.394 \cdot CN + 0.0001 \cdot \frac{L_{rz} L_{cA}}{\sqrt{i_r}} \qquad (55)$$

$$C_t = 6.886 - 0.038 \cdot CN - 132.014 \cdot i_r \qquad (56)$$

$$C_p = 0.568 + 0.0046 \cdot T_L - 16.342 \cdot i_r \tag{57}$$

Where: L_{rz} – river length from source to gauged cross-section, L_{cA} – distance from gauged cross-section to catchment's center of gravity, i_r – equalized slope of watershed calculated from the dependence: $i_r = 0.6 \cdot (H_{zr} - H_{przd})/L_{rz}$, H_{zr} – source abscissa, H_{przd} – gauged cross-section ordinate, and CN – CN parameter from SCS method.

The multiple correlation coefficient, R, of the established relationships varied between 0.587 and 0.90.

SCS-UH

In this method, in order to determine the SUH shape from the non-dimensional q/q_p versus the t/t_p hydrograph—Fig. 10, the peak discharge q_p and the time to peak t_p are computed, as:

$$q_p = \frac{2.08 \cdot A}{t_p} \tag{58}$$

$$t_p = \frac{t_r}{2} + T_l \tag{59}$$

Where: A is the area, t_r is the duration of rainfall and T_l is the lag time from centroid of rainfall to peak discharge. The T_l can be calculated from watershed characteristics using main stream length L, watershed slope s, and curve number (CN):

$$T_l = \frac{(3.28 \cdot L \cdot 1000)^{0.8} \left(\dfrac{1000}{CN} - 9 \right)^{0.7}}{1900 \cdot s^{0.5}} \tag{60}$$

SCS UH is used in direct runoff calculations for catchments up to 16 km^2 in defining T_l by means of formula (60) or above 8.0 km^2, if lag time is estimated by the hydraulic method.

Methods of low-flow estimation in ungauged catchments

Most of low-flow measures and calculation methods require adequate observed stream flow records, which can only be provided for gauged catchments. Ungauged catchments pose a different problem (Smakhtin 2001). Methods to determine minimum flows in ungauged catchments are divided into three groups: empirical methods, statistic methods, and rainfall-runoff models (WMO 2008). The empirical methods use simple mathematical Eqs. (32) and (33) are based on the transfer of flow rates data from the analog catchment to the test catchment (WMO 2008).

Statistic methods based on regression techniques

Regression can be used to derive equations to predict the values of various hydrologic statistics as a function of physiographic characteristics and other parameters. Such relationships are needed when little or no flow data are available at or near a site. Regional regression models have long been used to predict flood quantiles at ungauged

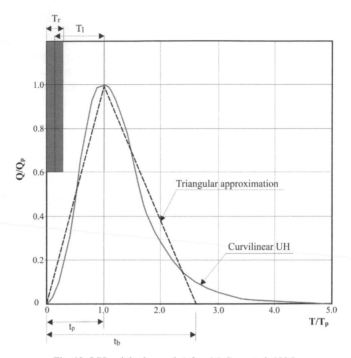

Fig. 10. SCS unit hydrograph (after: McCuen et al. 1996).

sites (Maidment 1993). This method is perhaps the most widely used technique in low-flow estimation at ungauged sites (Smakhtin 2001).

The regionalization of streamflow characteristics in general is based on the premise that catchments with similar climate, geology, topography, vegetation, and soils would normally have similar stream flow responses, for example, in terms of unit runoff from the catchment area, average monthly flow distribution, duration of certain flow periods, frequency, and magnitude of high and low-flow events in similar sized catchments (Smakhtin 2001).

For the regional regression approach, regions need to be identified that are homogenous and the same regression equation can be expected to apply to all catchments within a region. Different methods can be used to determine hydrologically homogeneous regions, for example, data clustering, regression tree, or low flow seasonality (Laaha and Blöschl 2007). The regression model is a relationship between dependent low-flow characteristics, independent catchment, and climatic variables. The stream flow data used should represent natural flow conditions in the catchments: the approach will most probably not work or will be misleading if flow regimes analyzed are continually changing under man-induced impacts. Technically, regression model is constructed by means of a multiple regression analysis (Laaha and Blöschl 2007):

$$q = \beta_0 + \beta_1 \cdot x_1 + \beta_2 \cdot x_2 + \dots + \beta p_{-1} \cdot xp_{-1} \qquad (61)$$

Where x_i are morphoclimatic parameters of a catchment, β_i is regression coefficient.

Laaha and Blöschl (2007) proposed regression equations (Table 1) for Austria. Among other things, they used the seasonal nature of low flows for isolating hydrologically homogeneous regions. Data set covers about 60% of the national territory of Austria. They used 31 physiographic parameters in their analysis.

Table 1. Component regressions of specific low flows q_{95} (l s^{-1} km^{-2}) in Austria (Laaha and Blöschl 2007).

Group	Region	R^2	Model q_{95} =
A-C	Alps< 1200 m	51%	$0.67 + 0.40P + 0.17GQ - 0.01GC + 6.43LWE + 0.14SM - 0.04LR - 0.20H0$
A-C	High Alps> 1800 m	68%	$-8.21 + 2.64P + 0.74HM - 0.02GC - 0.12PHM$
D	Pre-Alps (Styria)	89%	$-7.99 + 1.08P + 0.04LF$
E	Pre-Alps (Vorarlberg)	60%	$18.20 - 0.18SMO$
1	Flatland and hilly terrain (N, E of Austria)	71%	$-0.12 + 0.11SM + 0.05GGS + 0.02GC$
2	BohemianMassif	64%	$-3.31 + 1.96PW$
3	Foothills of Alps (Upper Austria)	68%	$-10.04 - 0.76D + 3.27P - 2.22H0$
4	Flyschzone	63%	$-6.17 + 0.06GL + 2.07PS - 0.06LF$

Where: P – average annual precipitation, PS – average summer precipitation, GQ – Quaternary sediments, GC – crystalline rock, GL – limestone, GGS – shallow groundwater table, LF – forest, LR – wasteland-rocks, LWE – wetland, SM – the mean slope, SMO – slope, $H0$ – the altitude of the stream gauge, HM – mean altitude, PW – average winter precipitation, and D – stream network density.

Engeland and Hisdal (2009) defined regression models for Norway. They carried out their analysis for 51 gauging stations, catchments below 2000 km^2, and 24 physiographic parameters describing the catchments. They tested six alternative models (Table 2) and proposed regression relationships for each model (Table 3).

Table 2. The regression model (Engeland and Hisdal 2009).

Name	Model
M1	Untransformed variables
M2	Model from Væringstad et al. (Engeland and Hisdal 2009)
M3	All variables log-transformed
M4	Only the QC is log transformed
M5	The QC is log-transformed, the model chooses between untransformed and log-transformed independent variables
M6	Like M5, but the winter and summer regions are merged

Table 3. The estimated regression coefficients (Engeland and Hisdal 2009).

Model	Equation
M1-Winter M1-Summer	$Q_c = 6.289 + 0.0484Q_M - 0.0312R_L + 0.873T_W$ $Q_c = 30.802 + 0.0766Q_M + 0.115L_\% - 0.192B_\%$
M2-Winter M2-Summer	$Q_c = \exp[-0.570 + 0.770 \ln (Q_M) - 0.202 \ln (H_{min})]$ $Q_c = \exp[-2.08 + 1.166 \ln (Q_M) - 0.534 \ln (R_G) - 0.368\ln(B_\%) + 0.1) + 0.153(M_\% + 0.1)]$
M3-Winter M3-Summer	$Q_c = \exp[-6.387 + 0.835 \ln (Q_M) - 0.391 \ln(M_\% + 0.1) - 0.175\ln(F_\% + 0.1) + 0.274\ln (T_A + 10)]$ $Q_c = \exp[-4.288 + 1.282\ln(Q_M) + 0.379\ln(L_\% + 0.1) - 0.272\ln(B_\% + 0.1)]$
M4-Winter M4-Summer	$Q_c = \exp[-0.00758 + 0.00767C_L + 0.0204Q_M - 0.00609M_\% + 0.0894T_A]$ $Q_c = \exp[-2.983 + 0.00285P_S + 0.0976L_\% + 0.0116M_\% + 0.0150C_L)]$
M5- Winter M5-Summer	$Q_c = \exp[3.3325 + 0.0102C_L + 0.03298\ln(L_{eff}) \_ 0.026485Q_M + 1.601\ln(T_S) - 0.215(F_\%) + 0.1) - 0.0173M_\%]$ $Q_c = \exp[-4.734 + 1.301\ln(Q_M) - 0.448\ln(B_\%) + 0.1) + 0.102L_\% + 0.0130C_L)]$
M6-All	$Q_c = \exp[-1.355 + 0.0238(Q_M) + 0.380\ln(C_W) + 0.243\ln(L_\% + 0.1)]$

Where : R_L – length of main river from the outlet to the most distant river string, C_L – catchment length from outlet to the most distant point at the water divide, C_W – catchment width, Q_M – mean annual runoff, Ps – summer precipitation, R_G – river gradient, H_{min} – minimum elevation, $F_\%$ – forested area, $B_\%$ – bogs, $M_\%$ – mountainous areas, $L_\%$ – lake percentage, L_{eff} – effective lake percentage, T_A – average annual temperature, T_S – average summer temperature, and T_W – average winter temperature.

A regional equation for selected catchments in the Upper Vistula basin was also proposed by Cupak (2012):

$$q_{95} = -10.3876 + 0.0103H_{me} - 4.1183N + 0.0186L + 0,798T_l = 0.0041P \cdot z. \tag{62}$$

Where: H_{me} – mean height of catchment, N – soils, L – length of watercourse, T_l – air temperature in the summer half-year, and P.z. – gauge zero level.

As opposed to the estimation of a single low-flow characteristic for which regression model has been constructed, regional prediction curves are suitable. Flow duration curves, low-flow frequency curves, and low flow-spell curves from a number of gauged catchments of varying size in a homogeneous region can be converted to a similar scale, superimposed and averaged to develop a composite regional curve. To make curves from different catchments comparable, all flows are standardized by catchment area, mean or median flow, or other 'index' flow. A curve for ungauged site may then be constructed by multiplying back the ordinates of a regional curve by either catchment area or an estimate of the index low-flow depending on how the flows for the regional curve were standardized. The index flow is estimated either by means of regression equation or from regional maps (Smakhtin 2001).

Rainfall-runoff models

Described the interaction between catchment structure, rainfall inputs, evaporative outputs, and streamflow outputs by representing hydrological processes by mathematical equation (WMO 2008).

Rainfall–runoff model which converts actual rainfall data in a simulated catchment into a continuous flow time series. The difficulties with rainfall–runoff simulation method are associated with the reliability or representativeness of the model employed and the ability of the user to satisfactorily quantify the parameter values for the specific catchment under investigation. Methodological aspects of model calibration with specific regard to low flows are currently not well developed. The rainfall-runoff models were used, for instance, by Duba and Pitchen in their studies (1983) in which they used the Streamflow Synthesis and Reservoir Regulation (SSARR) for a Bolivian river basin, Clausen and Rasmussen (1993) in Denmark, by Lanen et al. (1993), by Smakhtin and Watkins (1997) in South Africa, and by Engeland and Hisdal (2009) who used the HBV model in Norway (Smakhtin 2001).

Keywords: Stage, flow curve, frequency discharges, rainfall-runoff model.

References

Application of Hydrologic Methods in Maryland. State Highway Administration 2010.

Atiem, I.A. and Harmancioglu, N.B. 2006. Assessment of regional floods using L-moments approach: the case of the River Nile. Water Resources Management 20: 723–747.

Bahmani, R., Radmanesh, F., Eslamian, S., Khorsandi, M. and Zamani, R. 2013. Proper rainfall for peak flow estimation by integration of L-moment method and a hydrologic model. International Research Journal of Applied and Basic Sciences 4(10): 2959–2967.

Banasik, K., Rutkowska, A. and Kohnová, S. 2014a. Retention and curve number variability in a small agricultural catchment: the probabilistic approach. Water 2014 6: 1118–1133; doi:10.3390/w6051108.

Banasik, K., Krajewski, A., Sikorska, A. and Hejduk, L. 2014b. Curve number estimation for a small urban catchment from recorded rainfall-runoff events. Archives of Environmental Protection 40(3): 75–86.

Bedient, P.B., Huber, W.C. and Vieux, B.E. 2013. Hydrology and Floodplain Analysis. Pearson, England, 813 pp

Belete, M.A. 2009. Synthetic Unit Hydrographs in the Upper Awash and Tekeze Basins. Methods, Procedures and Models. VDM Verlag Dr Müller.

Butler, D. and Davies, J.W. 2011. Urban Drainage. Third Edition. Spon Press, London and New York.

Byczkowski, A. 1979. Hydrologiczne podstawy projektów wodnomelioracyjnych. Przepływy charakterystyczne. PWRiL, Warszawa.

Chauhan, M.S., Kumar, V. and Rahul, A.K. 2013. Modelling and quantifying water use efficiency for irrigation project and water supply at large scale. International Journal of Advanced Scientific and Technical Research 3(5): 617–639.

Ciepielowski, A. 1985. Wybrane zagadnienia hydrologiczne regulacji rzek. Wyd.SGGW-AR, Warszawa.

Clausen, B. and Rasmussen, K.R. 1993. Low flow estimation and hydrogeology in a chalk catchment. Nordic Hydrology 24(5): 297–308.

Cupak, A. 2012. Określenie przepływów niskich w zlewniach niekontrolowanych z wykorzystaniem regionalizacji. Gaz, Woda i Technika Sanitarna 2, pp. 62–64.

De Paola, F., Ranucci, A. and Feo, A. 2013. Antecedent moisture condition (SCS) frequency assessment: a case study in southern Italy. Irrigation and Drainage 62: 61–71.

Deshmukh, D.S., Chaube, U.C., Hailu, A.E., Gudeta, D.A. and Kassa, M.T. 2012. Estimation and comparison of curve numbers based on dynamic land use land cover change, observed rainfall-runoff data and land slope. Journal of Hydrology 492: 89–101.

Duba, D. and Pitchen, J.R.G. 1983. Low flow synthesis by computer. Geological Survey, Oklahoma City, OK.

DVWK. 1985. Niederschlag—Starkegenauswertung nach Wiederkehrzeit und Dauer. Regeln 124, Verlag Paul Parey, Hamburg.

Engeland, K. and Hisdal, H. 2009. A comparison of low flow estimates in ungauged catchments using regional regression and the HBV-model. Water Resources Management 23: 2567–2586.

FLOODFREQ COST Action ES0901. 2013. Review of Applied—Statistical methods for flood-frequency analysis in Europe. Center for Ecology and Hydrology. ISBN: 978-1-906698-32-4.

Gustard, A., Young, A.R., Rees, G. and Holmes, M.G.R. 2004. Operational hydrology. pp. 455–484. *In*: Tallaksen, L.M. and Van Lanen, H.A.J. (eds.). Hydrological Drought Processes and Estimation Methods for Streamflow and Groundwater. Developments in Water Sciences, Vol. 48. Elsevier, The Netherlands.

Handbook on good practices for Floyd map ping in Europe. 2007. EXIMAP.

Hosking, J.R.M. 1990. L-moments: analysis and estimation of distributions using linear combinations of order statistics. Journal of the Royal Statistical Society. Series B (Methodological) 52(1): 105–124.

Huff, F.A. 1990. Time distributions of Heavy Rainstorms in Illinois. Illinois State Water Survey. Champaign ISWS/CIR-173/90.

Institute of Hydrology. 1980. Low flow studies report. Institute of Hydrology, Wallingford, UK.

Izinyon, O.C. and Ehiorobo, J.O. 2015. L-moments method for flood frequency analysis of River Owan at Owan in Benin Owena River Basin in Nigeria. Curent Advances in Civil Engineering 3(1): 1–10.

Jungwirth, M., Haidvogl, G., Moog, O., Muhar, S. and Schmutz, S. 2003. Angewandte Fischökologiean Fließgewässern. FacultasUniversitätsverlag, Vienna, Austria. ISBN 3-8252-2113-X.

Kaczmarek, Z. 1970. Metody statystyczne w hydrologii i meteorologii. Wydawnictwo Komunikacji i Łączności, Warszawa.

Kaczor, G. and Wałęga, A. 2011. Przebieg wybranych epizodów opadowych na obszarze aglomeracji krakowskiej w aspekcie modelowania sieci kanalizacji deszczowej. Gaz, Woda i Technika Sanitarna 10: 364–366.

Kowalik, T. and Wałęga, A. 2015. Estimation of CN parameter for small agricultural watersheds using asymptotic functions. Water 7(3): 939–955.

Kupczyk, E. and Suligowski, R. 1997. Statystyczny opis struktury opadów atmosferycznych jako elementu wejścia do modeli hydrologicznych. [W:] Predykcja opadów i wezbrań o zadanym czasie powtarzalności. Red. Prof. Urszula Soczyńska. Wyd.UniwersytetuWarszawskiego, Warszawa 21–86.

Laaha, G. and Blöschl, G. 2007. A national low flow estimation procedure for Austria. Hydrological Sciences Journal 52(4): 635–644.

Lambor, J. 1971. Hydrologia inżynierska. Arkady, Warszawa.

Lanen, H.A.J., van Tallaksen, L.M., Kasparek, L. and Querner, E.P. 1997. Hydrological drought analysis in Hupsel basin using different physically based models. *In*: FRIEND'97—Regional Hydrology: Concepts and Models for Sustainable Water Resource Management, IAHS Publication 246: 189–196.

Limantara, L.M. 2009. The limiting physical parameters of synthetic unit hydrograph. World Applied Sciences Journal 7(6): 802–804.

Maidment, D.R. 1993. Handbook of Hydrology. McGraw-Hill, Inc. New York.

Materiały do ćwiczeń z hydrologii. 1994. Wyd. SGGW, Warszawa.

McCuen, R.H., Johnson, P.A. and Ragan, R.M. 1996. Highway hydrology. Hydraulic design series no. 2. Report No FHWA-SA-96-067, University of Maryland.

McCuen, R.H. and Levy, B.S. 2000. Evaluation of peak discharge transposition. Journal of Hydrologic Engineering, ASCE 5(3): 278–289.

McEnroe, B.M. and Gonzalez, P. 2003. Storm duration and antecedent moisture conditions for flood discharge estimation. Report Nb: K-TRAN: KU-02-4, University of Kansas.

McMahon, T.A. and Mein, R.G. 1986. River and Reservoir Yield. Water Resource Publication, CO., 368 pp.

Metodyka obliczania przepływów i opadów maksymalnych o określonym prawdopodobieństwie przewyższenia dla zlewni kontrolowanych i niekontrolowanych oraz identyfikacji modeli transformacji opadu w odpływ. Raport końcowy. 2009. Stowarzyszenie Hydrologów Polskich, maszynopis, Warszawa.

Mishra, S.K. and Singh, V.P. 2002. SCS-CN-based hydrologic simulation package. pp. 391–464. *In*: Singh, V.P. and Frevert, D.K. (eds.). Mathematical Models in Small Catchment Hydrology and Applications. Water Resources Publications, Littleton, CO.

Mishra, S.K., Jain, M.K., Bhunya, P.K. and Singh, V.P. 2005. Field applicability of the SCS-CN-based Mishra–Singh general model and its variants. Water Resources Management 19: 37–62.

Modarres, R. 2008. Regional maximum wind speed frequency analysis for the arid and semi-arid regions of Iran. Journal of Arid Environments 72: 1329–1342.

Morrill, R.A. 1975. A technique for estimating the magnitude and frequency of floods in Maine. U.S. Geological Survey Open File Report No. 75-292.

National Highway Institute. U.S. Department of Transportation. 2002. Highway Hydrology. Hydraulic Design Series No 2. Second Edition.

Oliveira, F. and Stolpa, D. 2003. Effect of the storm hyetograph duration and shape on the watershed response. *In*: Proc 82nd Annual Meeting of the Transportation Research Board. Washington DC, USA.

Ponce, V.M. 1989. Engineering Hydrology: Principles and Practices. Prentice Hall, Upper Saddle River, New Jersey.

Ponce, V.M. and Hawkins, R.H. 1996. Runoff curve number, has it reached maturity? Journal of Hydrologic Engineering – ASCE 1(1): 11–19.

Punzet, J. 1978. Zasoby wodne dorzecza Górnej Wisły (przepływy maksymalne). Materiałybadawcze IMGW.Instytut Meteorologii i Gospodarki Wodnej, Warszawa.

Rosbjerg, D., Madsen, H. and Rasmussen, P.F. 1992. Prediction in partial duration series with generalized Pareto-distributed exceedances. Water Resources Research 28(11): 3001–3010.

Sahu, R.K., Mishra, S.K. and Eldho, T.I. 2012. Performance evaluation of modified versions of SCS curve number method for two catchments of Maharashtra. India, ISH Journal of Hydraulic Engineering, 18(1): 27–36, doi:10.1080/09715010.2012.662425.

Salami, A.W., Bilewu, S.O., Ayanshila, A.M. and Oritola, S.F. 2009. Evaluation of synthetic unit hydrograph methods for the development of design storm hydrographs for Rivers in South-West, Nigeria. Journal of American Science 5(4): 23–32.

Sauer, V.B. 1973. Flood characteristics of Oklahoma stream. WRI Report 52-73, U.S.

Searcy, J.C. 1959. Flow duration curves. United States Geological Survey, Washington, DC, Water Supply Paper 1542A.

Shaw, E.M., Beven, K.J., Chappel, N.A and Lamb, R. 2011. Hydrology in Practice. Spon Press, London.

Smakhtin, V.U. 2001. Low flow hydrology: a review. Journal of Hydrology 240: 147–186.

Smakhtin, V.Y. and Watkins, D.A. 1997. Low-flow estimation in South Africa. Water Research Commission Report No. 494/1/97 Pretoria, South Africa.

Stachý, J. and Fal, B. 1986. Zasady obliczania maksymalnych przepływów prawdopodobnych. Prace Instytutu Badania Dróg i Mostów, nr 3-4, 91–149.

Straub, T.D., Melching, C.S. and Kocher, K.E. 2000. Estimation of Clark's Instantaneous Unit Hydrograph for small rural catchments in Illinois, Water-Resources Investigations Report 00-4184 USGS.

Sygut, M., Wałęga, A., Cupak, A. and Michalec, B. 2014. Weryfikacja współczynników redukcji przepływów maksymalnych rocznych o określonym prawdopodobieństwie przewyższenia w wybranych rzekach karpackiego dorzecza górnej Wisły. Infrastruktura i Ekologia Terenów Wiejskich. III/1, 971–983.

Szkutnicki, J., Kadłubowski, A. and Chudy, Ł. 2010. Metody wyznaczania przepływu w warunkach zmiennego spadku zwierciadła wody. "Hydrologia w inżynierii i gospodarce wodnej", t.1 (red. B. Więzik), Monografie Komitetu Inżynierii Środowiska, PAN, 68, Warszawa.

Tallaksen, L.M. and van Lanen, H.A.J. (eds.). 2004. Hydrological Drought: Processes and Estimation Methods for Streamflow and Groundwater. Developments in Water Science no. 48, Elsevier, The Netherlands.

U.S. Department of the Interior Geological Survey. Interagency Advisory Commitee of Water Data. Guidelines for Determining Flood Flow Frequency. Bulletin 17B of the Hydrology Subcommitee 1982.

Váňová, V. and Langhammer, L. 2011. Modelling the impact of land cover changes on flood mitigation in the upper Lužnice basin. J Hydrol Hydromech 59(4): 262–274 doi:10.2478/v10098-011-0022-8.

Ven Te Chow, Maidment, D.K. and Mays, L.W. 1988. Applied of Hydrology. McGraw Hill Book Company, New York.

Wałęga, A. and Cupak, A. 2011a. Informacja hydrologiczna na potrzeby określania stref zagrożenia powodziowego w dolinach rzecznych. [W:] Woda i surowce odnawialne a ich oddziaływanie na środowisko naturalne. Monografia pod redakcją Janusza R. Raka. Wyd. Muzeum Regionalnego im. Fastnachta w Brzozowie, Brzozów, 101–120.

Wałęga, A., Grzebinoga, M. and Paluszkiewicz, B. 2011b. On using the Snyder and Clark unit hydrograph for calculations of flood waves in a highland catchment (the Grabinka River example). ActaActa Scientarium Polonorum, Formatio Circumiectus 10(2): 47–56.

Wałęga, A., Cupak, A. and Miernik, W. 2011c. Wpływ parametrów wejściowych na wielkość przepływów maksymalnych uzyskanych z modelu NRCS-UH. Infrastruktura i Ekologia Terenów Wiejskich 7, PAN o/Kraków, 85–95.

Wałęga, A. 2012. Próba opracowania zależności regionalnych do obliczania parametrów syntetycznego hydrogramu jednostkowego Snydera. Infrastruktura i Ekologia Terenów Wiejskich 2/III, 5–16.

Wałęga, A., Drożdżal, E., Piórecki, M. and Radoń, R. 2012. Wybrane problemy związane z modelowaniem odpływu ze zlewni niekontrolowanych w aspekcie projektowania stref zagrożenia powodziowego. Acta Scientiarum Polonorum, Formatio Circumiectus 11(3): 57–68.

Wałęga, A., Michalec, B., Cupak, A. and Grzebinoga, M. 2015. Comparison of SCS-CN determination methodologies in a heterogeneous catchment. Journal of Mountain Science 12(5): 1084–1094.

Węglarczyk, S. 1998. Wybrane problem hydrologii stochastycznej. Seria Inżynieria Sanitarna i Wodna, Monografia 235. Wyd. Politechniki Krakowskiej, Kraków.

Węglarczyk, S. 2010. Statystyka w inżynierii środowiska. Politechnika Krakowska, Krakow.

WMO. 2008. Manual on low-flow estimation and prediction. Operationa hydrology report no. 50.

CHAPTER 12

Selected Principles of Fluvial Geomorphology

Jan Hradecký* and Václav Škarpich

INTRODUCTION

Fluvial geomorphology is an interdisciplinary branch of science that has been experiencing, in the last decade, a relatively significant increase in the knowledge regarding the functioning and development of fluvial systems. Increased exploitation of river systems by human society calls for the need to solve not only negative effects but also anticipate changes within basins, river channels or individual channel, and floodplain segments and fluvial landforms. Archetypal attraction of human society to rivers has assumed vast dimensions and the human impact on river morphology no longer takes place locally but has acquired regional and global dimensions. As Grill et al. (2015) point out, human society currently affects 48% of the river volume via river regulation and fragmentation. It is assumed that owing to further regulation and construction of water works, this influence is going to reach 93% of the river volume. This fact alone makes it necessary to implement the knowledge of fluvial geomorphology into practice by applying its principles in the management of rivers that in many countries are in a critical condition and that, if not helped, will, in return, limit the development of human society as well as the functioning of fluvial systems and related plant communities and organisms.

On the basis of the study conducted by Newson and Sear (1998), Dollar (2000) summarises: "fluvial geomorphology seeks to investigate the complexity of behaviour of river channels at a range of scales from cross sections to catchments; it also seeks to investigate the range of processes and responses over a very long timescale but usually within the most recent climatic cycle" (other definitions are also given by Gregory 2004). This chapter sums up the main principles of fluvial geomorphology that are the starting point of further research and application of results into practice. These include,

Department of Physical Geography and Geoecology, Faculty of Science, University of Ostrava, Chittussiho 10, 710 00 Ostrava, Czech Republic.
* Corresponding author: jan.hradecky@osu.cz

in particular: (i) the principle of fluvial system and complexity, (ii) the principle of fluvial continuum, and (iii) the principle of time structure. The chapter finishes with a case study from the area of the Czech Carpathians.

Fluvial System Approach and Geomorphic Complexity

System approach to the research of fluvial units is nothing new. On the contrary, along with the development of new methodological approaches, it deepens bringing completely new knowledge about the processes and reactions to changes in partial components. Fluvial geomorphology studies not only the genesis of fluvial landforms and processes but also other river basin factors that play a role in the functioning, changes, and dynamics of fluvial landforms that are not primarily linked to river channels (e.g., climate or geological structure). The study of these phenomena is only possible by using the knowledge of more branches of science, which is why fluvial geomorphology is an interdisciplinary branch that has transformed from descriptive science into science that uses modern technologies and modelling, and that tries to predict changes in fluvial processes with regard to changing conditions in the environment and human activities. If we work with such a concept of fluvial geomorphology, it is evident that system approach is a paradigm that has fully established itself in fluvial geomorphic research. The implementation of system theory stems from the study of Chorley and Kennedy (1971). The theory was further developed in a number of subsequent studies (Schumm 1977; Piégay and Schumm 2003; Brierley and Fryirs 2005, etc.). According to the theory of Chorley and Kennedy (1971), fluvial system can be understood as a "complex adaptive process-response system" composed of two sub-systems, namely the morphological system and cascade system. The morphological portion includes landforms such as channel, floodplain, or slopes, while the cascade system is understood as the flow of water, sediments, and energy" (Piégay and Schumm 2003). Despite considerable scientific progress and ability to model complex fluvial systems, we are still unable to describe and define the parameters and properties of fluvial systems perfectly. One of the key issues that fluvial geomorphology deals with is, among others, the detection of feedback interactions within fluvial systems. Another view of the fluvial system can be that of a hierarchic structure according to Frissell et al. (1986), in which the system is formed by valley segments, reaches of water streams, and channel units. This concept was expanded by Thomson et al. (2001), Poole (2002), and Brierley and Fryirs (2005), in which the fluvial system hierarchic structure is described by means of units of specific measures that reflect evolutionary adjustment changes and that have a specific frequency of disturbances. These units are ecoregion, river network, basin, landscape unit, reach, geomorphic unit, hydraulic unit, and microhabitat. Each unit is further studied for its system elements. Montgomery and Buffington (1997) present one more important fluvial system element—valley segment, which comprises colluvial and alluvial valleys, and bedrock valleys. Within individual valley segments Montgomery and Buffington (1997) distinguish different types of channel reaches according to the channel morphology that reflects transport capacity in relation to sediment supply. A conceptual model of the fluvial system presented by Piégay and Schumm (2003) is shown in Fig. 1.

Besides purely geomorphic aspects considered in the study of the fluvial system, the understanding of the processes affecting the fluvial morphology within individual measures can be helped by other elements—geological structure, basin morphology,

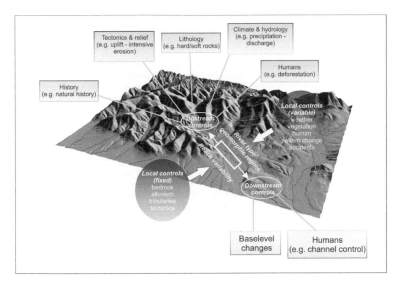

Fig. 1. Conceptual model of the fluvial system introduced by Piégay and Schumm (2003).

hydroclimatic conditions, soil conditions, vegetation, and land use changes. In this respect, a classical model is brought by Piégay and Schumm (2003) who define fluvial system as a set of variables, some of which are characterised as control – external – independent variables (e.g., sediment load or peak discharge), while others as adjustable – internal – dependant variables (e.g., channel pattern or channel slope). Areas of the study of fluvial systems within individual landscape components or fluvial segments are presented by Brierley et al. (2013) using typical questions. Examples are given regardless of the scale of the study area in Table 1.

System approach introduces another aspect of fluvial systems referred to as complexity—according to Wohl (2016), the so-called river geomorphic complexity. This phenomenon is amply discussed in the literature, however many of the principles are summarised in the above mentioned study. Complexity is understood both as (i) complex behaviour or as a property of a system (this complexity refers to the simultaneous presence of simple and complicated behaviours or the presence of non-linear dynamics, self-organization, and emergent properties) and (ii) as involving spatial heterogeneity. Wohl (2016) mentions several reasons why it is necessary to focus on geomorphic complexity, that is important not only in the study of the very landforms themselves but can also be widely applied in the problems of fluvial landscape ecology. Complexity in the form of habitats/locations (given by landforms) is reflected in biodiversity. Complexity has an impact on the attenuation of downstream fluxes such as water, sediments, dissolved substances and organic matter, and organisms. The third display of complexity is related to the resistance and resilience of fluvial systems towards natural and man-induced disturbances. Fourth, complexity reflects and affects the character of processes in streams. The fifth aspect is the influence of spatial heterogeneity on human ability to describe its form and detect changes in time and space. The last point is the application aspect since complexity is increasingly being taken into account in the management and restoration of water streams. However, what is problematic in the research is how to "measure" complexity. According to Wohl (2016), there is not

Table 1. Examples of typical scientific questions related to fluvial geomorphology (selection from Brierley et al. 2013).

Subject area	Question
Geological controls	→ How does the setting influence the erodibility/erosivity of this landscape, and associated flow–sediment fluxes?
Climatic/hydrological settings	→ What is the character of the discharge regime (annual flow, seasonal variability, interannual variability)?
	→ What is the flood history of the catchment?
Morphometry	→ What is the drainage pattern of the (sub)catchment?
	→ What is the shape of the longitudinal profile?
	→ How do slope and discharge affect patterns of gross stream power along the longitudinal profile?
Position in the catchment	→ Are you in a source, transfer, or accumulation zone?
Valley setting	→ What is the valley width/confinement?
	→ What is the valley width/confinement?
	→ What sediments make up the valley floor (type, size, and distribution in channels and on floodplains)?
Vegetation and wood conditions	→ How does vegetation affect instream and floodplain roughness?
Anthropogenic controls	→ How have direct human activities modified the river?

one single parameter of geomorphic complexity that would express the complexity of fluvial system in a simple way. At present, it is clear that one of the possibilities of detecting complex links in the fluvial system is using multidimensional statistics (Fig. 2), which is observed in relation to the studies dealing with the issues of river patterns or variability of channel reaches, or looking into the origin of specific channel morphology (e.g., Montgomery and Buffington 1997; Galia and Hradecký 2014; Güneralp and Marston 2012).

Fluvial Continuum—Problems of Connectivity

Rivers usually begin in higher elevations, and on their way to the base level, which is the level of the sea, they display variegated morphology that is reflected not only in the overall character, the so-called river pattern, but also in the varied morphology of the channel bed, banks, and floodplain. Fluvial systems are typical examples of paradynamic systems that display their functionality and integrity by the processes of mass and energy movement in various directions. As Ward (1989) states, lotic systems can be understood as four-dimensional entities. The longitudinal dimension is understood in the direction spring—mouth. The lateral dimension represents the relations between the channel and the floodplain or the channel and the slopes. The interaction between the channel and groundwater, or the channel and alluvium is the vertical dimension. Since all the three directions are exposed to changing development, the fourth dimension is time.

Within the basin area, relations are observed not only between the channel and the floodplain, but there is a close interconnection between geomorphic subsystems of the landscape, namely the channel and the adjacent slopes (Fig. 3). These relations are very important in terms of the supply of slope sediment into the channel, or the

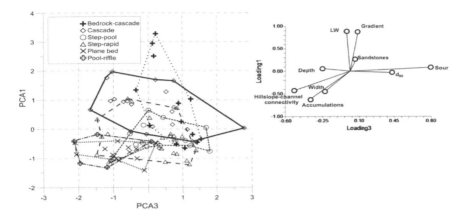

Fig. 2. Principal component analysis - plotting PC1 against PC3 reveals some arrangement of the channel-reach morphologies from positive to negative values along both components (diagonal direction). The lower gradient ones (plane beds and pool-riffles) are concentrated in the lower-left corner of the plot, and negative values for PC1 and PC3 prevail. These morphologies often contain channel accumulations, and an at-a-reach sediment supply is usually presented only in the case of incised channel reaches. In contrast, the inverse area of the plot mostly contains bedrock-cascades, cascades, step-pools, and step rapids, which in terms of PC1 and PC3 suggests that such channel reaches usually lack accumulations of sediment in an active channel (except remains of debris-flow accumulations), and these channel reaches are often supplied directly at-a-reach (Galia and Hradecký 2014, with permission of Elsevier).

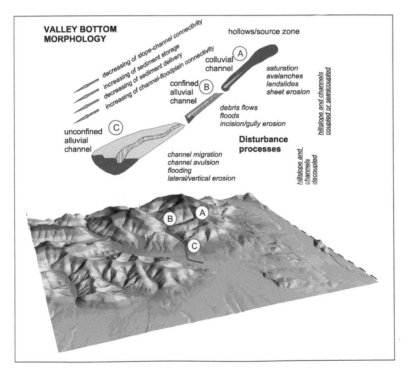

Fig. 3. Coupling and decoupling segments along mountain stream—examples of various confinements and disturbances along channel segments (inspired by Montgomery and Buffington (1998)).

impact of vegetation, mainly large woody debris, on the functioning of river systems (Wohl 2014). The key landscape parameter for the resulting character of river is valley setting (e.g., Brierley et al. 2002). Valley segments within the river network are characterised by similar morphology and typical geomorphic processes (Montgomery and Buffington 1998). An important factor is the form and intensity of slope-channel coupling. Valley segments are divided into three basic types: colluvial, alluvial, and bedrock. The colluvial segment is very specific for its shallow and ephemeral fluvial transport in headwater valleys. Sediments that enter the segment from the adjacent slopes are not effectively transported and therefore they accumulate. Channel beds in this segment are highly undeveloped and can hardly be termed as "channel beds". Bedrock valleys lack valley bottom sediments; floodplain is undeveloped as well. Channel beds, which are usually narrow, gully-like, and straight, are characterised by distinct incision and degradation trend. Bedrock landforms may be covered with a very thin or discontinuous layer of alluvium (Montgomery and Buffington 1998), which is given by high transport capacity. Bedrock valley segments closely adjoin the neighbouring slopes. These valley segments are called confined valley segments (Fryirs et al. 2016). In wider mountain or foothill valleys whose bottom is filled with alluvium, we study alluvial valley segments. Lower energy of the flow leads to a decrease in transport capacity and sediment deposition. Channels are more or less lined by floodplain that reduces direct contact between the channel and the valley slopes. These segments are called partly-confined or laterally-unconfined (Fryirs et al. 2016). Alluvial channels are characterised by a high variability in morphology and a different level of confinement, which is the reflection of local channel slope, sediment supply, and hydraulic discharge (Montgomery and Buffington 1998).

Connectivity of rivers is closely related to the concept of fluvial continuum. Within fluvial geomorphology, fluvial continuum and connectivity refer mainly to the movement of sediments in basin, valley segments, and river reaches. Connectivity has an important application potential since current quality water management projects (e.g., dam constructions) can hardly be carried out without a proper connectivity analysis. Many studies (Fryirs 2012; Fryirs and Brierley 1999; Bracken et al. 2015; Baartman et al. 2013, etc.) see fluvial continuum as interconnection of individual valley segments of a given basin through energy and material transport. Individual segments could then be linked to one of the three zones of a river system (e.g., Kondolf 1994), namely the source zone (zone of erosion) that generates sediments for lower-lying zones, the central zone dominated by the transport of sediments (zone of transport), and the lowest-lying accumulation zone (zone of deposition). A river may thus be perceived as a conveyor belt that provides sediment transport from the place of erosion into the place of prevailing accumulation. Connectivity in geomorphic systems is defined by Fryirs (2013) as the water-mediated transfer of sediment between two different compartments of the catchment sediment cascade. Sediments that enter the river come not only from the zone of erosion of higher-lying, usually mountain channel reaches but from other zones as well. Since the sediment conveyor belt may also be cut off, in order to understand its functionality, we have to understand the following components: (1) sediment delivery from sources (slope or valley floors); (2) entrainment at critical shear stress; (3) transport downstream (whether suspended load, bedload, or mixed load); (4) deposition in a temporary store or a more permanent sink (Fryirs 2013). Long-term weathering and a wide range of other processes produce a significant amount of sediment within the basin, only a part of which, however, is carried as far as the seacoast. This fact points to many limitations in sediment supply and transport in river channels. In this respect, we can speak about (dis)connectivity defined by Fryirs et al. (2007): "Landscape decoupling can be expressed as the degree to which any limiting factor constrains the efficiency of sediment transfer relationships". In the previous part, directions of mutual interconnection of fluvial system units were defined and similarly factors limiting sediment movement can be studied as well. Fryirs et al. (2007) defines three forms of landscape (dis)connectivity: buffers, barriers, and blankets. All of the mentioned groups of connectivity disruption entail a different level of limitation related to sediment entry in the fluvial system and its subsequent movement through the fluvial system. Buffers represent the limitation of sediments in entering the channel network in the lateral and longitudinal directions. This may concern alluvial fans or alluvial pockets of the floodplain that disrupt lateral connectivity, namely the direct contact between the channel and valley slopes or their foot. Longitudinal connectivity may be disrupted by intact valley fills and floodouts that have ephemeral or absent streams. Critical disruption in longitudinal connectivity is caused by barriers. These appear in the channel due to resistant bedrock that forms a rocky terrace above which the sediments may get trapped. Typical examples of barriers are landslide accumulations representing channel barriers of differently long duration (e.g., Korup 2005; Pánek et al. 2007; Kuo and Brierley 2014, etc.) or accumulations of large woody debris or beaver dams. Higher-lying channels of a higher gradient can experience sediment transport triggered by debris flows (e.g., Montgomery and Buffington 1997). In case the transport capacity has been exceeded, accumulations of these forms may fill the valley bottom making a barrier, which is only eroded at extremely high discharges (e.g., Pánek et al. 2010, 2013). The last variant of decoupling is represented by the so-called blankets that disrupt vertical connectivity through their effects on surface-subsurface sediment structures. Blankets can be found both in the channel bed and the floodplain. In the channel bed, these often concern armouring layers that decrease hydraulic roughness and thus inhibit channel bed reworking (Fryirs et al. 2007). Apart from natural buffers, barriers, and blankets, there are also anthropogenic variants (embankments, dams, artificial channel beds, etc.). All factors limiting connectivity have different durations and different effects on sediment entry into the channel and sediment transport through the channel. If an anthropogenic dam remains in existence until it is removed by man again, a reinforcing layer or accumulation of large woody debris may exert influence on the movement of sediments down the channel for a few tens of years (Fryirs et al. 2007). Landslide dams can be broken during a few days or weeks having extreme effects on channel development (Pánek et al. 2010).

Changing in Time

Fluvial systems are not unchanging; they are dynamic elements of landscape and therefore they undergo changes as any other element of the natural system. Some changes are common displays of the system functioning (sediment transport, restructuring of channel bed forms). In some cases, the system is experiencing a shift to another level without any changes in the river pattern (e.g., development of a meander leading to meander neck rupture or fast changes in the layout of anabranching rivers). Sometimes frequent changes of lower magnitude are observed, sometimes there is a fatal event (disturbance) that shifts the state of the fluvial system into a totally different, new evolutionary trajectory. Such changes are usually rare but can be of a large magnitude.

A typical example is a situation in which volcanic eruption causing massive snow melt in the summit parts of the volcano creates a flooding wave that transports millions of tons of sediment that affect the character of the channel and floodplain for a long time. Fluvial disturbances change the nature of the functioning of the river that subsequently changes it geomorphic regime. Werritty (1997) describes short-term changes and long-term changes that have an influence on channel stability. The changes are consequences of processes that take place outside or inside the fluvial system and are generally displays of the environmental change (Macklin and Lewin 1997). Many disturbances occur due to displays of the variability in physical geographical environment (e.g., climate oscillations, tectonics) or they represent essential man-induced transformations (land-use changes, direct interventions into the channel morphology, etc.). Schumm (1973, 1980) elaborated a concept of geomorphic thresholds that he defines as "a threshold of landform stability that is exceeded either by intrinsic change of the landscape itself, or by a progressive change of an external variable". Extrinsic threshold is defined as "one that is exceeded by the application of a force or process external to the system" and intrinsic threshold could be defined as "one in which change occurs without a change in an external variable" (Schumm 1980). A typical example of the first type of geomorphic threshold is a change in basin use or base level change, whereas the second type represents meander cut-off development (Werrity 1997). The development of fluvial systems is broadly discussed in the studies of Brierley and Fryirs (e.g., 2005 or 2016) that deal with the issues of evolutionary trajectories. This concept that is based on the mutual interaction between flow, sediment, and resistant elements along the valley floor is adequately described by the river evolution diagram in Fig. 4. The evolution diagram is the visual presentation of the fluvial system adapting together with its changes within geomorphic timeframes. A characteristic feature reflecting river behaviour is stream power and, as it is generally known, stream power analysis belongs to key tools in fluvial geomorphology since it can be interpreted in relation to sediment transport, deep erosion intensity, floodplain typology, and time-space variability of landforms and processes (Brierley and Fryirs 2016). If total stream power Ω (W m^{-1}) (calculated as $\Omega = \gamma Qs$, where γ is the unit weight of water 9800 N m^{-3}, Q is discharge m^3 s^{-1}, and s is energy slope m/m) defines a full range of energy conditions that can be reached within given valley conditions, then unit stream power ω (W m^{-2}) can be interpreted in relation to channel morphology (this power is calculated as $\omega = \Omega / w$, where w is width of flow in m). Mentioned authors define the so-called imposed boundary conditions within which the river functions, which determines energy situation affected by upstream catchment area, slope and valley confinement, and morphology. These boundary conditions specify a potential range of variability within which certain river morphology can exist under given valley conditions. The outer band is given by the highest and lowest calculated value of total stream power for the given valley segment (an important aspect is valley confinement). As it was mentioned above, it is necessary to distinguish between the terms river behaviour and river change. The term river behaviour describes how the river adapts to a certain morphology, while the river change comprises abrupt shift of the system into a different state represented by different fluvial processes and landforms. In the river evolution diagram, behavioural regime is represented by means of the inner band of the flux boundary conditions. Different river types are within flux boundary conditions affected by the amount of water, sediments, and vegetation character and these conditions easily change in time. The position of the inner band within the outer band is based on channel and floodplain adjustments towards typical flux conditions. The

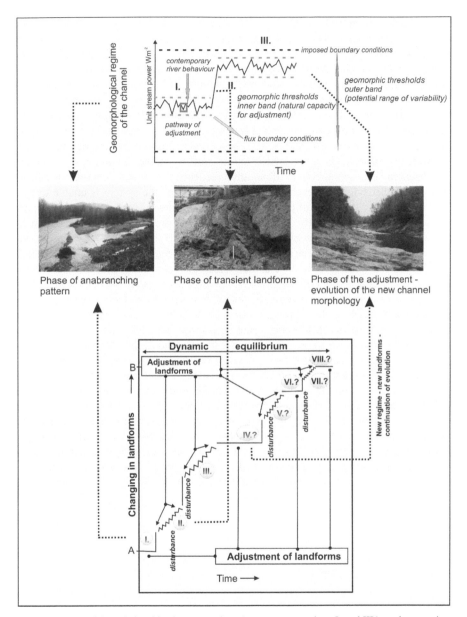

Fig. 4. Scheme of the relationships between robust (upper part, numbers I. and III.) and responsive (upper part, number II.) behaviour of the fluvial system, geomorphic thresholds and three phases of dynamic equilibrium-based on Werritty (1997) and Fryirs et al. (2012).

upper and lower boundaries are delimited by the highest and lowest modelled value of unit stream power. Each river type has its own behavioural regime, which is defined by its expected capacity for adjustment and characterised by its own pathway of adjustment (Brierley and Fryirs 2005). Evolution trajectories present a trend in applied fluvial geomorphology that should become a part of modern management of water streams.

The functioning of fluvial processes reveals a wide range of examples showing that the system can accelerate the changes (positive feedback), or that the system is maintained is the steady-state (negative feedback). Intensity of the display of the change in the fluvial system is then also a function of positive and negative feedbacks (e.g., Phillips 2014). Disruption-induced deflection of the fluvial system leads to the displays of a certain type of behaviour. The character of the behaviour is related to the fact whether the system is deflected to such an extent that its limiting thresholds are exceeded. Robust behaviour corresponds to the situation is which the system continues functioning within extrinsic thresholds. The system is exposed to numerous deflections and intrinsic thresholds are exceeded but no shift to another evolutionary trajectory is observed because the system of negative feedback works well. On the other hand, responsive behaviour corresponds to the state in which the system is unable to bear the disruption or the effect of positive feedback becomes evident. As a result, a major change may occur (Werritty 1997). Phillips (2014) presents a feedback set on an example of three fluvial system elements: surface runoff convergence, shear stress, and fluvial erosion (Fig. 5). All the cases represent positive feedback. Increase in the runoff leads to raised water level in the channel, which increased shear stress. If the water is considerably concentrated in the channel, however, its capacity is not exceeded, flow velocity and stream power increase. Exceeded flow velocity and increased shear stress in alluvial channels leads to fluvial erosion. Deep bottom erosion causes channels to cut deeper and the runoff to concentrate even more. This feedback loop has its end since each of the involved elements has important negative self-effects. Convergence flow is limited by precipitation amount, while resistance flow and gradient are limited by the depth and velocity. Fluvial erosion is limited by local base level, whereas shear stress is dependent on the gradient, and water depth and amount. It is clear that if the character of the channel shifts towards bedrock channel or local base level is reached, the system enters the steady-state phase. In such a case, a limiting factor and negative feedback start to manifest themselves and the system stabilises. On the other hand, the unsteady-state phase begins if shear stress is bigger than shear strength and precipitation is not limited. Therefore, the channel bed of anabranching rivers can thus change very intensively and very fast in the humid climatic phase.

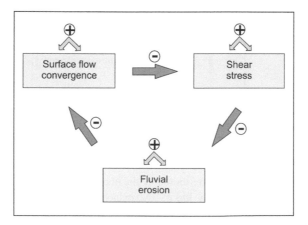

Fig. 5. Selected negative and positive feedbaecks within fluvial system based on Phillips (2014).

OLŠE—GRAVEL-BED RIVER—SHORT CASE STUDY

This section points to practical aspects of the fluvial system concept, namely on an example of the Carpathian gravel-bed river—the Olše River (Škarpich et al. in press). Long-term integration of man in the catchment area and directly in the channel has led to such changes that nowadays these are considered to be so fundamental that the river has exceeded the liming conditions—extrinsic thresholds—and the fluvial morphology has been transformed.

Rivers draining the Czech part of the Flysch Carpathians have deeply incised over the last 60 years. Originally, the gravel-bed character of anabranching rivers with a high amount of transportable material has changed to an incised single channel pattern. The incision resulted from a decrease in the sediment supply into the river continuum system and an accompanying increase in energy potential of the river channels. Sediment supply was particularly reduced due to afforestation and bank stabilisation. Focusing on the Olše River in the Czech part of flysch Carpathians (Moravian-Silesian Beskids, MSB), the results of energy potential aspects of contemporary Carpathian river channels are summarised. The average width of the studied reach of the Olše River active channel was narrowed from 49 m in 1836–1852 and 35 m in 1955 to 24 m in the year 2010. At some parts, the original river bed lowered as much as by 2.3 m between the years 1960 and 2003, which indicates an incision rate of 5.2 cm per year. The average value of incision in the studied reach is about 0.54 m (1.25 cm per year). Therefore, new internal conditions of the channel affected the acceleration of erosion processes. The main reason is the adjustment of water flow dynamics. The unit stream power and hydraulic radius values increased twice or three times between the years 1960 and 2003.

The Olše River is 99 km long. It belongs to the Odra River basin. The highest and lowest sites within the Olše River basin are Ropice Mt, at 1083 m a.s.l., and the mouth to the Odra River at 190 m a.s.l. The study focuses on the river reach between 55.0–70.0 r. km (see Fig. 6), with an average 0.006 channel slope. In the last century, the studied channel reach has developed from the originally anabranching gravel carrying stream into a single channel and, more concretely, a single channel with the occurrence of bedrock. The river bed, especially the gravel bars, is mainly composed of gravels of the medium particle size (d_{50}) of approximately 70 mm in the studied reach. Drainage area of the Olše River basin is 1,118 km^2. Mean annual discharge of the river amounts to 13.7 m^3.s^{-1} at the Věřňovice gauging station located near the mouth to the Odra River (see Fig. 6), where the basin area is 1,071 km^2 (data source: Czech Hydrometeorological Institute). At the Jablunkov-Olše gauging station, located in the study area (see Fig. 6), the mean annual discharge of the river amounts to 1.83 m^3.s^{-1} (the basin area is 92.85 km^2, data source: Czech Hydrometeorological Institute). The Olše River is characterised by high discharge variation, approximately 1:2000 (data source: Povodí Odry Enterprise). It is caused by frequent snowmelt-triggered floods of moderate magnitude and rare floods occurring due to summer rains connected to cyclones. At the Jablunkov rain gauging station, mean annual precipitation in the basin ranges from 800 to 1,400 mm (for location see Fig. 6, data source: Czech Hydrometeorological Institute).

Channelization works in the studied reach of the Olše River have been conducted since the 19th century. As a result, the river is currently characterised by simple channel geometry. The reason was mainly timber floating for iron furnaces and coal mining, construction of water and lumber mills (Polášek 2006), and protection against

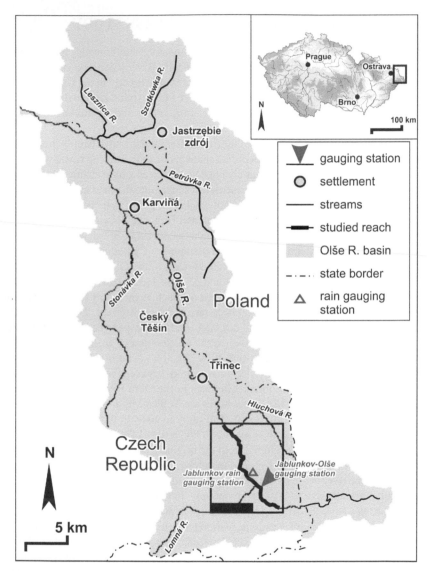

Fig. 6. Location of the Olše R. basin and the studied river reach in the Czech Republic (based on Škarpich et al. 2016).

floods. The main works were connected with: (i) concentrating spur dykes (from the 19th century to the first half of the 20th century); (ii) transversal constructions (weirs and anthropogenic steps, from the 19th century to the present, e.g., see Fig. 7); (iii) sediment mining (from the 19th century to the present); (iv) channel regulation by bank stabilising (from the second half of the 20th century to the present). Towards the end of the first half of the 20th century, the channelization works had unsystematic character (Škarpich et al. 2013). A few short regulated reaches (regulated by concentrating spur dykes especially) had been of short existence before being damaged or totally destroyed

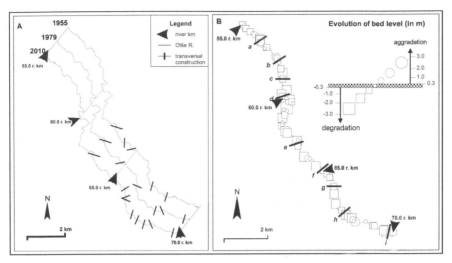

Fig. 7. Transverse constructions and their effects on channel processes. (A) Development of the Olše River course and transverse construction from the year 1836 to the year 2010 in the studied reach. The river km location is from the year 2010. (B) Vertical development of the bed level of the Olše R. studied reach (based on the scheme Landon et al. 1998). A longitudinal profile (low flow waterline altitude) was surveyed in year 1960 (data source: Provincial Archives in Opava) and in year 2003 (data source: Povodí Odry Enterprise Archives); analysed cross profiles h and i see in Fig. 8 (based on Škarpich et al. 2016).

during single flood events. In the second half of the 20th century, massive systematic regulation of the studied reach started including bank stabilising, channelization works via channel straightening, and gradual construction of weirs (Škarpich et al. 2013).

Arable lands, pasture and grasslands, and agricultural areas underwent rapid reduction between the years 1845 and 2000. This area had predisposing factors to gully and sheet erosion and sediment supply into the channels. Areas of forests were expanded about 21% in this period. It helped to reduce sheet and gully erosion and sediment supply deficit into the channels. During the period of 1836–2003, the study reach of the Olše River was shortened by about 11%. The channel was narrowed and the anabranching channel pattern was transformed to a single-thread channel. The active channel of the Olše River study reach displays a distinctive trend of narrowing in the years from 1836–1853 to 2010. The highest aggradation of the river bed in the studied reach is at the 68.85 r. km, were the value of aggradation is 2.50 m (5.81 cm per year). The rate of the deepest incision is observed at the 60.83 r. km, were the value of incision is 2.30 m (5.2 cm per year). The average value of incision in the studied reaches with deep erosion trends is about 0.54 m (1.25 cm per year). During the period of 1960–2003, morphological transformation of the Olše River channel changed the hydraulic parameters. Figure 8 shows a more increasing trend in the values of the cross-sectional area during the discharge in the year 1960 than in 2003. Analysis of wetted perimeter shows relatively similar values in the years 1960 and 2003. These facts are given by the morphology of the wider and shallower channel in the year 1960 (see Fig. 9). Figure 10 shows values and trends in the hydraulic radius in the studied reach of the Olše River channel. In fact, the higher the hydraulic radius is, the higher the velocity of the flow is. In the studied reach, lower values of the hydraulic radius are rather observed in the year 1960 than in 2003. Incision and narrowing of the study

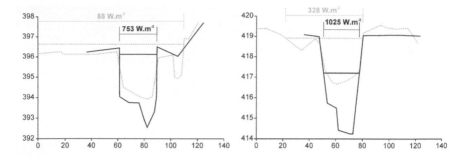

Fig. 8. Cross profiles, water level stage and ω during the 100-recurrence interval discharge in years 1960 (grey colour, upper value of the ω) and 2003 (black colour, lower value of the ω) in the studied river reach of the Olše R.: h (left) – 67.426; i (right) – 69.850 r. km; vertical axes – elevation (in m a.s.l.); horizontal axes – distance (in m) (based on Škarpich et al. 2016).

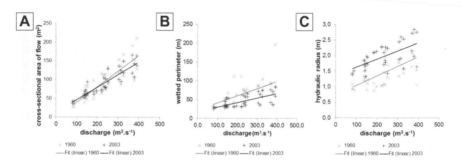

Fig. 9. Cross-sectional area of a flow (A), wetted perimeter (B) and hydraulic radius (C) during the 5, 20, 50, and 100-recurrence interval discharge of the studied cross profiles of the Olše R. channel in the years 1960 (grey color) and 2003 (black color) (based on Škarpich et al. 2016).

reach of the Olše River channel was markedly by the increase of ω (unit stream power) (Fig. 10). Comparing ω in the year 1960 and 2003, some cross profiles analyses show multiple increased values. The highest increase is observed at the 69.85 r. km where the value of ω increased from 328.58 W.m^{-2} in the year 1960 to 1025.25 W.m^{-2} in the year 2003 (during 100-recurrence interval discharge). The increase took place about three times. Mean value of the unit stream power in the studied reach increased from 188 W.m^{-2} in the year 1960 to 482 W.m^{-2} in the year 2003 (during 100-recurrence interval discharge).

Before the 19th century, the channel development was affected by external factors in the basin. Firstly, deforested slopes and little resistance of flysch lithology of the MSB (especially in the Olše River basin) predisposed a high amount of sediment supply into the channel. Hypothetically, an increase in incision factors in this period was identified, namely on the basis of the climatic setting analysis compared to land use changes linked with the deforestation. Deforestation can affect the acceleration of outflow in the basin (*sensu* Unucka and Adamec 2008 or Bobáľ et al. 2010). Higher discharges (flood flows) in this period affected the increase in energy potential of channels. Especially higher discharges of more than 1–2.5 years´ occurrence have a significant impact on the development and morphodynamics of channels (*sensu*, e.g., Wolman and Leopold

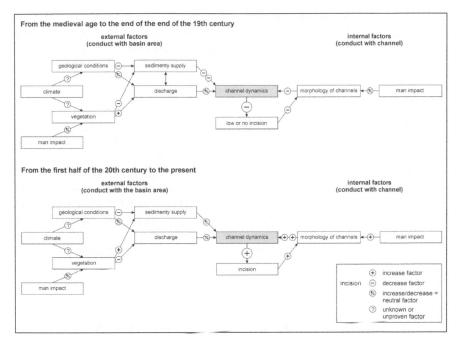

Fig. 10. Conceptual model of channel processes and morphology influencing incision in connection with main factors operates in the Olše R. basin.

1957; Leopold et al. 1964). Yet, this factor can only be identified with difficulties in the fluvial system (from a short observation period of discharges and precipitation in the area of the MSB), similar to climatic fluctuations (especially the period of the Little Ice Age) in the area of the MSB. In conclusion, as based on the above defined causes and the morphological setting of a wide and anabranching channel pattern, the channels were of accumulation character (*sensu* Škarpich et al. 2011, 2013).

In the 19th century, afforestation of slopes started in the MSB, as based especially on the analysis in the Olše River basin. This state reflected a decrease in sediment supply into the channel network and increase factor affected incision. Neutral factor of sediment supply affected the channel dynamics presented in the scheme (see Fig. 10) was defined. The main reason was still existing gravel-bed character of the Olše River channel (mainly the channel bed was built by gravel and gravel bars occurred in the channel). Hypothetically, (analogously on the ground described above), afforestation had a slightly decreasing effect on the incision through the outflow in the basin and higher magnitude discharges (*sensu* Škarpich et al. 2013). Some of the main factors influencing the channel morphology of the 19th century is human impact, and incision and narrowing impact. The occurrence of barriers (e.g., weirs) and buffers (e.g., bank stabilising structures) affected the morphology of channels and sediment supply into the channel continuum system (*sensu* Fryirs et al. 2007). This state caused a decrease in the amount of sediment material in channels, inducing the effect of hungry water and incision of channels (*sensu* Kondolf 1997). Contemporary state of an incised and narrowed channel leads to increased incision. Energy potential of a narrow and incised or regulated channel is higher than in the past. In conclusion, when we exclude external factors (especially sediment deficit) in the basin, internal factors of the channel morphology affected by man and processes operating since the 19th century (incision, narrowing) play an important role in the channel development.

Conclusion

Fluvial geomorphology of the last decades has significantly shifted the level of understanding the development and functioning of rivers. This is supported convincingly by the studies of Dollar (2000, 2002), Hardy (2006), and Scott (2010, 2011, 2013). With respect to application potential, fluvial geomorphology brings an interdisciplinary view of the functioning of rivers, which used to be ignored greatly by water managers in the past and which is why many rivers all over the world have suffered due to either direct or indirect effect impact on the geomorphic regime induced by man. Typical examples are accelerated sediment accumulations in channels or river bed incision monitored by, for example, Australian geomorphologists (Brizga and Finlayson 2000). Nowadays, solution and remedy for the contemporary state of such rivers are impossible without fluvial geomorphic analysis. Therefore, fluvial geomorphology cooperating hand in hand with hydrology and ecology (e.g., De Waal et al. 1998; Brierley and Fryirs 2005) but also technical and social sciences (Lee and Choi 2012; Lah et al. 2015, etc.) seeks ways to take sustainable care of rivers and minimise economic costs on their management. Complex restoration solutions then require accepting the concept of fluvial system, by which the care of the basic element of our landscape—rivers—can be optimised.

Acknowledgement

This chapter was supported by SGS05/PřF/2017-2018. Thanks are extended to Monika Hradecka for English translation and English style corrections.

Keywords: Fluvial system, complexity, fluvial continuum, time changes, Olše River case study.

References

Baartman, J.E.M., Masselink, R., Keesstra, S.D. and Temme, A.J.A.M. 2013. Linking landscape morphological complexity and sediment connectivity. Earth Surface Processes and Landforms 38: 1457–1471.
Bobál, P., Šír, B., Richnavský, J. and Unucka, J. 2011. Analysis of the impact of land cover spatial structure change on the erosion processes in the catchment. Acta Montanistica Slovaca 15: 269–276.
Bracken, L.J., Turnbull, L., Wainwright, J. and Bogaart, P. 2015. Sediment connectivity: a framework for understanding sediment transfer at multiple scales. Earth Surface Processes and Landforms 40: 177–188.
Brierley, G., Fryirs, K., Outhet, D. and Massey, C. 2002. Application of the river styles framework as a basis for river management in New South Wales, Australia. Applied Geography 22: 91–122.
Brierley, G., Fryirs, K., Cullum, C., Tadaki, M., Huang, H.Q. and Blue, B. 2013. Reading the landscape: integrating the theory and practice of geomorphology to develop place-based understandings of river systems. Progress in Physical Geography 37: 601–621.
Brierley, G.J. and Fryirs, K.A. 2005. Geomorphology and the River Management: Applications of the River Styles Framework. Blackwell Publishing, Malden.
Brierley, G.J. and Fryirs, K.A. 2016. The use of evolutionary trajectories to guide 'moving targets' in the management of river futures. River Research and Applications 32: 823–835.

Brizga, S. and Finlayson, B. (eds.). 2000. River Management: The Australasian Experience. Wiley and Sons.

Chorley, R.J. and Kennedy, B.A. 1971. Physical Geography: A Systems Approach. Prentice-Hall International, London.

De Waal, L.C., Large, A.R.G. and Wade, P.M. (eds.). 1998. Rehabilitation of Rivers: Principles and Implementation. Wiley and Sons, Chichester.

Dollar, E.S.J. 2000. Fluvial geomorphology. Progress in Physical Geography 24: 385–406.

Dollar, E.S.J. 2002. Fluvial geomorphology. Progress in Physical Geography 26: 123–143.

Frissell, C.A., Liss, W.J., Warren, C.E. and Hurley, M.D. 1986. A hierarchical framework for stream habitat classification: viewing streams in a watershed context. Environmental Management 10: 199–214.

Fryirs, K. and Brierley. G.J. 1999. Slope-channel decoupling in Wolumna catchment, New South Wales, Australia: the changing nature of sediment sources following European settlement. Catena 35: 41–63.

Fryirs, K. 2013. (Dis)Connectivity in catchment sediment cascades: a fresh look at the sediment delivery problem. Earth Surface Processes and Landforms 38: 30–46.

Fryirs, K.A., Brierley, G.J., Preston, N.J. and Kasai, M. 2007. Buffers, barriers and blankets: the (dis) connectivity of catchment-scale sediment cascades. Catena 70: 49–67.

Fryirs, K.A., Wheaton, J.M. and Brierley, G.J. 2016. An approach for measuring confinement and assessing the influence of valley setting on river forms and processes. Earth Surface Processes and Landforms 41: 701–710.

Galia, T. and Hradecký, J. 2014. Channel-reach morphology controls of headwater streams based in flysch geologic structures: an example from the Outer Western Carpathians, Czech Republic. Geomorphology 216: 1–12.

Gregory, K.J. 2004. Fluvial geomorphology. pp. 392–398. In: Goudie, A.S. (ed.). Encyclopedia of Geomorphology. Routledge, London, New York, USA.

Grill, G., Lehner, B., Lumsdon, A.E., MacDonald, G.K., Ch. Zarfl and Liermann, C.R. 2015. An index-based framework for assessing patterns and trends in river fragmentation and flow regulation by global dams at multiple scales. Environmental Research Letters 10: 015001.

Güneralp, I. and Marston, R.A. 2012. Process-form linkages in meander morphodynamics: bridging theoretical modeling and real world complexity. Progress in Physical Geography 36: 718–746.

Hardy, R.J. 2006. Fluvial geomorphology. Progress in Physical Geography 30: 553–567.

Kondolf, G.M. 1994. Geomorphic and environmental effects of instream gravel mining. Landscape and Urban Planning 28: 225–243.

Kondolf, G.M. 1997. Hungry water: Effects of dams and gravel mining on river channels. Environmental Management 21: 533–551.

Korup, O. 2005. Geomorphic hazard assessment of landslide dams in South Westland, New Zealand: fundamental problems and approaches. Geomorphology 66: 167–188.

Kuo, Ch.W. and Brierley, G. 2014. The influence of landscape connectivity and landslide dynamics upon channel adjustments and sediment flux in the Liwu Basin, Taiwan. Earth Surface Processes and Landforms 39: 2038–2055.

Lah, T.J., Park, Y. and Cho, Y.J. 2015. The Four major rivers restoration project of South Korea: an assessment of its process, program, and political dimensions. Journal of Environment and Development 24: 375–394.

Landon, N., Piégay, H. and Bravard, J.P. 1998. The Drôme river incision (France): from assessment to management. Landscape and Urban Planning 43: 119–131.

Lee, S. and Choi, G.W. 2012. Governance in a river restoration project in South Korea: the case of Incheon. Water Resources Management 26: 1165–1182.

Leopold, L.B., Wolman, M.G. and Miller, J.P. 1964. Fluvial Processes in Geomorphology. Freeman, San Francisco.

Macklin, M.G. and Lewin, J. 1997. Channel, floodplain and drainage response to environmental change. pp. 15–45. In: Thorne, C.R., Hey, R.D. and Newson, M.D. (eds.). Applied Fluvial Geomorphology for River Engineering and Management. Wiley and Sons, Chichester.

Montgomery, D.R. and Buffington, J.M. 1997. Channel-reach morphology in mountain drainage basins. Geologica Society of America Bulletin 109: 596–611.

Montgomery, D.R. and Buffington, J.M. 1998. Channel processes, clasification and response. pp. 13–42. In: Naiman, R. and Bilby, R. (eds.). River Ecology and Management. Springer-Verlag, New York.

Newson, M.D. and Sear, D. 1998. The role of geomorphology in monitoring and managing river sediment systems. Water and Environment Journal 12: 18–24.

Pánek, T., Smolková, V., Hradecký, J. and Kirchner, K. 2007. Landslide dams in the northern part of Czech Flysch Carpathians: geomorphic evidence and imprints. Studia Geomorphologica Carpatho-Balcanica 41: 77–96.

Pánek, T., Hradecký, J., Smolková, V., Šilhán, K., Minár, J. and Zernitskaya, V. 2010. The largest prehistoric landslide in northwestern Slovakia: Chronological constraints of the Kykula long-runout landslide and related dammed lakes. Geomorphology 120: 233–247.

Pánek, T., Smolková, V., Hradecký, J., Sedláček, J., Zernitskaya, V., Kadlec, J., Pazdur, A. and Řehánek, T. 2013. Late-Holocene evolution of a floodplain impounded by the Smrdutá landslide, Carpathian Mountains (Czech Republic). The Holocene 23: 218–229.

Phillips, J.D. 2014. Thresholds, mode switching, and emergent equilibrium in geomorphic systems. Earth Surface Processes and Landforms 39: 71–79.

Piégay, H. and Schumm, S.A. 2003. System approaches in fluvial geomorphology. pp. 105–134. *In*: Kondolf, G.M. and Piégay, H. (eds.). Tools in Fluvial Geomorphology. John Wiley and Sons, Chichester.

Polášek, J. 2006. Tradice výroby a zpracování železa v Beskydech a Pobeskydí: Plavení dřeva a zaniklé výrobní objekty v oblasti Moravskoslezských a Slezských Beskyd. Beskydian Museum, Frýdek Místek.

Poole, G.C. 2002. Fluvial landscape ecology: addressing uniqueness within the river discontinuum. Freshwater Biology 47: 641–660.

Schumm, S.A. 1973. Geomorphic thresholds and the complex response of drainage systems. pp. 299–310. *In*: Morisawa, M. (ed.). Fluvial Geomorphology. State University of New York, Binghapton.

Schumm, S.A. 1977. The Fluvial System. John Wiley and Sons, Chichester.

Schumm, S.A. 1980. Some applications of the concept of geomorphic thresholds. pp. 473–485. *In*: Coates, D.R. and Vitek, J. (eds.). Thresholds in Geomorphology. George Allen and Unwin, London.

Scott, T. 2010. Fluvial geomorphology. Progress in Physical Geography 34: 221–245.

Scott, T. 2011. Fluvial geomorphology 2008–2009. Progress in Physical Geography 35: 810–830.

Scott, T. 2013. Review of research in fluvial geomorphology 2010–2011. Progress in Physical Geography 37: 248–258.

Škarpich, V., Hradecký, J. and Tábořík, P. 2011. Structure and genesis of the quaternary filling of the Slavič River valley (Moravskoslezské Beskydy Mts., Czech Republic). Moravian Geographical Report 19: 30–38.

Škarpich, V., Hradecký, J. and Dušek, R. 2013. Complex transformation of the geomorphic regime of channels in the forefield of the Moravskoslezské Beskydy Mts.: case study of the Morávka River (Czech Republic). Catena 111: 25–40.

Škarpich, V., Galia, T. and Hradecký, J. 2016. Channel bed adjustment to over bankfull discharge magnitudes of the flysch gravel-bed stream—case study from the channelized reach of the Olše River (Czech Republic). Zeitschrift für Geomorphologie 60: 327–341.

Thomson, J.R., Taylor, M.P., Fryirs, K.A. and Brierley, G.J. 2001. A geomorphological framework for river characterization and habitat assessment. Aquatic Conservation: Marine and Freshwater Ecosystems 11: 373–389.

Unucka, J. and Adamec, M. 2008. Modeling of the land cover impact on the rainfall-runoff relations in the Olse catchment. Journal of Hydrology and Hydromechanics 56: 257–271.

Ward, J.V. 1989. The four-dimensional nature of lotic ecosystems. Journal of the North American Benthological Society 8: 2–8.

Werritty, A. 1997. Short-term changes in channel stability. pp. 47–65. *In*: Thorne, C.R., Hey, R.D. and Newson, M.D. (eds.). Applied Fluvial Geomorphology for River Engineering and Management. Wiley and Sons, Chichester.

Wohl, E. 2014. A legacy of absence: wood removal in US rivers. Progress in Physical Geography 38: 637–663.

Wohl, E. 2016. Spatial heterogeneity as a component of river geomorphic complexity. Progress in Physical Geography 40: 598–615.

Wolman, M.G. and Leopold, L.B. 1957. River flood plains: some observation on their formation. U.S. Geological Survey Professional Paper 282: 87109.

CHAPTER 13

On Measurements of Fluvial Geomorphology Parameters in the Field

Václav Škarpich

INTRODUCTION

The fluvial system is a process-response system where the processes are involved in the movement, storage, and transformation of energy and matter between system elements (fluvial forms). It includes the morphologic component of fluvial forms and the cascading component of water and sediment as fluvial processes (Schumm 1977). These two main components are viewed as main geomorphic agents observed by fluvial-geomorphic research. The first one, the morphologic component, is described as a three-dimensional system (Brierley and Fryirs 2005) where the basins are given by geographical space of the Earth and each place is characterised by altitude. Morphology of the basin is a main factor for water and energy movement that transports, erodes, and deposits sediment within the basin (Kondolf 1997). The relation between the processes that create fluvial landforms and the resulting fluvial morphology is complex and varies over time and space (Schumm 1977; Leopold et al. 1995; Kondolf 1997).

Based on this theoretical approach of interaction between morphologic and cascading components of the fluvial system, fluvial geomorphic research includes analysis of fluvial forms and fluvial processes. It is devoted to understanding rivers, both in their natural setting as well as how they respond to human-induced changes in the basin.

In fact, the main parameters monitored in the fluvial system are: (i) morphology characteristics of the channel and floodplain, (ii) channel bed sediment composition, (iii) discharge of water, defined as the volume rate of flow, and (iv) discharge of sediment, defined as rate of sediment movement. The main goal of this chapter is to show methods for measuring these basic parameters of the fluvial system.

Department of Physical Geography and Geoecology, University of Ostrava.
  E-mail: vaclav.skarpich@osu.cz

Basic Methodologies for Morphology Characteristics Measurement

The basic methodology approach of morphology analysis in fluvial research is fluvial-geomorphological field mapping. Sketches and maps of fluvial landscapes and landforms have been and are fundamental methods to analyse and visualise surface and near-surface features (Kondolf and Piégay 2003). The main problem in field mapping, is that by its nature, it is subjective and affected by the skills of the mapper. The level of these skills determines the quality of field mapping (Cooke and Doornkamp 1990).

The physical character of the morphology of a fluvial system is described by topography, and in detail, by longitudinal and transversal profiles. A longitudinal profile (as a two dimensional view along a channel) illustrates the type of bed forms and changes in slope. A transversal (cross-sectional) profile is a two dimensional view across a channel (and across secondary channels) and the adjacent floodplain perpendicular to stream (river) centerline. Parameters drawn from cross sections include width, depth, area of section flow, wetted perimeter, and the width to depth ratio (as hydraulic radius). Topography shows a three-dimensional view of a fluvial system and illustrates patterns of erosion and accumulation. Longitudinal and transversal profiles and topography are often surveyed by using electronic distance meters, automatic or hand levels, or total stations (TST, total station theodolite) (Shilling et al. 2005). The important parameter monitored in a river system is the bankfull stage. The bankfull stage is associated with the flow that just fills the channel to the top of its banks and at a point where the water begins to overflow onto a floodplain (Leopold et al. 1995). This parameter is necessary for determining, for example, of channel width, depth or hydraulic radius. Nevertheless, the topography of a river bed can be determined by sounding measurements from the water surface. The sounding measurements are also referred to as bathymetry. Bathymetry is the study of underwater depth of channels (can be used for lakes or oceans too), which gives information about the bottom configuration (Boiten 2008). Bathymetric surveys are generally conducted with transducers and echosounders. Transducers transmit a sound pulse from the water surface and record the same signal when it reflects from the bottom of the water body. An echosounder attached to the transducer records time interval between the emission and return of a pulse. This time interval is used to determine the depth of water along with the speed of sound in water at the time (Boiten 2008; Thurman 1997). A swath-sounding sonar system is one that is used to measure the depth in a line extending outwards from the sonar transducer. This system acquires data in swath at right angles to the direction of motion of the transducer head.

Except for field measurements, archive maps or remote sensing data are used in fluvial-geomorphic research. In the last years, LIDAR (Light Detection and Ranging) is used to examine the surface of the floodplain. LIDAR is a remote sensing technology that measures distance by illuminating a target with a laser and analysing the reflected light (Olsen 2007). Subsequently, data derived from LIDAR gives information about landforms occurring in fluvial landscapes. In the last few years, unmanned aerial vehicles (UAVs) are used for remote sensing of the fluvial landscape.

Bed Sediment Sampling Procedures and Equipment

Bed sediment analysis is frequently used, for example, in fluvial-geomorphic research or water management. Information on bed-material is needed for a variety of purposes, for example, streambed monitoring for detecting watershed impacts, analysing stream habitat, computations of flow hydraulics (as roughness of beds and banks), etc. (Bunte and Abt 2001).

Bed material (especially in gravel and cobble-bed streams) can be sampled by two different methods: (i) surface sampling and (ii) volumetric sampling. Surface sampling is based on obtaining samples of a pre-selected number of surface particles from a pre-defined sampling area (Wolman 1956), while volumetric sampling is based on obtaining samples of a pre-selected sediment mass/volume from a predefined sedimentary layer (Bunte and Abt 2001).

Surface sampling collects bed-surface particles that are exposed to the top of the streambed. The vertical extent of the surface sediment is equal to the diameter of one particle, that is, the particle that is exposed on the surface at any given point. The sampling of particles is based on the analysis of the three particle axes that define the three-dimensional shape of a particle: (i) the longest (a-axis), (ii) the intermediate (b-axis), and (iii) the shortest (c-axis) axis (see Fig. 2). However, only the intermediate (b-axis) is frequently used for grain-size characterization of particle.

Three fundamental methodological approaches that are used for the bed-surface sediment sampling: (i) line counts, (ii) grid counts, and (iii) areal sampling procedures (Bevenger and King 1995; Bunte and Abt 2001; Wolman 1956). The line count (or pebble count) sampling procedure is based on the select and hand-picked pre-set number of

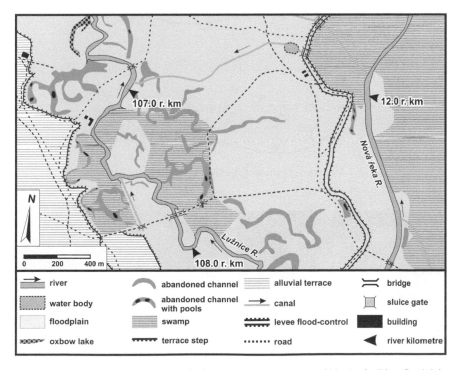

Fig. 1. Example of thematic map of fluvial landscape of the Lužnice and Nová řeka River floodplain in Czech Republic.

surface particles at even-spaced increments along transect. Grid counts select a hand-picked pre-set number of surface particles at even-spaced grid points. Areal sampling procedure includes analysis of all surface particles contained within a small pre-set area of the river-bed. It is used to ensure that small particles are included representatively in the sample. Photographs of the bed (see Fig. 3) are frequently used determination for bed sediment-size areal sampling (Bridge and Demicco 2008).

Volumetric samples extract a pre-defined volume of sediment from the bed. Volumetric samples are three-dimensional and may be taken from various strata of the sediment column. In general, the volume of material has to be large enough to be independent from the maximum particle size. The minimum depth of a volumetric sample should be at least twice the diameter of the maximum particle size and the minimum weight should be 200 times the weight of the largest particle of interest (based on study Diplas and Fripp 1992). Volumetric sampling of bed material in dry beds doesn't need special sampling equipment. Difficulties arise from sampling under water. The main problem is water flow, which washes fine particles during the sampling (Blomqvist 1991). Thus, bed material should generally be sampled during lowest flows when much of the bed is exposed. Nevertheless, several procedures and equipment for taking volumetric samples under water are possible to use, for example, McNeil samplers, barrel samplers, freeze-core samplers, etc. (Bunte and Abt 2001).

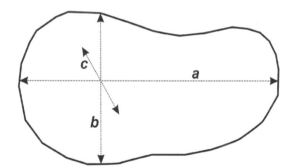

Fig. 2. Definition of particle a-, b-, and c-axes (Bunte and Abt 2001).

Grains selected

Greyscale image overlaid on grains selected

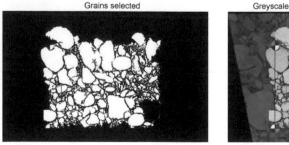
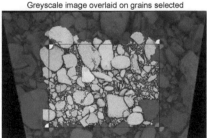

Fig. 3. Particle-size analysis procedures with using of photographs computed in Sedimetrics Digital Gravelometer 1.0 software.

Discharge Measurements

The main driving factor of (morphological and sedimentological) changes in river systems is discharge. Discharge is the volume rate of water flow, more precisely the volume of water in a river per unit time. The SI (International System of Units) unit is m^3/s (cubic meters per second). In the United States, Customary Units or British Imperial Units are used, and volumetric flow rate is often expressed as ft^3/s (cubic feet per second) or gallons per minute (either U.S. or imperial definitions). Discharge is highly variable in natural river systems. It can be placed within a continuum from low flow to extremely large floods. Quantifying discharge is necessary in order to put measurements of energy and material transport into a basin context.

For determination of discharge in open channel, various methods can be applied. The flow measuring methods are based on a simple concept. In open channels, the discharge (Q) in any cross-section and at any moment is given by relation:

$$Q = \overline{v} \cdot A, \tag{1}$$

where A is the area of flow and $\overline{v}$ is average stream velocity calculated on the basis of velocity measurements carried out in that particular section (Boiten 2008). We identified several methods allowing for access to the flow regime of the river. These can be divided into: (i) single measurement methods and (ii) methods of continuous measurement.

Single measurement methods are based on a measurement over a short period. It is often performed to calibrate one of the methods of continuous measurement or in ungauged rivers. Velocity area method is one of them. Discharge is determined as the product of the cross-sectional area of flow and flow velocities measured by using current-meters, or electromagnetic sensors. Hydrometric propeller (Woltmann 1790) is very often used for identification of discharge in the measured area of flow. A hydrometric propeller measures the water speed, just like a wind speed sensor. Many measurements at different depths across the river make an average stream velocity. Together with the cross-section area, the discharge can be calculated using of Eq. (1).

In recent years, advances in technology have allowed discharge measurements to be made, by the use of an Acoustic Doppler Current Profiler (ADCP, see Fig. 4). An ADCP uses the principles of the Doppler Effect (Doppler 1903) to measure the velocity

Fig. 4. Discharge measurements by use of an Acoustic Doppler Current Profiler (ADCP) on the Bílovka River (Czech Republic).

of water more precisely. It determines water velocity by sending a sound pulse into the water and measuring the change in frequency of that sound pulse reflected back to the ADCP by sediment or other particulates being transported in the water. The change in frequency that is measured is translated into water velocity (Griffiths and Flatt 1987).

Other examples of single measurement discharge methods are dilution methods (Boiten 2008). They are based on the suitable selected tracers which are injected at an upstream section and subsequently measured in a downstream sampling section. The discharge can be calculated from the concentration of tracer injected at the upstream section and the concentration measured at the downstream section (Collins and Wright 1964). This method is especially suitable for mountain streams where the discharge is highly affected by turbulent processes of flow (Schuster 1970).

The discharge can be calculated from the measurement of the main morphological parameters of channels and water surface slope. This methodological approach is called the slope-area method (Boiten 2008). For example, Manning formula (Manning 1891) is useful for discharge estimation. It is an empirical formula estimating the average velocity of a liquid flowing (water flowing) in a conduit that does not completely enclose the liquid (water), that is, open channel flow. The Manning equation, written in terms of discharge, is:

$$Q = \frac{1}{n} \cdot R^{2/3} \cdot S^{1/2} \cdot A, \tag{2}$$

where S is water surface slope, A is cross-sectional area of flow, R is hydraulic radius, and n is estimating a roughness coefficient for the channel boundaries. This methodological approach is suitable, for example, for peak discharge estimating during the floods (Manning 1891). This methodology is very inaccurate and the main problem is determination of roughness coefficient.

Discharge in river channels are normally measured indirectly. It is based on relying on the conversion of a record of water level (e.g., by water level measurement sensor) to flow using a stage-discharge relation. The stage discharge relation (rating curve) is obtained either by installing a gauging structure with known hydraulic characteristics (channel slope, cross section shape), or by measuring the stream velocity using propeller-type current meters or other methods.

Other continuous methods can be regarded, for example, using flow measuring structures, where discharges are derived from measurements of the upstream water level, which are continuously measured at a certain distance upstream of the structure. As modern methods for continuous discharge measurement are: (i) acoustic methods, where discharges are calculated from measurements of the velocity computed from the difference in running time of a sound wave, which is transmitted diagonally across the channel in upstream and downstream direction or (ii) electro-magnetic method, where the flow velocity is determined by measuring the voltage induced by a moving conductor (streamflow) in a magnetic field (Boiten 2008).

Sediment Transport Measurements—Procedures and Equipment

Fluvial sediment transport is the movement of solid particles. Typically, is it due to a combination of gravity acting on the sediment and the movement of the fluid in which the sediment is entrained (Schumm 1977; Leopold et al. 1995). Sediment transport in the basin determines the character of the fluvial landscape, as well as erosion or deposition processes. A reason for measuring these parameters is to predict processes and behaviours of the fluvial system. Typical questions are: "how far will banks erode?", "how much time will pass before a reservoir fills with sediment?", etc.

The mode of sediment transport moves by rolling or saltating or in suspension (Abbott and Francis 1977; Kondolf and Piégay 2003). Based on the mode of sediment movement, we distinguished methods for measuring bed load transport and sediment suspension transport (suspended load).

A suspended load gauging requires the spatially distributed measurement of both sediment concentration and water velocity (Kondolf and Piégay 2003). Similar to discharge measurement, we can identify several methods within the single measurement methods and methods of continuous measurement. The simplest way of taking a sample of suspended sediment is to dip a bucket or other container into the stream, while subsequently discharge is measured. The sediment contained in a measured volume of water is filtered, dried, and weighed. This gives a suspended sediment discharge as concentration of sediment during the measured discharge. Also, samples can be manually pumped from a stream, where the water with suspended sediment sample is pumped through a filter (Rijn 1979). For continuous measuring, automatic samplers are available, which can pump a sample through series of filters into a series of bottles, either at pre-determined times and intervals, or as triggered by predetermined flow conditions. An optical turbidimeter is often used for continuous measuring. An optical turbidimeter passes a beam of light through sediment-laden water from a source on one side of a channel to a sensor at the other side. The sensor may either measure the extent through which light is absorbed by the sediment particles or measure the extent through which the light is scattered by suspended particles (Grobler and Weaver 1981). A similar principle is used, for example, in nuclear gauges, which measure either the absorption or the scattering of gamma radiation instead of light (Lal 1994).

The bottom of alluvial river channels is composed of granular material. This sediment material can be transported by flowing water if the flow velocities are sufficiently high (Boiten 2005). It is defined as a bed-load transport.

The basic principle of mechanical trap-type bed-load samplers is used to measure the interception of the sediment particles. These sediment particles are in transport close to the bed over a small incremental width of the channel bed. Most of the particles close to the bed are transported as bed-load, but the sampler will inherently collect a small part of the suspended-load (related to vertical size of intake mouth) too. The bed-load transport measured by a mechanical sampler is dependent on its efficiency (instrumental errors), on its location with respect to the bed form geometry (spatial variability), and on the near-bed turbulence structure (temporal variability) (Rijn 1986). We distinguish mobile or stationary trap-type bed-load samplers. Mobile trap-type bedload samplers, for example, the Helley-Smith sampler (HS), bed-load transportmeter Arnhem (BTMA), etc., can be used only for a limited time periods during sufficiently high flow velocities (Rijn 1986). The simplest way to measure the bed-load is to dig a hole in the streambed (which is stationary) and remove and weigh the material that drops into it (Lenzi et al. 1999).

Other methodology for bed-load transport measuring is to record the bed profile as a function of time (bed form tracking). The basic principle of bed form tracking is to compute the bed-load transport from bed form profiles measured at successive time intervals under similar flow conditions (Engel and Lau 1980).

A number of studies report the use of radio-active tracers to monitor the bed-load movement (e.g., Osterberg et al. 1965; Nelson et al. 1966). The technique is to insert a radio-active tracer into the stream in a form similar to the bed-load; it is should have the same shape, size, and weight as the natural sediment. The movement downstream can then be monitored using portable detectors. Another similar technique used to investigate sediment transport is the application of passive integrated transponders (PIT) tags (Kondolf and Piégay 2003). It traces pebble movements by inserting PIT tags into individual clasts. The movement is registered by antennas capable of sampling and detecting PIT tags. Each PIT tag has its own signal identification and this technique is ideal for tracking the individual movements of episodically transported particles, for example, in gravel-bed rivers. In the last few years, the using of piezoelectric or geophone sensors have provided information about continuous bed-load transport rates during high flow events also (Lenzi et al. 1999).

Safe Field Operations

Promoting a safe environment and preventing injury in the workplace is a top priority for all fluvial geomorphologists, water engineers, researchers, etc. It is paramount that each person must be aware of safety procedures and operational policies to ensure avoiding or minimizing personal injury in the vicinity of the field activity, for example, during the fluvial-geomorphic research. Preparation before beginning field work will help in avoiding accidents and safe returns (Yobbi et al. 1996). A list of required tools, instruments, and supplies should be made when the researcher is planning research activities. Also, determination of any potential risks should be noted, for example, the maximum velocity and depths of rivers that may be encountered in studied river reach.

In preparing, the researcher should consider some of the following elements (after Yobbi et al. 1996):

- Learn how to swim or be prepared to wear a personal floatation device.
- Maintain a schedule of physical examinations, as appropriate for your age or with your personal physician.

Seek medical advice if you have a limiting medical condition; have an adequate supply of medication on hand for field trips.

- If you have severe allergic reactions to insect bites or stings, carry appropriate serum and instruct colleagues on administering it in case you are incapacitated.
- Learn modern first aid procedures.
- Learn about potential hazards of the field area.
- Learn wilderness survival techniques.
- Know where the nearest emergency medical facilities are located.
- Be prepared for everything (e.g., unexpected events)!

Appropriate personal gear may include (after Yobbi et al. 1996):

- Adequate clothing for weather conditions.
- Proper foot gear (including hiking boots, steel-toed safety shoes, hip boots or waders).

- Hats appropriate for weather conditions (e.g., wool stocking cap for cold weather).
- Life jacket, safety goggles, work gloves.
- Orange fluorescent vest is required, for example, while working on bridges or surveying on or adjacent to roads.

112 is the common emergency telephone number that can be called free-of-charge from any fixed or mobile phone in order to reach emergency services in numerous European Countries, including all member states of the European Union, as well as several other countries in the world. 911 is emergency telephone number used in the United States, Canada, as well as in some Latin American countries.

Acknowledgements

The author sincerely thanks Zdeněk Máčka (Masaryk University in Brno) and Stanislav Ruman (University of Ostrava) for their informal review of the early draft. I would like to extend thanks to the anonymous reviewer for inspirational comments and recommendations.

Keywords: Hydrometry, fluvial-geomorphological field research, morphology of rivers measurement, bed sediment sampling, discharge measurement, sediment transport measurement, safe field operations.

References

Abbott, J.E. and Francis, J.R.D. 1977. Saltation and suspension trajectories of solid grains in a water stream. Philosophical Transactions of the Royal Society a-Mathematical Physical and Engineering Sciences 284(1321): 225–254. doi:10.1098/rsta.1977.0009.

Bevenger, G.S. and King, R.M. 1995. A pebble count procedure for accessing watershed cumulative effect. U.S. Dept. of Agriculture, Forest Service, Rocky Mountain Forest and Range Experiment Station, Fort Collins, Co., USA, 23 p.

Blomqvist, S. 1991. Quantitative sampling of soft-bottom sediments: problems and solutions. Marine Ecology Progress Series 76: 295–304.

Boiten, W. 2008. Hydrometry: A Comprehensive Introduction to the Measurement of Flow in Open Channels, 3rd Ed. CRC Press, Boca Raton.

Bridge, J. and Demicco, R.V. 2008. Earth Surface Processes, Landforms and Sediment Deposits. Cambridge University Press, Cambridge, 815 pp.

Brierley, G.J. and Fryirs, K.A. 2005. Geomorphology and River Management: Applications of the River Styles Framework. Blackwell Publishing, Malden, Mass, 398 pp.

Bunte, K. and Abt, S.R. 2001. Sampling Surface and Subsurface Particle-Size Distributions in Wadable Gravel- and Cobble-Bed Streams for Analyses in Sediment Transport, Hydraulics, and Streambed Monitoring. U.S. Dept. of Agriculture, Forest Service, Rocky Mountain Research Station, Fort Collins, Co., USA, 428 pp.

Collins, M.R. and Wright, R.R. 1964. Application of fluorescent tracing techniques to hydrologic studies. Journal of the American Water Works Association 56(6): 748–754.

Cooke, R.U. and Doornkamp, J.C. 1990. Geomorphology in Environmental Management: A New Introduction. Oxford University Press, New York, 410 pp.

Diplas, P. and Fripp, J.B. 1992. Properties of various sediment sampling procedures. Journal of Hydraulic Engineering 118: 955–970. doi:http://dx.doi.org/10.1061/(ASCE)0733-9429(1992)118:7(955).

Doppler, C. 1903. Über das farbige Licht der Doppelsterne und einiger anderer Gestirne des Himmels: Versuch einer das Brodley›sche Aberrations-Theorem. Prague: Verlag der königlichen böhmischen Gesellschaft der Wissenschaften, 25 p. (in German).

Engel, P. and Lau, Y.L. 1980. Computation of bed load using bathymetric data. Journal of the Hydraulics Division 106(3): 369–380.

Griffiths, G. and Flatt, D. 1987. A self-contained acoustic Doppler current propeller: design and operation. 5th International Conference on Electronics for Ocean Technology, Edinburgh. London: IERE: 41–47.

Grobler, D.C. and Weaver, A.B. 1981. Continuous measurement of suspended sediment in rivers by means of a double beam turbidity meter. Erosion and Sediment Transport Measurement. Proceedings of the Florence Symposium, IAHS Publ 133: 97–103.

Kondolf, G.M. 1997. Hungry water: effects of dams and gravel mining on river channels. Environmental Management 21: 533–551. doi:10.1007/s002679900048.

Kondolf, G.M. and Piégay, H. 2003. Tools in Fluvial Geomorphology. Wiley, Chichester (in England, Great Britain) 688 pp.

Lal, R. 1994. Soil Erosion Research Methods. Soil and Water Conservation Society, Ankeny, IA, USA, 340 pp.

Lenzi, M.A., Agostino, V.D' and Billi, P. 1999. Bedload transport in the instrumented catchment of the Rio Cordon: part I. Analysis of bedload records, conditions and threshold of bedload entrainment. Catena 36(3): 171–190. doi:10.1016/S0341-8162(99)00016-8.

Leopold, L.B., Wolman, M.G. and Miller, J.P. 1995. Fluvial Processes in Geomorphology. Dover Publications, New York, 544 pp.

Manning, R. 1891. On the flow of water in open channels and pipes. Transactions of the Institution of Civil Engineers of Ireland 20: 161–207.

Nelson, J.L., Perkins, R.W. and Haushild, W.L. 1966. Determination of Columbia River flow times downstream from Pasco, Washington, using radioactive tracers introduced by the Hanford reactors. Water Resources Research 2(1): 31–39. doi:10.1029/WR002i001p00031.

Olsen, R. 2007. Remote Sensing from Air and Space. SPIE Press, Bellingham, Washington, 255 pp.

Osterberg, C., Cutshall, N. and Cronin, J. 1965. Chromium-51 as a radioactive tracer of Columbia River water at sea. Science 150: 1585–1587. doi:10.1126/science.150.3703.1585.

Rijn, L.C. van. 1979. Pump Filter Sampler. Delft: Delft Hydraulics Laboratory, Report S4CW.

Rijn, L.C. van. 1986. Manual sediment transport measurements. Delft Hydraulics Laboratory, Delft.

Schumm, S.A. 1977. The Fluvial System. Blackburn Press, Caldwell, NJ, USA, 338 pp.

Schuster, J.S. 1970. Water measurement procedures-irrigation operators. Workshop, Report No. REC-OCE-70-38. Denver: Bureau of Reclamation.

Shilling, F., Sommarstrom, S., Kattelmann, R., Washburn, B., Florsheim, J. and Henly, R. 2005. California Watershed Assessment Manual: Volume I. Prepared for the California Resources Agency and the California Bay-Delta Authority (Web site: http://cwam.ucdavis.edu/), 230 pp.

Thurman, H.V. 1997. Introductory Oceanography. Prentice Hall College, New Jersey.

Wolman, M.G. 1954. A method of sampling coarse bed material. American Geophysical Union 36: 951–956. doi:10.1029/TR035i006p00951.

Woltmann, R. 1790. Theorie und Gebrauch des hydrometrischen Flügels oder eine zuverlässige Methode die Geschwindigkeit der Winde und strömenden Gewässer zu beobachten. Hamburg, 60 p. (in German).

Yobbi, D.K., Yorke, T.H. and Mycyk, R.T. 1996. A guide to safe field operations. U.S. Geological Survey Open-File Report 95-777. Tallahassee: Water Resources Division Safety Officer, U.S. Geological Survey, 34 p.

CHAPTER 14

Geomorphology and Hydraulics of Steep Mountain Channels

Tomáš Galia

INTRODUCTION

Geomorphological processes, bed morphology, and hydraulics of steep mountain channels significantly differ from typical gravel bed or meandering rivers. For our purpose, we can describe steep mountain channels by the criteria of their channel gradient (≥ 0.03 m/m) *sensu* Comiti and Mao (2012), drainage area (≤ 10 km^2) and they are often channels of I or II order following Strahler classification. They link slopes and the fluvial system, supplying low gradient rivers with water and sediments. These parts of drainage networks show extremely rapid rainfall-runoff response (Rickenmann 1997). Confined or semi-confined valley configurations with limited spaces for the floodplain development are conducive for strong hillslope-channel coupling, when colluvial sediments are transported directly from hillslopes to channels (Fig. 1). Steep mountain channels are usually sediment supply limited, so the resulting intensity of sediment transport during flood events is rather than dependent on direct sediment delivery from adjacent hillslopes and individual sediment inputs (e.g., bank failures, gullies) than absolute value of flow discharge. Therefore, sediment transport has pulse character. On the highest channel gradients (> 0.2 m/m), debris flows act as significant transport agents.

Bed material contains high variety of particle sizes from sand to large boulder fractions. Coarser fractions are responsible for the origin of typical vertical bed oscillations in forms of alluvial steps on channel gradients ≥ 0.05 m/m. On the other hand, pool-riffle morphology on lower gradients represents alternation of coarser particles in riffles and finer particles deposited in pools. Bedrock outcrops and large wood debris are also often presented on a stepped-bed profile in mountain channels, increasing their total flow resistance. In fluvial systems, individual grains of bed sediments downstream decrease in size during fluvial transport by selective sorting, particle abrasion, breaking, and weathering (Knighton 1984; Gomez et al. 2001). This

Department of Physical geography and geoecology, University of Ostrava.
E-mail: tomas.galia@osu.cz

Fig. 1. Strong hillslope-channel coupling in a steep headwater channel indicated by the arrow. "A" denotes small alluvial accumulation which originate in a limited space of the channel (the Lubina Stream, Western Carpathians, Czech Republic).

is not the case of steep headwaters, where downstream coarsening of bed sediments is often observed, reflecting the continual sediment supply from the adjacent hillslopes or range of debris flows occurrence (Brummer and Montgomery 2003; Johnson et al. 2010).

Although steep streams are usually not human-affected by the same intensity as the lower gradient rivers, interactions of forest or water management can alter their processes and morphology. For example, check-dam constructions decrease biota and sediment connectivity within the longitudinal profile and intensive forest management directly influences supply of woody material into the channels. Nevertheless, indirect impacts of reforestation or deforestation have been documented, when both activities lead to significant changes of sediment supply into stream networks, later resulting in variability of bedload transport intensity and transformation of channel morphology.

Morphological Aspects

The cross-sectional geometry of low-gradient rivers usually reflects discharge value of bankfull flow with 1–2.5y recurrence interval (Leopold et al. 1964). On the other hand, steep streams adjust their geometry and morphology to the complexity of several internal and external factors, including discharge regime (frequency, magnitude, and duration of floods), sediment supply (frequency, magnitude and calibre of individual sediment inputs), bedrock resistance controlling channel gradient and local knickpoints, land cover (supply of sediments and woody material), history of landscape (e.g., glaciation, deforestation), direct human interactions (presence of check-dams, embankments, sediment mining) and others. By contrast to lowland rivers, mountain steep streams are generally adjusted to these agents rather by vertical bed oscillations rather than lateral migration of the channel, owing to valley confinement by surrounding hillslopes. Channel-reaches and channel units are the basic components of mountain streams in a hierarchical point of view. A channel-reach is typically defined as at least 10–20 channel widths long in a longitudinal stream profile. Every channel-reach is characterised by the occurrence of typical bedforms and is governed by relatively uniform geomorphic processes (Frissell et al. 1986; Montgomery and Buffington 1997). Channel units play an important role as a subsystem of channel-reaches; the length of a channel unit varies from 1 to several channel widths. They are usually described as single bedforms representing an area of macroscopically uniform flow (e.g., step, pool, riffle) and simultaneously separating regions of non-uniform flow (Grant et al. 1990; Halwas and Church 2002).

It has been recognized for decades that channel-reach morphology varies with channel gradient, also reflecting a balance between transport capacity and sediment supply. The most recently used process-based classification of mountain channels is that of Montgomery and Buffington (1997), which distinguishes channel-reaches into the main groups of colluvial, bedrock, and alluvial channels. Alluvial channels are later divided into cascades, step-pools, plane beds, pool-riffles, and dune-ripples (Fig. 2), where intermediate or forced morphologies can also occur. Bedrock channels can be defined as those which lack continuous cover of alluvial sediments and exist only where transport capacity exceeds sediment flux over the long term (Whipple 2004). Colluvial channels occur in headwater parts, lack permanent flow, and sediments are

Fig. 2. Examples of channel-reach morphologies from the Western Carpathians (Czech Republic): A – bedrock, B – bedrock-cascade, C – cascade, D – step-pool, E – plane bed, and F – pool-riffle.

mobile usually during episodic debris-flow events or extraordinary flood events. Cascade channels occupy the highest channel gradients ≥ 0.10 m/m, and are characterized by longitudinally and laterally disorganized coarse bed material with small, partially channel-spanning pools. Step-pools consist of step and pool units, where pools are longer than the channel width. Steps acting as important flow resistance elements are created by large boulders or logs and they represent zones of supercritical flows. Pools with subcritical flows contain finer material and they provide refuges for benthos organisms during low-flow periods. Both cascades and step-pools are recognised as so-called stepped-bed morphology (Comiti and Mao 2012) due to their similar hydraulic parameters, difficult to distinguish morphologies, and relatively rare occurrence of regular step-pools in mountainous landscape. Plane bed morphology, sometimes referred to as rapids, occupy lower channel gradients about 0.05 m/m, lack vertically oscillating bedforms, and usually represent transport-balanced channel-reaches. Pool-riffles include alternation of zones with relatively rapid shallow flows and slower flows in pools with finer sediments. Also, rhythmically-spaced gravel bars are usually present, resulting from local flow convergences and divergences. Pool-riffles represent lowest channel gradients in mountainous headwaters with transitions from supply-limited to transport-limited conditions. Dune-ripples are typical for low-gradient sand bed rivers, but they can occasionally appear in mountain streams in transport-limited zones with fine bed material.

As an example of intermediate channel-reach morphology, one can mention bedrock-cascade channels in limited sediment-supply conditions where protruding bedrock outcrops act as steps and intervening pools are filled with alluvial sediments (Galia and Hradecký 2014a). Forced morphologies are typically related to the higher occurrence of river wood in steep channels. For instance, individual logs span the channel width instead of boulders and trap sediments in forced step-pool channel morphology. In the case of removal of these logs, the channel may transform to cascade or bedrock morphology due to decrease of roughness elements and thus, increase in transport capacity. Similarly, on lower channel gradients, large wood jams force flow divergence and sediment impoundment and can lead to the origin of forced pool-riffles instead of plane bed morphology (Fig. 3).

Fig. 3. Woody jam inducing forced pool-riffle morphology (the Libotínský Stream, Western Carpathians, Czech Republic).

Flow Velocity and Resistance

In terms of hydraulics, steep streams are characterised by shallow turbulent flows and high relative roughness owing to protruding boulders, bedrock, and woody material. In contrast to lower gradient rivers, these large roughness elements account for up to 80–90 percentage of total flow resistance and therefore, total bed roughness is practically divided into skin (grain) roughness and form roughness in steep channels (Reid and Hickin 2008; Chiari and Rickenmann 2011). Flow velocity varies in relatively short longitudinal distances due to irregularity of morphological elements with slower flows in the pools, and higher flows in the rapids and on the steps. The largest variations are observed in stepped-bed channels, where supercritical flows typically occur from the step crests to the impingement points, whereas sub-critical conditions establish thereafter in the pools (Comiti et al. 2007). Moreover, the flow resistance varies considerably with the depth of flow, much more so than in large rivers. Reid and Hickin (2008) demonstrated over six orders of magnitude of the Darcy–Weisbach resistance factor for a mean-depth range that averages 20 cm and is less than 36 cm. One also should note difficulties in the mean flow depth estimations in shallow flows with large roughness elements.

There exist several approaches for the calculation of mean flow velocity in steep channel-reaches, which were developed in laboratorial flumes and by field experiments. These are represented by the traditional Manning formula, equations based on the Darcy–Weisbach friction factor, power–law empirical equations, and empirical relationships between the dimensionless parameters of discharge and flow velocity. When the discharge is known, calculations of flow velocity v (m s^{-1}) based on the unit discharge q (m^2 s^{-1}) or relationship between dimensionless parameters of discharge $q* = q/\sqrt{(g\,D_{84}{}^3)}$ and velocity $v* = v/\sqrt{(g\,D_{84})}$ provide best results (Comiti et al. 2007; Zimmermann 2010), for example:

$$v = 1.3g^{0.2}S^{0.2}q^{0.6}D_{90}{}^{-0.4} \tag{1}$$

$$v* = 0.92q^{*0.66} \tag{2}$$

where g is the acceleration of gravity (9.81 m^2 s^{-1}), S is the channel gradient (m m^{-1}), and D_x denotes bed grain-size percentile (m). Power–law Eq. (1) has been developed by Rickenmann (1990) in a flume with $0.03 \leq S \leq 0.40$, Eq. (2) has been derived by field tests in an Alpine torrent with $0.08 \leq S \leq 0.21$ (Comiti et al. 2007).

The Manning formula

$$v = 1/n\,R^{0.67}S^{0.5} \tag{3}$$

is generally not recommended for calculations of shallow flows and large roughness elements, except the case when n roughness parameter is determined reflecting measured discharge, for example, by Rickenmann's approach (1996):

$$n = (0.97g^{0.41}Q0.19)/(S^{0.19}D_{90}{}^{0.64}) \tag{4}$$

The Darcy–Weisbach friction factor f is defined as:

$$f = (8g\,d\,S)/v^2 \tag{5}$$

where d denotes mean flow depth (m). Standard power function for estimation of the channel roughness based on d/D_{84} ratio and the Strickler relationship takes form:

$$\sqrt{(8/f)} = 8.2(d/D_{84})^{1/6} \qquad (6)$$

Logarithmic–law Keulegan equation is proposed as:

$$\sqrt{(8/f)} = (1/k)\, ln(11d/k_s) \qquad (7)$$

where k is von Kármán constant equal to 0.4 and k_s stands for the height of bed roughness element, usually expressed as some multiple of D_{50} or D_{84} or standard deviation of bed taken from geodetic measurements.

Recent comprehensive reviews of flow velocity calculations in steep streams are provided by Zimmermann (2010), Rickenmann and Recking (2011), and Comiti and Mao (2012). In case the discharge is not known, regular occurrence of bedrock outcrops or river wood in channels, which cannot be described by grain-size percentiles in roughness coefficients, represent additional uncertainty in flow velocity calculations. One should take into account the fact that the flow resistance significantly differs with changes of water stage owing to submergence of roughness elements (e.g., individual boulders) by increasing discharge (Yochum et al. 2012). For stepped-bed streams, shift from nappe to transitional or skimming flow regime occur with increasing discharge, submergence of alluvial boulder steps and smoothing relative heights between steps and pools (Fig. 4) (Chin 2003; Comiti et al. 2007).

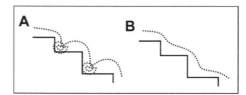

Fig. 4. Schematization of hydraulics in step-pool channel-reach during low (A) and high (B) flows.

Sediment Transport

Bedload transport, that is, the transport of larger particles (coarse sand, gravel, cobble, and boulder–size fractions) along the channel bed, is directly related to channel morphology and its maintenance in steep streams and thus, this type of transport is more important in these kinds of streams than suspended transport. Massive bedload transport during extraordinary floods also represents one of the natural hazards in mountainous areas for human society. Lenzi et al. (2006) documented two thresholds of bedload effective discharges in a steep Alpine torrent: relatively frequent floods with 1.5–3 year recurrence interval were responsible for maintaining the channel form (e.g., pool depth) and more infrequently, higher flows with recurrence interval in tens of years were responsible for macro-scale channel shaping (e.g., channel width and the destruction and origin of alluvial steps). Similar recurrence intervals were introduced for step-pool channels, when finer material stored in pools travels as bedload over stable bed-forming clasts during more frequent events, whereas steps are remobilized during high-magnitude floods (Grant 1990; Chin 2003).

Supply-limited character and limited amounts of accessible sediments in the channel together with increased flow resistance are the main reasons for the fact that observed bedload transport in steep channels may be considerably smaller than that predicted by conventional bedload transport equations (Yager et al. 2007; Chiari and Rickenmann 2011). Bedload transport in these streams is dependent on the frequency and magnitude of direct sediment supply and the relaxing time between individual high-flow events rather than the absolute value of discharge, and significant differences in bedload transported volumes may occur between individual headwater watersheds (Mao et al. 2009; Yu et al. 2009). The specific volume of sediment supply and resulted bedload transport during ordinary floods is also crucial for the origin of well-developed step-pools or other channel-reach morphologies (Recking et al. 2012; Galia and Hradecký 2014b). One should also mention irreplaceable role of debris flows as non-Newtonian fluids, which are able to transport massive volumes of sediment in the steepest parts of mountainous watersheds and produce massive incision of valleys in their transport zones. Source zones of debris-flow material are related to channel gradients exceeding 0.20 m/m, but accumulation zones of these processes can occur on relatively low channel gradients about 0.05 m/m (Stock and Dietrich 2003).

The beginning of bedload transport during high-flow events is also dependent on the character of the supplied material and bed sediments; especially their grain-size characteristics which play an important role. The parameters of the critical shear stress τ_{ci} (8-9) with $\tau_b \approx \tau_{ci}$, the unit stream power ω (10) or the unit discharge g as the ratio Q/W for the beginning of motion of individual grain with D_i diameter are usually evaluated:

$$\tau_b = S R \rho g \tag{8}$$

$$\tau_{ci} = \theta(\rho_s - \rho) \, g \, D_i \tag{9}$$

$$\omega = (Q S \rho g)/W, \tag{10}$$

where τ_b denotes the bed shear stress (N m^{-2}), θ means the dimensionless shear stress or Shield's number, and W the channel width (m). The parameter of θ is dependent on the character of the bed sediment, for example, its sorting or imbrication. Buffington and Montgomery (1997) reported a wide range of $0.03 \geq \theta \geq 0.086$ for lower gradient gravel-bed rivers. For the case of the stepped-bed Alpine torrent, Lenzi et al. (2006) reported the following relationship:

$$\theta = 0.054(D_i/D_{90})^{-0.737} \tag{11}$$

Galia et al. (2015) documented very low critical values of the unit stream power and discharge as the incipient of grain motion in flysch-based headwater streams due to missing boulder fraction, implying low-flow resistance. On the other hand, high values of the critical unit discharge and bed shear stress were obtained for bedload transported particles in a limited sediment-supplied Alpine torrent (Lenzi et al. 2006).

Modelling of sediment transport in steep channels represents a challenge for hydraulic engineering and fluvial geomorphology owing to the huge complexity of the topic. In past years, two free-available transport models were designated for the computation of bedload transport in torrential Alpine watersheds: Tomsed (formerly known as Setrac) developed by the University of Natural Resources and Applied Life Sciences, Vienna (Friedl and Chiari 2011, www.bedload.at) and Sedflow developed by the WSL Swiss Federal Institute for Forest, Snow and Landscape Research (Heimann et al. 2014, www.wsl.ch/sedFlow). The first one represents more detailed calculations of bedload transport based on real-time changes in channel cross-sectional geometry during simulated flood events, whereas Sedflow model is more suitable for estimations of sediment routing in an entire mountain watershed for longer time periods. Although both models use a variety of modern approaches for flow velocity and bedload discharge calculations, it is always necessary to validate calculated bedload transport volumes with real observed transport. Because of a lack in direct bedload transport measurements in steep streams, a rough model validation could be done via estimations of intensity of erosion or deposition processes in examined streams by field inspections or comparison of pre and post-flood aerial photographs or LiDAR data if available (Chiari and Rickenmann 2011; Galia and Hradecký 2014b; Heimann et al. 2014).

Human Interactions in Steep Channels

Constructions of grade-control structures on the one side and restoration of fluvial systems on the other side can be mentioned as the main contradiction in direct human interventions into steep channels. The management of mountain torrents brings tasks about protection against erosional processes at steep channel gradients and consequent massive bedload transport. Thus, a widely used method is stabilisation of channel beds by staircase-like sequences of cemented, boulder, or wooden grade-control structures. Grade-control structures are called check-dams (Fig. 5) when they are at least 1.5–2 m high above the channel bed level. If their height is less, the term "bed sill" is used (Lenzi et al. 2003a,b). The main idea is the reduction of initial bed slope S to lower equilibrium bed slope S_{eq} between individual constructions, when the value of S_{eq} generally represents equilibrium between potential erosion and aggradation processes in assessed channel-reach. As a result, the channel bed is locally stabilised and the bedload transport rate in downstream channel-reaches decreases. However, decreased connectivity of sediment transport can lead into incision, narrowing and armouring processes in unregulated downstream reaches and high artificial steps decreases potential for biota migration, which may affect local populations of aquatic invertebrates and fishes.

Fig. 5. Wooden and concrete check-dam structures during higher flow event with *ca.* 2 year recurrence interval (the Malá Ráztoka Stream, Western Carpathians, Czech Republic).

Restorations of steep streams are primarily concerned with the increase of longitudinal connectivity for aquatic organisms and transfer of "safe" volumes of coarse sediments. In past decades, open check-dams have been developed in Alpine countries, enabling a certain degree of such connectivity. From the 1990s, uncemented boulder grade-control structures in forms of low artificial steps have been tested in moderate-gradient streams and today, this approach represents a main restoration technique for torrential channels with bed gradients up to $S = 0.15$ m/m and usually reflecting a lower degree of safety up to 20–30 flood recurrence interval. Construction parameters of these artificial steps follow the geometry of natural streams with step-pool morphology reported by Abrahams et al. (1995):

$$1 < (H/L)/S < 2 \tag{12}$$

where H denotes the step height (m) and L means the step wavelength (longitudinal distance between individual steps) (m). For the pilot restoration of an Italian torrent, Lenzi (2002) reported field wavelength $L = H/(1.1–1.3S)$ and $H/D_{90} = 2$. Experiences from a few step-pool restorations in USA summarized by Chin et al. (2009), who clearly demonstrated that natural adjustments of these structures occurred toward geomorphic stability and ecological improvement and also discussed future strategy of step-pool restorations. At lower channel gradients with plane beds and pool-riffles, rapid hydraulic structures can also be applied (Radecki-Pawlik 2013).

Conclusion Remarks

Steep mountain channels function as very dynamic system. In contrast to lowland rivers, these streams (i) usually adjust their bed to flow conditions vertically by formations of stepped-bed morphology or pool/riffle alternations instead of changes in planar geometry, (ii) bedloads with its pulse character represent an important part of transported material, and (iii) steep channels often interact with adjacent hillslopes due to valley confinement with limited space for floodplains. High-flow variations with quick response to precipitation events and additional roughness elements (bedrock and alluvial steps, large wood) strongly affect hydraulics in these channels. Together with a variety in frequency, magnitude, and calibre of sediment supply, this makes it difficult to calculate or simulate bedload transport during high-flows. Restorations of torrents are usually concerned with substitutions of check-dams by more ecological friendly bed-stabilisation elements.

Acknowledgements

The author sincerely thanks Anna Kidová (Institute of Geography of Bratislava) and Karel Šilhán (University of Ostrava) for informal reviews of the early draft and an anonymous reviewer of the later version. The chapter was supported by the project of the University of Ostrava Foundation SGS18/PřF/2015-2016.

Keywords: Steep channel, mountain stream, hillslope-channel coupling, channel-reach morphology, flow resistance, bedload transport, sediment transport modelling, check-dam, stream restorations, step-pool, torrent.

References

Abrahams, A.D., Gang, L. and Atkinson, J.F. 1995. Step-pool streams: adjustment to maximum flow resistance. Water Resources Research 31: 2593–2602.

Brummer, C.J. and Montgomery, D.R. 2003. Downstream coarsening in headwater channels. Water Resources Research 39: 1–14.

Buffington, J.M. and Montgomery, D.R. 1997. A systematic analysis of eight decades of incipient motion studies, with special reference to gravel-bedded rivers. Water Resources Research 33: 1993–2029.

Chiari, M. and Rickenmann, D. 2011. Back-calculation of bedload transport in steep channels with a numerical model. Earth Surface Processes and Landforms 36: 805–815.

Chin, A. 2003. The geomorphic significance of step–pools in mountain streams. Geomorphology 55: 125–137.

Chin, A., Anderson, S., Collison, A., Ellis-Sugai, B.J., Haltiner, J.P., Hogervorst, J.B., Kondolf, G.M., O'Hirok, L.S., Purcell, A.H., Riley, A.L. and Wohl, E. 2009. Linking theory and practice for restoration of step-pool streams. Environmental Management 43: 645–661.

Comiti, F., Mao, L., Wilcox, A., Wohl, E.E. and Lenzi, M.A. 2007. Field-derived relationships for flow velocity and resistance in high-gradient streams. Journal of Hydrology 340: 48–62.

Comiti, F. and Mao, L. 2012. Recent advances on the dynamics of steep channels. pp. 353–387. *In*: Church, M., Pascale, M.B. and Roy, A.G. (eds.). Gravel-Bed Rivers: Processes, Tools, Environments. John Wiley & Sons, Chichester, UK.

Friedl, K. and Chiari, M. 2011. A one-dimensional transport model for steep slopes. Manual Tomsed version 0.1 beta. Accessible online at http://www.bedload.at.

Frissell, C.A., Liss, W.J., Warren, C.E. and Hurley, M.D. 1986. A hierarchical framework for stream habitat classification: viewing streams in a watershed context. Environmental Management 10: 199–214.

Galia, T. and Hradecký, J. 2014a. Morphological patterns of headwater streams based in flysch bedrock: examples from the Outer Western Carpathians. Catena 119: 174–183.

Galia, T. and Hradecký, J. 2014b. Channel-reach morphology controls of headwater streams based in flysch geologic structures: an example from the Outer Western Carpathians, Czech Republic. Geomorphology 216: 1–12.

Galia, T., Hradecký, J. and Škarpich, V. 2015. Sediment transport in headwater streams of the Carpathian Flysch belt: its nature and recent effects of human interventions. pp. 13–26. *In*: Heininger, P. and Cullmann, J. (eds.). Sediment Matters. Springer. Heidelberg.

Gomez, B., Rosser, B.J., Peacock, D.H., Hicks, D.M. and Palmer, J.A. 2001. Downstream fining in a rapidly aggrading gravel bed river. Water Resources Research 37: 1813–1823.

Grant, G.E., Swanson, F.J. and Wolman, M.G. 1990. Pattern and origin of stepped-bed morphology in high-gradient streams, Western Cascades, Oregon. Geological Society of America Bulletin 102: 340–352.

Halwas, K.L. and Church, M. 2002. Channel units in small, high gradient streams on Vancouver Island, British Columbia. Geomorphology 43: 243–256.

Heimann, F.U.M., Rickenmann, D., Böckli, M., Badoux, A., Turowski, J.M. and Kirchner, J.W. 2014. Recalculation of bedload transport observations in Swiss mountain rivers using the model sedFlow. Earth Surface Dynamics Discussions 2: 773–822.

Johnson, R.M., Warburton, J., Mills, A.J. and Winter, C. 2010. Evaluating the significance of event and post-event sediment dynamics in a first order tributary using multiple sediment budgets. Geografiska Annaler: Series A, Physical Geography 92: 189–209.

Knighton, A.D. 1984. Fluvial Forms and Processes. Edward Arnold, London.

Lenzi, M.A. 2002. Stream bed stabilization using boulder check dams that mimic step-pool morphology features in Northern Italy. Geomorphology 45: 243–260.

Lenzi, M.A., Marion, A. and Comiti, F. 2003a. Interference processes on scouring at bed sills. Earth Surface Processes and Landforms 28: 99–110.

Lenzi, M.A., Marion, A. and Comiti, F. 2003b. Local scouring at grade-control structures in alluvial mountain rivers. Water Resources Research 39: 1176.

Lenzi, M.A., Mao, L. and Comiti, F. 2006. When does bedload transport begin in step boulder-bed streams? Hydrological Processes 20: 3517–3533.

Leopold, L.B., Wolman, M.G. and Miller, J.P. 1964. Fluvial Processes in Geomorphology. Freeman, San Francisco.

Mao, L., Cavalli, M., Comiti, F., Marchi, L., Lenzi, M.A. and Arattano, M. 2009. Sediment transfer processes in two Alpine catchments of contrasting morphological settings. Journal of Hydrology 364: 88–98.

Montgomery, D.R. and Buffington, J.M. 1997. Channel-reach morphology in mountain drainage basins. Geological Society of America Bulletin 109: 596–611.

Radecki-Pawlik, A. 2013. On using artificial rapid hydraulic structures (RHS) within mountain stream channels—some exploitation and hydraulic problems. pp. 101–115. *In*: Rowiński, P. (ed.). Experimental and Computational Solutions of Hydraulic Problems. GeoPlanet: Earth and Planetary Sciences, Springer-Verlag Berlin Heidelberg.

Recking, A., Leduc, P., Liébault, F. and Church, M. 2012. A field investigation of the influence of sediment supply on step-pool morphology and stability. Geomorphology 139–140: 53–66.

Reid, D.E. and Hickin, E.J. 2008. Flow resistance in steep mountain streams. Earth Surface Processes and Landforms 33: 2211–2240.

Rickenmann, D. 1990. Bedload transport capacity of slurry flows at steep slopes. Mitteilung 103 der Versuchsanstalt für Wasserbau, Hydrologie Glaziologie, ETH Zürich.

Rickenmann, D. 1996. Fliessgeschwindigkeit in Wildbächen und Gebirgsflüssen. Wasser Energie Luft 88: 298–304.

Rickenmann, D. 1997. Sediment transport in Swiss torrents. Earth Surface Processes and Landforms 22: 937–951.

Rickenmann, D. and Recking, A. 2011. Evaluation of flow resistance in gravel-bed rivers through a large field data set. Water Resources Research 47: W07538.

Stock, J. and Dietrich, W.E. 2003. Valley incision by debris flows: evidence of a topographic signature. Water Resources Research 39: 1089. doi:10.1029/2001WR001057.

Whipple, K.X. 2004. Bedrock rivers and the geomorphology of active orogens. Annual Review of Earth and Planetary Sciences 32: 151–185.

Yager, E.M., Kirchner, J.W. and Dietrich, W.E. 2007. Calculating bed load transport in steep boulder bed channels. Water Resources Research 43: W07418.

Yochum, S.E., Bledsoe, B.P., David, G.C.L. and Wohl, E.E. 2012. Velocity prediction in high-gradient channels. Journal of Hydrology 424–425: 84–98.

Yu, G., Wang, Z., Zhang, K., Chang, T. and Liu, H. 2009. Effect of incoming sediment on the transport rate of bed load in mountain streams. International Journal of Sedimentary Research 24: 260–273.

Zimmermann, A. 2010. Flow resistance in steep streams: an experimental study. Water Resources Research 46: W09536.

CHAPTER 15

Slope Stability
Basic Information

Tymoteusz Zydroń,[1,]* *Anna Bucała-Hrabia,*[2] *Andrzej T. Gruchot*[1]
and Veronika Kapustova[3]

INTRODUCTION

Mass movements are one of the geodynamic processes that occur worldwide and cause significant changes in landscape. In many cases they cause substantial economic damage and sometimes even fatalities. It is estimated that damages caused by mass movements are more than ten times greater than the ones caused by earthquakes (Graniczny and Mizerski 2007). Schuster (1996) states that annual damages (direct and indirect) caused by landslides in Japan are estimated at 4 billion dollars and in the USA, Italy and India range from 1 billion to 2 billion dollars.

Landslides have occurred for a long time and they concern both natural slopes and man-made engineering structures. The causes for mass movements are generally known, but the place and time of their occurrence is hard to predict because of randomness of the natural phenomena that initiate these processes. Mass movements are often triggered by two or more factors occurring at the same time (Thiel 1989). Due to the complexity of mass movements and their influence on human activity they are the subject of research in many disciplines.

Therefore the methods and scope of work are very diverse in mass movement research. This chapter presents mass movements issues, their typology and general test methods. Special attention was given to engineering tools used in slope stability evaluation. A description of chosen physical processes that influence stability conditions was also included.

[1] Agricultural University of Cracow, Dept. of Hydraulic Engineering, and Geotechnics.
   E-mail: rmgrucho@cyf-kr.edu.pl
[2] Institute of Geography and Spatial Organization Polish Academy of Sciences, Department of Geomorphology and Hydrology of Mountains and Uplands.
   E-mail: abucala@zg.pan.krakow.pl
[3] University of Ostrava, Department of Physical Geography and Geoecology.
   E-mail: veronika.kapustova@osu.cz
* Corresponding author: t.zydron@ur.krakow.pl

Mass Movements

Typology

Mass movement is the downward and outward movement of slope-forming material under the influence of gravity (Varnes 1978). It is different from 'mass transport', as it does not require transporting medium such as water, air, or ice (Brunsden 1984). Popular expression 'landslide' is often used for mass movement, but in a pure sense, a landslide is only one of the mass movement types (here called slide). Mass movements are common processes in the landscape, as they may affect slopes with gradient higher than 2°. We can study mass movements from different points of view: (1) geomorphology and geology, (2) geotechnics and engineering geology, (3) geoecology, and (4) economy and crisis management, etc.

Numerous classifications of mass movements were created, based on mass movement mechanism, activity, rate of movement, morphology and morphometry, etc. (e.g., Varnes 1978; Hutchinson 1988; Cruden and Varnes 1996; Dikau et al. 1996). The most often used classification is that introduced by Cruden and Varnes (1996), based upon mass movement mechanism and type of material (Table 1).

Table 1. Classification of mass movements according to Cruden and Varnes (1996).

Type of movement		Type of material		
		Bedrock	**Engineering soils**	
			Predominantly coarse (debris)	**Predominantly fine (earth)**
Fall		Rock fall	Debris fall	Earth fall
Topple		Rock topple	Debris topple	Earth topple
Slide	Rotational	Rock slide	Debris slide	Earth slide
	Translational			
Lateral spread		Rock spread	Debris spread	Earth spread
Flow		Rock flow (Sackung)	Debris flow	Earth flow
Complex		e.g., rock avalanche	e.g., flow slide	e.g., slump - earth flow

According to mechanisms of movement, Cruden and Varnes (1996) distinguished six basic types of mass movements: fall, topple, slide (rotational and translational), lateral spread, flow, and complex.

Fall is a sudden free movement of material away from very steep slopes (Dikau et al. 1996). Detached material fragments lose contact with the slope and move by free falling at least in a part of their trajectory. Subsequent material deformation involves break-up, roll, bounce, slide, or dry flow onto the slopes bellow (Dikau et al. 1996). Falls are extremely rapid processes. The velocity of falls reaches an order of 100 km·h^{-1}. Falls occur in various sites as mountain faces, coastal cliffs, steep river banks, and road outcrops, etc. (Dikau 2004). Typical fall features are planar sheets forming wedge-shaped hollows of the scar and accumulation forms such as talus cones and screes (Fig. 1a).

Topples consist of a long-term forward rotation of a rock, debris, or soil column about a pivot or hinge on a hillslope (Dikau et al. 1996). The type of topple movement is tilting without collapse (Dikau 2004). Toppling is a relatively slow process. Initial velocity is in order of 10^{-3}–10^2 mm·year^{-1}. Exponential increase in the rate of movement may take

Fig. 1. Examples of selected mass movement types: (a) Large rock fall on W slopes of Demerdzhi Yaila, Crimean Mountains (photo: T. Pánek); (b) Block topple in massive limestones above Foros Town, southern slope of the Crimean Mountains (photo: T. Pánek); (c) Large simple rotational landslide on northern slopes of Petros Mt., Chornohora, Ukraine (photo: T. Pánek); (d) Small translational slide in a gully, the shear plane is predisposed by colluvium-bedrock boundary, Košařiska, Jablunkovská Brázda Furrow, Czech Republic (photo: L. Kubiszová); (e) Rock spread on the Kráľova studňa Mt., Veľká Fatra, Slovakia (photo: V. Kapustova); (f) Rock flow (sackung) on Veľká Kamenistá Mt., Western Tatras (photo: T. Pánek); (g) Debris flow accumulation in one of the gullies on the northern slope of the Smrk Mt., Moravskoslezské Beskydy Mt., Czech Republic (photo: J. Hradecký); (h) Rock avalanche 'Seit', Tien Shan, Kyrgyzstan (photo: A. Strom).

thousands of years often culminating in an abrupt fall or slide. It is also possible, that the movement will stop and does not result in the destruction by fall or slide (Dikau et al. 1996). Common types of toppling failures are: (1) flexural toppling of slabs of rock, dipping steeply into face, (2) block toppling (Fig. 1b) of columns of rock containing widely spaced orthogonal joints, and (3) block flexure toppling characterized by pseudo-continuous flexure of long columns through accumulated motions along numerous cross-joints (Goodman and Bray 1976).

Slide is a mass movement along single or multiple recognizable shear surfaces (Dikau 2004). Movement rates of slides can vary by several orders of magnitude, from several cm year^{-1} to m s^{-1}. Slides can develop in a wide range of slope gradients. According to the shear surface type, two groups of slides are distinguished: (1) rotational and (2) translational slides (Fig. 1c,d). Rotational slides move along circular or spoon-shaped shear surface. In the head area, sliding blocks tilt backwards while moving downhill (flattening or slope reversal occurs). Moving material has a small degree of internal deformation, so the slide body will essentially be the same as the surrounding slope. Sometimes, sliding material liquefies and turns into a flow at its toe. Rotational rock slides often develop in bedrock consisting of interbedded stronger and weaker materials, like flysch, marls, and limestones. Soil slides often develop in cohesive fine-textured materials, such as consolidated clays or weathered marls (Dikau et al. 1996). Translational slide is a non-circular slope failure along near-planar slip surface. The slip surface is typically some structural surface of weakness within the slope-forming material (fault, bedding, etc.). Translational slides will either stay as a discrete block or break into debris depending on the slope angle and the slide velocity (Dikau 2004).

Lateral spreading is a lateral extension of a cohesive rock or soil mass over a deforming mass of softer underlying material in which the controlling basal shear surface is often not well-defined (Dikau et al. 1996). Rock spreading (Fig. 1e) is a slow, deep-seated, plastic deformation in a rock mass, leading to extension at the surface. Masses move predominantly under gravitational stresses either in homogenous rock or in fractured capping (Dikau 2004). Due to extensional conditions, typical features such as tension cracks, trenches, double ridges, uphill facing scarps, pseudo-karst, and graben-like depressions evolve. Often pseudo-karst caves occur under the surface. Soil/debris spreading is the collapse of a sensitive soil layer at a certain depth, followed by either adjustment of the overlying more resistant soil layers, or progressive failure of the whole sliding mass. Lateral movement along the basal nearly horizontal mobile zone occurs. The duration is typically only a few minutes. Quickclay slides and sand liquefaction slides belong in this category. General morphology of the soil spread is similar to that of rock spreading (alternating horst and graben structures), but on a smaller scale (Dikau et al. 1996).

Flow is a mass movement, in which the individual particles travel separately within a moving mass. They involve different types of material, such as highly fractured rock, clastic debris in a fine matrix, or only fine grains (Dikau 2004). Velocity of flowing mass decreases with depth due to friction. There are different types of flows according to material in which they evolved. Rock flow (Fig. 1f) is a creeping flow-type deep-seated gravitational deformation affecting homogenous rock masses (Dikau 2004). For a rock flow, high volumes of the rock mass (10^3–10^6.m^3) and small total displacement rates are typical. In the upper part of affected slopes, high angle extensional shear planes produce typical features, such as graben-like depressions, double ridges, troughs, etc. Due to the compressional conditions on the slope foot, often bulging and low-angle shear planes occur. Rock flows occur only on valley slopes in mountain areas and high coastal cliffs, where slopes are high enough to induce strong gravitational stress (Dikau et al. 1996). Usually, rock flows display a very slow constant creep deformation, but sometimes they include short rapid reactivation phases related to critical events (extreme rainfall, earthquake, etc.). Debris flow (Fig. 1g) is a rapid destructive gravitational movement of a mixture of fine material, coarse material, and variable admixture of water. The debris flow moves in surges and involves the erosion of the channel bottom and bank failures. Main features of debris flows are source area, main track, lateral ridges, and depositional toe. Source areas of debris flows are slopes covered by unconsolidated colluvial mantle (Dikau 2004). Earth flow in its wet form is a special category of debris flow, where the material is of well sorted fine grain size (rare coarse grains may occur). They are very rapid. Similarly as the debris flows, earth flows tend to follow existing drainage ways and in certain conditions they may travel for very long distances (Dikau et al. 1996). While debris flows accumulate where the channel gradient decreases, earth flows are mobile even on flat surfaces (according to their viscosity). When earth flows reach low gradients, they spread into a flat fan or thin sheet (Dikau 2004). Velocity may change from slow creep into 10 m·s^{-1}.

Complex mass movement is a combination of two or more principal types of movements (Fig. 1h). Nearly all mass movements develop complex behavior moving downslope (Dikau 2004). Typical complex mass movement is a rock avalanche or flow slide. Rock avalanche is a highly damaging high volume of dry rock debris, originating as a rock fall or other type of slope collapse and moving at a high velocity and for a long distance, even on a gentle slope (Hsü 1975 in Dikau et al. 1996). Velocity of rock avalanche may reach several m·s^{-1}. Flow slide is a highly damaging form of debris flow, evolving as a rotational slide in unconsolidated material (very often in man-made tips and spoil heaps, but also in debris of geological origin). The material loses its cohesion with a reduction in strength, becoming a fluidized mass. Flow slides are high magnitude events (in terms of their velocity and destruction). They not only fluidize very quickly, but also rapidly become solid when they cease moving (Dikau 2004).

For example, in Polish Carpathians, all the types of mass movements described by Varnes (1978) can be found; these movements can occur in rock or earth material (in mantle, alluvial deposits, soils), debris material, and also in complex (e.g., rocky mantle) material (Margielewski 2008).

Causes of Mass Movements

Passive and active factors are distinguished as causes of mass movement generation (Thiel 1989). Passive factors include lithologic and structural conditions of rock formations and soils (dip of beds and their relation to morphological surfaces, tectonic faults), and morphological conditions. On the other hand, active factors include morphological, climatic, hydrological, and anthropogenic conditions (loading of slopes by structures, a change in slope inclination), tectonic, and floristic movements. Morphological factors are considered to be both passive and active since mass movements lead to slope transformation, which cause formation of new morphological conditions. Among the most common causes of mass movements, saturation of a rock massif with rain water or snowmelt, as well as lateral and bottom erosion of mountain streams at slope feet, are particularly significant ones. Therefore, when investigating the conditions of slope stability, it is very common to analyze the amount and intensity of precipitation as the main factor that initiates mass movements.

Numerous works on the subject of geomorphology provide threshold values for precipitation that initiates morphogenetic processes (among others, Caine 1980; Govi et al. 1982; Kotarba 1994; Glade 1998; Gil and Długosz 2006; Guzzetti et al. 2007).

Taking into account the region of Polish Carpathians, Starkel (1996) distinguished three types of precipitation, during which threshold values are exceeded. Brief downpours of a stormy character and intensity of $1-3$ mm·min^{-1} lead to intense surface runoff and wash-down. Precipitation amounting to 200–600 mm within 2–5 days and with the intensity of up to 10 mm·h^{-1} brings about formation of shallow mantle slides, rejuvenation of old landslide surfaces, and transformation of the channel system. On the other hand, constant precipitation with monthly totals of 100–500 mm causes saturation of the subgrade and favors activation of deep landslides. A frequently observed phenomenon is when various types of precipitation overlap, especially when downpours of a stormy character overlap with heavy rainfall, several-day or daily-occurring downpours. Such a situation took place in Poland in July 1997, and it caused mass activation of landslides, mainly in the western part of the Carpathians (Poprawa and Rączkowski 1998; Mrozek et al. 2000; Gorczyca 2004; Bucała 2009). Values of rainfall threshold compiled for different parts of the world and various types of mass-movements are widely presented in the Guzzetti et al. (2007).

Mass movements may also be initiated by saturation of mass deposited on slopes by snowmelt (Starkel 1960; Gil 1997; Cardinali et al. 2000; Guzzetti et al. 2002; Rączkowski and Mrozek 2002). For example, in the Czech part of the Flysch Carpathians (predominantly in non-forested areas) in 2006, a fast thawing of an unusually thick snow cover in conjunction with massive rainfall caused evolution of numerous shallow landslides. Bíl and Müller (2008) calculated a Total Cumulative Precipitation (TCP) index of snow thaw periods for the last 20 years before 2006. TCP is combining the amount of water from both thawing snow and rainfall. From this data, they calculated the TCP safe threshold value without landslides for the study area, which is ca. 100 mm of water delivered to the soil during the spring thaw (which is *ca.* 11 mm day^{-1}). In 2006, 10% of the landslides occurred under or at 100 mm of TCP. The upper value of 155 mm covered all of the landslides (Bíl and Müller 2008).

The direct factors that lead to landslide formation also include slope undercuttings by lateral erosion (which are particularly intensive during high-water conditions) (Swanson et al. 1985; Kotarba 1986; Gorczyca 2004; Liu et al. 2004) and by headward erosion (developing mainly in lateral tributaries of main watercourses) in headwater areas (Starkel 1960; Bober et al. 1977; Margielewski 1994). Loss of stability of river channel banks can have various behaviors. Rotational slides are usually less common than other types of mass movements (Simon et al. 2000) and usually occur along high banks. Shallow slides (Fig. 2a) occur more often in the vicinity of river channels and are a result of bank undercutting (Fig. 2b) caused by water erosion and/or by formation of fissures (joints) in the upper parts of banks. Slope failure may take place at the foot of the bank or at the border between media varying significantly in permeability.

Fig. 2. Examples of mass movements near river channels: (a) translational slide, Targanice, Beskid Mały Mts., Poland (photo: T. Zydroń), (b) erosion of river bank, Zdynia, Beskid Niski Mts., Poland (photo: T. Zydroń).

Examples of typical models of loss of stability of channel banks are presented in the work by Langendeon and Simon (2008).

Formation of mass movements takes place also as a result of seismic tremors (Schenk et al. 2001; Owen et al. 2008; Khattak et al. 2010). Based on aerial photographs and field investigations, catastrophic landslides have been extensively mapped and analyzed in California, El Salvador, Japan, and Italy (e.g., Bommer and Ridriguez 2002; Sassa et al. 2007; Harp et al. 2011). In the Polish Carpathians, the formation of a landslide near Dukla, described by Gerlach et al. (1976), and formation of small landslides in the Silesian Beskids in the 1990s were thought to be connected with seismic tremors (Bajgier 1993).

Next to natural factors, changes in land use are one of the main factors which influence the occurrence of mass movements (Selby 1976; Marden and Rowan 1993; Bergin et al. 1995; Glade 2003). Land use and vegetation cover have an effect on the magnitude of the processes of water runoff and soil washout, and on the magnitude of shallow slope processes. For example, slope wash on arable land is by 3–4 orders of magnitude higher than in forest areas (Gerlach 1976; Gil 1976). Overloading of slopes by construction is another example of human activities that lead to activation of landslides.

Mass movements are rarely brought about by only one initiating factor. It usually takes a series of factors which lead to disturbing the slope natural equilibrium.

Methods of mass movement investigations

Mass movements play an enormous role in shaping the relief of mountainous regions. They are processes which are a natural hazard to life and human economic activity. Terrain mapping of landslides, which can be enriched with geodetic methods, is the fundamental method for studying mass movements (Kirchner and Krejči 2002). For much larger regions, these methods can be substituted by analysis of aerial photographs. Geographic Information Systems (GIS) spatial analysis techniques help to analyze and locate landslide hazards (Campus et al. 2000). These techniques, supported by the use of proper cartographic materials, calculation models, and statistical techniques, find application in research on determining zones of landslide susceptibility and on determining the landslide hazard (Bonham-Carter et al. 1989; Clerici et al. 2002; Van Westen et al. 1997, 2003; Ayalew et al. 2004). Landslide susceptibility zones are estimated by considering the correlation of environmental conditions (lithology, tectonics, relief, land use) and maps of landslides in a given region.

Currently, data from Airborne Laser Scanning (ALS) and from Light Detection and Ranging (LIDAR) can be a great support in studying landslides. It is one of the important techniques used for gathering information on land surface (Borkowski 2006; Vosselman and Mass 2010). These data are characterized by high vertical accuracy (0.1–0.3 m) and allow researchers to create a detailed terrain model. Laser scanning is also done using terrestrial laser scanners, which, when compared with airborne scanners, give more accurate measurements.

Methods of mass movement investigations can include geo-engineering, hydrogeological, geophysical, and geotechnical works. Range of geological and geotechnical research of landslides are largely regulated by appropriate regulations, standards, and instructions. Typical geo-engineering tests involve preparation of geo-engineering documentation. Geo-engineering documentation covers activities related to the execution of field work (which includes geodetic and photogrammetric measurements, as well as geological and laboratory measurements). Geological works cover geophysical, hydrogeological, and geo-engineering works. The scope of laboratory tests of soils, rocks, and water meets the requirements of Eurocode 7 (2004), and therefore, in relation to landslides, it is analogical to the scope for objects of the 3rd geotechnical category. Field works are of significant importance; they usually involve making boreholes (at cross-sections perpendicular to the landslide) and collecting samples of soils and rocks from them in order to determine their physical-mechanical parameters. The scope of field works also involves geophysical tests, particularly helpful in determination of the depth range of mass movements and lithologic structure of slopes.

Typical geotechnical tests *in situ* include static and dynamic probings, dilatometric tests, and pressuremeter tests. However, they are of limited applicability if rock formations occur in the soil (Bednarczyk 2007). On the other hand, the scope of laboratory tests include determination of basic geotechnical parameters of soils, but the shear strength tests are of particularly great importance. Geo-engineering tests are the base for the creation of a geological model of the analyzed slope, which can be used for estimation of the safety factor of the analyzed fragment of orogene. These calculations allow for the determination of the causes of landslides, and they can also help to design procedures which will mitigate the analyzed landslide or the slope fragment that is threatened by landslide movements.

Monitoring of mass movements (which usually involves a system of surface and vertical measurements of landslide displacements, as well as hydrogeological measurements of the level or pressure of underground water) can also be included in the studies on mass movements. Inclinometer columns, which, apart from determining the degree of activity, make it also possible to establish the location of the failure zone, are a relatively common tool that allows for observation of landslide activity. Other measuring devices used for studying landslides may include extensometers, tensiometers, tiltmeters, and meteorological network (Grabowski et al. 2008). For instance, in Poland, such monitoring is conducted by the Polish Geological Institute within the framework of SOPO project (Anti-Landslide Protection System) (Poprawa and Rączkowski 2003), and consists in conducting surface and in-depth monitoring on 100 selected landslides.

Estimation of displacements of landslide colluvia is also carried out using the techniques of Global Positioning System (GPS) (Gili et al. 2000). Assessment of slope stability may also involve recording the traces of present-day and historical landslide activity (Crozier 1986). This assessment assumes that slopes on which there are no zone of depletion or colluvia can be considered stable. Data from historical archives on landslide activity are helpful in such studies (Ibsen and Brunsden 1996). Dendrochronological tests are also used when studying landslide activity (Wistuba et al. 2013); based on the analysis of deconcentricity of tree-ring increments, it is possible to reconstruct the history of mass movements on a particular area.

Slope Stability

Shear strength

Landslide processes involve forces which have an impact on soil or rock mediums. Therefore, in order to understand the principles of conducting slope stability calculations it is necessary to give the notion of stress, which represents the limit value of the ratio of the force that acts on an infinitely small part of the cross-sectional area of a body to the size of this area (Glazer 1985):

$$\sigma = \lim_{\Delta A \to 0} \frac{\Delta N}{\Delta A} \tag{1}$$

Where: σ – stress, N – force, A – the cross-sectional area.

The above definition, which is commonly used in mechanics, needs to be expanded in the case of soils due to their internal structure. This is because soil is a multi-phase system in which we can distinguish solid particles or mineral grains that constitute the soil skeleton, as well as spaces between particles – pores, filled with liquid, gas, or ice.

Stresses (**total stresses** σ) occur in each point of the soil medium. These stresses result from loading soil layers lying above, and from an external loading. If we assume that an additional load is applied to saturated soil, then at first it is taken over by water, causing an increase in **pore water pressure (u_w)**. As a result of this, outflow of water to places with lower pressure will occur, soil will compact, and the additional load will be entirely taken over by the soil skeleton. This stress is called **effective stress (σ')**:

$$\sigma' = \sigma - u_w \tag{2}$$

Soil pores in the case of unsaturated soil are partially filled with air, and in this case effective stress can be written down as:

$$\sigma' = (\sigma - u_a) + \chi \cdot (u_a - u_w) \tag{3}$$

Where: σ – normal stress (kPa), u_a – air pressure in pores (kPa), u_w – pore water pressure in pores (kPa).

Assuming that χ is the function of water saturation S_r, then if soil is watered $S_r = 1$, $\chi = 1$, and in the case of completely dry soil ($S_r = 0$) parameter $\chi = 0$.

In the case of soils, shear strength is the fundamental force responsible for keeping soil in balance, and it prevents formation of mass movements. Shear strength represents the value (usually maximum) of resistance that soil provides to shear stresses, and it is expressed by Coulomb's equation:

$$\tau_f = (\sigma - u_w) \cdot \tan \phi' + c' \tag{4}$$

Where: σ, u_w – as above, ϕ' – effective angle of internal friction (°), c' – effective cohesion (kPa).

The above equation applies when soil pores are filled with water. However, soil pores are usually filled both with water and air, which is why matrix suction forces occur in soil; these forces cause an increase in soil strength (Fredlund et al. 1978):

$$\tau_f = c' + (\sigma - u_a) \cdot \tan \phi' + (u_a - u_w) \cdot \tan\phi^b \tag{5}$$

Where: σ, u_w, u_a, ϕ', c' – as above, ϕ^b – angle that determines the increase in shear strength in connection with an increase in suction pressure, σ - u_a – net stress, u_a - u_w – soil suction.

Theoretical bases for stability calculations

In soil mechanics, evaluation of slope stability is the most often described notion of the factor of safety (Cornforth 2005):

$$FS = \frac{total\ available\ shear\ (or\ moment)\ resistance}{shear\ force\ (or\ moment)\ needed\ for\ static\ equilibrium} \tag{6}$$

According to the provided definition, stability calculations involve determination of the ratio of holding moments (or forces) to turning moments (or sliding forces). If the factor of safety FS = 1.0, then conditions of limit equilibrium occur. On the other hand, a slope is stable if FS>1.0.

Table 2. General characteristics of selected limit equilibrium methods (based on Abramson et al. 2002; Slope stability 2002; Cornforth 2005).

Name of method	Shape of the failure plane	Calculation conditions	Notes
Infinite model	Shallow failure surface, parallel to land surface	equilibrium of horizontal and vertical force projections	interslice forces are negligible; method used for landslides with a relatively high length to their depth
Fellenius method (Ordinary method of slices)	circular	equilibrium of moments	interslice forces neglected in the analysis
Simplified Bishop method	circular	equilibrium of vertical force projections and moments	interslice forces neglected in the analysis
Janbu generalized method	non-cylindrical	equilibrium of horizontal and vertical force projections	inclination of interslice forces results from the position of pressure line
Morgentern-Price	non-cylindrical	equilibrium of horizontal and vertical force projections, and moments	interslice forces have different inclinations
Spencer method	non-cylindrical	equilibrium of horizontal and vertical force projections, and moments	interslice forces have identical inclinations in each slice
Corps of Engineers (Modified Fellenius method)	any	equilibrium of horizontal and vertical force projections	inclination of interslice forces is identical in each slice and corresponds to the average slope inclination

In the case of slope stability analysis, two calculation approaches are possible, taking into account total or effective stresses (Cornforth 2005). The analysis carried out with respect to effective stresses covers conditions under which the value of shear strength takes into account stresses transferred onto the soil skeleton, and values of shear strength parameters are termed effective parameters (ϕ', c'). The advantage of this analysis is that it allows for the determination of the effects of changes in the level of groundwater table or the effect of infiltration on the state of stresses in the soil,

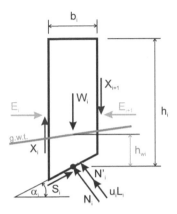

Legend:
b_i – width of slice i; h_i – height of slice i; h_{wi} – height of groundwater table within slice i; α – angle of inclination in respect to the base of slice i; L_i – length of the base of slice i; W_i – weight of slice i; N'_i – value of the normal reaction at the base of slice i; E_i, E_{i+1} – horizontal components of the interslice forces; X_i, X_{i+1} – vertical components of the interslice forces; S_i – shear strength force at the base of slice i.

Fig. 3. Distribution of forces in a slice in the Bishop method (modified from Sozański 1977).

and further consequences on the stability conditions. On the other hand, in the case of analysis carried out with respect to total stresses, the value of shear strength takes into account the effect of all the factors on the value of measured force. In this type of analysis, one can take into account undrained shear strength or total parameters of shear strength (ϕ, c).

In order to determine the factor of stability, a number of calculation methods were elaborated. These methods can be divided into limit stress methods, limit equilibrium methods, and numerical methods. Stability analysis is usually carried out for two-dimensional models of profile slopes prepared based on geo-engineering documentation. The most commonly used stability analyses include limit equilibrium methods which, in general terms, involve division of the existing (or potential) landslide mass into slices and determination of the size of forces occurring within individual slices, but particular methods differ in terms of adopted calculation assumptions. Table 1 shows general characteristics of selected limit equilibrium methods, and below is a detailed description of two of them.

Bishop method (1955) belongs to a group of methods which assume a circular failure surface (Fig. 3). The method assumes that forces of interaction between slices are unknown, and their values are determined by a method of successive approximations using internal equilibrium equations. The factor of safety is calculated from the moment equilibrium equation in relation to the center of a potential failure surface. The equation does not take into account forces of interaction between slices due to the internal character of these forces. That is why the moment caused by them for the entire landslide mass, in relation to any given point, should equal zero.

The value of shear resistance forces on slice bases is determined from the limit state criterion of the Coulomb-Mohr hypothesis, and amounts to:

$$T = \frac{1}{FS}\left(N_i \cdot \tan\phi + c_i \cdot L_i\right) \tag{7}$$

The equation of projections of all forces on the vertical direction allows researchers to obtain the following:

$$W_i + \left(X_i - X_{i+1}\right) - N_i \cdot \cos\alpha_i - T_i \cdot \sin\alpha_i = 0 \tag{8}$$

Assuming that

$$\Delta X_i = X_i - X_{i+1} \tag{9}$$

the equation for the value of the normal force at the slice base is obtained:

$$N_i = \frac{W_i + \Delta X_i - \left(c_i \cdot L_i \cdot \sin\alpha_i\right)/FS}{\cos\alpha_i + \left(\tan\phi_i \cdot \sin\alpha_i / FS\right)} \tag{10}$$

On the other hand, assuming:

$$m_{i(\alpha)} = \cos\alpha_i + \left(\tan\phi \cdot \sin\alpha_i\right)/FS \tag{11}$$

the following is obtained:

$$N_i = \frac{W_i + \Delta X_i - \left(c_i \cdot L_i \cdot \sin\alpha_i\right)/FS}{m_{i(\alpha)}} \tag{12}$$

Equation of moments for the whole landslide mass in relation to the center of a potential failure surface has the following form:

$$R\sum W_i \cdot \sin\alpha_i = R\sum S_i \tag{13}$$

where: R – radius of failure plane,
and so:

$$\sum W_i \cdot \sin\alpha_i = \sum\left(N_i \cdot \tan\phi + c_i \cdot L_i\right)/FS \tag{14}$$

Assuming that the factor of safety for all slices has the same value (*FS=const*), the following is obtained:

$$FS = \frac{1}{\sum W_i \cdot \sin\alpha_i} \sum \frac{\left(W_i + \Delta X_i\right) \cdot \tan\phi_i + c_i \cdot L_i \cdot \cos\alpha_i}{m_{i(\alpha)}} \tag{15}$$

The factor of safety for a saturated medium take the form:

$$FS = \frac{1}{\sum W_i \cdot \sin\alpha_i} \sum \frac{\left(W_i + \Delta X_i - u_i \cdot L_i \cdot \cos\alpha_i\right) \cdot \tan\phi_i + c_i \cdot L_i \cdot \cos\alpha_i}{m_{i(\alpha)}} \tag{16}$$

The above equation allows for the determination of the factor of safety using the method of successive approximations. The calculations start with the highest slice on which

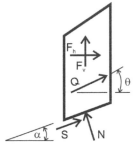

Fig. 4. Generalized distribution of forces in a slice in the Spencer method (modified from Duncan and Wright 2005).

internal forces act only from one side, and their value matches the increase in force on the slice width. Due to the complicated nature of the equations, these calculations are laborious.

That is why the Bishop method is used in practice. The method assumes that vertical components of interslice forces equal zero. This assumption implies that interslice forces are horizontal. The factor of safety equation then assumes the following form:

$$FS = \frac{1}{\sum W_i \cdot \sin\alpha_i} \sum \frac{(W_i - u_i \cdot L_i \cdot \cos\alpha_i) \cdot \tan\phi_i + c_i \cdot L_i \cdot \cos\alpha_i}{m_{i(\alpha)}} \qquad (17)$$

The factor of safety is determined by the iteration method. In the first step, the FS = 1.0 value, or the value determined from the previous use of another method, is assumed on the right side of the equation. Iterative calculations are done until the following condition is met:

$$\left| FS_o - FS_z \right| \leq \varepsilon \qquad (18)$$

where: FS_0 – calculated value of the factor of safety in the next iterative step; FS_z – the assumed value of the factor of safety in the previous iterative step; ε – the assumed accuracy of calculations.

The $m_{i(\alpha)}$ factor in the Bishop method depends on the angle of inclination of the slice base. At low values of this angle, this coefficient may assume very low or even negative values, which causes a disproportionately high increase in the factor of safety value. This is why this method may give incorrect estimates of the factor of safety, particularly in the case of failure surfaces that pass below the lower edge of the slope, which may happen in the case of gently inclined slopes or when weak soils with low strength parameters occur at the slope base.

Spencer method (1976) belongs to the limit equilibrium methods where any failure surface is assumed. This method requires satisfying equilibrium of forces and moments acting on individual blocks (Fig. 4). Interslice forces (Z_L and Z_R) are taken into account in calculations; the forces have identical inclination for all slices. Inclination of these forces (θ) is not known and computed as one of the unknowns in the equilibrium equation.

By summing up the forces acting perpendicularly and parallel to the slice base (Fig. 4), two equilibrium equations are obtained:

$$N - F_V \cdot \cos\alpha + F_h \cdot \sin\alpha + Q \cdot \sin(\alpha - \theta) = 0 \qquad (19)$$

$$S - F_V \cdot \sin\alpha - F_h \cdot \cos\alpha - Q \cdot \cos(\alpha - \theta) = 0 \qquad (20)$$

Value of force Q represents the interslice forces ($Q=Z_R-Z_L$), whereas values (F_h) and (F_v) represent all known forces (horizontal and vertical) acting on a slice, including slice weight, and they can also take into account seismic loads, dispersion, and concentrated surface forces.

By joining two equilibrium equations for forces (1) and (2) with the Coulomb-Mohr equation which describes shear forces (S), the equation for force (Q) is obtained:

$$Q = \frac{-F_V \cdot \sin\alpha - F_h \cdot \cos\alpha + \left(\dfrac{c' \cdot L}{FS}\right) + \left(F_v \cdot \cos\alpha - F_h \cdot \sin\alpha - u \cdot L\right) \cdot \left(\dfrac{\tan\phi'}{FS}\right)}{\cos(\alpha - \theta) + \left[\sin(\alpha - \theta) \cdot \dfrac{\tan\phi'}{FS}\right]} \quad (21)$$

Two equations need to be solved in order to determine the factor of stability (FS) and the unknown inclination of interslice forces. Assuming that interslice forces are parallel and are in the state of limit equilibrium, equations of general equilibrium of forces in the horizontal and vertical direction are as follows:

$$\sum Q_i = 0 \quad (22)$$

In the case of moment equilibrium, moments may be summed about any arbitrary point. Assuming that the moments are determined in the Cartesian coordinate system in relation to the point located at the beginning of the coordinate system (x=0, y=0), the moment equilibrium equation can take the following form:

$$\sum Q \cdot \left(xb \cdot \sin\theta - yQ \cdot \cos\theta\right) = 0 \quad (23)$$

where: xb – horizontal coordinate of the center of the slice base, yQ – vertical coordinate of a point on the line of action of force, Q, is located above the center of the slice base.

By substituting Eq. (21) into Eqs. (22–23), one can obtain two equilibrium equations with two unknowns: factor of safety (FS) and inclination of interslice forces (θ). In order to solve them, the method of successive approximations is used. This method allows researchers to determine FS and θ at the assumed and permissible error level.

In recent years, there has been a growing interest in numerical methods (FEM) with respect to solving tasks connected with stability of slopes. This is connected with development of computer science, wide availability of computers with high computing power, and programs which allow researchers to construct complex models of slopes. However, possibilities of a wide use of numerical methods for solving specific engineering problems are limited, mainly due to weak knowledge of strength and deformation properties in models used (Cała 2007). The advantage of numerical methods is that there are no assumptions regarding the failure surface or distribution of forces between blocks. In addition, these methods allow taking into account factors which influence the stability conditions and which are associated with the geological structure and properties of soil.

The effect of active factors on slope stability

The factor of safety is a dynamic value that changes as a result of the action of active factors which bring about changes in the state of stresses in the soil medium.

In the case of slopes, water circulation is the basic factor which determines changes in the state of stresses. This factor contributes to the change in pore pressure or suction pressure. In most cases, water movement in soil is described by Darcy's law (1956):

$$Q = k_s \cdot I \cdot F \tag{24}$$

where: Q – amount of water flowing through soil, k_s – hydraulic conductivity of saturated soil (permeability coefficient), I – hydraulic gradient representing the difference in height of water columns or the difference in pressures along water flow paths, F – cross-sectional area of soil.

In the case of unsaturated soils, the equation of water flow in the basic part of soil takes the following form (Richards 1931):

$$\frac{\partial \theta}{\partial t} = \frac{\partial}{\partial x}\left(k_x(h) \cdot \frac{\partial h}{\partial x}\right) + \frac{\partial}{\partial y}\left(k_y(h) \cdot \frac{\partial h}{\partial y}\right) + \frac{\partial}{\partial z}\left(k_z(h) \cdot \frac{\partial h}{\partial z}\right) + Q \tag{25}$$

where: θ – soil moisture content, t – time, h – suction height, x, y, z – coordinates of the coordinate system, $k_x(h)$, $k_y(h)$, $k_z(h)$ – hydraulic conductivities of unsaturated soil in directions x, y and z, Q – water inflow.

Depending on complexity of the investigated problem, the above equation is reduced to a two-dimensional or one-dimensional form, and is solved by numerical methods which use appropriate computer programs (e.g., Geostudio, Z_Soil, Geo5). In the case of simple soil conditions (e.g., vertical flow in a uniform medium), the Richards' equation may be substituted by simplified models of infiltration, for example, the Iverson's model (2000) or Green-Ampt (1911). The Iverson's model is a linearized version of the Richards' equation and makes it possible to determine the changes in pore pressure caused by rainfall, which is used for determining the factor of safety:

$$FS = \frac{\tan\phi'}{\tan\beta} + \frac{c' - \gamma_w \cdot \psi(z,t) \cdot tg\phi'}{\gamma \cdot z \cdot \sin\beta \cdot \cos\beta} \tag{26}$$

where: β – slope inclination, ϕ' – effective angle of internal friction, c' – effective cohesion, z – depth, $\psi(z,t)$ – pore water pressure head as a function of depth and duration of rainfall.

In the case of the Green-Ampt's model, calculations involve determining the infiltration capacity of soil and comparing it with the intensity of precipitation. The law of conservation of mass allows for the calculation of the amount of water accumulated in the soil profile and the depth of wetting front in the soil profile. In its simplest form, the factor of safety can be determined from the equation (Pradel and Raad 1993):

$$FS = \frac{\tan\phi'}{\tan\beta} + \frac{c' - \gamma_w \cdot z \cdot tg\phi'}{\gamma \cdot z \cdot \sin\beta \cdot \cos\beta} \tag{27}$$

where: β, ϕ', c' – as above, and z – location of wetting front.

In the above equation it is assumed that precipitation causes saturation of the soil surface in and above the level of wetting front. The above equation makes physical sense when the intensity of precipitation is higher than soil water permeability. Otherwise, precipitation does not cause soil saturation, which means that negative pore pressure dominates in soil (soil suction). In such a case, in order to determine the factor of safety, knowledge of the retention characteristic of soil is necessary.

Vegetation, whose effect on soil can be examined both in hydrological and mechanical terms, is another factor influencing slope stability (Greenway 1987). However, this effect is not unambiguous. On one hand, vegetation takes up water from soil (reducing its moisture); the roots form natural soil reinforcement (also reducing the intensity of the erosion process). On the other hand, roots increase soil water permeability and contribute to an increase in water infiltration into the soil profile, and high trees are an additional dynamic load exerted on a slope during strong winds.

The effect of root systems on soil shear strength is usually characterized by the so-called root cohesion (c_r). This parameter is associated with the transfer of tensile stresses by roots. According to Waldron model (1977) and Wu et al. (1979), the size of this force depends greatly on the tensile strength of individual roots and on their share in soil:

$$c_r = T_r \cdot \frac{A_r}{A} \cdot (\cos\theta \cdot \tan\phi' + \sin\theta)$$

(28)

where: T_r – tensile strength (kPa), Ar – cross-sectional root area (m²), A – the soil area being studied (m²), θ – angle of root deviation from the perpendicular to the shear plane at the moment of breaking (°).

If there are roots with various diameters in soil, the above equation takes the following form:

$$c_r = k' \cdot \sum_{i=1} T_{ri} \cdot \left(\frac{A_r}{A}\right)$$

(29)

The model sensitivity analysis conducted by Wu et al. (1979) showed that for typical values of soil, Wu et al. (1979) showed that for typical values of the angle of internal friction of soils the value of parameter k' is 1.0–1.3 (most often 1.2). Example values of tensile strength and the root area ratio of typical species of European trees are presented in the works by Bischetti et al. (2005, 2009) and Genet et al. (2007).

The shortcoming of the Wu/Waldron model is the assumption that maximum tensile stresses in all roots occur at the moment of the soil shear. As shown by the results of research by, among others, Docker and Hubble (2008) and Pollen and Simon (2005), the Wu-Waldron model overestimates the values of root cohesion, in extreme cases even more than two-fold. Therefore, Pollen and Simon (2005) proposed the fiber bundle model. This model assumes that in the initial phase of shearing, tensile stresses in soil are evenly transferred to all roots (n). An increase in tensile stresses, which takes place during shearing at some point, leads to breaking of the weakest root. As a result, tensile force is distributed to other roots (n-1), whereas the value of stresses increases due to the fact that the same load is transferred to a lower number of roots. Furthermore, in consequence of the increase in external load, progressive destruction (breaking) of successive roots takes place. This leads to destruction of the root system.

The hydrological effect of vegetation on slope stability results indirectly from the effect of plant roots on soil water permeability, which primarily influences the degree of saturation of the soil profile. Monitoring of pore pressure changes, which was used by, among others, Simon et al. (2000) and Simon and Collison (2002), is a useful tool in such a type of assessment. The authors showed that, in the case of trees, the hydrological effect influences the increase in stability of channel banks, whereas in the case of soil covered with grasses a reverse effect was found.

In the case of slopes located in the vicinity of river channels, fluvial processes which occur in the channel of watercourses are an important factor influencing the stability of banks. Based on the research on stability of channel banks of the Sieve River, Rinaldi and Casagli (1999) distinguished several stages (periods) in the evolution of channel banks. In periods of low levels of surface waters, stability of river channel banks is generally high, and their inclination can reach even up to 90°. The upper part of bank is usually exposed to atmospheric conditions, and soil material which can be found in this part of bank is in unsaturated condition. In such conditions, soil profile saturation changes depending on meteorological conditions. Reduction of soil suction is generally associated with the occurrence of precipitation, and the increase in value of this parameter is a result of evapotranspiration and changes in height of underground water level. Disturbance of river bank stability in this period may result from leaching of particles and grains of loose non-cohesive soils.

During the time when the water level in a watercourse rises (due to high-water condition), saturation of soils in the coastal zone takes place. The amount and intensity of water inflow into soil deposited on the bank depend on a lot of factors, including its water permeability, its moisture, value of suction pressure, and also on the rate at which water level in a watercourse rises. As a result of this process, a reduction in suction pressure and in soil shear strength takes place, and the range of soil saturation may correspond to the water level in a watercourse. Flood water exerts pressure on river banks, and this pressure causes an increase in bank stability (Simon et al. 2000), but in turn, bottom erosion is a destabilizing factor. The next period of transformations of channel banks is associated with decreasing water level in a watercourse, and this process activates changes in the underground water level near the bank and contributes to activation of forces of flow-off pressure. As shown by the results of research (by, among others, Rinaldi and Casagli 1999; Simon et al. 2000; Chiang et al. 2011), the period when water level in a watercourse decreases is critical in terms of stability of channel banks. As a result of transformation (failure), new conditions of bank equilibrium are formed. The bank remains stable until material accumulated in its foot gets eroded.

The above information indicates that **erosion** is an important factor which determines the stability of channel banks. According to the proposition of Simon and Pollen (2006), the intensity of erosion ($m \cdot s^{-1}$) of channel banks may be determined from the equation:

$$E = k_d \cdot t \cdot (\tau_0 - \tau_c) \tag{30}$$

where: k_d – erodibility coefficient ($m^3 \cdot N^{-1} \cdot s^{-1}$), τ_0 – average boundary shear stress (Pa), τ_c – critical shear stress (Pa), t – time (s).

The mean shearing stress can be determined from the equation:

$$\tau_0 = \gamma_w \cdot R_h \cdot S \tag{31}$$

where: γ_w – unit weight of water ($N \cdot m^{-3}$), R_h – hydraulic radius (m), S – channel slope (–).

On the other hand, critical shear stress can be calculated from the following formula:

$$\tau_c = 0.06 \cdot (\rho - \rho_w) \cdot g \cdot D \tag{32}$$

where: ρ – soil volumetric density ($kg \cdot m^{-3}$), ρ_w – water volumetric density ($kg \cdot m^{-3}$), g – gravitational acceleration ($m \, s^{-2}$), D – representative particle diameter (m).

Tests conducted by Hanson and Simon (2001) on a dozen or so sections of rivers in the USA showed that values of erodibility factor reached $0-1.3$ m^3·N^{-1}·s^{-1} (on average 0.07 m^3·N^{-1}·s^{-1}). These authors, based on the results of their observations, found that erodibility factor for silty soils, silty-clayey soils, and clays can be defined as a function of shear stresses:

$$k_d = 2 \cdot 10^{-7} \cdot \tau_c^{-0,5} \tag{33}$$

The calculations allow for determining changes in the geometry of river channel banks, which lead to the formation of new equilibrium conditions.

To determine the stability of river channel banks, calculation methods appropriate for the assumed or actual failure plane need to be used. In the case of rotational slides, methods described in Table 2 can be applied, and in the case of a shallow failure plane, the following equation will be of use (Simon et al. 2000):

$$FS = \frac{\sum c_i' \cdot L_i + S_i \cdot \tan\phi_i' + \left[W_i \cdot \cos\beta - U_i + P_i \cdot \cos(\alpha - \beta)\right] \cdot \tan\phi_i'}{\sum W_i \cdot \sin\beta - P_i \cdot \sin(\alpha - \beta)} \tag{34}$$

where: W_i – weight of block "i", L_i – length of the failure plane in block "i", U_i – hydrostatic-uplift force on the saturated portion of the failure surface, P_i – hydrostatic-confining force due to external water level, ϕ' – effective angle of internal friction, c' – effective cohesion, α – angle of inclination of the watercourse bank, β – angle of inclination of the failure plane, and S_i – shear strength of soil in the unsaturated zone.

Other methods of calculating the stability of scarps of watercourse channels for shallow failure surfaces are presented in the work of Simon and Pollen (2006), and for the practical needs a calculation program called BSTEM has been created (http://www.ars.usda.gov). BSTEM takes into account the influence of vegetation and fluvial processes on the stability of channel banks.

Keywords: Mass movements, instability factors, shear strength, slope stability, factor of safety, erosion of riverbanks, root cohesion.

References

Abramson, L.W., Lee, T.S., Sharma, S. and Boyce, G.M. 2002. Slope Stability and Stabilization Methods. John Wiley & Sons, Inc., New York, USA.

Ayalew, L., Yamagishi, H. and Ugawa, N. 2004. Landslide susceptibility mapping using GIS-based weighted linear combination, the case in Tsugawa area of Agano River, Niigata Prefecture, Japan. Landslides 1: 73–81.

Bajgier, M. 1993. Rola struktury geologicznej w ewolucji rzeźby wschodniego skłonu Beskidu Śląskiego i zachodniej części Kotliny Żywieckiej. Kwart. AGH, Geologia 19(1): 1–69.

Bednarczyk, Z. 2007. Ocena przydatności wybranych testów In situ i badań laboratoryjnych dla identyfikacji i przeciwdziałania procesom geodynamicznym. Górnictwo i Geoinżynieria 31(3/1): 65–79.

Bergin, D.O., Kimberley, M.O. and Marden, M. 1995. Protective value of regeneration tea tree stands on erosion-prone hill country, East Coast, North Island, New Zealand. New Zealand Journal of Forestry Science 25: 3–19.

Bíl, M. and Müller, I. 2008. The origin of shallow landslides in Moravia (Czech Republic) in the spring of 2006. Geomorphology 99: 246–253.

Bischetti, G.B., Chiaradia, E.A., Simonato, T., Speziali, B., Vitali, B., Vullo, P. and Zocco, A. 2005. Root strength and root area ratio of forest species in Lombardy (Northern Italy). Plant and Soil 278: 11–22.

Bischetti, G.B., Chiaradia, E.A., Epis, T. and Morlotti, E. 2009. Root cohesion of forest species in the Italian Alps. Plant and Soil 324: 71–89.

Bishop, A.W. 1955. The use of the slip circle in the stability analysis of slopes. Geotechnique 5: 7–17.

Bober, L., Chowaniec, J., Oszczypko, N., Witek, K. and Wójcik, A. 1977. Geologiczne warunki rozwoju osuwiska w Brzeżance koło Strzyżowa. Przegląd Geologiczny 7: 372–376.

Bommer, J.J. and Ridriguez, C.E. 2002. Earthquake-induced landslides in Central America. Engineering Geology 63: 189–220.

Bonham-Carter, G.F., Agteberg, F.P. and Wright, D.F. 1989. Weight of evidence modelling: a new approach to mapping mineral potential. *In*: Agteberg, F.P. and Bonham-Carter, G.F. (eds.). Statistical Applications in the Earth Sciences. Geological Survey of Canada Paper 89: 171–183.

Borkowski, A. 2006, Lotniczy skaning laserowy jako metoda pozyskiwania danych dla potrzeb modelowania hydrodynamicznego. pp. 129–136. *In*: Kotecko A. (ed.) Aktualny problemy rolnictwa, gospodarki żywnościowej i ochrony środowiska, Wydawnictwo Akademii Rolniczej we Wrocławiu.

Brunsden, D. 1984. Mudslides. pp. 363–418. *In*: Brunsden, D. and Prior, D.B. (eds.). Slope Instability. Wiley, Chichester, UK.

BSTEM. 2013. USDA-ARS National Sedimentation Laboratory, Oxford.

Bucała, A. 2009. Rola opadów nawalnych w kształtowaniu stoków i koryt w Gorcach na przykładzie zlewni Jaszcze i Jamne. Przegląd Geograficzny 81(3): 399–418.

Caine, N. 1980. The rainfall intensity-duration control of shallow landslides and debris flows. Gegrafiska Annaler 62A(1-2): 23–27.

Cała, M. 2007. Numeryczne metody analizy stateczności zboczy. Rozprawy, Monografie. AGH, Uczelniane Wydawnictwo. Naukowo-Dydaktyczne, Kraków, Poland.

Campus, S., Fortati, F. and Scavia, C. 2000. Preliminary study for landslide hazard assessment: GIS techniques and a multivariate statistical approach. pp. 215–220. *In*: Bromhead, E.N., Dixon, N. and Ibsen, M.L. (eds.). Landslides in Research, Theory and Practice. Proceedings of the 8th International Symposium on landslides, Cardiff 2000, Thomas Telford, London.

Cardinali, M., Ardizzone, F., Galli, M., Guzzetti, F. and Reichenbach, P. 2000. Landslides triggered by rapid snow melting: the December 1996–January 1997 event in Central Italy. pp. 439–448. *In*: Proceedings 1st Plinius Conference on Mediterranean Storms, Bios Publisher, Cosenza, Italy.

Chiang, S.-W., Tsai, T.-L. and Yang, J.-C. 2011. Conjunction effect of stream water level and groundwater flow for riverbank stability analysis. Environmental Earth Sciences 62: 707–715.

Clerici, A., Perygo, S., Tellini, C. and Vescovi, P. 2002. A procedure for landslide susceptibility zonation by the conditional analysis method. Geomorphology 48(4): 349–364.

Cornforth, D.H. 2005. Landslides in Practice. Investigations, Analysis, and Remedial/Preventative Options in Soils. John Wiley & Sons, Inc., New Jersey, USA.

Crozier, M.J. 1986. Landslides: Causes, Consequences and Environment. Croom Helm, London, United Kingdom.

Cruden, D.M. and Varnes, D.J. 1996. Landslide types and processes. pp. 36–75. *In*: Turner, A.K. and Schuster, R.L. (eds.). Landslides. Investigation and Mitigation, National Academy Press, Washington, USA

Darcy, H. 1856. Les Fontaines Publiques de la Ville de Dijon, Dalmont, Paris, France.

Dikau, R., Brundsen, D., Schrott, L. and Ibsen, L.M. (eds.). 1996. Landslide Recognition: Identification, Movement and Causes. Wiley & Sons Inc., New York, 251.

Dikau, R. 2004. Mass movement. pp. 696–698. *In*: Goudie, A.S. (ed.). Encyclopedia of Geomorphology. Routledge, London, UK.

Docker, B.B. and Hubble, T.C.T. 2008. Quantifying root-reinforcement of river bank soils by four Australian tree species. Geomorphology 100: 401–418.

Duncan, J.M. and Wright, S.G. 2005. Soil Strength and Slope Stability. John Wiley & Sons, New Jersey, USA.

Eurocode 7: Geotechnical design. European Committee for Standardization, Brussels, 2004.

Fredlund, D.G., Morgenstern, N.R. and Widger, R.A. 1978. The shear strength of unsaturated soils. Canadian Geotechnical Journal 15(3): 313–321.

Genet, M., Stokes, A., Salin, F., Mickovski, S.B., Forcaud, T., Dumail, J.-F. and van Beek, R. 2007. The influence of cellulose content on tensile strength in tree roots. pp. 3–11. *In*: Stokes, A., Spanos, I., Norris, J. and Cammeraat, E. (eds.). Eco- and Ground Bio-Engineering. The Use of Vegetation to Improve Slope Stability. Springer, Dordrecht, The Netherlands.

GEO-SLOPE. GEO-SLOPE International Ltd., Seepage Modeling with SEEP/W 2007: An Engineering Methodology (4th ed.). Alberta: GEO-SLOPE International Ltd., 2010.

Geo5. 2015. Fine Spol. s r.o., Praha, Czech Republic.

Gerlach, T. 1976. Współczesny rozwój stoków w polskich Karpatach fliszowych. Prace Geograficzne, 122, Wydawnictwo IGiPZ PAN, Wrocław, Poland.

Gil, E. 1976. Spłukiwanie gleby na stokach fliszowych w rejonie Szymbarku. Dokumentacja Geograficzna, 2, Wydawnictwo IGiPZ PAN, Warszawa, Poland.

Gil, E. 1997. Meteorological and hydrological conditions of landslides in Polish Flysch Carpathians. Studia Geomophologica Carpatho-Balcanica 31: 143–158.

Gil, E. and Długosz, M. 2006. Threshold values of rainfalls triggering selected deep-seated landslide in the Polish Flysch Carpathians. Studia Geomorphologica Carpatho – Balcanica 40: 21–43.

Gili, J.A., Corominas, J. and Rius, J. 2000. Using global positioning system techniques in landslide monitoring. Engineering Geology 55: 167–192.

Glade, T. 1998. Establishing the frequency and magnitude of landslides – triggering rainstorm events in New Zealand. Environmental Geology 35(2/3): 160–174.

Glade, T. 2003. Landslide occurrence as a response to land use change: a review of evidence from New Zealand. Catena 51: 297–314.

Glazer, Z. 1985. Mechanika gruntów. Wydawnictwa Geologiczne, Warszawa, Poland.

Goodman, R.E. and Bray, J.W. 1976. Toppling of rock slopes. Proceedings of Special Conference on Rock Engineering for Foundation 2: 201–234.

Gorczyca, E. 2004. Przekształcenie stoków fliszowych przez procesy masowe podczas katastrofalnych opadów (dorzecze Łososiny). Wyd. Uniwersytetu Jagiellońskiego, Kraków, Poland.

Govi, M., Sorzana, P. and Tropeano, D. 1982. Landslide mapping as evidence of extreme regional events. Studia Geomorphologica Carpatho-Balcanica 15: 43–61.

Grabowski, D., Marciniec, P., Mrozek, T., Neścieruk, P., Rączkowski, W., Wójcik, A. and Zimnal, Z. 2008. Instrukcja opracowania Mapy osuwisk i terenów zagrożonych ruchami masowymi w skali 1:10 000. Państwowy Instytut Geologiczny, Warszawa, Poland.

Graniczny, W. and Mizerski, W. 2007. Katastrofy przyrodnicze. Wydawnictwo Naukowe PWN, Warszawa, Poland.

Green, W.H. and Ampt, G.A. 1911. Studies of soils physics I. The flow of air and water through soils. Journal of Agricultural Science 4: 1–24.

Greenway, D.R. 1987. Vegetation and slope stability. pp. 187–230. *In*: Anderson, M.G. and Richards, K.S. (eds.). Slope Stability. Wiley and Sons, New York, USA.

Guzzetti, F., Malamud, B.D., Turcotte, D.L. and Reichenbach, P. 2002. Power-law correlations of landslide areas in central Italy. Earth and Planetary Science Letters 195(3): 169–183.

Guzzetti, F., Peruccacci, S., Rossi, M. and Stark, C.P. 2007. Rainfall thresholds for the initiation of landslides in central and southern Europe. Meteorology and Atmospheric Physics 98: 239–267. doi:10.1007/s00703-007-0262-7.

Hanson, G.J. and Simon, A. 2001. Erodibility of cohesive streambeds in the loess area of the Midwestern USA. Hydrological Processes 15: 23–28.

Harp, E.L., Keefer, D.K., Sato, H.P. and Yagi, H. 2011. Landslide inventories: the essential part of seismic landslide hazard analyses. Engineering Geology 122(1-2): 9–21.

Hutchinson, J.N. 1988. General Report: Morphological and geotechnical parameters of landslides in relation to geology and hydrogeology. pp. 3–35. *In*: Bonnard, C. (ed.). Proceedings of the 5th International Symposium on Landslides. Balkema, Rotterdam, The Netherlands.

Ibsen, M.L. and Brunsden, D. 1996. The nature, use and problems of historical archives for the temporal occurrence of landslides, with specific reference to the south coast of Britain, Ventnor, Isle of Wight. Gemorphology 15: 241–258.

Iverson, R.M. 2000. Landslide triggering by rain infiltration. Water Resources Research 36(7): 1897–1910.

Khattak, G.A., Owen, L.A., Kamp, U. and Harp, E.L. 2010. Evolution of earthquake-triggered landslides in the Kashmir Himalaya, northern Pakistan. Geomorphology 115(1): 102–108.

Kirchner, K. and Krejči, O. 2002. Slope deformations and their significance for relief development in the middle part of the Outer Western Carpathians in Moravia. Moravian Geographical Reports 10(2): 10–19.

Kotarba, A. 1986. Rola osuwisk w modelowaniu rzeźby beskidzkiej i pogórskiej. Przegląd Geograficzny, LVIII, 1-2: 119–129.

Kotarba, A. 1994. Geomorfologiczne skutki katastrofalnych letnich ulew w Tatrach wysokich. Acta Universitatis Nicolai Copernici, Geographia 27(92): 21–34.

Langendoen, E.J. and Simon, A. 2008. Modeling the evolution of incised streams II: Streambank Erosion. Journal of Hydraulic Engineering 134(9): 905915.

Liu, J.G., Mason, P.J., Clerici, N., Chen, S., Davis, A., Miao, F. and Liang, L. 2004. Landslide hazard assessment in the Three Gorges area of the Yangtze river using ASTER imagery: Zigui–Badong. Geomorphology 61(1): 171–187.

Marden, M. and Rowan, D. 1993. Protective value of vegetation on tertiary terrain before and during Cyclone Bola, East Coast, North Island, New Zealand. New Zealand Journal of Forestry Science 23: 255–263.

Margielewski, W. 1994. Typy sukcesji ruchów masowych na przykładzie osuwisk pasma Jaworzyny Krynickiej. Sprawozdania z czynności i posiedzeń PAU, Kraków 58: 110–114.

Margielewski, W. 2008. Wpływ ruchów masowych na współczesną ewolucję rzeźby Karpat fliszowych. pp. 69–80. *In*: Starkel, L., Kostrzewski, A., Kotarba, A. and Krzemień, K. (eds.). Współczesne przemiany rzeźby Polski. Instytut Geografii i Gospodarki Przestrzennej Uniwersytetu Jagiellońskiego, Kraków, Poland.

Mrozek, T., Rączkowski, W. and Limanówka, D. 2000. Recent landslides and triggering climatic conditions in Laskowa and Pleśna regions, Polish Carpathians. Studia Geomorphologica Carpatho-Balcanica 34: 89–109.

Owen, L.A., Kamp, U., Khattak, G.A., Harp, E., Keefer, D.K. and Bauer, M. 2008. Landslides triggered by the October 8, 2005, Kashmir earthquake. Geomorphology 94: 1–9.

Pollen, N. and Simon, A. 2005. Estimating the mechanical effects of riparian vegetation on stream bank stability using a fiber bundle model. Water Resources Research 41: W07025, doi:10.1029/2004WR003801.

Poprawa, D. and Rączkowski, W. 1998. Geologiczne skutki powodzi 1997 r. na przykładzie osuwisk województwa nowosądeckiego. pp. 119–131. *In*: Starkel, L. and Grela, J. (eds.). Powódź w dorzeczy górnej Wisły w lipcu 1997 r. Wydawnictwo PAN, Kraków, Poland.

Poprawa, D. and Rączkowski, W. 2003. Osuwiska Karpat. Przegląd Geologiczny 51(8): 685–692.

Pradel, D. and Raad, G. 1993. Effect of permeability on surficial stability of homogeneous slopes. Journal of Geotechnical Engineering 119(2): 315–332.

Rączkowski, W. and Mrozek, T. 2002. Activating of landsliding in the Polish Flysch Carpathians by the end of the 20th century. Studia Geomorphologica Carpatho–Balcanica 36: 91–111.

Richards, L.A. 1931. Capillary conduction of liquids through porous mediums. Physics 1(5): 318–333.

Rinaldi, M. and Casagli, N. 1999. Stability of streambanks formed in partially saturated soils and effects of negative pore water pressure: the Sieve River (Italy). Geomorphology 26: 253–277.

Sassa, K., Fukuoka, H., Wang, F. and Wang, K. 2007. Landslides induced by a combined effect of earthquake and rainfall. pp. 193–207. *In*: Sassa, K., Fukuoka, H., Wang, F. and Wang, K. (eds.). Progress in Landslide Science. Springer-Verlag, Berlin, Germany.

Schenk, V., Schenkova, Z., Kottnauer, P., Guterch, B. and Labak, P. 2001. Earthquake hazards maps for the Czech Republic, Poland and Slovakia. Acta Geophys. Polonica 49: 287–302.

Schuster, R.L. 1996. Socioeconomic significance of landslides. pp. 12–35. *In*: Turner, K.A. and Schuster, R.L. (eds.). Landslides. Investigations and Mitigation. Special Report 247, National Academy Press, Washington, USA.

Selby, M.J. 1976. Earth's Changing Surface, an Introduction to Geomorphology. Clarendon Press, Oxford, UK.

Simon, A., Curini, A., Darby, S.E. and Langendoen, E.J. 2000. Bank and near-bank processes in an incised channel. Geomorphology 35: 193–217.

Simon, A. and Collison, A.J.C. 2002. Quantifying the mechanical and hydrological effects of riparian vegetation on streambank stability. Earth Surface Processes and Landforms 27: 527–546.

Simon, A. and Pollen, N. 2006. A model of streambank stability incorporating hydraulic erosion and the effects of riparian vegetation. Proceedings of the Eighth Federal Interagency Sedimentation Conference (8th FISC), April 2–6, Reno, NV, USA.

Sozański, J. 1977. Stateczność wykopów, hałd i nasypów. Wydawnictwo "Śląsk", Katowice, Poland.
Spencer, E. 1967. A method of analysis of the stability of embankments assuming parallel interslice forces. Géotechnique: 17(1): 11–26.
Starkel, L. 1960. Rozwój rzeźby Karpat fliszowych w holocenie. Prace Geograficzne IGiPZ PAN 22: 239.
Starkel, L. 1996. Geomorphic role of extreme rainfalls in the Polish Carpathians. Studia Geomorphologica Carpatho–Balcanica 30: 21–39.
Swanson, F.J., Graham, R.L. and Grant, G.E. 1985. Some effects of slope movements on river channels. International Symposium on Erosion, Debris Flow and Disaster Prevention. Tsukuba. Japan 273–278.
Thiel, K. (ed.). 1989. Kształtowanie fliszowych stoków karpackich przez ruchy masowe na przykładzie badań na stoku Bystrzyca w Szymbarku. Wydawnictwo PAN IBW, Gdańsk, Poland.
U.S. Army Corps of Engineers. 2003. Slope stability. Engineering manual No. 1110-2–1902. Washington, USA.
Van Westen, C.J., Ranwers, N., Terlin, M.T.J. and Soeters, R. 1997. Prediction of the occurrence of slope instability phenomena through GIS-based hazard zonation. Geologishe Rundschau 86: 404–414.
Van Westen, C.J., Rengers, N. and Soeters, R. 2003. Use of geomorphological information in indirect landslide susceptibility assessment. Natural Hazards 30: 399–419.
Varnes, D.J. 1978. Slopes movement types and processes. pp. 12–33. *In*: Schuster, R.L. and Krizek, R.J. (eds.). Landslides: Analysis and Control, Special Rep. 176, Transportation Research Board, National Academy of Science, Washington, USA.
Vosselman, G. and Mass, H.G. 2010. Airborne and Terrestrial Laser Scanning. Whittles Publishing, Dunbeath, Scotland.
Waldron, L.J. 1977. The shear resistance of root-permeated homogeneous and stratified soil. Soil Science Society of America Journal 41(5): 843–849.
Wistuba, M., Malik, I., Gärtner, H., Kojs, P. and Owczarek, P. 2013. Application of eccentric growth of trees as a tool for landslide analyses: the example of Picea abies Karst. in the Carpathian and Sudeten Mountains (Central Europe). Catena 111: 41–55.
Wu, T.H., McKinnell, W.P. and Swanston, D.N. 1979. Strength of tree roots and landslides on Prince of Wales Island, Alaska. Canadian Geotechnical Journal 16: 19–33.
Z_SOIL PC. 2014. Instruction Manual. ZACE Services, Lausanne, Switzerland.

CHAPTER 16

Changes of Mountain River Channels and Their Environmental Effects

Elżbieta Gorczyca, * *Kazimierz Krzemień,*[a] *Maciej Liro*[b] and
Mateusz Sobucki[c]

INTRODUCTION

When geomorphologists began using the system approach, they started looking more closely at the links between the river channel and the environment of its basin (Chorley and Kennedy 1971; Klimek 1979; Kaszowski and Krzemień 1979, 1999). Their studies mention features of lowland, mountain, and high-mountain channels (Miller 1958; Gardner et al. 1983; Krzemień 1992; Montgomery and Buffington 1997; Knighton 1998; Wohl and Merritt 2005). A significant step forward was made when researchers widened their focus into entire channel systems. Since that time, they started focusing on the fluvial system, beginning with headwater areas, and on the regularity of the fluvial system structure in the longitudinal river profile. It is typically assumed that stream channels begin at the end of their headwaters, as fluvial processes take over from slope processes, and the sheet outflow is channelled (Montgomery and Buffington 1997; Wrońska-Wałach 2009, 2014; Płaczkowska et al. 2014). Some researchers also began looking for characteristic features of mountain river channels (e.g., Heede 1972; Krzemień 1992; Whiting and Bradley 1993; Montgomery and Buffington 1997; Wrońska-Wałach 2009). They highlighted a set of features (steps, rib-steps, bedrock bottoms, and riffle-pool sequences) characteristic of mountain rivers. Others also added: (1) the channel gradient as a significant factor influencing the stream's power,

Jagiellonian University, Institute of Geography and Spatial Management, ul. Gronostajowa 7, 30-387 Kraków, Poland.
[a] E-mail: kazimierz.krzemien@uj.edu.pl
[b] E-mail: maciej.liro@uj.edu.pl
[c] E-mail: mateusz.sobucki@uj.edu.pl
* Corresponding author: elzbieta.gorczyca@uj.edu.pl

and therefore its capacity for erosion and transport, (2) the size of the bedload material that influenced the location of debris ribs, (3) the average distance between ribs, and (4) the average rib width (McDonald and Banerjee 1971; Heede 1972, 1977, 1981; Krzemień 1992; Montgomery and Buffington 1997; Chartrand and Whiting 2000). Some studies investigated the relief energy in a river system, and the environmental conditions influencing the functioning of river channels (Klimek 1979). These authors typically underlined the fact that mountain river channels were either cut in bedrock, or in thick layers of gravel.

Mountain river channels may follow a number of patterns, and they can be divided into straight, sinuous, meandering, braided, and wandering types, depending on the volume and rate of sediment supply as well as channel stability (Leopold and Wolman 1957; Schumm 1981). Considering their longitudinal channel section, they are typically assumed to begin at the end of their headwaters, as fluvial processes take over from slope processes, and the sheet outflow is channelled (Montgomery and Buffington 1997; Wrońska-Wałach 2009, 2014; Płaczkowska et al. 2014). The characteristics of mountain fluvial systems are assumed to lose their mountain characteristics along with the disappearance of the riffle-pool sequence (Montgomery and Buffington 1997; Wohl and Merritt 2005).

Mountain rivers can be defined based on a variety of criteria proposed in published studies: high channel slope (e.g., Dębski 1970; Gardner et al. 1983; Wohl 2000; Bartnik 2006), coarse bed material and its relatively regular supply, and a hydrological regimen that permits bedload transport and fluctuating water discharge during the year (Fig. 1A). Mountain rivers may have either bedrock or alluvial channels, each associated with a different set of typical features. While bedrock channels characteristically feature a bedrock bottom, steps, and evorsion hollows, alluvial channels involve step-pool and

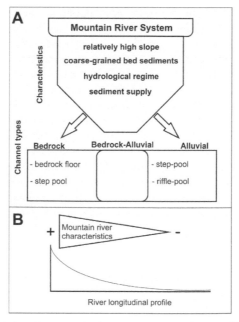

Fig. 1. Main characteristics of a mountain river system (A) and its potential occurrence along the longitudinal section (B).

riffle-pool sequences (Fig. 1A). These identification criteria are typically found in mountain or mountain-foreland sections of rivers. Downstream from these sections, rivers tend to gradually lose the energy coming from the mountain relief, and their characteristic mountain features disappear (Fig. 1B).

Several types of natural mountain channels have been identified, that is: (1) rocky channels with their bottom and banks cut in the bedrock (Fig. 2A), (2) rocky channels with their bottom cut in the bedrock and the banks cut in alluvia (Fig. 2B), and (3) fully alluvial channels (Fig. 2C and D). Mountain channels can be straight, sinuous, braided (Leopold and Wolman 1957), or wandering (Fig. 2D). Rivers flowing through foothill areas tend to have either sinuous or meandering channel patterns (Fig. 2E) cut in alluvial covers. Channels of high-mountain rivers are cut in bedrock (Fig. 2A), in slope covers (Fig. 2F), moraine deposits (Fig. 2G), or fluvioglacial-fluvial deposits (Fig. 2H).

The structure of mountain river channels in their longitudinal profile. Historically, the direct human impact on river channels involved the development of transversal (Fig. 3B) or longitudinal (Fig. 3A) training structures (Fig. 3A–D) (Radecki-Pawlik 2012). Such regulated reaches often alternate with natural and semi-natural ones within the same fluvial system (Kaszowski and Krzemień 1979; Gorczyca and Krzemień 2010; Gorczyca et al. 2011).

The high complexity and high levels of energy in mountain fluvial systems (i.e., Kaszowski and Krzemień 1977; Chełmicki and Krzemień 1999) cause difficulties in understating river response to human intervention. With their complexity of biotic and abiotic conditions, mountain river channels are highly sensitive to change and degradation under human influence both in the valley bottom and across the basin (Lach and Wyżga 2002; Krzemień 2003; Zawiejska and Krzemień 2004; Wohl 2006; Korpak 2007; Korpak et al. 2008; Gorczyca and Krzemień 2010; Wyżga et al. 2012; Krzemień et al. 2015). The most common effects of mountain river training include channel narrowing and incision. Ultimately, river energy is concentrated in a very narrow zone (Fig. 3C and D).

The Structure of Mountain River Channels in Their Longitudinal Profile

Fluvial systems are characterized by a certain segmental structure, which allows us to assume the degree of their complexity associated with their natural and anthropogenic conditions (Chorley and Kennedy 1971; Kaszowski and Krzemień 1977; Montgomery and Buffington 1997; Krzemień 2003). In the longitudinal profile of mountain river channels, there are different kinds of morphodynamic segments (Figs. 4 and 5). Their existence is a result of the evolution of channels, and indicates the stage which this evolution has reached (Kaszowski and Krzemień 1977). Two segments belonging to different systems can be distinguished in the longitudinal profile of a valley, that is, a denudation segment and a fluvial one. The boundary between them is usually not sharp, and has the characteristics of a transition or border zone (Kaszowski and Krzemień 1977; Wrońska-Wałach 2009; Krzemień 2012). The denudation segment can be of the type of a slope channel, lined with sharp-edged debris which, during catastrophic rainfall, can be cleaned out (Kaszowski 1973; Wrońska-Wałach 2009). The fluvial segment is usually characterized by a complex segmental structure referring to the diversity of the geological structure, the changing energy of the river, or the anthropopression related to the regulation of river channels, the exploitation of debris, or land use changes in individual basins.

Fig. 2. Examples of mountain channels: A – Bedrock channel (Ribeira do Seixal, Madera), B – Bedrock-alluvial channel (Wisłoka River, Poland), C – Braided channel (Dades River, Morocco), D – Wandering channel (Czarny Dunajec River, Poland), E – Sinuous channel in foothills area (stream in Mecsek Mts., Hungary), F – Channel cut in slope material (stream in Quilian Mts., China), G – Channel cut in moraine material (Starorobociański Stream, Poland), H – Channel cut in glacifluvial material (Chochołowski Stream, Poland).

Fig. 3. Examples of mountain channel regulation structures and other anthropopressure: A – Longitudinal regulations (Czarny Dunajec River, Poland), B – Transversal regulations (Ribeira do Grande, Madera), C – Artificial trough (Krośnica Stream, Poland), D – Channel narrowed due to regulation (Liyuan-He, China), E – Large boulders used for building purposes (Podhale region, Poland), F – Channel with remaining elements of training structures (Skawa River, Poland).

The segmental structure of river channels may change over time (Krzemień 1999, 2003). As confirmed by previous studies, the knowledge of whole fluvial systems and their segmental structure is essential to get to know the stages of their development, to protect them, and to carry out any properly planned possible regulations of channels (Krzemień 2012). In order for the possibility of getting to know whole fluvial systems to exist in a uniform manner, studies proposing to collect uniform information about segments of river channels, or entire channel systems, were developed on a world scale (Fig. 5). For this purpose, ready-made instructions with diaries and charts including ready-made proposals of answers were presented (Kamykowska et al. 1975; Mosley et al. 1987; Wasson et al. 1993; Rosgen 1994; Thorne 1998; Kamykowska et al. 1999, 2012). Studies made using these methodological proposals were conducted, among others, in the Polish Western Carpathians (Kaszowski and Krzemień 1979; Krzemień

1981, 1992; Rączkowska 1983; Gorczyca 2012), in New Zealand (Mosley 1987), in France (Wasson et al. 1993; Krzemień 1999), in Scotland (Chełmicki and Krzemień 1999), in the Italian Alps (Krzemień 1999), and in Morocco (Izmaiłow and Krzemień 2008) as well as works summarizing the results from USA, Canada, and New Zeland (Rosgen 1994).

In the 1970s, based on the studies of gravel-bed rivers in the Western Carpathians, the first instruction for mapping river channels was developed at the Jagiellonian University (Kamykowska et al. 1975). The studies made on its basis were developed in subsequent years (Kaszowski and Krzemień 1977, 1979; Krzemień 1981; Rączkowska 1983). This method was perfected, verified, and prepared for printing (Kamykowska et al. 1999, 2012). Aerial photographs from different periods, as well as archival maps, materials relating to the regulation of river channels, expert opinions, and hydrological data were used as supplementary materials during the study of the segmental structure of river channels.

Based on the conducted field research and cameral studies, simple and complex channel systems were distinguished. A simple structure characterizes those channels within which the same type of the morphodynamic segment is not repeated twice (Fig. 4b). Channels with repeating sequences of morphodynamic segments were recognized as well (Fig. 4). The more morphodynamic segments in a system, the greater its complexity. This attests to the diversified development of the erosional curve (Fig. 4). The diversity of bedrock resistance is an important cause of conditioning the borders of morphodynamic segments; however, boundaries between segments, which cannot be explained with differences in resistance, are common. The genesis of erosion segments occurring in the lower and middle sections of channels often dates back just a few decades. In addition, morphodynamic segments may serve a similar function, but at the same time, they may vary in their dynamic condition, that is, the intensity of processes occurring in them. These differences stem not only from the structural

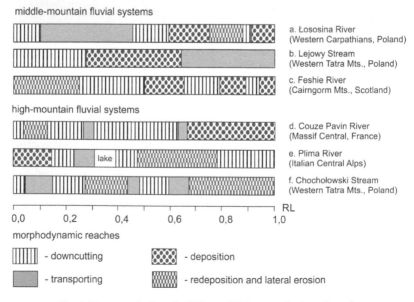

Fig. 4. Structure of selected middle- and high-mountain river channels.

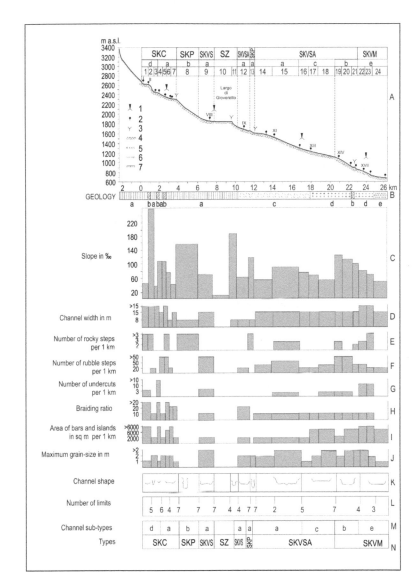

Fig. 5. Plima Stream typology and structure (A): (1) – recessive moraines, (2) – bed rubble measurement sites, (3) – artificial reservoir dams, (4) fluvioglacial and moraine covers, (5) – rocky outcrops, (6) – lacustrine covers, (7) – glacier (Vedretta Lunga). Channel types: SKC – glacial cirque type, SKP – glacial thresholds, SKVS – glacial trough – moderately stable, SKVM – glacial trough – mobile; Channel subtypes (channels displaying a tendency to): (a) – weak deep erosion, (b) – intensive deep erosion, (c) – moderate lateral migration, (d) – intensive lateral migration, (e) – intensive deep and lateral erosion, SZ – the Lago Gioveretto dam system. Geology (B): (a) – quartz phyllites and chlorine shales, (b) – marbles, (c) – orthogneisses, granitoides and granodiorites, (d) – paragneises and chlorine shales, (e) – alluvial fan fluvioglacial drift, (C) – slope in%, (D) – channel width in m, (E) – number of rocky steps per 1 km, (F) – number of rubble thresholds per 1 km, (G) – number of undercuts per kilometre, (H) – braiding ratio (number of medial bars and islands per 1 km), (I) – area of middle bars and islands per 1 km, (J) – largest grain–size in m, (K) – channel shape, (L) – number of limits, (M) – channel subtype, (N) – channel type (Krzemień 1999).

and functional differences of the channel system itself, but also of the whole basin (Kaszowski and Krzemień 1977).

A diagram of the structure of selected channel systems in medium and high mountains has been presented in Fig. 4. In mid-mountain areas, in semi-natural conditions, river channels are usually cut into solid rock in the upper reaches. At lower altitudes, these channels become more winding with the dominance of downward erosion, and then alluvial with the dominance of deposition. Such a development model of channels was disturbed by human activity (Fig. 4a, Gorczyca 2012). In this way, for example, new erosion segments appeared at lower positions. The complexity of fluvial systems is also affected by the complex history of the development of a given area. In the case of the Feshie River, the complex structure of the channel system was also affected by the Pleistocene glacier in the Cairngorm Mountains, whose activity caused the upper part of the Geldie River to take over, and the depositing of moraine covers in the lower reaches of the Feshie River (Fig. 4c, Chelmicki and Krzemień 1999). In high mountain areas, the complex structure of channels was influenced by the impact of Pleistocene and contemporary mountain glaciers, for example, in the Plima valley in the Italian Alps (Figs. 4e and 5, Krzemień 1999), or in the Chochołowska Valley in the Western Tatras (Fig. 4f, Krzemień 1991). The complex structure was also influenced by the impact of volcanism in the Holocene, for example, in the Monts Dore Massif in France (Fig. 4d, Krzemień 1999). Thus, the comparison of features of morphodynamic segments of various channels points to the individual properties of not only each of the channel subsystems, but also of their basins (Fig. 4).

Human Impact on River Channels on the Examples of Rivers in the Polish Carpathians

For more than 200 years, dynamic and highly diverse mountain river systems all over the world have been subject to direct human impact (Gregory 2006). Their channels have been degraded, the ecological condition has been weakened, and erosion processes have intensified (Klimek 1987; Krzemień 2003; Wohl 2006; Korpak 2007). Most major European rivers have been trained. Yet, these huge efforts aimed at preventing floods by involving retention reservoirs, embankments, and other river training structures have failed to provide full flood protection. Instead, the isolation of channels from the rest of their valleys resulted in a considerable decrease in flood water retention, and in accelerated flood flow velocity. For example, the turn of the 20th century marked the beginning of river training efforts in the Polish Carpathian Mountains (Kędzior 1929; Gregory 2006). Their general objective was to protect urban and rural communities in the area from floods. More direct objectives included improving channel capacity (by straightening) and stability, limiting the sediment supply by bank undercutting, slope processes, and through linear erosion and mass movements. These projects typically started from the lowest reaches, and gradually progressed upstream (Zawiejska 2006; Korpak 2007; Starkel and Łajczak 2008). Thus, sections trained featured gentle bends and long straights (Kościelniak 2004; Korpak et al. 2008) in the middle of the former wide riverbed. The overall history of the Carpathian river training can be divided into four stages:

a. From the late 19th century until the outbreak of the Second World War;
b. From the end of the Second World War until the 1970s;

c. Between 1971 and 1990;

d. After 1990.

Between the beginning of the 20th century and the outbreak of the Second World War, training schemes on major Carpathian rivers proceeded from the mountain foreland into the upper river reaches. The measures primarily included step structures, built of timber or concrete with stone cladding. The objective was to provide protection for property, and to prevent lateral erosion. One of the effects of these schemes was the reclamation of new land that could be put to various economic uses. Some of the projects, however, were gradually lost in floods, for example, in 1934. In the late 1930s, engineers added several-metre-high check-dams to these projects, built as isolated structures. Also, at that time, plans were made for systematic training projects along extensive channel sections. The Second World War put a stop to these endeavours, and only a few structures were completed (Korpak 2007). Modelled on Alpine structures, the check-dams were ill-suited to the Carpathian environment as they only trapped fine suspended matter. Post-war training efforts gradually started in the late 1950s. That stage was characterised by a continuation of high dam-structures, including weirs and check-dams, plus groynes and training banks. After 1970, training programmes included the large-scale use of step-structures (0.5–1.0 m) and bank protection, typically made of rock boulders, and the high check-dam design was abandoned. The most extensive river training schemes were carried out in the 1970s when Poland enjoyed a period of economic revival fuelled by new international loans. The projects stopped abruptly after 1989, as the country embarked on its political and economic transformation. The work restarted later in the 1990s, especially after the 1997 flood (Korpak et al. 2008). By then, a European Union (EU) associated country, Poland, received aid to address the damage from a series of extreme floods, including 1997, 1998, and 2001. The structures involved step-structures and protection of river banks mainly using rock boulders (Korpak et al. 2008). Subsequently, the scale of river training schemes shrank again and concentrated primarily along reaches near buildings and other structures. New "close-to-nature" training principles, as well as the guidelines of the Framework Water Directive, were also beginning to make their mark in Poland. A compromise between environmental and human needs became a factor to be considered (Korpak et al. 2008).

Some reaches of Carpathian rivers have been subject to more than one training scheme. Indeed, as training structures were destroyed, renewed efforts were often launched to bring a given reach back under control. Each subsequent project tended to be implemented using a different technology and approach, which altered the scope and structures involved (Fig. 3). The combined effect of generations of training projects included straightening, narrowing, and deepening of the channels, as well as changes to their longitudinal profiles. The dominant process triggered by these works was excessive downcutting (Kościelniak 2004; Zawiejska 2006; Korpak et al. 2008). As a result, some channel reaches have cut into bedrock (Wyżga 1991; Krzemień 2003; Gorczyca and Krzemień 2010).

Transformations that involved a shift from a braided to a single-thread channel were caused by engineering projects, but were aided by changes to the land use within the basin which limited the supply of sediment. The training approach in Carpathian rivers involved carving out of a single narrow channel, and cutting off other branches by using longitudinal constructions, such as groynes and training banks (Korpak et al.

2008; Radecki-Pawlik 2012). The most characteristic feature of the structure found today in many trained rivers is their considerable uniformity of channel geometry and its morphodynamic function (Zawiejska and Krzemień 2004; Zawiejska 2006). Thus straightened, shortened, and narrowed channels became steeper, and offered higher flow energy, which in turn increased the transport of bedload, leading to channel incision.

Dams also play a significant role in the transformation of mountain river morphology. Since the beginning of the 20th century, more than 40,000 large dams, which affect about 15% of annual runoff globally (Nilsson et al. 2005), have been built worldwide (see for review Brandt 2000; Petts and Gurnell 2005). Fourteen such structures operate in the Polish Carpathians (Sroczyński 2004), and more are under construction. Dam construction in a river valley disturbs the natural water and sediment transport by the river. In a gravel-bed mountain river, this causes deposition of the bedload upstream of the reservoir and of the suspended sediments inside the dam reservoir (Łajczak 2005). The reduction of sediment supplies into channels caused significant channel incision downstream of the dam (for example, Williams and Wolman 1984; Brandt 2000; Petts and Gurnell 2005; Skalak et al. 2013). This geomorphological effect is often observed over several tens of kilometres below dams (Williams and Wolman 1984), and can be qualitatively and quantitatively predicted (Petts and Gurnell 2005; Schmidt and Wilcock 2008; Grant et al. 2012). The deposition of sediments upstream of the dam generally leads to the silting of reservoirs, the rapid reduction of their capacity (Łajczak 1996, 2005), and the development of deltas near the reservoir inlet (Klimek et al. 1989). The evolution of gravel-bed mountain rivers in their fluctuating backwater zones is poorly understood, and existing models only allow qualitative predicting (Skalak et al. 2013; Liro 2014). Recent studies of gravel-bed river channels upstream of reservoirs suggest that the forced deposition of bedload in that zone has led to: a considerable increase in the size of channel bars, intensification of lateral erosion (Liro 2016), and a considerable widening of the active channel width (Liro 2015), which may contribute to improved hydro-morphological quality of watercourses in these zones, in comparison to the reaches downstream of the dams (Wiejaczka and Kijowska-Strugała 2015). Such tendencies were observed in the gravel-bed Dunajec River upstream of the Czorsztyn Reservoir, where the channel underwent significant adjustments in the 15 years from the commissioning of the reservoir (Fig. 6). These observations suggest that channels with sediment deficit, and affected by previous regulation works, may, at least relatively, quickly return to the transport capacity—sediment supply equilibrium. This is the result of the artificially forced deposition of bedload upstream from the reservoirs, where the base-level rising occurs. The direction of these changes can be predicted by taking into account the evolution of a given channel system, local conditions, and the reservoir's mode of operation (see Liro 2014).

In recent years, increasing numbers of large dams are being decommissioned, especially in the USA (see for review East et al. 2015; O'Connor et al. 2015). The removal of a dam allows the river channel to recover its continuity, including water and sediment transport, and the migration of animals, thus potentially contributing to an improvement in the valley's ecological conditions (O'Connor et al. 2015). However, the geomorphological and ecological impact of such operations on the river valley remains understated, and still poses a challenge and demands further interdisciplinary research (O'Connor et al. 2015).

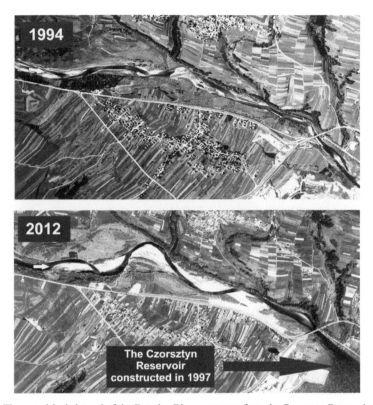

Fig. 6. The gravel-bed channel of the Dunajec River upstream from the Czorsztyn Reservoir (Polish Carpathians).

Bedload Extraction

The evolution of mountain rivers, both in Poland and worldwide, has been influenced by gravel mining, a common practice during the last one hundred years (e.g., Augustowski 1968; Osuch 1968; Wyżga 1991; Kondolf 1994, 1997; Marston et al. 2003; Krzemień 2003; Rinaldi et al. 2005). In Poland, the exploitation of river bedload material for construction purposes resulted in higher rates of channel deepening, especially after the Second World War (Klimek 1983; Rinaldi et al. 2005). Large scale operations of this type used bulldozers, while private individuals tended to only extract large boulders for construction purposes (Fig. 3E). The depletion of the channel bottom armouring exposed the channels to downcutting, and the affected reach would go through a deepening phase. This development also impacted upon adjacent reaches, mainly through headward erosion. These large-scale and long-term operations accelerated the deepening of some Carpathian river reaches to the point where the bedrock would be uncovered (Krzemień 2003; Kościelniak 2004; Zawiejska 2006; Rinaldi et al. 2005). The Ropa river channel is a case in point, where an industrial-scale gravel extraction operation paved the way for a dramatic modification of the affected reach, as an initial 1.5 metre downcutting continued up and downstream converting tens of kilometres of an alluvial channel into a bedrock channel (Augustowski 1968). Similar developments were observed in other Carpathian rivers, particularly in the 1970s (Klimek 1987; Wyżga

1991), where downcutting would lead to the lowering of the groundwater level and to channel geometry changes resulting in greater downstream flood hazard levels (Wyżga 1997, 2001). In the past, similar cycles involving bedload extraction and its consequent adverse effects were also observed in other European countries (Marston et al. 2003; Rinaldi et al. 2005). Today, EU member states do not allow such activities, although illegal operations continue undermining this consensus in the Polish Carpathians. Education of local communities and strict enforcement of the law are necessary if these processes are to be stopped.

Channel Degradation and the Disappearance of Braided Channels

Braided rivers are often found in mountain areas with a glacial history, or where glaciers are present. Their channels display features typical of mountain rivers. They include a relatively steep valley, frequent and large fluctuations in the water stage and discharge, a low proportion of cohesive sediments building the floodplain, coarse bed material, and a large supply of sediment into the channel (Leopold and Wolman 1957; Schumm 1960; Knighton 1998). This set of conditions typically produces high stream power. This high stream power is responsible for high rates of erosion and bedload transport (Leopold and Wolman 1957; Schumm 1960).

Braided as well as wandering rivers typically feature some of the most complex channel systems, and these have been proved to be the most susceptible to the adverse effects of human influence (Surian and Rinaldi 2003; Piégay et al. 2006; Surian 2006). Indeed, these gravel-bed channels degraded as channel bars merged with each other, thus increasing in size, but decreasing in number. Next, the large bars and islands merged with the floodplain, and cut off some of the side streams (Germanowski and Schumm 1993). Finally, multi-stream channels would be replaced by single, but deep-cutting channels.

A good example of the degradation and disappearance of braided channel systems is provided by three rivers running out of the high mountain range of the Tatra Mountains. Braided channels in the Podhale region of the Polish Carpathian Mountains are known to have degraded due to a combination of river training, gradual environmental change, and increasing human impact (Zawiejska and Krzemień 2004; Wyżga 2007; Gorczyca and Krzemień 2010). Nearly 90% of the Biały Dunajec, and 80% of the Czarny Dunajec channel systems were trained in the past (Fig. 7). In the 20th century, the training gradually constricted these expansive braiding systems to narrow and straight channels. Over time, however, some of the training structures were destroyed by floods. This effect was the strongest in the lower course of the Białka, the third river, where nearly all training structures were destroyed in the 1970s and 1980s. Its channel returned, if not entirely, to its old pattern (Baumgart-Kotarba 1983). Today, the Białka River channel is one that is the least transformed by human influence of all the river channels crossing the Podhale region in the foreland of the Tatras, and some of its reaches provide a good example of a braided river channel in a river running out of the Tatra Mountains. Even this channel, however, has been observed to be gradually losing its braided character, although this is due to factors other than extensive training schemes, or bedload extraction. Large alluvial forms are found in the channel, a result of the merging (by vegetation expansion) of bars and islands. These merged bars have been preserved at the higher elevation remains of formerly active channels. In the next phase, these large channel forms are transformed into sidebars, while islands gradually become a floodplain. This process combines with the downcutting of the high stream

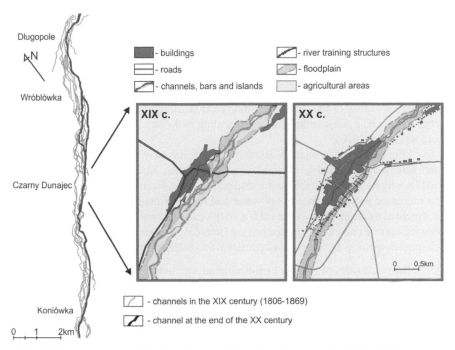

Fig. 7. Planform changes of the Czarny Dunajec River channel between the XIX and XX century.

power river and, partly supported by local human impacts, produces a downcut main stream that carries most of the water; the channel tends to be a single-threaded one, and the braided nature of the system disappears.

The problem of the degradation and destruction of braided river channels was also studied in the French Alps (e.g., Piegay et al. 2009). The length of braided channels was estimated to be 1.214 km, of which approximately 53% was completely destroyed during the last two centuries. Training, including that with levees, was the main reason for the disappearance of the braiding feature along the course of 437 kilometres. For the remaining 17% of the channels, the causes of degradation are unclear, but they probably also include the extraction of riverbed material.

Among the multitude of the channel types found in the Polish Carpathian Mountains, the proportion of regulated to unregulated reaches varies (Fig. 3). In 1991, 621 kilometres, or about 28%, of major mountain river channels were trained (including the Vistula, Soła, Raba, and Dunajec rivers) (Hennig 1991). This proportion was lower among smaller rivers, such as the Łososina, Ropa, and Jasiołka at 465 kilometres, or about 4.3%. The proportion of trained channels varied widely between channel systems, from 3 to 60% (Hennig 1991). In the period following 1991, there were plans to train more than 1.600 kilometres of channels, but these have been reduced to a necessary minimum.

The Environmental Effect of River Channel Incision

In areas with strong human impact, the functioning of rivers and streams is adjusted to changes in the valley bottom and sides resulting mainly from channel regulation works, dam constructions, gravel mining, and land use changes. As the channel systems adjust their structures to these changes, they display varying rates of change. During the last half of a century, major rivers in the Polish Carpathians deepened their channels by up to 2–8 metres (Łapuszek and Ratomski 2006). This process also caused backward erosion in some of their side valleys. The intensified erosion processes were caused by human activities in channels, including river training, bedload extraction, and excavation in the central section of the channel (Gorczyca et al. 2011), large-scale land use change after 1989, and a considerable degree of sealing in the basin (Bartnik 2006; Gorczyca and Krzemień 2010).

After 1989, the Western Carpathian Mountains saw an increase in grasslands and forests at the expense of arable land (Soja 2002; Kozak 2003; Gorczyca et al. 2011), which helped to stabilise the slopes. However, river channels began to receive an increased portion of energy as water had less sediment to carry. This effect was augmented by an increase in the sealing of the basin (Bartnik 2006). This extra energy carried into the channels triggered processes leading to their deepening. As the scale of channel deepening increases, environmental and economic problems begin, including:

- Increased rates of wear in hydrotechnical structures.
- Loss of channel bank stability.
- Over-drainage of certain sections of valley bottoms.
- Death of old trees in valley bottoms.
- Lowering the ground water level, and the loss of water in wells.
- Deterioration of water quality due to more rapid discharge.
- Deterioration of biodiversity in rivers.

Post-Regulation River Adjustments

One positive aspect to the deepening of Carpathian rivers and the destruction of training structures is that the channels return to their pre-training character (Fig. 8). This restoration, however, takes place in a zone that is narrower than before, and takes place at a level 2–8 metres lower. The resulting channels have fewer branches and streams, and might be dubbed as post-training channels. They resemble semi-natural channels with some remaining elements of the training structures. The channels tend to be deeper and wider than the trained ones (Fig. 8). They are found in various parts of fluvial systems, and are clearly an effect of channels being left to natural processes.

A good example is a section of the upper Wisłoka River (Polish Carpathians), which underwent channel training in the 1960s. Prior to the project, this semi-natural reach had a meandering course (Fig. 9). Its dominant processes included lateral erosion on the outside of the meanders, and accumulation on bars on the inside of the meanders. The training project straightened the channel, shortened its course, increased the gradient, and narrowed the stream. Some of the cut-off meanders were filled in and, decades later, they would become overgrown by a riparian forest. Until the end of the 20th century, the little change that occurred along the channel was due to lateral erosion. In the early 21st century, this process accelerated as a result of a series of floods in 2004, 2006, 2010, 2011, and 2014 (Fig. 9, Gorczyca et al. 2013), and half a century after its training the river was spontaneously restored to its natural condition. Vegetation gradually entered the newly formed accumulation zones. In the expanding active zone of the channel, this included a mosaic of herbaceous plants, shrubs, and a riparian forest (Fig. 9). The

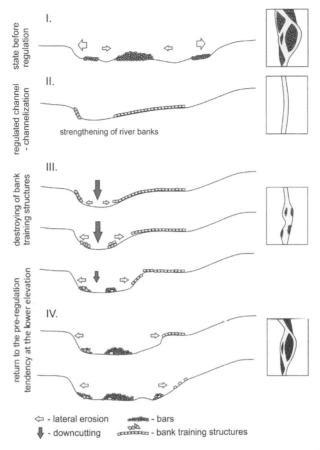

Fig. 8. Conceptual model of channel adjustments from regulation to self-renaturisation.

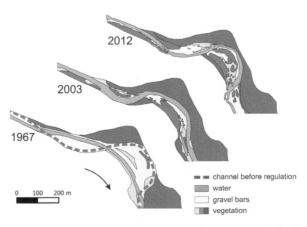

Fig. 9. Channel renaturisation of the upper Wisłoka River reach in Krempna following regulation in 1960s.

course of the reach is quite similar to the original one. Meanders developed in the same places, rendering the original training scheme a wasted expense. A better alternative for a river running through sparsely populated areas would be to provide a corridor for its migration, where it could freely shape its course (Wyżga 2008; Gorczyca and Krzemień 2010).

One obstacle to such river restoration is bedload extraction, which may be legally prohibited, but sometimes continues in local communities. Gradually, however, there are more and more examples of activists who oppose these operations (Bojarski et al. 2005). There are also collective actions, such as the one initiated by the municipality of Pcim, which blocked all access roads to river channels with lifting barriers. Other activities intended to prevent this involve planting trees and shrubs along the banks of river channels.

Conclusions

River and stream channels in the mountain regions and in their forelands are undergoing significant transformation processes. Rivers and streams have been adjusting the structure of their channel systems to the changing conditions in the valley bottoms and slopes, particularly in areas with strong human impact. Numerous natural channel systems have undergone various types of training carried out independently in different reaches, which artificially created different morphodynamic reaches. Both training and channel bedload extractions upset the equilibrium of entire fluvial systems, and limit their continuity. The most apparent human-induced changes visible across entire fluvial systems include channel narrowing, straightening, and deepening, as well as transformations of multi-channel systems into single channels. For this reason, even adjacent reaches can function in very different ways. Long-term periods without proper maintenance of training structures gradually lead to their deterioration, and natural morphogenetic processes restore the channels to their pre-training form. The channels spontaneously restore their natural status, improving the geo-ecological conditions of the valley bottom and the channel itself. Over long periods of time, channel systems can gradually regain their semi-natural equilibrium in this way. Far from being stable, mountain river channels continue to change considerably and their systems are sensitive to natural and human impacts on the geographical environment of their basins.

Keywords: Mountain rivers, gravel-bed channel changes, channel structure, human impact, renaturisation.

References

Augustowski, B. 1968. Spostrzeżenia nad zmianami antropogenicznymi w korycie rzeki Ropy w Karpatach. Zeszyty Geograficzne WSP w Gdańsku 10: 161–168.
Bartnik, W. 2006. Charakterystyka hydromorfologiczna rzek i potoków górskich. Infrastruktura i Ekologia Terenów Wiejskich 4(1): 143–174.
Baumgart-Kotarba, M. 1983. Kształtowanie koryt i teras rzecznych w warunkach zróżnicowanych ruchów tektonicznych: na przykładzie wschodniego Podhala. Prace Geograficzne PAN 145.
Brandt, S.A. 2000. Classification of geomorphological effects downstream of dams. Catena 40: 375–401.
Chartrand, S.M. and Whiting, P.J. 2000. Alluvial architecture in headwater streams with special emphasis on step-pool topography. Earth Surface Processes and Landforms 25: 583–600.
Chełmicki, W. and Krzemień, K. 1999. Channel typology for the River Feshie in the Cairngorm Mts., Scotland. Prace Geograficzne Instytut Geografii UJ 104: 57–68.

Chorley, R.J. and Kennedy, B.A. 1971. Physical Geography. A Systems Approach. Prentice Hall, London.

Dębski, K. 1970. Hydrologia. Arkady, Warszawa.

East, A.E., Pess, G.R., Bountry, J.A., Magirl, C.S., Ritchie, A.C., Logan, J.B., Randlec, T.J., Mastin, M.C., Minearf, J.T., Duda, J.J., Liermann, M.C., McHenryh, M.L., Beechie, T.J. and Shafrothi, P.B. 2015. Large-scale dam removal on the Elwha River, Washington, USA: River channel and floodplain geomorphic change. Geomorphology 228: 765–786.

Gardner, J.S., Smith, D.J. and Desloges, J.R. 1983. The dynamic geomorphology of the Mt. Rae Area: a high Mountain Region in Southwestern Alberta. Department of Geography Publication Series No. 19 University of Waterloo 1–237.

Germanowski, D. and Schumm, S.A. 1993. Changes in braided river morphology resulting from aggradation and degradation. J Geol 101: 451–466.

Gorczyca, E. and Krzemień, K. 2010. Channel Structure Changes in Carpathian Rivers, pp. 185–198. *In*: Radecki-Pawlik, A. and Hernik, J. (eds.). Cultural Landscapes of River Valleys. Wydawnictwo Uniwersytetu Rolniczego w Krakowie.

Gorczyca, E., Krzemień, K. and Łyp, M. 2011. Contemporary trends in the Białka River channel development in the Western Carpathians. Geographia Polonica 84. Special Issue Part 2: 39–53.

Gorczyca, E., Krzemień, K., Wrońska-Wałach, D. and Sobucki, M. 2013. Channel changes due to extreme rainfalls in the Polish Carpathians. pp. 23–35. *In*: Lóczy, D. (ed.). Geomorphological Impacts of Extreme Weather. Springer, Netherlands.

Grant, G.E. 2012. The geomorphic response of gravel-bed rivers to dams: perspectives and prospects. pp. 165–181. *In*: Church, M., Biron, P.M. and Roy, A.G. (eds.). Gravel Bed Rivers VII: Processes, Tools, Environments. John Wiley & Sons. Chichester, UK.

Gregory, K.J. 2006. The human role in changing river channels. Geomorphology 79: 172–191.

Heede, B.H. 1972. Flow and channel characteristics of two high mountain streams. USDA Forest Service Research Paper RM-96. 1–12.

Heede, B.H. 1977. Influence of forest density on bedload movement in a small mountain stream. Hydrology and Water Resources in Arizona and Southwest 7: 103–107.

Heede, B.H. 1981. Dynamics of selected mountain streams in the Western United States of America. Zeitschrift für Geomorphologie Neue Folge 25. 1.

Hennig, I. 1991. Zabudowa potoków i rzek górskich. pp. 158–161. *In*: Dynowska, I. and Maciejewski, M. (ed.). Dorzecze górnej Wisły II. PWN. Warszawa-Kraków.

Jeleński, J., Jelonek, M., Litewka, T., Wyżga, B. and Zalewski, J. 2005. Zasady dobrej praktyki w utrzymaniu rzek i potoków górskich. Ministerstwo Środowiska. Departament Zasobów Wodnych.

Kaszowski, L. and Krzemień, K. 1977. Structure of mountain channel systems as exemplified by chosen Carpathians streams. Studia Geomorphologica Carpatho-Balcanica 11: 111–125.

Kaszowski, L. and Krzemień, K. 1979. Channel subsystems in the Polish Tatra Mts., Studia Geomorphologica Carpatho-Balcanica 8: 149–161.

Kaszowski, L. and Krzemień, K. 1999. Classification systems of mountain river channels. Prace Geograficzne IG UJ 104: 27–40.

Kędzior, A. 1929. Roboty wodne i melioracyjne w południowej Małopolsce [Regulation and draining works in southern Little Poland]. 2. Lwów.

Klimek, K. 1979. Geomorfologiczne zróżnicowanie koryt karpackich dopływów Wisły [Morphodynamic channel types of the Carpathian tributaries to the Vistula]. Folia Geographica. Series Geographica-Physica 12: 35–47.

Klimek, K. 1987. Man's impact on fluvial processes in the Polish Western Carpathians Geografiska Annaler 69A: 221–226.

Klimek, K., Łajczak, A. and Zawilińska, L. 1990. Sedimentary environment of the modern Dunajec delta in artifical Lake Rożnów, Carpathian Mts., Poland. Quaestiones Geographicae 11/12: 81–92.

Knighton, D.A. 1998. Fluvial Forms and Processes. Edward Arnold, London.

Kondolf, G.M. 1994. Geomorphic and environmental effects of instream gravel mining. Landscape and Urban Planning 28: 225–243.

Kondolf, G.M. 1997. Hungry water: effects of dams and gravel mining on river channels. Environmental Management 21: 533–551.

Korpak, J. 2007. The influence of river training on mountain channel changes (Polish Carpathian Mountains). Geomorphology 92: 166–181.

Korpak, J., Krzemień, K. and Radecki-Pawlik, A. 2008. Wpływ czynników antropogenicznych na zmiany koryt cieków karpackich [Influence of anthropogenic factors on changes of Carpathian stream channels]. Infrastruktura i ekologia terenów wiejskich. Monografia 4 PAN Kraków 1–88.

Kościelniak, J. 2004. Influence of river training on functioning of the Biały Dunajec River channel system. Geom Slovaca 4(1): 62–67.

Kozak, J. 2003. Forest cover changes in the Western Carpathians over the past 180 years: a case study from the Orawa region in Poland. Mountain Research and Development 23(4): 369–375.

Krzemień, K. 1992. The high-mountain fluvial system in the Western Tatras—perspective. Geographia Polonica 60: 51–65.

Krzemień, K. 2003. The Czarny Dunajec River, Poland, as example of human-induced development tendencies in a mountain river channel. Landform Analysis 4: 57–64.

Krzemień, K., Gorczyca, E., Sobucki, M., Liro, M. and Łyp, M. 2015. Effects of environmental changes and human impact on the functioning of mountain river channels, Carpathians, southern Poland. Ann. Warsaw Univ. of Life Sci – SGGW, Land Reclam 47(3): 249–260.

Lach, J. and Wyżga, B. 2002. Channel incision and flow increase of the upper Wisłoka River, southern Poland, subsequent to the reforestation of its catchment. Earth Surface Processes and Landforms 27(4): 445–462.

Łajczak, A. 1996. Modelling the long-term course of non-flushed reservoir sedimentation and estimating the life of dams. Earth Surface Processes and Landforms 21: 1091–1107.

Łajczak, A. 2005. Deltas in dam-retained lakes in the Carpathian part of the Vistula drainage basin. Prace Geograficzne 116: 99–109.

Łapuszek, M. and Ratomski, J. 2006. Metodyka określania i charakterystyka przebiegu oraz prognoza erozji dennej rzek górskich dorzecza górnej Wisły (The method of establishing, characteristics and forecasting of the riverbed erosion of the upper Vistula basin mountainous river). Inżynieria Środowiska. Monografia 332: 1–122.

Leopold, L.B. and Wolman, M.G. 1957. River channel patterns: braided, meandering, and straight. U.S. Geol Survey Prof Paper 282B: 39–85.

Liro, M. 2014. Conceptual model for assessing the channel changes upstream from dam reservoir. Quaestiones Geographicae 33: 61–74.

Liro, M. 2015. Gravel-bed channel changes upstream of a reservoir: the case of the Dunajec River upstream of the Czorsztyn Reservoir, southern Poland. Geomorphology 228: 694–702.

Liro, M. 2016. Development of sediment slug upstream from the Czorsztyn Reservoir (Southern Poland) and its interaction with river morphology. Geomorphology 253: 225–238.

Marston, R.A., Bravard, J.-P. and Green, T. 2003. Impacts of reforestation and gravel mining on the Malnant River, Haute-Savoie, French Alps. Geomorphology 55: 65–74.

McDonald, B.C. and Banerjee, I. 1971. Sediments and bedforms on a braided outwash plain. Can J Earth Sciences 8: 1281–1290.

Miller, J.P. 1958. High mountain streams: effects of geology on channel characteristics and bed material, State Bureau of Mines and Mineral Resources, New Mexico Institute of Mining and Technology, Socorro, New Mexico, Memoir 4: 1–52.

Montgomery, D.R. and Buffington, J.M. 1997. Channel-reach morphology in mountain drainage basins. Geological Society of America Bulletin 109(5): 596–611.

Nilsson, C., Reidy, C.A., Dynesius, M. and Revenga, C. 2005. Fragmentation and flow regulation of the world's large river systems. Science 308(5720): 405–408.

O'Connor, J.E., Duda, J.J. and Grant, G.E. 2015. 1000 dams down and counting. Science 348: 496–497.

Osuch, B. 1968. Problemy wynikające z nadmiernej eksploatacji kruszywa rzecznego na przykładzie rzeki Wisłoki. Zeszyty Naukowe AGH w Krakowie 219(15): 283–301.

Piégay, H., Grant, G., Nakamura, F. and Trustrum, N. 2006. Braided river management: from assessment of river behaviour to improved sustainable development. pp. 257–275. *In*: Sambrook-Smith, G.H., Best, J.L., Bristow, C.S. and Petts, G.E. (eds.). Braided Rivers: Process, Deposits, Ecology and Management. Special publication 36 of the International Association of Sedimentologists 25.

Petts, G.E. and Gurnell, A.M. 2005. Dams and geomorphology: research progress and future directions. Geomorphology 71: 27–47.

Płaczkowska, E. 2014. Geological aspects of headwater catchments development in the Lubań Range (the Outer Carpathians, Poland). Zeitschrift für Geomorphologie 58(4): 525–537.

Radecki-Pawlik, A. 2012. Budowle hydrotechniczne w korytach rzek górskich. pp. 55–77. *In*: Krzemień, K. (ed.). Struktura koryt rzek i potoków (studium metodyczne). IGiGP UJ. Kraków.

Rinaldi, R., Wyżga, B. and Surian, N. 2005. Sediment mining in alluvial channels: physical effects and management perspectives. River Research and Applications 21: 805–828.

Rosgen, D.L. 1994. A classification of natural rivers. Catena 22: 169–199.

Schmidt, J.C. and Wilcock, P.R. 2008. Metrics for assessing the downstream effects of dams. Water Resources Research 44(4): W04404, doi:10.1029/2006WR005092.

Schumm, S.A. 1960. The shape of alluvial channels in relation to sediment type. U.S. Geol Survey Prof Paper 352-B: 17–30.

Schumm, S.A. 1981. Evolution and response of the fluvial system, sedimentologic implications. Soc Econ Paleontol Mineral Spec Publ 31: 19–29.

Skalak, K.J., Benthem, A.J., Schnek, E.R., Hupp, C.R., Galloway, J.M., Nustad, R.A. and Wiche, G.J. 2013. Large dams and alluvial rivers in the Anthropocene: the impacts of the Garrison and Oahe Dams on the Upper Missouri River. Anthropocene 2: 51–64.

Soja, R. 2002. Hydrologiczne aspekty antropopresji w polskich Karpatach [Hydrological aspects of anthropopropression in the Polish Carpathians. Prace Geograficzne IGiPZ PAN 186: 1–130.

Sroczyński, W. 2004. Jeziora zaporowe w krajobrazie Karpat (Barrier lakes in the Carpathians – selected problems in spatial organization). pp. 87–98. *In*: Myga-Piątek, U. (ed.). Przemiany Krajobrazu Kulturowego Karpat. Wybrane aspekty. Komisja Krajobrazu Kulturowego PTG. Sosnowiec (in Polish, with English summary).

Starkel, L. and Łajczak, A. 2008. Kształtowanie rzeźby den dolin w Karpatach (koryt i równin zalewowych). pp. 95–108. *In*: Starkel, L., Kostrzewski, A., Kotarba, A. and Krzemień, K. (eds.). Współczesne przemiany rzeźby Polski. IGiGP UJ. Kraków.

Surian, N. and Rinaldi, M. 2003. Morphological response to river engineering and management in alluvial channels in Italy. Geomorphology 50(4): 307–326.

Surian, N. 2006. Effects of human impact on braided river morphology: examples from Northern Italy. pp. 327–338. *In*: Sambrook-Smith, G.H., Best, J.L., Bristow, C. and Petts, G.E. (eds.). Braided Rivers, IAS Special Publication 36, Blackwell Science.

Thorne, C.R. 1998. Stream reconnaissance handbook. Geomorphological investigation and analysis of river channels. John Wiley and Sons, Chichester, New York, Weinheim, Brisbane, Singapore, Toronto.

Whiting, P.J. and Bradley, J.B. 1993. A process-based classification system for headwater streams. Earth Surface Processes and Landforms 18: 603–612.

Wiejaczka, Ł. and Kijowska-Strugała, M. 2015. Assessment of the hydromorphological state of Carpathian rivers above and below reservoirs. Water and Environment Journal 29: 277–287.

Williams, G.P. and Wolman, M.G. 1984. Downstream effects of dams on alluvial rivers. United States Geological Survey, Professional Paper 1286.

Wohl, E. 2000. Mountain rivers revisited. Water Resources Monograph Series 114, AGU, Washington, D.C.

Wohl, E. and Merritt, D. 2005. Prediction of mountain stream morphology. Water Resources Research 41. W08419.

Wohl, E. 2006. Human impacts to mountain streams. Geomorphology 79: 217–248.

Wrońska-Wałach, D. 2009. Dendrogeomorphological analysis of a headwater area in the Gorce Mountains. Studia Geomorphologica Carpatho-Balcanica 63: 97–114.

Wrońska-Wałach, D. 2014. Differing responses to extreme rainfall events in headwater areas recorded by wood anatomy in roots (Gorce Mountains, Poland). Catena 118: 41–54.

Wyżga, B. 1991. Present-day downcutting of the Raba River channel (Western Carpathians, Poland) and its environmental effects. Catena 18: 551–566.

Wyżga, B. 1997. Methods for studying the response of flood flows to channel change. Journal of Hydrology 198: 271–288.

Wyżga, B. 2001. Impact of the channelization-induced incision of the Skawa and Wisłoka Rivers, southern Poland, on the conditions of overbank deposition. Regulated Rivers: Research and Management 17: 85–100.

Wyżga, B. 2007. A review on channel incision in the Polish Carpathian rivers during the 20th century. Developments in Earth Surface Processes 11: 525–553.

Zawiejska, J. and Krzemień, K. 2004. Man-induced changes in the structure and dynamics of the Upper Dunajec River channel. Geografický Časopis 56(2): 111–124.

Zawiejska, J. 2006. Struktura i dynamika koryta Dunajca. Rozprawa doktorska. Arch IGiGP UJ 1–179.

CHAPTER 17

Sediment Transport and Channel Morphology

Implications for Fish Habitat

Marwan A. Hassan,* Carles Ferrer-Boix, Piotr Cienciala and
Shawn Chartrand

INTRODUCTION

This chapter reviews the basic principles of sediment transport in streams, channel classification, mountain channel morphology, and associated implications for fish spawning habitats. Local sediment transport processes construct the differing channel morphologies, which are often opportunistically used by spawning fish. At the watershed scale, the various channel morphologies are spatially organized, as represented by channel classifications. Thus, sediment transport, channel morphology, and freshwater fish ecology are critically linked, and appreciation for the characteristic associations between these natural attributes is crucial to our understanding of (i) river evolution, (ii) responses to natural and human-induced disturbances, and (iii) implications for riverine ecosystems and the distribution of fish habitats. This chapter reviews exiting links between physical and biological processes and identifies the needs for future research for channel ecosystems in mountain streams.

Sediment Transport

Modes of transport and sediment sources

Sediment transport can be classified according to the transport mode, or by focusing on sediment source zones.

Department of Geography, The University of British Columbia, Vancouver, British Columbia, Canada
    V6T 1Z2.
* Corresponding author: mhassan@geog.ubc.ca

Modes of transport

Sediment load in streams can be transported in various ways. Dissolved load corresponds to material transported in solution with the fluid, whose concentration depends primarily on the geochemical character of basin geologic materials, and the dissolution of these materials by groundwater flow. Since dissolved load is a part of the streamflow, it bears little significance for channel processes, despite comprising a majority of the river basin sediment load. On the other hand, clastic load refers to discrete particles or groups of particles carried by the streamflow, either in suspension in the water column or as bedload within a relatively thin layer located at the streambed surface. In the first case, particles are sustained in the fluid by buoyancy and flow turbulence. Conversely, bedload consists of particles rolling, saltating, bouncing, and sliding on the bed surface, continuously interacting with other particles resting on the bed. Taken together, clastic sediment transport and continued exchange between the moving particles and the grains resting on the bed determines the morphology of rivers (see Section 3).

Sediment sources

Sediment sources in river basins can be distinguished as two types for timescales ranging from individual floods to many flood seasons. Sediment derived from the river bed is an internal source, whereas sediment entering rivers from tributaries, river banks, hillslopes, and adjacent floodplains are external sources. In general, external sediment sources for rivers are spatially concentrated in headwater regions where mountain streams occur. It is important to note that upon reaching the channel, sediment originally derived from external sources will become an internal source. The timescale for external sources to become an internal source varies. For example, in mountain streams fluvial and hillslope processes are often strongly coupled, as exemplified by a landslide which directly enters a stream.

Sediment entraintment

Sediment resting on the streambed will move, or become entrained, when the overlying fluid pressure, or shear stress, exceeds some threshold value. In laboratory studies with streambed sediments of uniform size, it is relatively straightforward to identify the threshold stress which will result in measurable sediment transport. Generalizing the threshold stress for natural streams which exhibit many different sediment sizes on the channel bed is, however, a difficult task. Mathematical expressions, or sediment transport formulae, used to simulate sediment entrainment and transport have been completed with field and laboratory studies. Sediment transport formulae either use a critical value of shear stress below which sediment transport is zero (e.g., Meyer-Peter and Müller 1948; Ashida and Michiue 1972), or a reference shear stress below which sediment transport can be considered insignificant (e.g., Parker 1990; Wilcock and Crowe 2003). Modeling of sediment transport in streams is central to river evolution, channel morphology, and stream ecology. It is therefore important to review how the threshold shear stress is commonly computed.

Uniform sediment

Shields (1936) found that a minimum shear stress is required to initiate the motion of streambed particles. Based on dimensional analysis and flow similarity, Shields derived an empirical critical shear stress curve for the onset of motion. Dimensional shear stress τ was expressed in dimensionless form (Shields number) τ^* as

$$\tau^* = \frac{\tau}{\rho R g D} \tag{1}$$

where $R = (\rho_s - \rho)/\rho$ is the submerged specific gravity (ρ_s and ρ are the density of sediment and water, respectively), g is the gravitational acceleration, and D is the bed surface characteristic grain size. When streambed sediments begin to move, the dimensional and dimensionless stresses are termed the critical values, represented symbolically as τ_c and τ_c^*, respectively. By dimensional analysis, Shields found that τ_c^* is a function of the Reynolds particle number Re_* defined as

$$Re_* = \frac{u_* D}{\upsilon} \tag{2}$$

where $u_* = \sqrt{\tau/\rho}$ is the shear velocity and is the kinematic viscosity of water. The Shields number expresses the ratio of critical driving (expressed in terms of τ_c) to resisting forces (submerged weight of the particle). The Reynolds particle number, which represents the ratio of inertial to viscous forces acting near the bed, is used as a measurement of flow turbulence (the higher Re_*, the more turbulent the flow). Interestingly, the Shields' diagram shows that τ_c^* becomes independent of Re_* for high Reynolds values. In fact, for values of $Re_* > 400$ (i.e., within the fully rough flow regime), typically found in rivers, the value of the dimensionless critical shear stress approaches a constant value ≈ 0.06. However, the original Shields diagram depicts a relatively broad region reflecting initial motion conditions, suggesting that bed sediment initial motion is an inherently statistical process (e.g., Shields 1936; Paintal 1971). Random fluctuations of fluid motions in the flow blur a clear definition of conditions for the onset of motion. Besides, differences in methodologies used to determine the threshold for initial bed sediment motion were identified by Buffington and Montgomery (1997) as a source of potential scatter of the critical shear stress. It is known that, in general, the Shields' diagram overpredicts the values of the τ_c^* (Garcia 2008; Parker 2008). In this regard, Neill (1968) determined that τ_c^* is 0.03. Figure 1 illustrates the Shields' diagram by means of the equation fitted by Brownlie (1981) and by other researches.

Unlike empirical methodologies which aim to provide averaged values of the shear stress responsible for initiation of motion, a number of researchers have attempted to provide theoretical estimates of τ_c^* (e.g., Ikeda 1982; Wiberg and Smith 1987). Ikeda (1982) obtained an expression of the critical shear stress based on a time-averaged force balance that included the weight of the particle and the lift and the drag forces caused by the flow acting at the center of the particle:

$$\tau_c^* = \frac{4\mu C_f}{3(c_D + \mu c_L)} \tag{3}$$

where μ is the dynamic Coulomb friction, C_f is the dimensionless friction coefficient ($\tau = \rho C_f u^2$, is the mean flow velocity) and c_D and c_L are the drag and lift force coefficients.

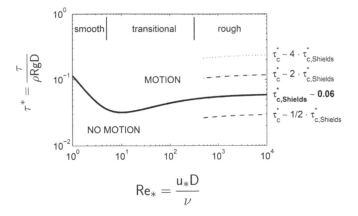

Fig. 1. Shields' curve for the initiation of motion from the interpolated curve fitted by Brownlie (1981). Also included, values of the critical shears stress obtained by Neill (1968) ($\tau_c^* \sim 1/2 \cdot \tau_{c,Shields}^*$), those associated to armored and clustered beds ($\tau_c^* \sim 2 \cdot \tau_{c,Shields}^*$), and to highly structured beds ($\tau_c^* \sim 4 \cdot \tau_{c,Shields}^*$) (Church et al. 1998).

Parker and Andrews (1985) extended this analysis to sediment mixtures. Kirchner et al. (1990) included the effects of grain protrusion and friction angle with respect to surrounding grains.

Equation (3) does not include gravitational effects due to local bed topography. Thus, Eq. (3) can be used under the hypothesis of small longitudinal and transverse bed slopes. Classical expressions have been derived to explicitly account for either non-negligible longitudinal or transverse slopes (e.g., Wiberg and Smith 1987). These latter expressions have been generalized to arbitrary non-negligible slopes (e.g., Kovacs and Parker 1994; Seminara et al. 2002). As bed slope approaches the submerged dynamic angle of repose for bed sediments, all of these critical stress formulations gradually reduce the maximum critical shear stress to zero. Contrary to these findings, Lamb et al. (2008) found that the critical shear stress increases with bed slope.

Sediment mixtures

For hydraulically rough flows, Fig. 1 indicates that the critical shear stress is directly proportional to grain size. As a result, larger particles should require a higher shear stress for entrainment. As pointed out by Parker (2008), the Brownlie (1981) fit of the original Shields' data overpredicts the critical shear stress data at bankfull flow. This suggests that there must be a physical process that reduces the critical shear stress for initial motion in streambeds composed of poorly-sorted mixtures typical of gravel-bed rivers.

Coarse particles surrounded by finer grains generally protrude into the fluid column more versus protrusion associated with a uniform coarse particle mixture. Thus, drag and lift forces acting on coarse grains in a poorly sorted mixture are relatively higher. This leads to a reduction of the critical shear stresses. By the same token, finer particles, sheltered by coarse grains present on the bed, are less exposed to the flow than if they were surrounded by similar grain sizes. Therefore, unlike what is predicted by the Shields' diagram, an increase in the critical shear stress associated with the lower range of the grain size distribution is expected when they are part of poorly-sorted mixtures. Two extreme conditions thus can be defined as far as bedload transport is concerned: selective transport and equal mobility (Parker and Klingeman 1982): while the former takes place when the mobility (and entrainment) of a grain is directly proportional to its diameter (Shields' diagram), the latter occurs when particle mobility (and entrainment) is the same for all grain sizes of the streambed mixture, irrespective of their grain size.

Egiazaroff (1965) was the first to propose a sound hiding/exposure function to account for the grain size effects discussed above. The hiding function derived by Egiazaroff (1965) was later included in the sediment transport formula derived by Ashida and Michiue (1972), and other transport functions have also included the effects of grain exposure and hiding (e.g., Parker 1990). Additionally, the Wilcock and Crowe (2003) transport formula accounts for grain size effects according to the percent sand content of a poorly sorted mixture.

Further complexities on the critical shear stress determination arise under the presence of poorly-sorted mixtures. Gravel-bed rivers usually exhibit coarse armored surfaces formed by (i) sorting processes by which fine material is winnowed vertically and in the downstream direction, or by (ii) overrepresentation of coarse particles on the bed surface so that equal mobility is achieved (Parker and Klingeman 1982). Further, surface armor development is characterized both by the accumulation of coarse particles on the surface, and due to formation of spatially organized bed structures and particle clusters that increase flow resistance, reduce the mobility of surface grains (Parker 2008), and enhance bed stability (Church et al. 1998). These effects, not taken into consideration in any sediment transport formulations, may lead to significant under-predictions of critical shear stress in some cases (Hassan and Church 2000), and thus to substantial inaccuracies in sediment transport calculations. Notably, Church et al. (1998) reported values of the critical shear stress for armored and clustered beds ~2 times that defined by Shields, and up to ~ 4 times that for highly structured surfaces (Fig. 1).

Sediment transport

Sediment transport rates q_b are defined as the sediment fluxes across a certain channel station for a given period of time per unit channel width. The following discussion focuses on bedload transport, which in gravel-bed rivers is usually observed to be small in magnitude (Church and Hassan 2002). Under these conditions, bedload transport is inherently a discrete process by which individual grains on the bed are entrained and transported by the flow. Few attempts have been proposed to evaluate sediment transport rates in a discrete way in the laboratory (e.g., Roseberry et al. 2012). However, as pointed out by Furbish et al. (2012), these methodologies are not practically appropriate since high-speed cameras are needed to obtain the discrete velocity of each moving grain. Discrete measurements of bedload transport rates in the laboratory (Wong et al. 2007) and in the field (e.g., Hassan et al. 1991) have been successfully carried out by the use of tracers. Tracer-based sediment transport rates are computed as:

$$q_b = \frac{(1-\lambda)L_a}{N} \sum_{i=1}^{N} V_{si} \tag{4}$$

where V_{si} is the virtual velocity of the ith tracer (measured as the travel distance of the particle L_{si} for a given period of time), N is the total number of tracers, and λ is the active layer bed porosity, whose thickness L_a encompasses fluctuations of bed elevation for a given transport event.

Bedload transport capacity formulae

Lumped empirical or semi-empirical expressions (e.g., Meyer-Peter and Müller 1948) have been widely used to compute bedload transport rates. Unlike predictions obtained with discrete formulations, these expressions provide a global estimate of the maximum (capacity) sediment transport rate at a particular stream location based on streamflow mean characteristics (typically the mean boundary shear stress τ) and a characteristic grain size for the bed material distribution. Given some of the difficulties discussed above, a universal bedload transport formula has not been obtained. Conversely, a plethora of empirical or semi-empirical expressions have been developed from field and/or laboratory observations. These sediment transport expressions inherently reflect the associated and particular sediment supply and transport conditions.

Sediment transport formulae typically reflect observational conditions of (i) uniform steady flows and (ii) a high relative sediment supply (transport limited conditions). When these conditions do not prevail, lower bedload transport rates compared to those predicted by the formulae occur. Inspired by the seminal work of Bell and Sutherland (1983), a number of corrections based on the assessment of the spatial length needed to attain equilibrium conditions have been proposed to overcome the fact that steady uniform conditions are rarely achieved in nature. However, as pointed out by Canelas et al. (2013), results rendered by these formulations are sensitive to the selected correction factors. Details of bulk bedload transport capacity formulae proposed for uniform and poorly-sorted sediment mixtures can be found in the literature (e.g., Garcia 2008; Parker 2008, respectively).

Channel Classification and Morphology

Stream channel morphology reflects complex interactions between several governing factors including flow character, boundary conditions, sediment supply, sediment transport mode and intensity, and the geological/geomorphological history and character of the basin. This web of spatially and temporally-varying interactions manifest as stream networks which exhibit tremendous physical diversity at broad scales of view. Stream classifications frequently attempt to collapse diversity through identification of commonly observed and measured streambed shapes, as well as the local processes which effect development of these shapes. In this section we will review some of the more popular stream classifications, as well as the morphologies these classifications attempt to organize.

Channel classification

Over the last few decades, river classification has emerged as a major interdisciplinary topic that has fostered collaboration among scientists, managers, and stakeholders in the study, monitoring and restoration of streams (Buffington and Montgomery 2013). Land use changes, rising demands for natural resources, and climate change have created an urgent need for better understanding of interactions among the hydrologic, geomorphic, ecological, and anthropomorphic processes affecting riverine ecosystems. Whereas channel classifications reflect a range of styles and contextual applicability, channel classification is a tool that can be used to address these needs (Buffington and Montgomery 2013). Unfortunately, a universal channel classification that can suit all needs is lacking. Furthermore, few channel classification systems have been tested, and comparability of classification systems is difficult because they are descriptive. This is because the different classification systems lack a common objective, reflect the wide range of methods that has been used to develop and test them, and the regional nature of some of them (Buffington and Montgomery 2013). Fundamentally, each classification system will yield what it was designed to yield. Accordingly, readers should consult previous reviews of channel classifications (e.g., Rosgen 1994; Montgomery and Buffington 1997; Buffington and Montgomery 2013).

Based on the domains of various geomorphic processes, Whiting and Bradley (1993) developed a process-based classification of headwater streams. The Whiting and Bradley classification (1993) is based on system attributes that most significantly influence stream ecosystems and sediment mobilization in headwater streams: (1) hillslope gradient, which determines slope stability, mass movement type, magnitude, and frequency; (2) channel gradient which determines stream power and the mode of fluvial sediment transport; (3) valley bottom width which determines the potential for mass movements to directly enter channel segments; and (4) sediment calibre, which determines movement mode and magnitude of sediment transport. They included two response elements: (a) channel width, which indicates the degree to which hillslopes contribute material directly to the channel (coupling/decoupling); and (b) channel depth, which, with channel gradient, determines shear stress.

The Rosgen (1994, 1996) channel classification is one of the most widely used systems for mountain basins. The Rosgen classification's main tenet is to support design of *natural* channels. Indeed the Rosgen classification serves as the basis for many river restoration projects. However, the scientific merits of the Rosgen system remain controversial (e.g., Simon et al. 2007). The Rosgen classification is focused on reach-scale delineation of the channel morphology and boundary conditions, resulting in eight major and 94 minor stream types. The classification is based on channel boundary conditions, entrenchment ratio (the degree of lateral confinement), sinuosity (the degree of meandering), width/depth ratio (channel shape), and channel slope. The eight major channel types are divided between single-thread (A, G, F, B, E, and C), and multiple channels (D and DA).

Another widely used channel classification for mountain streams was developed by Montgomery and Buffington (1997). They proposed a process-based framework for classifying mountain streams based on the fundamental relation that exists between bed surface roughness, sediment supply, and sediment transport capacity, which links processes to form in stream channels. Furthermore, they recognized that channel morphology is likely to be affected by external controls such as coupling and LWD, and therefore these external conditions with the sediment transport regime were combined to classify channel units into reach-scale aggregations. Montgomery and Buffington tested their classification framework using data collected from Oregon and Washington streams, in addition to published data in prior studies. Confirming earlier finding by Grant et al. (1990), Montgomery and Buffingtons' field test indicates that channel types occur within a limited range of slope classes. A unique aspect of Montgomery and Buffingtons' classification is identification of morphogenic processes and response potential associated with each channel type (Buffington and Montgomery 2013). Another critical aspect of this classification is the emphasis on the role of geomorphic history on channel types. Subsequent research identified the role of glacial history (e.g., Brardinoni and Hassan 2006, 2007; Moir et al. 2006) and lithology (e.g., Thompson et al. 2006), and the classification has been used to study fish habitat and its spatial distribution in a wide range of fluvial environments (Miller et al. 2008).

Church (2002, 2006) developed a channel classification based on channel bed mobility in association with channel stability, channel morphology, sediment transport regime, and sediment type. He defined bed mobility as the bankfull Shields stress (τ_*bf) that describes the relative mobility of the median size of the bed material. Six type of channels were described: (1) jammed channel (cobble-bolder channel; $S > 3°$; $\tau_*bf = 0.04$); (2) threshold channel (cobble-gravel; $\tau_*bf = 0.04$); (3) threshold channel (sandy-gravel to cobble-gravel; τ_*bf up to 0.15); (4) transitional channel (sand to gravel; $\tau_*bf = 0.15–1.0$); (5) labile channel (sandy channel bed, fine sand to silt banks, $\tau_*bf > 1.0$); and (6) labile channel (silt and sandy channel bed, silty to clay-silt bans, cobble-gravel; τ_*bf up to 10). Buffington and Montgomery (2013) pointed out that Church's six channel types are comparable to the primary reach scale morphologies suggested by other classifications (e.g., Rosgen 1994, 1996; Montgomery and Buffington 1997, 1998).

Morphological Types of Mountain Streams

Customarily, channel morphology of streams is studied at the reach scale, which are typically several to many channel widths in length (Montgomery and Buffington 1998). Based on the boundary material (bed and banks), we distinguish between three headwater stream morphology types: bedrock, colluvial, and alluvial channels (e.g., Grant et al. 1990; Montgomery and Buffington 1997; Hassan et al. 2005; Church 2013). Mountain landscapes exhibit a variety of channel types (Grant et al. 1990), over which there is some confusion of terminology. As a result, the following discussion adopts the most commonly used terms.

Colluvial reaches are channels that contain sediments and wood generally derived from hillslopes, which are delivered through mass movements (e.g., debris flow), and of which is typically immobile under the prevailing hydrologic regime (Montgomery and Dietrich 1988; Church 2013; Zimmermann 2013). As a result, colluvial channel material can remain in storage for a relatively long period of time (Montgomery and Buffington 1998). Whereas colluvial channels are dominant features of headwater systems, less attention had been paid to their morphology and dynamics (Church 2013).

Bedrock reaches lack continuous alluvial deposits, are typically constrained by rock outcrops and for which sediment supply is lower than the sediment transport capacity, i.e., supply-limited streams (e.g., Church 2013). Bedrock reaches are characteristically steep and are usually located in the uppermost part of drainage basins. Their morphology is determined by the prevailing flow and sediment transport capacity—which influence abrasion of the rock boundary—rock jointing and/or bedding, and bedrock resistance to abrasion and weathering processes, which influence the rate and style of erosion (see Tinkler and Wohl 1998). Elevated rates of sediment and wood supply from upstream reaches can change bedrock reaches into alluvial ones due to transient storage of the delivered material (Benda 1990). Notably, Wohl (1998) suggested a classification system for bedrock reaches based on basic morphologic characteristics.

Alluvial reaches has been the focus of extensive research over the last few decades. They differ from other channel types in that their boundary consists of sediment that has been reworked by flood flows. Alluvial reaches are considered transport-limited channels (i.e., sediment supply rate is larger than transport rate) and characterised by a wide range of bed morphologies, which the remainder of this section will review.

Boulder cascade

Montgomery and Buffington (1997) described boulder cascades as steep channel reaches that are characterized by longitudinally and laterally disorganized boulders and cobbles, with small pools upstream and downstream of the larger sediments (Fig. 2a). The framework boulders are thought to be generally stable and mobile only during extreme flood events (Grant et al. 1990; Church 2006, 2013). Much of the larger sediments in cascade reaches could be part of lag deposits laid down by glaciers, debris flows, or extreme floods during wetter climatic periods. As a result, cascade reaches are commonly regarded as colluvial in nature (Hassan et al. 2005; Church 2013). Upstream supplied gravels and sand can also be found in cascade reaches; this finer material is generally mobile during most flood events. Flow in cascade reaches is described as tumbling flow

Fig. 2. Photographs of common channel morphologies that occur in mountain streams: (a) cascade channel, (b) step-pool channel, (c) rapid channel and (d) pool-riffle channel. Photographs a and c courtesy of Michael Church and photographs b and d courtesy of Shawn Chartrand.

due to the typically large proportion of boulders within the bed substrate. Although cascade reaches constitute a significant proportion of the fluvial system, little attention has been paid to the study of these morphologies.

Step-pool

Step-pool morphology is characterized by alternating channel-spanning steps and intervening pools (Fig. 2b), of which steps are depositional features of interlocking boulders, cobbles, and/or streamwood, and pools are areas of focused bed erosion and consist of fine sediments overlying boulders, cobbles, and/or bedrock (Keller and Swanson 1979; Montgomery and Buffington 1997; Chartrand and Whiting 2000). Step-pools occur over a wide range of stream gradients (S) ranging from about 3% (Whittaker 1987; Chin 1989; Abrahams et al. 1995; Chartrand and Whiting 2000) to upwards of 30% (Montgomery and Buffington 1997; Chartrand and Whiting 2000; Lenzi 2001; Church and Zimmermann 2007; Church 2013), and are described by several morphologic components including step height (H), step-pool wavelength (L), step drop height, and scour or residual depth. Step-pool occurrence is generally thought to reflect a low upstream sediment supply rate, and channels with a low width-to-depth ratio (Grant et al. 1990; Church and Zimmermann 2007), for which the largest sediments may measure 6 or less times the channel width (termed the jamming ratio) (Zimmermann et al. 2010). Step-pools are thought immobile except during relatively large events (recurrence interval of 30–50 years) (Grant and Mizuyama 1991; Lenzi 2001; Zimmerman and Church 2001).

Abrahams et al. (1995) used field data and flume experiments to demonstrate that step-pools impart a maximum resistance to flow under conditions defined as: $1 < H/L/S < 2$. Abrahams et al. (1995) inferred that maximum resistance to flow implies maximum stability, and hence they suggested step-pool geometry will evolve toward the maximum resistance organization. Data from a wide range of studies indicates that the evolution process likely occurs at a wide range of timescales, and that step-pools may not universally achieve maximum resistance conditions (Chartrand and Whiting 2000; Zimmerman and Church 2001; Comiti et al. 2005; Chartrand et al. 2011). Zimmerman et al. (2010) showed that the question of step-pool channel stability is probabilistic in nature, with some dependence on the Shields ratio (a measure of force exerted on the streambed) and the jamming ratio, as hypothesized by Church and Zimmerman (2007), and a further hypothesized implicit dependence on the exact nature of how steps form (Zimmermann 2013).

Step forming clasts are generally accepted to determine step height; however the question of explicit controls on step wavelength remains open. In general, step wavelength and slope are negatively related (Whittaker 1987; Chin 1989; Grant et al. 1990; Wohl et al. 1997; Chartrand and Whiting 2000). However, data from many different studies shows that data do not collapse along a specified trend (Chartrand et al. 2011), suggesting that slope is best viewed as a first order control on step-pool occurrence (Chartrand and Whiting 2000). Many studies highlight a link between the step wavelength and channel width (see Chin and Wohl 2005), with Church and Zimmerman (2007) proposing the first clear hypothesis of any direct linkage. Chartrand et al. (2011) showed that step wavelength is well described by channel width, when both attributes are scaled by step drop height. Channel width scaled by step drop height is termed the step-pool aspect ratio, which reflects the intensity and rate of potential energy release from step to pool (Chartrand et al. 2011), providing a basic linkage between step wavelength and the nature of flow hydrodynamics along step-pools (Marston 1982; Wilcox and Wohl 2007; Chartrand et al. 2011). Notably, channel form scaling by width is consistent across many studies (e.g., Whittaker 1987; Chin 1989; Grant 1990), and specifically with respect to the jamming theory of step-pool development (Church and Zimmerman 2007; Zimmerman et al. 2010).

Step-pool formation has garnered tremendous interest amongst researchers (e.g., Judd 1963; Grant et al. 1990; Abrahams et al. 1995; Chin 1999; Chartrand and Whiting 2000; Zimmermann and Church 2001; Curran and Wilcock 2005b; Wohl and Chin 2005; Church and Zimmermann 2007; Zimmermann et al. 2010). Central to all of the previous work is localized boulder stability for durations sufficient to accommodate step construction via jamming of other boulders, cobbles; and wood to the initial keystone. Formative ideas and theories range from a process that is in some manner controlled by the flow field (Whittaker 1987; Chin 1999; Chartrand and Whiting 2000; Comiti et al. 2005; Curran and Wilcock 2005b), to ideas which reflect a strong dependence on the location of key stones within the channel (Zimmermann and Church 2001; Curran and Wilcock 2005; Church and Zimmerman 2007), with each set of ideas garnering support from controlled laboratory experiments (Curran and Wilcock 2005b; Zimmermann et al. 2010). Flow field driven step-pool formation differs from key stone driven formation in that the location of key stones is described as random, and likely to depend on their entry location into the channel from the channel banks, adjacent slopes, and non-fluvial processes.

The general interest in step-pools has surged over the last decade because river restoration and enhancement projects commonly emulate step-pools in order to stabilize steep stream reaches and provide passage corridors for salmonids (Lenzi 2001; Roni et al. 2006b; Chin et al. 2008; Chartrand et al. 2011), because step-pools are a critical habitat for a number of aquatic species (Gregory et al. 1991; Roni et al. 2006a, 2008; Meyer et al. 2007), because step-pools are important sources of nutrients and organic matter to the larger downstream portions of the stream network (Wipfli et al. 2007), and because mountain streams pose a significant risk to people and infrastructure (Zimmermann 2013).

Rapid (plane bed)

To avoid confusion with the plane-bed regime well established in sand-bed hydraulics, we use the term rapid as suggested by Grant et al. (1990) to describe the plane bed morphology that has been used by Montgomery and Buffington (1997). The relatively featureless gravel channel units with moderate gradients in which pools are typically absent has been described by Montgomery and Buffington (1997) as plane beds (rapid) (Fig. 2c). The typical range of gradients is 2% to 10% and relative roughness is of order 0.5 to 1.0. The bed surface of the rapid is well-armored which is associated with features such as clusters, stone lines, and stone nets (Church et al. 1998; Montgomery and Buffington 1998; Church 2013). Such bed surface structures indicate relative stability and a supply-limited sediment transport regime (cf. Dietrich et al. 1989). At relatively high flow (~ bankfull discharge), partial destruction of the bed surface may increase the transport rate and the extent of sediment mobility within the channel (Church et al. 1998; Hassan and Church 2001). Rapid reaches are considered a transition between supply- and transport-limited reaches (Montgomery and Buffington 1997). Depending on sediment availability and flow conditions these reaches may function as sediment storage or sediment source sites (Hassan et al. 2005).

Pool and riffle

Pool-riffles can be observed across a wide range of mountain and valley landscapes (Fig. 2d), comprising straight to sinuous stream reaches, across mild (0.001%) to moderate (3.0%) bed slopes (Montgomery and Buffington 1997; Chartrand and Whiting 2000), and reflecting perhaps the most common bedform found in gravel-bed streams (Leopold et al. 1964). From the most general perspective, pools reflect locations of focused bed erosion, and riffles locations of focused deposition. Given that pool-riffles occur in sequence of one or more bedforms, pool-riffle persistence across many floods can be thought to reflect a general stability of riffle sediments. Accordingly, surficial riffle sediments are commonly reported to be larger than that observed within pools, and hence less mobile. Freely-formed pool-riffles commonly exhibit spacing intervals or wavelengths that measure 5–7 times the channel width (Leopold et al. 1964; Keller and Melhorn 1978), and forced pool-riffles reflect shorter width-scaled intervals (Grant et al. 1990; Montgomery et al. 1995; Hogan et al. 1998). Forced pool-riffles are driven by some type of physical obstruction within a channel.

Along straight channel reaches, sinuosity ≤ 1.07, pool-riffles can be situated transverse to the dominant flow direction such that pool and riffle occur in an alternating fashion in the downstream direction, and can express weakly-developed to strongly-developed bars, which are generally situated as part of the riffle. Along sinuous reaches pool-riffles are more commonly expressed with bars which are bank-attached and occupy laterally-varying positions from one-side of the channel to the next. Bars are built adjacent to pools, and bar-tails extend into riffles which can be situated at varying angles to the dominant flow direction (Church and Jones 1982). In general, bars will not develop at width/depth ratios < 10 (Colombini et al. 1987), and bar growth will be inhibited for conditions of low-relative sediment supply (e.g., Dietrich et al. 1989). As a result, bars can be observed as loci of sediment storage, provided there is sufficient accommodation space for sediment to be stored.

Pool-riffles are theorized to form through several different mechanisms, characterized as processes governed by (1) channel or boundary obstructions, (2) valley or channel conditions which drive alternating zones of flow-convergence and divergence, and (3) bank roughness conditions which drive a flow-field response. Pool-riffle formation due to channel or boundary obstructions was formalized by Clifford (1993b) by linking the work of Yalin (1971), Richards (1976), Church and Jones (1982), and Lisle (1986) and utilizing results of his own fieldwork. From this body of information Clifford proposed a systematic 3-step process of pool-riffle formation that is both probabilistic and autogenic in nature. The three steps include: (1) pool scour via a local obstruction which leads to upstream and downstream deposition of a riffle (probabilistic); (2) pool-riffle maturation from the initial bedform through loss of the obstruction and development of a pool-riffle unit; and (3) further pool-riffle pair creation via local flow perturbations driven by the initial pool-riffle unit (autogenic). The key to Clifford's work is that the obstruction persists long enough to permit the flow to deform the bed to a degree whereby an autogenic behaviour can emerge.

The flow convergence hypothesis (MacWilliams et al. 2006) supposes that flow constrictions create a pool through flow concentration and increased capacity of sediment entrainment. Some volume of material eroded from the pool deposits downstream to form a riffle. The hypothesis was developed based on a re-evaluation of Keller's (1971) original data with use of 2-dimensional and 3-dimensional hydrodynamic models. A key finding of the simulations revealed flow-steering interactions between a small bank protrusion and a prominent bar located alongside the subject pool, such that flows converged through the bar centre, and not through the pool. The model simulations further suggest that the converged flows drive secondary circulation cells that may be responsible for pool scour. Sawyer et al. (2010) and White et al. (2010) subsequently demonstrated that the general idea of MacWilliams et al. (2006) appears to provide an explanation for pool-riffle reaches along the Yuba River in northern California, and de Almeida and Rodríguez (2011, 2012) have demonstrated that the occurrence of several width contractions and expansions along a tributary to the Buffalo River in Arkansas is sufficient to drive pool-riffle formation from an initially flat bed. MacVicar and Rennie (2012) and MacVicar and Best (2013) have further explored the dynamics of convergent and divergent flow patterns for an idealized fixed pool-riffle along a straight walled flume. The former work indicates that transitions from shallow to deep and shallow water regions coupled with no-slip conditions at the flume walls is sufficient to drive somewhat robust zones of convergent and divergent flow.

Einstein and Shen (1964) completed a series of flume experiments on emergent meander formation in a straight flume. Their work included a series of runs where bank roughness was accentuated to explore its effect on patterns of bed erosion and deposition. The scale of bank roughness to flume width was ~35, assuming the bank roughness length was equivalent to 0.5D. Results from the bank roughness experiments resulted in alternating bar-pool-riffle (Church and Jones 1982), with diagonal riffles (Church and Jones 1982) separating pools. The alternating features wavelength scaled as the flume width, which is much shorter than the results of most field studies. Einstein and Shen (1964) repeated the experiment without the rough banks, keeping all other conditions identical, and found that alternating bar-pool-riffle morphology did not emerge.

Implications of Sediment Transport and Channel Morphology for Spawning Habitat of Salmonid Fish

Sediment transport and channel morphology are important components of physical habitat for aquatic biota in rivers. We focus the last part of this chapter on salmonid fish (salmon, trout, and charr), which belong to the taxonomic family Salmonidae, because salmonids play a critical role in riverine-riparian ecosystems, strongly influencing their structure and function (e.g., Naiman et al. 2002; Gende et al. 2002), and because they are fairly widespread across large parts of temperate and subarctic regions of North America, Europe, and Asia (Quinn 2005; Thorstad et al. 2011).

The way salmonid fish are influenced by sediment transport and channel morphology varies with life stage and season, as fish engage in their basic life history activities (spawning and incubation, foraging, dispersal, and migration; Quinn 2005; Thorstad et al. 2011). We restrict the scope of the following discussion to geomorphic controls on spawning and incubation habitat only, but note that water temperature, food abundance, as well as off-channel habitat, are also fundamental constituents of salmonid habitat (e.g., Bjornn and Reiser 1991; Quinn 2005).

Spawning substrum availability

Reproductive success of salmonid fish is directly affected by channel morphology, sedimentology, and sediment transport. Salmonid females deposit eggs in depressions excavated in the bed using flexing body motions (Esteve 2005). After the eggs are fertilized by males, spawning females cover an "egg pocket" with a layer of material derived from similar pits dug upstream (Groot and Margolis 1991; Esteve 2005). Through this process, salmonids winnow fine sediment from the nest and, as a result, substratum in which the eggs are ultimately buried is coarser than the ambient bed material (Kondolf et al. 1993). Nesting sites, termed redds, may contain one or more egg pockets (e.g., Groot and Margolis 1991; Quinn 2005).

The availability of unconsolidated bed material for redd construction and egg burial is a fundamental requirement for salmonid reproduction. Spawning can be also limited by the calibre of bed material. In order for fish to excavate and cover their nests, at least part of the sediment particles must be small enough to be mobilized and displaced by hydraulic forces generated by a digging female. The maximum size of sediment particles that a spawning female can displace appears to be correlated to body size (Kondolf and Wolman 1993; Riebe et al. 2014). Based on an envelope curve fitted to empirical data, Kondolf and Wolman (1993) proposed that, as a rule of thumb, salmonids can excavate redds if the bed material median grain size does not exceed 10% of the fish body length. New data presented by Riebe et al. (2014) seem to suggest that this threshold may be somewhat underestimated. However, given reports that the ability of fish to construct a redd may be hampered by bed compaction (Burner 1951), it would seem that the relationship between body size and limiting threshold of spawning gravel calibre may be more complex due to variability in bed structure described by several authors (e.g., Church et al. 1998; Hassan et al. 2007; Hodge et al. 2013).

Channel morphology and sediment transport influence the distribution of potentially available spawning substrate at the morphologic unit scale through topographic sorting (Keller 1971; Lisle and Hilton 1992; Thompson et al. 1999; Nelson et al. 2010). Field observations suggest however that in pool-riffle channels, salmonids most frequently utilize the pool-riffle transition as their spawning sites (e.g., Baxter and Hauer 2000; Moir and Pasternack 2008; Bean et al. 2015), and this morphological association is believed to be partly related to hyporheic downwelling that promotes intergravel flow (Baxter and Hauer 2000; see below). Bars are other morphological units frequently chosen for spawning, although redds can be also constructed in riffles (e.g., Peterson and Quinn 1996; Moir et al. 2004; Moir and Pasternack 2008; Gottesfeld et al. 2004). In addition, non-alluvial channel features—such as large wood—can strongly influence spatial organization of textural patches (e.g., Buffington and Montgomery 1999) and salmonids have been reported to spawn in hydraulically sheltered positions downstream of flow obstructions (Shellberg et al. 2010).

In coarse-grained channels, such as step-pool or rapid reaches, topographic sorting of bed material may constitute a more critical constraint because of limited availability of potentially movable sediment. Indeed, under such conditions distribution of spawning sites has been observed to reflect patches of suitable substratum (e.g., Moir et al. 2008), or transient shadow deposits in the lee of large boulders, called "pocket gravel" (e.g., Kondolf et al. 1991). As a result, small-bodied salmonids (resident fish) might be forced to spawn in hydraulically sheltered locations, where finer bed material accumulates (Cienciala and Hassan 2013).

Given their dependence on availability of unconsolidated bed material, the occurrence of salmonid spawning in mountain drainage basins tends to be most frequent in alluvial, lower gradient reaches, especially those with pool-riffle morphology (Montgomery et al. 1999; Moir et al. 2004). Because of a relatively high supply of sediment with respect to transporting capacity (Montgomery and Buffington 1997), these reaches tend to have abundant spawning gravel. However, these reaches may also be favored because of "holding pools" that provide fish with cover from predators prior to spawning. Both of these factors tend to correlate highly with spawner density (Magee et al. 1996; Braun and Reynolds 2011; Anlauf-Dunn et al. 2014).

Disturbance due to fine sediment

The survival of incubating embryos and hatched fish larvae to the juvenile "fry" life stage is partly dependent upon whether or not the streambed is disturbed and/or reorganized after spawning has taken place (Chartrand et al. 2015). Salmonid embryos develop within the redd for several weeks, the exact duration depending on factors like temperature and species (Quinn 2005). Before emerging from the bed, newly hatched fish larvae, called alevins, remain within the bed material for a few more weeks until yolk sacs off which they feed are absorbed (Groot and Margolis 1991; Quinn 2005). The interstitial environment provides shelter from predators and environmental conditions (e.g., high velocities, bedload transport). Alevins and incubating embryos rely on intergravel flow for supply of oxygen, removal of waste, and maintenance of moisture (Groot and Margolis 1991). Moreover, at the time of emergence, connectivity of interstitial pores within the streambed is critical for alevin emergence from the bed.

Intergravel flow and interstitial passages within a redd are affected by local bed material porosity and permeability (e.g., McNeill and Ahnell 1964), which in turn depend on its packing and composition, in particular, the content of fine sediment within framework gravel (e.g., Kamann et al. 2007). Several studies linked low egg-to-fry salmonid survival to high content of fine sediment within the respective spawning sites (e.g., McNeill and Ahnell 1964; Hartman et al. 1996). More detailed observations indicated that decreases in intergravel velocity due to fine sediment and the associated reduction of oxygen flux was strongly correlated with embryo mortality (e.g., Bjornn and Reiser 1991; Greig et al. 2005). Other research demonstrated that, despite high oxygen availability, blockage of interstitial pathways by fine material also resulted in high in-redd mortality (Franssen et al. 2012). The relative importance of oxygen shortage and entombment as mechanisms driving embryo and alevin mortality is still unknown, and presumably, may vary from case-to-case. Furthermore, the relationship between egg-to-fry survival and the quality of substratum in spawning nests seems to vary with size of eggs (van den Berghe and Gross 1989). Research evidence indicates that size-dependent survival might stem from different physiological requirements regarding oxygen supply (Quinn et al. 1995) or restricted ability of larger alevins to move in small pore spaces (e.g., Crisp 1993).

Sedimentological and hydraulic conditions within the egg pockets change over time as a result of fish activity as well as sediment transport following redd construction. As noted above, a salmon female initially removes large quantities of fine material during redd construction, thus coarsening the substratum and improving the incubation conditions within the redd (McNeill and Ahnell 1964; Kondolf et al. 1993). For example, decline in redd substratum permeability may occur due to compaction of framework particles during the incubation period (Gustafson-Greenwood and Moring 1991). Moreover, the composition of bed material within the redd may change over time as fine sediment infiltrates following spawning, re-filling the interstitial spaces (Acornley and Sear 1999; Zimmermann and Lapointe 2005; Sear et al. 2008). This process is often referred to as redd siltation or sedimentation, and may occur immediately, as a result of local spawning activity. The rates at which fine sediment accumulates within the framework gravel depends on the rates of suspended sediment and bedload transport (e.g., Beschta and Jackson 1979; Lisle 1989; Sear et al. 2008). However, accumulation rates decline with increasing bedload rates, suggesting a reduction in the efficiency of this process (Lisle 1989).

Fine sediment penetration depth depends on shapes and relative sizes of infiltrating particles and interstitial spaces (e.g., Carling 1984; Frostick et al. 1984). Several studies reported formation of sand seals within the gravel framework close to the bed surface, which appeared to prevent accumulation of finer material deeper in the bed (e.g., Jackson and Beschta 1979; Lisle 1989; Allen and Frostick 1999). This process has been hypothesized to benefit survival of salmonid embryos (Meyer et al. 2005). More recently, experimental work (Wooster et al. 2008) and mathematical modeling (Cui et al. 2008) provided further support to such a decrease in fine sediment infiltration with depth in immobile gravel beds. However, Franssen et al. (2014) found that the effect of fine sediment seals was highly variable and dependent on the "filtering fraction", which they identified as very fine gravel and coarse sand.

Basin-wide patterns in fine sediment abundance and the associated risk of disturbance reflects regional geology, land cover and use (e.g., Cover et al. 2008) as well as the locations of sediment sources (e.g., Nistor and Church 2005) or lower-order confluences (Rice and Church 1998; Benda et al. 2004). Such impacts are also dependent on location of sedimentary sinks, often related to reduction in channel slope. Locally, sediment sinks may form due to presence of channel-spanning obstructions to flow in the form of log jams or mass wasting deposits (e.g., Hogan et al. 1998). For example, Cienciala and Hassan (2013) hypothesized that the efficient entrapment of sediment by flow obstructions may have large impacts on egg-to-fry survival in coarse-bed reaches. They reasoned that in such streams small-bodied salmonids may be forced to spawn in hydraulically sheltered zones to which areas of movable substrate are limited. Although the chronic and episodic inputs of fine material to streams can be related to both natural and anthropogenic causes, man-made landscape disturbance has been identified as particularly common (e.g., Lisle 1982; Roberts and Church 1986; Lisle and Hilton 1992).

Disturbance due to bed scour

Bed material mobilization during floods can affect egg-to-fry survival of salmonids. The risk of disturbance on mortality depends on depth of bed scour relative to the depth at which eggs are buried (DeVries 1997, 2008). Research into this subject established that the mean egg burial depth is positively correlated with female body size (e.g., Carling and Crisp 1989; Steen and Quinn 1999). The depth of scour in gravel and cobble bed channels may vary depending on specific driving mechanisms. "General scour", which can be defined as the lowering of the bed surface elevation within a channel reach that is associated with the passage of floods, occurs to depths equivalent to the active bedload transport layer (e.g., DeVries 2008). Thus, this type of scour usually scales with bed material calibre and reaches maximum depths of approximately $1.5-2D_{90}$ (DeVries

2002). In contrast, "local scour" occurs at the morphological unit-scale due to local flow conditions. A typical example if this kind of scour is that associated with flows convergence due to large wood or bank projection (Buffington et al. 2003). Local scour may be significantly deeper than the limit proposed for the general scour (DeVries 2008).

In addition, redd construction itself can influence bed mobility and, potentially, scour depth within the nesting site. Two contrasting effects have been identified in the literature. First, excavation of the egg pocket and covering it with loose material has been suggested to result in locally reduced entrainment thresholds due to disruption of bed structure; on the other hand, coarsening of bed material within the redd due to winnowing of fine sediment may act to increase entrainment thresholds (e.g., Montgomery et al. 1996; Hassan et al. 2008). The relative importance of these two factors is currently unknown.

Bed scour depth and, thus, scour disturbance risk, is highly variable in space (Hassan 1990; Rennie and Millar 2000), but its spatial pattern at the reach and morphological unit scale may be organized, to some extent, by the effect of bed topography on bedload transport (Lisle 1982; Sear 1996; Lisle et al. 2000). DeVries (2008) proposed a conceptual model of bed scour with reference to disturbance of salmonid spawning nests. Based on prior research he suggested that the common location of redds in the pool tail may be particularly prone to scour, which occurs due to temporary imbalance in sediment flux (e.g., Schuett-Hames et al. 2000; DeVries et al. 2001). Other research suggested that in a large gravel-bed river, deep scour was most likely in the thalweg and zones of high flow velocity (Lapointe et al. 2000). However, despite the intuitive appeal of these simple models, scour disturbance may be a more complex phenomenon. In particular, bed mobility associated with variation in local sediment storage (e.g., Lisle et al. 2000) may be an important factor governing scour. For example, Cienciala and Hassan (2013) found that in four reaches of a stable, coarse-bed stream with low sediment supply, spatial pattern of bed mobility was strongly conditioned by local sediment supply in the areas of recent deposition. Therefore, the pattern of morphological bed changes reflecting net scour varied significantly from year to year (Cienciala and Hassan 2013). Alternatively, the inherent non-uniform nature of sediment transport at ecologically meaningful timescales may be a governing factor of bed sediment mobility, and therefore redd scour risk (Chartrand et al. 2015).

The selection of a site for redd construction by a spawning female may strongly influence bed scour disturbance risk to salmonid eggs or alevins. For example, some researchers observed that salmonid females display preference for freshly deposited, loose gravel (e.g., Gottesfeld et al. 2004; Hassan et al. 2008). It appears that, given the preferential scour of recent deposits (Cienciala and Hassan 2013), such behavior of a fish would make its offspring more vulnerable to scour disturbance. However, in most cases the locations of redds have been reported to reflect the zones of low bed mobility relative to the ambient bed, such as in the channel margins or in the areas sheltered from hydraulic forces by flow obstructions (e.g., May et al. 2009; Shellberg et al. 2010). Correlation of selected spawning sites with low excess shear stress zones (Moir et al. 2009) seems to provide further, although indirect, evidence for avoidance of scour-prone locations.

Montgomery et al. (1999) have proposed that the bed scour regime is one of the key controls which defines the spatial (and temporal) pattern of salmonid spawning at the channel network scale. They suggested that spawning site distributions reflect the spatial pattern of channel gradient and the associated morphological channel types (representative of a suite of geomorphic processes; Montgomery and Buffington 1997; Buffington et al. 2003). This hypothesis has not been comprehensively tested but some of its components have garnered support from field evidence (e.g., Moir et al. 1998, 2004; Shellberg et al. 2010). Overall, because of logistical challenges, empirical basin-wide studies of bed scour disturbance at the scale of drainage network have been limited and have primarily relied on models (e.g., Goode et al. 2013). Consequently, the spatial patterns of bed scour disturbance across drainage basins is still poorly understood.

Sediment dynamics, channel morphology, and restoration of salmonid spawning habitat

Salmonid fish are well adapted to the natural range of variability in habitat conditions and disturbance regimes (Bisson et al. 2009). However, progressing loss, degradation, and fragmentation of freshwater habitat due to a combination of land use and climate change has imperiled many populations of these resilient fish, thus posing a challenge for river management and ecosystem conservation (Bisson et al. 2009; Rieman et al. 2006). Given the ecological, cultural, and economical role of salmonids, they are often target species of river restoration projects (e.g., Beechie and Bolton 1999).

In attempts to improve degraded habitat in regulated rivers, restoration actions have commonly focused on implementing flow regimes expected to support the target species (e.g., Palmer et al. 2014). However, in many cases, morphological changes due to flow regulation and channel engineering may prevent ecological benefits from these "environmental flows" (Brown and Pasternack 2008). Thus, the majority of restoration projects have adopted hard engineering approach, which involves designing and modifying the channel to achieve some reference or pre-disturbance morphological conditions (Lave 2009) or rescaling and reconfiguring the channel to match the altered flows (e.g., Harrison et al. 2011). Choice of restoration activities aimed specifically at improvement of bed material characteristics depends on the nature of disturbance. In cases of excessive bed coarsening due to limited sediment supply and storage, these activities typically involve artificial introduction of suitable spawning substrate (Wheaton et al. 2004). Such "gravel augmentation" may also be accompanied by installation of in-channel structures or large wood, which augments functionality of introduced gravels (e.g., Elkins et al. 2007), and enables their accumulation (e.g., Roni et al. 2014). If, on the other hand, elevated sediment supply results in accumulation of excess fine material in redds, spawning sites can be cleaned ("gravel enhancement"; Wheaton et al. 2004). In projects in which high sediment inputs are related to bank erosion, rip-rap construction has been a common remedial procedure. Other off-channel measures may also be adopted to control sediment production and delivery to stream networks. In addition to these local restoration operations, it may be necessary to enable or improve access of salmonid fish to their spawning sites by implementing fish passage solutions such as fish ladders at dams (Kemp 2015), or outright dam removal.

However, in-channel, structural approaches to channel restoration have had varied success rate and, typically, improvements of spawning habitat have been relatively short lived (but see Whiteway et al. 2010). For example, Pulg et al. (2013) evaluated gravel augmentation and cleaning efforts and noted that, although enhanced conditions increased spawning success for two years, the conditions quickly deteriorated and were predicted to reach pre-intervention level within 5–6 years. Even faster decline in intergravel conditions—within 1 year of restoration—was observed by Pander et al.

(2015). They concluded that augmentation was the most successful method of gravel improvement while substrate raking had little positive effect. In addition, the latter measure resulted in deposition of the mobilized fine material in downstream sites (Pander et al. 2015). This finding is in line with the contention of some researchers that restoration activities may themselves constitute a form of disturbance (e.g., Tullos et al. 2009). In another study focused on evaluation of restoration practices, Barlaup et al. (2008) reported that introduced gravels were indeed used by spawning salmonids but they were rapidly washed downstream during a flood event. Other limitations were reported by Mueller et al. (2014) who pointed out that improvements in spawning gravel quality varied with depth below the surface and that spawning success was improved only for some salmonid species. Similarly, efforts related to reconfiguration of channel morphology and placement of structures or wood have also frequently failed to achieve intended ecological benefits (e.g., Bernhardt and Palmer 2011; Roni et al. 2014).

Growing criticism of the reach-scale, structural approach to habitat restoration (e.g., Palmer et al. 2014; Simon et al. 2007) has led to increased recognition that a more holistic and process-based approach is needed to achieve long-term improvement in ecological function (Beechie et al. 2010; Palmer et al. 2014; Sear 2010; Wohl et al. 2005, 2015). For example, it has been argued that local restoration opportunities should be placed in the broader landscape context (e.g., Sear 2010; Wohl et al. 2005). Effectiveness of the reach scale restoration measures is often predicated upon addressing the root causes of disturbance and considering basin-scale controls on flow and sediment regimes as well as biogeochemical and ecological fluxes (e.g., Beechie et al. 2010; Palmer et al. 2014). Rather than focusing exclusively on spawning habitat, complementary types of habitat should be also taken into account, for example, those used at various life stages for foraging or refuge (e.g., Fausch et al. 2002; Lapointe 2012; White and Rahel 2008). The importance of food resources for fish highlights the advantages of ecosystem rather than species-oriented approach to restoration (e.g., Beechie et al. 2010; Lapointe et al. 2013; Wohl et al. 2015), even if keystone species such as salmonids are given special status. Moreover, multiscale approach enables incorporation of factors which define habitat connectivity and metapopulation dynamics (Fausch et al. 2002) into the restoration planning. While in many cases achieving the above objectives is certainty not feasible, some recent large-scale dam removals project in basins inhabited by salmonid populations may provide unique opportunities for such process-based ecosystem restoration (e.g., Brenkman et al. 2012). Finally, it has been increasingly recognized that restoration projects should be designed with careful consideration of various scenarios of regional climate change, including projected temperatures as well as flow and sediment regimes (Battin et al. 2007; Wohl et al. 2015).

Summary and Conclusions

This chapter has introduced the basic principles of sediment transport, channel morphology, and their implications for habitat spawning for fish. The chapter has mainly focused on mountain streams. A review of entrainment and transport processes of both uniform and heterogeneous sediment has been introduced. The different views of channel classification and morphology have been examined. Morphological units of mountain streams have been thoroughly reviewed. Finally, the implications of sediment transport and channel morphology for spawning habitat of fish are discussed.

Particularly, perturbations to fish habitats caused by fine sediment filling pools and bed elevation changes are reviewed.

Keywords: Fluvial geomorphology, sediment transport, channel classification, channel morphology, fish habitat.

References

Abrahams, A.D., Li, G. and Atkinson, J.F. 1995. Step–pool streams: adjustment to maximum flow resistance. Water Resources Research 31: 2593–2602.

Acornley, R.M. and Sear, D.A. 1999. Sediment transport and siltation of brown trout (*Salmo trutta* L.) spawning gravels in chalk streams. Hydrological Processes 13: 447–458.

Anlauf-Dunn, K.J., Ward, E.J., Strickland, M. and Jones, K. 2014. Habitat connectivity, complexity, and quality: predicting adult coho salmon occupancy and abundance. Canadian Journal of Fisheries and Aquatic Sciences 71: 1864–1876.

Ashida, K. and Michiue, M. 1972. Study on hydraulic resistance and bedload transport rate in alluvial streams. Transactions, Japan Society of Civil Engineering 206: 59–69.

Barlaup, B.T., Gabrielsen, S.E., Skoglund, H. and Wiers, T. 2008. Addition of spawning gravel—a means to restore spawning habitat of Atlantic salmon (*Salmo salar* L.), and Anadromous and resident brown trout (*Salmo trutta* L.) in regulated rivers. River Research and Applications 24: 543–550.

Battin, J., Wiley, M.W., Ruckelshaus, M.H., Palmer, R.N., Korb, E., Bartz, K.K. and Imaki, H. 2007. Projected impacts of climate change on salmon habitat restoration. Proceedings of the National Academy of Sciences 104: 6720–6725.

Bean, J.R., Wilcox, A.C., Woessner, W.W. and Muhlfeld, C.C. 2015. Multi-scale hydrogeomorphic influences on bull trout (*Salvelinus confluentus*) spawning habitat. Canadian Journal of Fisheries and Aquatic Sciences 72: 514–525.

Beechie, T. and Bolton, S. 1999. An approach to restoring salmonid habitat-forming processes in Pacific Northwest watersheds. Fisheries 24: 6–15.

Beechie, T.J., Sear, D.A., Olden, J.D., Pess, G.R., Buffington, J.M., Moir, H., Roni, P. and Pollock, M.M. 2010. Process-based principles for restoring river ecosystems. BioScience 60: 209–222.

Bell, R. and Sutherland, A. 1983. Non-equilibrium bedload transport by steady flow. Journal of Hydraulic Engineering 109: 351–367.

Benda, L. 1990. The influence of debris flows on channels and valley floors in the Oregon Coast Range, USA. Earth Surface Processes and Landforms 15: 457–466.

Benda, L., Poff, N.L., Miller, D., Dunne, T., Reeves, G., Pess, G. and Pollock, M. 2004. The network dynamics hypothesis: how channel networks structure riverine habitats. BioScience 54: 413–427.

Bernhardt, E.S. and Palmer, M.A. 2011. River restoration: the fuzzy logic of repairing reaches to reverse catchment scale degradation. Ecological Applications 21: 1926–1931.

Beschta, R.L. and Jackson, W.L. 1979. The intrusion of fine sediments into a stable gravel bed. Journal of the Fisheries Board of Canada 36: 204–210.

Bisson, P.A., Dunham, J.B. and Reeves, G.H. 2009. Freshwater ecosystems and resilience of Pacific salmon: habitat management based on natural variability. Ecology and Society 14: 45. [online] URL: http://www.ecologyandsociety.org/vol14/iss1/art45/.

Bjornn, T. and Reiser, D.W. 1991. Habitat requirements of salmonids in streams. *In*: Meehan, W.R. (ed.). Influences of Forest and Rangeland Management on Salmonid Fishes and their Habitats. American Fisheries Society Special Publication, Bethesda, MD 19: 83–138.

Brardinoni, F. and Hassan, M.A. 2006. Glacial erosion, evolution of river long profiles, and the organization of process domains in mountain drainage basins of coastal British Columbia. Journal of Geophysical Research 111: F01013.

Brardinoni, F. and Hassan, M.A. 2007. Glacially induced organization of channel-reach morphology in mountain streams. Journal of Geophysical Research 112: F03013.

Braun, D.C. and Reynolds, J.D. 2011. Relationships between habitat characteristics and breeding population densities in sockeye salmon (Oncorhynchus nerka). Canadian Journal of Fisheries and Aquatic Sciences 68: 758–767.

Brenkman, S.J., Duda, J.J., Torgersen, C.E., Welty, E., Pess, G.R., Peters, R. and Mchenry, M.L. 2012. A riverscape perspective of Pacific salmonids and aquatic habitats prior to large-scale dam removal in the Elwha River, Washington, USA. Fisheries Management and Ecology 19: 36–53.

Brown, R.A. and Pasternack, G.B. 2008. Engineered channel controls limiting spawning habitat rehabilitation success on regulated gravel-bed rivers. Geomorphology 97: 631–654.

Brownlie, W.R. 1981. Prediction of flow depth and sediment discharge in open channels, Report No. KH-R-43A, pp. 232, W. M. Keck Laboratory of Hydraulics and Water Resources, California Institute of Technology, Pasadena, California, USA.

Buffington, J.M. and Montgomery, D.R. 1997. A systematic analysis of eight decades of incipient motion studies, with special reference to gravel-bedded rivers. Water Resources Research 33: 1993–2029.

Buffington, J.M. and Montgomery, D.R. 1999. Effects of hydraulic roughness on surface textures of gravel-bed rivers. Water Resources Research 35: 3507–3521.

Buffington, J.M., Woodsmith, R.D., Booth, D.B. and Montgomery, D.R. 2003. Fluvial processes in Puget Sound rivers and the Pacific Northwest. pp. 46–78. *In*: Montgomery, D.R. and Bolton, S. (eds.). Restoration of Puget Sound Rivers. University of Washington Press, Seattle, WA.

Buffington, J.M. and Montgomery, D.R. 2013. Geomorphic classification of rivers. pp. 730–767. *In*: Shroder, J. (Editor in Chief), Wohl, E. (ed.). Treatise on Geomorphology. Vol. 9: Fluvial Geomorphology. Academic Press, San Diego, CA.

Burner, C.J. 1951. Characteristics of spawning nests of Columbia River salmon. US Department of Interior.

Canelas, R., Murillo, J. and Ferreira, R.M.L. 2013. Two-dimensional depth-averaged modelling of dam-break flows over mobile beds. Journal of Hydraulic Research 51: 392–407.

Carling, P.A. 1984. Deposition of fine and coarse sand in an open-work gravel bed. Canadian Journal of Fisheries and Aquatic Sciences 41: 263–270.

Chartrand, S.M. and Whiting, P.J. 2000. Alluvial architecture in headwater streams with special emphasis on step–pool topography. Earth Surface Processes and Landforms 25: 583–600.

Chartrand, S.M., Jellinek, M., Whiting, P.J. and Stamm, J.S. 2011. Geometric scaling of step-pool mountain streams: observations and implications. Geomorphology 129: 141–151.

Chartrand, S.M., Hassan, M.A. and Radic, V. 2015. Pool-riffle sedimentation and surface texture trends in a gravel bed stream. Water Resources Research 51, doi:10.1002/2015WR017840.

Chin, A. 1989. Step–pools in stream channels. Progress in Physical Geography 13: 391–408.

Chin, A. 1999. On the origin of step-pool sequences in mountain streams. Geophysical Research Letters 26: 231–234.

Church, M., Hassan, M.A. and Wolcott, J.F. 1998. Stabilizing self-organized structures in gravel-bed stream channels: field and experimental observations. Water Resources Research 34: 3169–3179.

Church, M. 2002. Geomorphic thresholds in riverine landscapes. Freshwater Biology 47: 541–557.

Church, M. and Hassan, M.A. 2002. Mobility of bed material in Harris Creek. Water Resources Research, 38: 1237, doi: 10.1029/2001WR000753.

Church, M. 2006. Bed material transport and the morphology of alluvial rivers. Annual Review of Earth and Planetary Sciences 34: 325–354.

Church, M. and Zimmermann, A. 2007. Form and stability of step-pool channels: research progress. Water Resources Research 43, W03415, doi:10.1029/2006WR005037.

Church, M. 2013. Steep headwater channels. pp. 528–549. *In*: Shroder, J. (Editor in Chief), Wohl, E. (ed.). Treatise on Geomorphology. Vol. 9: Fluvial Geomorphology. Academic Press, San Diego, CA.

Cienciala, P. and Hassan, M.A. 2013. Linking spatial patterns of bed surface texture, bed mobility, and channel hydraulics in a mountain stream to potential spawning substrate for small resident trout. Geomorphology 197: 96–107.

Colombini, M., Seminara, G. and Tubino. M. 1987. Finite-amplitude alternate bars. Journal of Fluid Mechanics 181: 213–232.

Comiti, F., Andreoli, A. and Lenzi, M.A. 2005. Morphological effects of local scouring in step–pool streams. Earth Surface Processes and Landforms 30: 1567–1581.

Cover, M.R., May, C.L., Dietrich, W.E. and Resh, V.H. 2008. Quantitative linkages among sediment supply, streambed fine sediment, and benthic macroinvertebrates in northern California streams. Journal of the North American Benthological Society 27: 135–149.

Crisp, D.T. 1993. The ability of UK salmonid alevins to emerge through a sand layer. Journal of Fish Biology 43: 656–658.

Cui, Y., Wooster, J.K., Baker, P.F., Dusterhoff, S.R., Sklar, L.S. and Dietrich, W.E. 2008. Theory of fine sediment infiltration into immobile gravel bed. Journal of Hydraulic Engineering 134: 1421–1429.

Curran, J.C. and Wilcock, P.R. 2005a. Effect of sand supply on transport rates in a gravel-bed channel. Journal of Hydraulic Engineering 131: 961–967.

Curran, J.C. and Wilcock, P. 2005b. Characteristic dimensions of the step–pool configuration: an experimental study. Water Resources Research 41: W0203, 1–15.

DeVries, P. 1997. Riverine salmonid egg burial depths: review of published data and implications for scour studies. Canadian Journal of Fisheries and Aquatic Sciences 54: 1685–1698.

DeVries, P., Burges, S.J., Daigneau, J. and Stearns, D. 2001. Measurement of the temporal progression of scour in a pool-riffle sequence in a gravel bed stream using an electronic scour monitor. Water Resources Research 37: 2805–2816.

DeVries, P. 2002. Bedload layer thickness and disturbance depth in gravel bed streams. Journal of Hydraulic Engineering 128: 983–991.

DeVries, P. 2008. Bed disturbance processes and the physical mechanisms of scour in salmonid spawning habitat. pp. 121–147. *In*: Sear, D.A. and DeVries, P. (eds.). Salmonid Spawning Habitat in Rivers: Physical Controls, Biological Responses, and Approaches to Remediation, American Fisheries Society Symposium (Vol. 65). American Fisheries Society. Bethesda, Maryland.

Dietrich, W.E., Kirchner, J.W., Ikeda, H. and Iseya, F. 1989. Sediment supply and the development of the coarse surface layer in gravel bedded rivers. Nature 340: 215–217.

Egiazaroff, I.V. 1965. Calculation of non-uniform sediment concentrations. Journal of Hydraulic Engineering 91: 225–248.

Elkins, E.M., Pasternack, G.B. and Merz, J.E. 2007. Use of slope creation for rehabilitating incised, regulated, gravel bed rivers. Water Resources Research 43: W05432, doi: 10.1029/2006WR005159.

Esteve, M. 2005. Observations of spawning behaviour in Salmoninae: Salmo, Oncorhynchus and Salvelinus. Reviews in Fish Biology and Fisheries 15: 1–21.

Fausch, K.D., Torgersen, C.E., Baxter, C.V. and Li, H.W. 2002. Landscapes to riverscapes: bridging the gap between research and conservation of stream fishes a continuous view of the river is needed to understand how processes interacting among scales set the context for stream fishes and their habitat. BioScience 52: 483–498.

Franssen, J., Blais, C., Lapointe, M., Bérubé, F., Bergeron, N. and Magnan, P. 2012. Asphyxiation and entombment mechanisms in fines rich spawning substrates: experimental evidence with brook trout (*Salvelinus fontinalis*) embryos. Canadian Journal of Fisheries and Aquatic Sciences 69: 587–599.

Franssen, J., Lapointe, M. and Magnan, P. 2014. Geomorphic controls on fine sediment reinfiltration into salmonid spawning gravels and the implications for spawning habitat rehabilitation. Geomorphology 211: 11–21.

Frostick, L.E., Lucas, P.M. and Reid, I. 1984. The infiltration of fine matrices into coarse-grained alluvial sediments and its implications for stratigraphical interpretation. Journal of the Geological Society 141: 955–965.

Furbish, D.J., Haff, P.K., Roseberry, J.C. and Schmeeckle, M.W. 2012. A probabilistic description of the bed load sediment flux: 1. Theory J Geophys Res 117, F03031, doi:10.1029/2012JF002352.

Garcia, M.H. 2008. Sediment transport and morphodynamics. pp. 21–163. *In*: García, M.H. (ed.). Sedimentation Engineering: Processes, Measurements, Modeling and Practice. ASCE, Reston, VA.

Gende, S.M., Edwards, R.T., Willson, M.F. and Wipfli, M.S. 2002. Pacific Salmon in Aquatic and Terrestrial Ecosystems: Pacific salmon subsidize freshwater and terrestrial ecosystems through several pathways, which generates unique management and conservation issues but also provides valuable research opportunities. BioScience 52: 917–928.

Goode, J.R., Buffington, J.M., Tonina, D., Isaak, D.J., Thurow, R.F., Wenger, S., Nagel, D., Luce, C., Tetzlaff, D. and Soulsby, C. 2013. Potential effects of climate change on streambed scour and risks to salmonid survival in snow-dominated mountain basins. Hydrological Processes 27: 750–765.

Gottesfeld, A.S., Hassan, M.A., Tunnicliffe, J.F. and Poirier, R.W. 2004. Sediment dispersion in salmon spawning streams: the influence of floods and salmon redd construction. Journal of the American Water Resources Association 40: 1071–1086.

Grant, G.E., Swanson, F.J. and Wolman, M.G. 1990. Pattern and origin of stepped-bed morphology in high-gradient streams, Western Cascades, Oregon. Geological Society of America Bulletin 102: 340–352.

Greig, S.M., Sear, D.A. and Carling, P.A. 2005. The impact of fine sediment accumulation on the survival of incubating salmon progeny: implications for sediment management. Science of the Total Environment 344: 241–258.

Groot, C. and Margolis, L. 1991. Pacific Salmon Life Histories. UBC Press, 608 p.

Gustafson-Greenwood, K.I. and Moring, J.R. 1991. Gravel compaction and permeabilities in redds of Atlantic salmon, *Salmo salar* L. Aquaculture Research 22: 537–540.

Harrison, L.R., Legleiter, C.J., Wydzga, M.A. and Dunne, T. 2011. Channel dynamics and habitat development in a meandering, gravel bed river. Water Resources Research 47: W04513. doi:10.1029/2009WR008926.

Hartman, G.F., Scrivener, J.C. and Miles, M.J. 1996. Impacts of logging in Carnation Creek, a high-energy coastal stream in British Columbia, and their implication for restoring fish habitat. Canadian Journal of Fisheries and Aquatic Sciences 53: 237–251.

Hassan, M.A. 1990. Scour, fill, and burial depth of coarse material in gravel bed streams. Earth Surface Processes and Landforms 15: 341–356.

Hassan, M.A., Church, M. and Schick, A.P. 1991. Distance of movement of coarse particles in gravel bed streams. Water Resources Research 27: 503–511.

Hassan, M.A. and Church, M. 2000. Experiments on surface structure and partial sediment transport on a gravel bed. Water Resources Research 36(7): 1885–1895.

Hassan, M.A. and Church, M. 2001. Rating bedload transport in Harris Creek: seasonal and spatial variation over a cobble-gravel bed. Water Resources Research 37: 813–825.

Hassan, M.A. and Woodsmith, R.D. 2004. Bed load transport in an obstruction-formed pool in a forest, gravel-bed stream. Geomorphology 58: 203–221.

Hassan, M.A., Church, M., Lisle, T.E., Brardinoni, F., Benda, L. and Grant, G.E. 2005. Sediment transport and channel morphology of small, forested streams. Journal of the American Water Resources Association 41: 853–876.

Hassan, M.A., Smith, B.J., Hogan, D.L., Luzi, D.S., Zimmermann, A.E. and Eaton, B.C. 2007. Sediment storage and transport in coarse bed streams: scale considerations. pp. 473–496. *In*: Habersack, H., Piegay, H. and Rinaldi, M. (eds.). Gravel-Bed Rivers VI: From Process Understanding to River Restoration. Elsevier. Amsterdam, The Netherlands.

Hassan, M.A., Gottesfeld, A.S., Montgomery, D.R., Tunnicliffe, J.F., Clarke, G.K., Wynn, G., Jones-Cox, H., Poirier, R., MacIsaac, E., Herunter, H. and Macdonald, S.J. 2008. Salmon-driven bed load transport and bed morphology in mountain streams. Geophysical Research Letters, 35, L04405, doi:10.1029/2007GL032997.

Hassan, M.A., Brayshaw, D., Alila, Y. and Andrews, E.D. 2014. Effective discharge in small formerly glaciated mountain streams of British Columbia: limitations and implications. Water Resources Research 50: 4440–4458.

Hodge, R.A., Sear, D.A. and Leyland, J. 2013. Spatial variations in surface sediment structure in riffle–pool sequences: a preliminary test of the Differential Sediment Entrainment Hypothesis (DSEH). Earth Surface Processes and Landforms 38: 449–465.

Hogan, D.L., Bird, S.A. and Hassan, M.A. 1998. Spatial and temporal evolution of small coastal gravel-bed streams: influence of forest management on channel morphology and fish habitat. pp. 365–392. *In*: Klingeman, P.C., Beschta, R.L., Komar, P.D. and Bradley, J.B. (eds.). Gravel-Bed Rivers in the Environment. Water Resources Publications. Highland Ranch, Colorado.

Ikeda, S. 1982. Incipient motion of sand particles on side slopes. J Hydraul Div Am Soc Civ Eng 108: 95–114.

Judd, H.E. 1963. A study of bed characteristics in relation to flow in rough, high-gradient natural channels. Ph.D. Thesis, Utah State University.

Kamann, P.J., Ritzi, R.W., Dominic, D.F. and Conrad, C.M. 2007. Porosity and permeability in sediment mixtures. Groundwater 45: 429–438.

Kemp, P.S. 2015. Impoundments, barriers and abstractions. pp. 717–769. *In*: Craig, J.F. (ed.). Freshwater Fisheries Ecology. John Wiley & Sons. Chichester, UK.

Keller, E.A. 1971. Areal sorting of bedload material: a hypothesis of velocity reversal. Geological Society of America, Bull 82: 753–756.

Kirchner, J.W., Dietrich, W.E., Iseya, F. and Ikeda, H. 1990. The variability of critical shear stress, friction angle, and grain protrusion in water-worked sediments. Sedimentology 37: 647–672.

Kondolf, G.M., Cada, G.F., Sale, M.J. and Felando, T. 1991. Distribution and stability of potential salmonid spawning gravels in steep boulder-bed streams of the eastern Sierra Nevada. Transactions of the American Fisheries Society 120: 177–186.

Kondolf, G.M. and Wolman, M.G. 1993. The sizes of salmonid spawning gravels. Water Resources Research 29: 2275–2285.

Kondolf, G.M., Sale, M.J. and Wolman, M.G. 1993. Modification of fluvial gravel size by spawning salmonids. Water Resources Research 29: 2265–2274.

Kovacs, A. and Parker, G. 1994. A new vectorial bedload formulation and its application to the time evolution of straight river channels. Journal of Fluid Mechanics 267: 153–183.

Lamb, M.P., Dietrich, W.E. and Venditti, J.G. 2008. Is the critical Shields stress for incipient sediment motion dependent on channel-bed slope? Journal of Geophysical Research 113: F02008, doi:10.1029/2007JF000831.

Lapointe, M., Eaton, B., Driscoll, S. and Latulippe, C. 2000. Modelling the probability of salmonid egg pocket scour due to floods. Canadian Journal of Fisheries and Aquatic Sciences 57: 1120–1130.

Lapointe, M. 2012. River geomorphology and salmonid habitat: some examples illustrating their complex association, from redd to riverscape scales. pp. 191–215. *In*: Church, M., Biron, P.M. and Roy, A.G. (eds.). Gravel-Bed Rivers:Processes, Tools, Environments. Wiley & Sons. Chichester, UK.

Lapointe, N.W.R., Cooke, S.J., Imhof, J.G., Boisclair, D., Casselman, J.M., Curry, R.A., Langer, O.E., McLaughlin, R.L., Minns, C.K., Post, J.R., Power, M., Rasmussen, J.B., Reynolds, J.D., Richardson, J.S. and Tonn, W.M. 2013. Principles for ensuring healthy and productive freshwater ecosystems that support sustainable fisheries. Environmental Reviews 22: 110–134.

Lave, R. 2009. The controversy over natural channel design: substantive explanations and potential avenues for resolution 1. Journal of the American Water Resources Association 45: 1519–1532.

Lenzi, M.A. 2001. Step-pool evolution in the Rio Cordon, Northeastern Italy. Earth Surface Processes and Landforms 26: 991–1008.

Leopold, L.B., Wolman, M.G. and Miller, J.P. 1964. Fluvial Processes in Geomorphology. W.H. Freeman and Company, San Francisco.

Lisle, T.E. 1982. Effects of aggradation and degradation on riffle-pool morphology in natural gravel channels, northwestern California. Water Resources Research 18: 1643–1651.

Lisle, T.E. 1986. Stabilization of a gravel channel by large streamside obstructions and bedrock bends, Jacoby Creek, northwestern California. Geological Society of America Bulletin 97: 999–1011.

Lisle, T.E. 1989. Sediment transport and resulting deposition in spawning gravels, north coastal California. Water Resources Research 25: 1303–1319.

Lisle, T.E. and Hilton, S. 1992. The Volume of fine sediment in pools: an index of sediment supply in gravel-bed streams. Journal of the American Water Resources Association 28: 371–383.

Lisle, T.E., Nelson, J.M., Pitlick, J., Madej, M.A. and Barkett, B.L. 2000. Variability of bed mobility in natural, gravel-bed channels and adjustments to sediment load at local and reach scales. Water Resources Research 36: 3743–3755.

MacVicar, B. and Rennie, C. 2012. Flow and turbulence redistribution in a straight artificial pool. Water Resources Research 48: W02,503, doi:10.1029/2010WR009374.

MacVicar, B. and Best, J. 2013. A flume experiment on the effect of channel width on the perturbation and recovery of flow in straight pools and riffles with smooth boundaries. Journal of Geophysical Research: Earth Surface 118: 1850–1863.

MacWilliams, M., Wheaton, J., Pasternack, G., Street, R. and Kitanidis, P. 2006. Flow convergence routing hypothesis for pool-rie maintenance in alluvial rivers. Water Resources Research 42: doi:10.1029/2005WR004391.

Magee, J.P., McMahon, T.E. and Thurow, R.F. 1996. Spatial variation in spawning habitat of cutthroat trout in a sediment-rich stream basin. Transactions of the American Fisheries Society 12: 768–779.

McNeil, W.J. and Ahnell, W.H. 1964. Success of pink salmon spawning relative to size of spawning bed materials. US Department of Interior, Fish and Wildlife Service.

Meyer, C.B., Sparkman, M.D. and Klatte, B.A. 2005. Sand seals in coho salmon redds: do they improve egg survival? North American Journal of Fisheries Management 25: 105–121.

Meyer, J.L., Strayer, D.L., Wallace, J.B., Eggert, S.L., Helfman, G.S. and Leonard, N.E. 2007. The contribution of headwater streams to biodiversity in river networks. Journal of the American Water Resources Association 43: 86–103.

Meyer-Peter, E. and Müller, R. 1948. Formulas for bed-load transport. Proc 2nd Congr Int Assoc Hydraul. Res., Stockholm, pp. 39–64.

Miller, D.J., Burnett, K.M. and Benda, L. 2008. Factors controlling availability of spawning habitat for salmonids at the basin scale. pp. 103–120. *In*: Sear, D.A. and DeVries, P. (eds.). Salmonid Spawning Habitat in Rivers. American Fisheries Society Symposium 65, Bethesda, MD.

Moir, H.J., Soulsby, C. and Youngson, A.F. 2002. Hydraulic and sedimentary controls on the availability and use of Atlantic salmon (*Salmo salar*) spawning habitat in the River Dee system, north-east Scotland. Geomorphology 45: 291–308.

Moir, H.J., Gibbins, C.N., Soulsby, C. and Webb, J. 2004. Linking channel geomorphic characteristics to spatial patterns of spawning activity and discharge use by Atlantic salmon (*Salmo salar* L.). Geomorphology 60: 21–35.

Moir, H.J. and Pasternack, G.B. 2008. Relationships between mesoscale morphological units, stream hydraulics and Chinook salmon (*Oncorhynchus tshawytscha*) spawning habitat on the Lower Yuba River, California. Geomorphology 100: 527–548.

Moir, H.J., Gibbins, C.N., Buffington, J.M., Webb, J.H., Soulsby, C. and Brewer, M.J. 2009. A new method to identify the fluvial regimes used by spawning salmonids. Canadian Journal of Fisheries and Aquatic Sciences 66: 1404–1408.

Montgomery, D.R. and Dietrich, W.E. 1988. Where do channels begin? Nature 336: 232–234.

Montgomery, D.R., Buffington, J.M., Smith, R.D., Schmidt, K.M. and Pess, G. 1995. Pool spacing in forest channels. Water Resources Research 31: 1097–1105.

Montgomery, D.R., Buffington, J.M., Peterson, N.P., Schuett-Hames, D. and Quinn, T.P. 1996. Stream-bed scour, egg burial depths, and the influence of salmonid spawning on bed surface mobility and embryo survival. Canadian Journal of Fisheries and Aquatic Sciences 53: 1061–1070.

Montgomery, D.R. and Buffington, J.M. 1997. Channel-reach morphology in mountain drainage basins. Geological Society of America Bulletin 109: 596–611.

Montgomery, D.R., Beamer, E.M., Pess, G.R. and Quinn, T.P. 1999. Channel type and salmonid spawning distribution and abundance. Canadian Journal of Fisheries and Aquatic Sciences 56: 377–387.

Mueller, M., Pander, J. and Geist, J. 2014. The ecological value of stream restoration measures: an evaluation on ecosystem and target species scales. Ecological Engineering 62: 129–139.

Naiman, R.J., Bilby, R.E., Schindler, D.E. and Helfield, J.M. 2002. Pacific salmon, nutrients, and the dynamics of freshwater and riparian ecosystems. Ecosystems 5: 399–417.

Neill, C.R. 1968. Note on initial movement of coarse uniform material. Journal of Hydraulic Research 6: 157–184.

Nelson, P.A., Dietrich, W.E. and Venditti, J.G. 2010. Bed topography and the development of forced bed surface patches. Journal of Geophysical Research (2003–2012), 115: 2003–2012.

Nistor, C.J. and Church, M. 2005. Suspended sediment transport regime in a debris-flow gully on Vancouver Island, British Columbia. Hydrological Processes 19: 861–885.

Paintal, A.S. 1971. Concept of critical shear stress in loose boundary open channels. Journal of Hydraulic Research 9: 91–113.

Palmer, M.A., Hondula, K.L. and Koch, B.J. 2014. Ecological restoration of streams and rivers: shifting strategies and shifting goals. Annual Review of Ecology, Evolution, and Systematics 45: 247–269.

Pander, J., Mueller, M. and Geist, J. 2015. A comparison of four stream substratum restoration techniques concerning interstitial conditions and downstream effects. River Research and Applications 31: 239–255.

Parker, G. and Klingeman, P. 1982. On why gravel bed streams are paved. Water Resources Research 18: 1409–1423.

Parker, G. and Andrews, E. 1985. Sorting of bed load sediment by flow in meander bends. Water Resources Research 21(9): 1361–1373.

Parker, G. 1990. Surface bedload transport relation for gravel rivers. Journal of Hydraulic Research 28: 417–436.

Parker, G. 2008. Transport of gravel and sediment mixtures. pp. 165–251. *In*: Garcia, M.H. (ed.). Sedimentation Engineering. Processes, Measurements, Modeling and Practice. ASCE, Reston, VA.

Peterson, N.P. and Quinn, T.P. 1996. Spatial and temporal variation in dissolved oxygen in natural egg pockets of chum salmon, in Kennedy Creek. Washington. Journal of Fish Biology 48: 131–143.

Pulg, U., Barlaup, B.T., Sternecker, K., Trepl, L. and Unfer, G. 2013. Restoration of spawning habitats of brown trout (*salmo Trutta*) in a regulated chalk stream. River Research and Applications 29: 172–182.

Quinn, T.P., Hendry, A.P. and Wetzel, L.A. 1995. The influence of life history trade-offs and the size of incubation gravels on egg size variation in sockeye salmon (*Oncorhynchus nerka*). Oikos 425–438.

Quinn, T.P. 2005. The Behavior and Ecology of Pacific Salmon and Trout. UBC Press.

Rennie, C.D. and Millar, R.G. 2000. Spatial variability of stream bed scour and fill: a comparison of scour depth in chum salmon (*Oncorhynchus keta*) redds and adjacent bed. Canadian Journal of Fisheries and Aquatic Sciences 57: 928–938.

Rice, S. and Church, M. 1998. Grain size along two gravel-bed rivers: statistical variation, spatial pattern and sedimentary links. Earth Surface Processes and Landforms 23: 345–363.

Richards, K.S. 1976. Channel width and the riffle-pool sequence. Geological Society of America Bulletin 87: 883–890.

Riebe, C.S., Sklar, L.S., Overstreet, B.T. and Wooster, J.K. 2014. Optimal reproduction in salmon spawning substrates linked to grain size and fish length. Water Resources Research 50: 898–918.

Rieman, B., Dunham, J. and Clayton, J. 2006. Emerging concepts for management of river ecosystems and challenges to applied integration of physical and biological sciences in the Pacific Northwest, USA. International Journal of River Basin Management 4: 85–97.

Roni, P., Beechie, T., Pess, G. and Hanson, K. 2014. Wood placement in river restoration: fact, fiction, and future direction. Canadian Journal of Fisheries and Aquatic Sciences 72: 466–478.

Roseberry, J.C., Schmeeckle, M.W. and Furbish, D.J. 2012. A probabilistic description of the bed load sediment flux: 2. Particle activity and motions, Journal of Geophysical Research 117: F03032, doi:10.1029/2012JF002353.

Rosgen, D.L. 1994. A classification of natural rivers. Catena 22: 169–199.

Rosgen, D.L. 1996. Applied River Morphology. Wildland Hydrology, Pagosa, Springs, CO.

Schuett-Hames, D.E., Peterson, N.P., Conrad, R. and Quinn, T.P. 2000. Patterns of gravel scour and fill after spawning by chum salmon in a western Washington stream. North American Journal of Fisheries Management 20: 610–617.

Sear, D. 2010. Integrating science and practice for the sustainable management of in-channel salmonid habitat. pp. 81–118. *In*: Kemp, P. (ed.). Salmonid Fisheries: Freshwater Habitat Management. Wiley-Blackwell. Chichester, UK.

Sear, D.A. 1996. Sediment transport processes in pool–riffle sequences. Earth Surface Processes and Landforms 21: 241–262.

Sear, D.S., Frostick, L.B., Rollinson, G. and Lisle, T.E. 2008. The significance and mechanics of fine-sediment infiltration and accumulation in gravel spawning beds. pp. 149–173. *In*: Sear, D.A. and DeVries, P. (eds.). Salmonid Spawning Habitat in Rivers: Physical Controls, Biological Responses, and Approaches to Remediation. American Fisheries Society Symposium (Vol. 65). American Fisheries Society. Bethesda, Maryland.

Seminara, G., Parker, G. and Solari, L. 2002. Bed load at low Shields stress on arbitrarily sloping beds: failure of the Bagnold hypothesis. Water Resources Research 38: 1249, doi: 10.1029/2001WR000681.

Shellberg, J.G., Bolton, S.M. and Montgomery, D.R. 2010. Hydrogeomorphic effects on bedload scour in bull char (*Salvelinus confluentus*) spawning habitat, western Washington, USA. Canadian Journal of Fisheries and Aquatic Sciences 67: 626–640.

Shields, A. 1936. Anwendung der Aehnlichkeitsmechanik und der Turbulenzforschung auf die Geschiebebewegung. der PreuBischen Versuchsanstalt fur Wassserbau und Schiffbau Berlin NW 87. Berlin. Ph.D.

Simon, A., Doyle, M., Kondolf, M., Shields, Jr. F.D., Rhoads, B. and McPhillips, M. 2007. Critical evaluation of how the Rosgen classification and associated "Natural Channel Design" methods fail to integrate and quantify fluvial processes and channel response. Journal of the American Water Resources Association 43: 1117–1131.

Steen, R.P. and Quinn, T.P. 1999. Egg burial depth by sockeye salmon (*Oncorhynchus nerka*): implications for survival of embryos and natural selection on female body size. Canadian Journal of Zoology 77: 836–841.

Thompson, D.M., Wohl, E.E. and Jarrett, R.D. 1999. Velocity reversals and sediment sorting in pools and riffles controlled by channel constrictions. Geomorphology 27: 229–241.

Thompson, C.J., Croke, J., Ogden, R. and Wallbrink, P. 2006. A morpho-statistical classification of mountain stream reach types in Southeastern Australia. Geomorphology 81: 43–65.

Thorstad, E.B., Whoriskey, F., Rikardsen, A.H. and Aarestrup, K. 2011. Aquatic nomads: the life and migrations of the Atlantic salmon. pp. 1–32. *In*: Klemetsen, A., Einum, S. and Skurdal, J. (eds.). Atlantic Salmon Ecology. Wiley-Blackwell, Chichester, UK.

Tinkler, K.J. and Wohl, E.E. (eds.). 1998. Rivers Over Rock. Geophysical Monograph, 107, American Geophysical Union, Washington, D.C., USA, 323 pp.

Tullos, D.D., Penrose, D.L., Jennings, G.D. and Cope, W.G. 2009. Analysis of functional traits in reconfigured channels: implications for the bioassessment and disturbance of river restoration. Journal of the North American Benthological Society 28: 80–92.

Van Den Berghe, E.P. and Gross, M.R. 1989. Natural selection resulting from female breeding competition in a Pacific salmon (coho: *Oncorhynchus kisutch*). Evolution 43: 125–140.

Wheaton, J.M., Pasternack, G.B. and Merz, J.E. 2004. Spawning habitat rehabilitation-I. Conceptual approach and methods. International Journal of River Basin Management 2: 3–20.

White, S.M. and Rahel, F.J. 2008. Complementation of habitats for Bonneville cutthroat trout in watersheds influenced by beavers, livestock, and drought. Transactions of the American Fisheries Society 137: 881–894.

Whiteway, S.L., Biron, P.M., Zimmermann, A., Venter, O. and Grant, J.W.A. 2010. Do in-stream restoration structures enhance salmonid abundance? Canadian Journal of Fisheries and Aquatic Sciences 67: 831–841.

Whiting, P.J. and Bradley, J.B. 1993. A process-based classification system for headwater streams. Earth Surface Processes and Landforms 18: 603–612.

Whittaker, J.G. 1987. Sediment transport in step–pool streams. pp. 545–570. *In*: Thorne, C.R., Bathurst, J.C. and Hey, R.D. (eds.). Sediment Transport in Gravel-Bed Rivers. Wiley, Chichester.

Wiberg, P.L. and Smith, J.D. 1987. Calculations of the critical shear stress for motion of uniform and heterogeneous sediments, Water Resources Research 23: 1471–1480.

Wilcock, P.R. and Crowe, J.C. 2003. Surface-based transport model for mixed-size sediment. Journal of Hydraulic Engineering 129: 120–128.

Wipfli, M.S., Richardson, J.S. and Naiman, R.J. 2007. Ecological linkages between headwaters and downstream ecosystems: transport of organic matter, invertebrates, and wood down headwater channels. Journal of the American Water Resources Association 43: 72–85.

Wohl, E., Angermeier, P.L., Bledsoe, B., Kondolf, G.M., MacDonnell, L., Merritt, D.M., Palmer, M.A., Poff, N.L. and Tarboton, D. 2005. River restoration. Water Resources Research 41: W10301, doi:10.1029/2005WR003985.

Wohl, E., Lane, S.N. and Wilcox, A.C. 2015. The science and practice of river restoration. Water Resources Research 51: 5974–5997.

Wohl, E.E., Madsen, S. and MacDonald, L. 1997. Characteristics of log and clast bed–step in step–pool streams of northwestern Montana, USA. Geomorphology 20: 1–10.

Wohl, E.E. 1998. Bedrock channel morphology in relation to erosional processes. pp. 133–152. *In*: Tinkler, K.J. and Wohl, E.E. (eds.). Rivers Over Rock: Fluvial Processes in Bedrock Channels. American Geophysical Union, Washington, D.C.

Wong, M., Parker, G., DeVries, P., Brown, T.M. and Burges, S.J. 2007. Experiments on dispersion of tracer stones under lower-regime plane-bed equilibrium bed load transport. Water Resources Research 43: W03440, doi:10.1029/2006WR005172.

Wooster, J.K., Dusterhoff, S.R., Cui, Y., Sklar, L.S., Dietrich, W.E. and Malko, M. 2008. Sediment supply and relative size distribution effects on fine sediment infiltration into immobile gravels. Water Resources Research 44, doi: 10.1029/2006WR005815.

Yalin, M. 1971. On the formation of dunes and meanders. pp. C101–C108. *In*: Proceedings of the 14th Congress of the International Association for Hydraulic Research, IAHR, Paris.

Zimmermann, A. and Church, M. 2001. Channel morphology, gradient profiles and bed stresses during spring runoff in a step-pool channel. Geomorphology 40: 311–328.

Zimmermann, A.E. and Lapointe, M. 2005. Intergranular flow velocity through salmonid redds: sensitivity to fines infiltration from low intensity sediment transport events. River Research and Applications 21: 865–881.

Zimmermann, A.E., Church, M. and Hassan, M.A. 2010. Step–pool stability: testing the jammed state hypothesis. Journal of Geophysical Research 115: F02008. http://dx.doi.org/10.1029/2009JF001365.

Zimmermann, A.E. 2013. Step–pool channel features. pp. 346–363. *In*: Shroder, J. (Editor in Chief), Wohl, E. (ed.). Treatise on Geomorphology. Vol. 9: Fluvial Geomorphology. Academic Press, San Diego, CA.

CHAPTER 18

Understanding the Past of Rivers

Reading the History of Rivers from Documents and Maps

Francisco da Silva Costa, * *António Avelino Batista Vieira*[a] and
António José Bento Gonçalves[b]

INTRODUCTION

An analysis of the routines of secular, public institutions raises several difficulties in location of precise sources, such as, for example, written documental sources that are dispersed throughout documentation centres and archives. The preservation and management of collections and information has yet to be accomplished in many documentation centres. Thus, any attempt to reconstruct the daily routines of an institution implies conducting new studies on documental sources (Martins 1997).

There are several documental approaches employed to access the historical context of a certain period. Written primary sources are testaments of the past, "footprints" left by humans in history, which research uses to reconstruct these memories that took place as realistically as possible (Carvalho 2009; Ketelaar 2004; Milligan 1979). Legal documents (constitutions, laws, decrees), sentences, correspondence, inventories, censuses, maps, graphs, etc., are some of the many interpretations researchers make of primary sources we can find in archives and departments in public bodies that need to archive documents (Prado 2010). However, it is important to understand that historical documents only become primary sources when used for the purposes of research (Camargo 2015).

Departamento de Geografia, Universidade do Minho; Centro de Estudos em Geografia e Ordenamento do Território-UMinho. Campus de Azurém, 4800 Guimarães (Portugal).
[a] E-mail: vieira@geografia.uminho.pt
[b] E-mail: bento@geografia.uminho.pt
* Corresponding author: costafs@geografia.uminho.pt

The Historical Archive of the Portuguese Environment Agency (APA)

The Portuguese Environment Agency (*Agência Portuguesa do Ambiente – APA*) holds an important archive resulting from the century-old activity of the Water Services in charge of water management and water resources planning, which should be studied and made public. The documental and technical collections of the former Douro Water Services (*Serviços Hidráulicos do Douro – SHD*) represent an opportunity to understand a public institution, where water and its use profoundly mark social relationships and economic activities. The technical and administrative documents the SHD produced are essential to explore the SHD's history and determine the role it played in the economy and society of northern Portugal, as well as all the correspondence and projects it received, in its capacity as competent body in charge of water management (Campelo 2011; Costa 2012; Costa and Cordeiro 2012, 2012a, 2015, 2015a; Costa et al. 2015). From the Minho to the Douro rivers, including the Lima and Cávado, their most important tributaries and other watercourses, shaping the water map of northern Portugal, we come across the art of engineers and designers/planners in the projects developed by hydroelectric power stations, *t*he petitions and complaints from citizens, the evaluation of projects proposed by private bodies, the stories of river guards, and detailed reports by clerks busy at their trade. All of this information is gathered through maps, projects, letters, registries, reports, and annotations on correspondence (Campelo 2011; Costa 2012, 2011, 2010; Costa and Cordeiro 2012, 2012a, 2015, 2015a; Costa et al. 2015).

The files in the APA's archive tell of case studies on the uses of public water for different purposes, the modalities of its use, more traditional techniques associated with irrigation as well as industrial productive processes, highlighting the importance of hydroelectric power in the development of the Ave River basin, without neglecting its public uses and the occupation of river banks for other activities (Costa 2012; Costa et al. 2015). Thus, several areas of scientific interest (historical, geographical, technical, etc.) are taken into account in the analysis of these processes, as well as the problems, impacts, and limitations of human activities in the management of the Public Waterways Domain (*Domínio Público Hídrico – DPH*).

Goals

This paper intends to highlight the importance of the historical sources in the SHD's archive for our understanding of the watercourses of a drainage basin in northwest Portugal – the Ave River basin. The choice of this basin for a case study derives from the need to employ a naturally delimited planning unit, as well as to understand and analyse the interrelations occurring in the river channel-river basin flows and vice versa (Costa 2008, 2008c, 2012).

Among the thousands of files found in the archive, we consulted and analysed those related to the Public Waterways Domain of the Ave River basin, from 1886 to the present day. They revealed the relevance of historical sources in producing information to broaden our knowledge of watercourses and riverside areas in this region.

We intend to present case studies of particular interest in historical terms related to the Ave River basin, located on the coastal region of northwest Portugal and the evolution of its watercourses' morphometric features.

Study area – the ave river basin

The Ave River basin is located in northwest Portugal between latitudes 41° 15' and 41° 40' North and longitudes 8° 00' and 8° 45' West, covering an area of approximately 1391 km². This basin is surrounded by the Cávado river basin to the north, the Douro river basin to the east, and the Leça river basin to the south (Fig. 1).

The Ave River, with its sources located at over 1050 metres in altitude on the Cabreira Mountain, running for about 100 km and with a river basin covering 1391 km², initially crosses the territory from northeast to southeast, turning later to west to flow to its mouth at Vila do Conde. According to the Ave River Basin Plan, the annual drainage at the Ave River mouth is, on average, is 1250 hm³, and it is estimated that annual average rainfall on the basin is 1791 mm, which corresponds to 2498 hm³. From this total, 1248 hm³ are lost to evaporation and 1203 hm³ infiltrates, recharging the aquifers, resulting in immediate surface runoff of 47 hm³. The 1203 hm³ that infiltrates appears at the surface, resulting in a total surface runoff of 1250 hm³ (DRAOT-Norte 2000). The Vizela River, on the left bank, and the Este River, on the right bank, are the main sub-drainage basins of the Ave River.

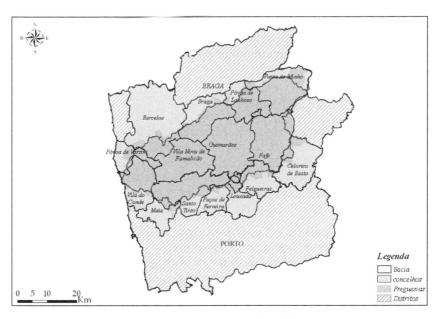

Fig. 1. Administrative framework of the Ave river basin. Source: IGEOE.

Legal water framework—the main diplomas at the beginning of the 20th century

According to Bacellar (2005), we need to understand how the administrative machine worked in the historical period under study. It was at the end of the 19th century that some of the reforms appeared that were to shape the organisational framework of public waters in Portugal (Costa 2008, 2011, 2012, 2012a; Costa et al. 2011). In 1884, the publication of *"Plano de organização dos serviços hidrográficos no continente de*

Portugal" ("*Plan to organise the hydrographic services on mainland Portugal*") is an important step in the first attempts to organise the territory. The proposal consisted in dividing the country into four hydrographic circumscriptions, employing as the main criterion for the grouping of the drainage basins of the respective rivers (Costa 2008, 2010, 2012; Costa et al. 2011; Pato 2008). In 1892, the Water Services were organised and regulated, based on two very important diplomas: the decree with the force of law No. 8, passed on the 1st of December 1892, which regulates the Organisation of Water Services and respective staff; the Decree of 19th December 1892, which establishes the regulations of the Water Services. These diplomas set the bases for the organisation and operation of the Water Services, as well as established the definition of water uses and property and management principles, among other aspects (Costa 2008, 2010; Costa et al. 2011).

Practically all issues related with water were regulated in Decree No. 5787–IIII, passed on 10th May 1919, with exceptions made for a few dispositions found in the 1892 regulations. Thus, the passing of the Water Law marks one of the highest points in Portuguese legislation on water, published in a recent past. These laws were already pointing to some of the essential principles of an appropriate management policy, such as considering water resources as a factor in national wealth, taking the river basin as the basic management unit, and the interdependent nature of the use of the different water resources. Water management is rooted in a centuries-old institutional and legal tradition which formulates concepts which are still relevant today (Costa 2004, 2008, 2010, 2010a).

Results—Discussion and Analyses

There are several physical structures associated with the use of public water for agricultural and industrial purposes, but our research will consider other types of intervention, such as works and actions that can be developed in a section transversally, or on the river bed and banks and, for that reason, longitudinally. The interventions of the DPH can be classified according to the nature of the works accomplished and the area covered (Costa 2008, 2012, 2010c; Wasson et al. 1998):

- Maintenance works – in this study, particularly focusing on clearing of river banks and the extraction of aggregates and boulders.
- Adjustment and management actions – particularly interventions related to water harnessing (weirs, tailraces, sluiceway channels, etc.) and adjusting or regulating currents (building walls, channellization alignment, etc.).
- Structural works – large-scale works such as, for example, building bridges and aqueducts.

Clearing and maintenance works and extraction of aggregates

Clearing, regulating, and overhauling works on river beds and banks was established in articles 276, 284, and 286 of the Water Services Regulations and in article 130 of the Water Law.

Clearing and extracting sand was the cause of hundreds of applications annually (Costa 2004a, 2008, 2010c) and were included in maintenance works, given that, in some situations, these procedures contributed to desilting the watercourses' beds.

Apart from the revenue generated by aggregate extraction, it also gave rise to conflict resulting from, in most cases, the lack of knowledge about the negative impacts of this type of activity. There are two references to the doubts about how extraction works could affect the local landscape and the consequences for leisure activities (Costa 2008):

- In 1931, in an internal memo, the section chief called the activity into question given the effects it had on the degradation of a very popular river beach for leisure and bathing. He thus stated that "(…) the extraction of thousands of m³ of sand from the Ave River on its northern bank and upstream from its rail bridge in Vila do Conde (…)" from a river beach described with the following characteristics – "(…) a regular surface of 7000 m² with an average height of 0.32 m above the current and a bend at a higher level near the bank, with a surface area of 800 m² and an average height of 1 m, giving an approximate volume of 3000 m³ (…)", to then conclude "(…) with the intended sand extraction, the public will be deprived of the utility the location offers for their domestic use (bathers) (…)";
- In 1946, a letter from the Santo Tirso City Hall denounced practices it rejected because "(…) those locations have been destroyed and disfigured due to sand extraction by contractors and master builders (…). They dig holes where the water stagnates in the summer; the sand is perforated, leaving only pebbles and coal residues the waters are dragging from factories. The situation in the park has thus become terribly distressful, unsightly and dirty".

The river guards had responsibilities in evaluating potential sites for aggregate extraction. They were in charge of analysing the applications and, in some cases, would have to visit the sites personally to determine the possibilities for such activities. This situation occurred in 1937, recorded in an internal memo. With regard to an application to extract 100 m³ of sand from the Pelhe River at four points in the parishes of Gavião, Calendário, and Antas, intense extraction was reported, and the river was described as follows at local level: "(…) the river is narrow and is three metres long. At the widest part of the river, and only where the bridges are located does the river reach its widest and the sands accumulate, is where extraction is possible. Even so, it is still doubtful, if the current does not bring new sands, because extraction has indeed been extraordinary (…)". In this memo, the conclusion unequivocally states that "(…) to extract a large amount of sand at only one location, the discharge will have to increase so that the river bed can be filled with new sands dragged in by the discharge. Consequently, they can extract at different locations, at which, in any case, I have seen that others have taken upon and continue to take (…).".

Apart from doubts resulting from the lack of information on the watercourses' capacity to provide enough sand to respond to the applications, other questions were raised, particularly with regard to the impacts of this activity for the stability of certain road structures. In 1939, the Póvoa de Varzim City Hall reported to the SHD that "(…) the fact that sand is being constantly extracted from the Este River, near the Central Pillar of the Vau Bridge (…), has clear detrimental effects on that bridge's stability (…)". The mayor thus alerted to the need to "(…) take measures to further stop the extraction of sand (…)" by intensifying inspections (Costa 2008).

Among the thousands of applications for this activity, there were also applications to extract boulders, in which the main purpose consisted of improving the watercourses. In most cases, the stones were used for works on the banks, particularly to build retaining walls. The destruction or cutting of boulders took place mostly in the Vizela River basin and is undoubtedly related with its mountainous morphology of steep slopes and large granite blocks and boulders (Costa 2004a, 2008).

Sand clearance and extraction are some of the examples of interventions by the DPH, with differentiated goals and impacts. Clearance served to regulate water drainage and reduce the effects of flooding, whereas extraction, although possibly also serving to make similar improvements, could result in the degradation of river beds and banks, with all the associated risks. The interventions were, however, of low environmental impact, given the local scale at which they were operated, in opposition to the large-scale works conducted on river beds and banks.

Regulating actions

Harnessing water became essential to the development of agricultural activities as well as to the different industries which require these types of structures. The works involved in water harnessing imply different types of intervention in the channels and their geometry. Among the different structures used in water harnessing, particularly relevant structures are weirs and dams, because they are the ones that provoke the highest impacts.

The first observation made from a reading of the DPH's files is that these two concepts are not clearly distinguished in the respective applications. Indeed, the applications for weirs and dams are of the same type and are referenced as small-scale hydraulic structures.

What is commonly called a dam falls outside the scope of this study, and although the distinction is made, we will focus on small-scale interventions, that is, those that come closer to the concept of weir. It is a water diversion structure built for several purposes on the river beds, in which the backwater zone does not surpass the river's banks (INE 2004). This type of structure creates artificial bodies of water upstream and requires the building of diversion or conveyance devices, so as to carry the water from where it is captured to the places where it is going to be used.

Licencing for the construction of weirs and dams is duly defined in the 1892 Water Services Regulations and in the 1919 Water Law. Article 265 of the Regulations establishes the technical criteria to be applied to this type of structure (Costa 2008, 2010, 2010d):

- Height – "(…) should be such that, in medium waters, the water level upstream is, at least, 0.20 m lower than the lowest part of the higher properties located throughout the extension of the reservoir and 0.10 m lower than the lowest part of the sewage ducts of the higher plots of land found in that extension (…)" and "(…) it will be clearly and visibility identified on its banks, in an easily accessible place, with a fixed sign or reference that cannot be easily destroyed (…).".
- The existence of a spillway and water sluices or gates in sufficient number and with sizes calculated such that, even if the water rises, the level of the reservoir maintains the necessary conditions so as not to cause damage. The construction of an inclined plane (Fig. 2) so as to ensure the necessary conditions for fish to pass (Fig. 3).

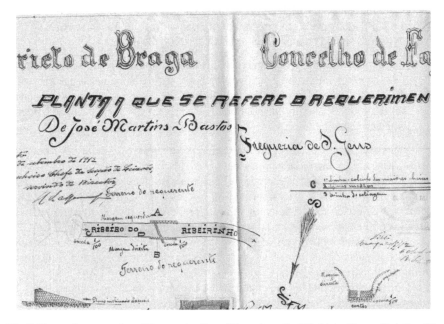

Fig. 2. Plan and cross-section of the construction of the weir on the Ribeirinho Stream (Ruivães, São Gens, Fafe 1912) (Source: APA).

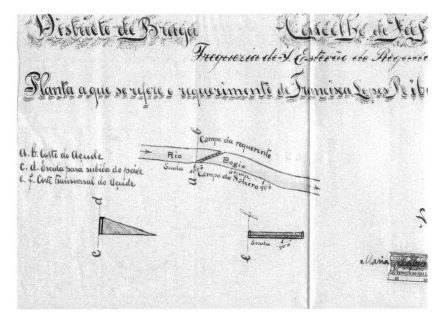

Fig. 3. Plan and cross-section of the construction of the weir on the Bugio River (Regadas, Fafe 1912) (Source: APA).

Under the terms of article 69 of the Regulations of the Water Law and articles 84 and 85 of the law itself, information of great interest from the perspective of river hydrodynamics should be made available (Costa 2008, 2010d):

- The harnessing area, with properties labelled and the names of their owners.
- The site where the weir or dam will be built, setting its height or the process to deviate the water. The direction and form of the diversion and return channels, canals, or ducts, identifying individually the properties which had to be burdened with curtailment.

Weirs have long been part of the river landscape of the Ave River basin (Costa 2008, 2010d). There are numerous examples that reveal the complexity associated with water harnessing in this type of structure. Most of the processes are related to properties for the building of agricultural irrigation weirs (Fig. 4) (Costa 2008, 2010, 2010d; INE 2004).

Water diversion for irrigation of adjacent properties is thus conducted due to the elevation of the backwater area, through structures built on the river bed (Fig. 5).

So as to control the diversion, floodgates were often used (Fig. 6) and intended to reduce impacts on the river bed as well as allow for a larger section of discharge during the flooding periods.

Operating the floodgates depended on the need for water from tributary flows and those requested downstream, as well as the optimisation of the weir's operation (Costa 2008, 2010, 2010d).

The weirs as fixed structures complied with their function to retain water to be diverted, but they could, in some cases, be obstacles to the high discharge flows during the rainy season (winter). In this situation, flooding of surrounding properties was frequent, causing damage to the soils because of less productive substrata deposits and the effects of intensive runoff. Most of the damage to riverside fields occurred in times of overflow, flooding agricultural fields and watermills which caused them to stop working normally (Costa 2008, 2008b). Proprietors were thus forced to introduce

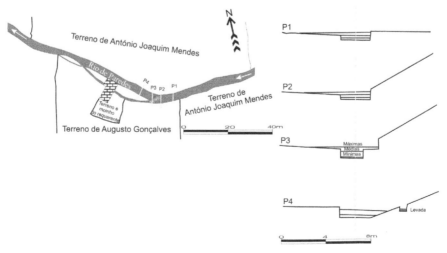

Fig. 4. Plan and cross-section of the construction of an irrigation weir on the Ferro River (Fraga, Armil, Fafe 1903) (Source: APA, adapted).

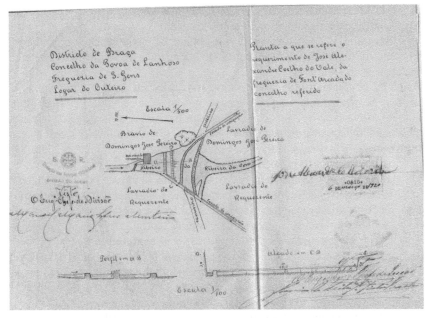

Fig. 5. Plan and cross-section of the construction of an irrigation weir on the Cova Stream (Outeiro, Fonte Arcada, Póvoa de Lanhoso 1920) (Source: APA).

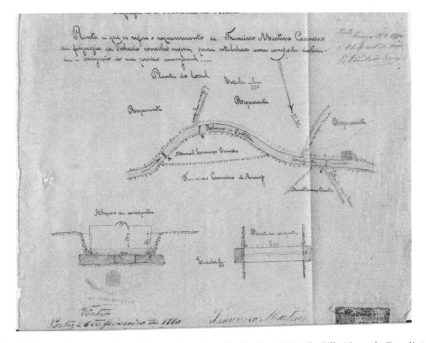

Fig. 6. Plan and cross-section of the construction of a floodgate (Portela, Vila Nova de Famalicão 1909) (Source: APA).

changes to these structures, so as to control the currents either by removing the floodgates (Fig. 7) or by changing the weir's threshold (Fig. 8).

Furthermore, the fixed weirs increase sedimentation on the river beds, impeding their capacity for regulation, which meant the floodgates frequently had to be used to regulate this situation (Fig. 9).

At the beginning of the 19th century, several hydraulic works, which had until then only been used to provide mechanical power, were rapidly converted to set up hydroelectric stations and for industrial power. In the Ave valley and its tributaries, where small industrial units and farms spread over time, small weirs were adapted to the production of hydroelectric power (Cordeiro 1995; Costa 2003, 2004, 2008, 2008a, 2009, 2010, 2010b, 2010d, 2010e; Providência 2003).

The applications for maintenance works on the weirs were quite common and tended to increase in years of flooding. Normally, the maintenance of dams and weirs did not significantly affect water flow. Nevertheless, there were certain types of intervention which changed the weirs' morphometric features, causing significant hydrological impacts. The most common and widely used intervention consisted of raising the weir's crest height, a change that could be permanent or only used in times of draught.

These morphometric changes brought about significant benefits, but could also have negative impacts for the landowners upstream from the weirs or dams. Complaints were thus frequent on the part of farmers and other factory owners located in the vicinity. One way in which to mitigate or avoid these problems was to place floodgates or spillways on the weirs or dams (Costa 2008, 2010d).

The concentration of weirs and dams is evident when we look at the distribution of these structures in 1973, along the Ave River's main course (Fig. 10).

The particular conditions of the Ave River, with one hundred seventeen weirs and six reservoirs, extending for about one hundred kilometres, provoke an altered and

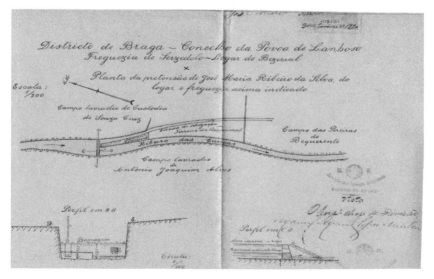

Fig. 7. Plan and cross-sections of the removal of a weir floodgate on the Cuncas Stream (Rego do Moinho, Serzedelo, Póvoa de Lanhoso 1920) (Source: APA).

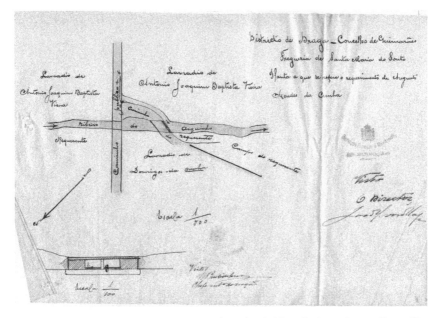

Fig. 8. Plan and cross-section of the process to raise a threshold on the Souto Stream (Souto (Santa Maria), Guimarães 1906) (Source: APA).

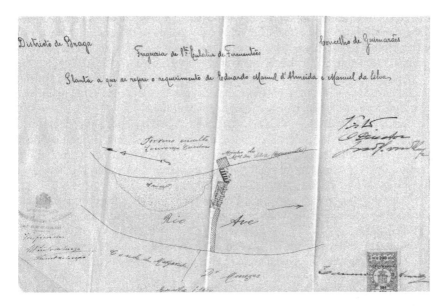

Fig. 9. Plan related to the opening of five floodgates on a watercourse weir on the Ave River (Silvares and Guimarães 1904) (Source: APA).

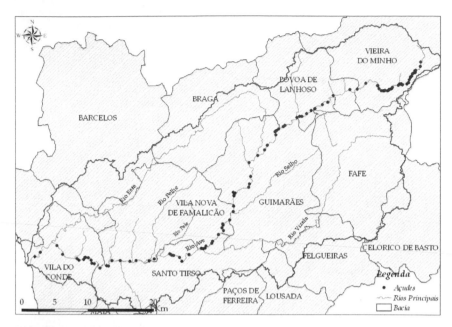

Fig. 10. Location of weirs and dams on the Ave River's main course in 1973 (Source: APA).

dynamic regime, essentially until the confluence of the Vizela River. The local scale of these many structures, associated with the low height of the waterfalls, favours the river's hydrodynamics, caused by a rise in the thalweg's gradient and consequently the currents' speed and their activity (Barbosa et al. 1992; Costa 2008, 2010d).

Apart from the weirs and dams, there are other forms of intervention, which we can define as works on the river banks and, as such, they are considered mainly from the perspective of the watercourse's longitudinal profile.

Based on an analysis of the DPH files, we can find a number of concepts which fall into this type of operation: channelization, building culverts, regulation, altering the river bed, rectifications, and watercourse alignment. Associated to these different forms of intervention, work on walls is fundamental in the relationship between the river banks and beds (Costa 2004a, 2008, 2010c).

The need to establish a physical boundary on the banks implied in most cases constructing retaining walls, built in a rudimentary manner and with local materials (Fig. 11).

Although different terms are used in the applications submitted, works on walls are intended to support and safeguard the banks against the force of the water's currents (Fig. 12).

Retaining walls are primarily intended to avoid landslides to the river bed, thus diminishing the silting process.

It is in the lower sector of the Ave River basin that we find the largest number of applications for retaining walls, as it is the area with the most drainage problems, whether

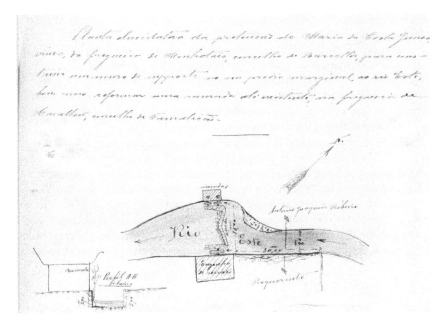

Fig. 11. Plan and cross-section of the construction of a retaining wall on the left bank of the Este River (Cestães, Cavalões, Vila Nova de Famalicão 1908) (Source: APA).

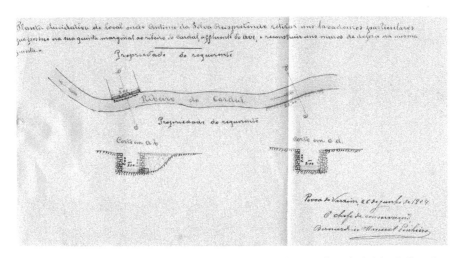

Fig. 12. Plan and cross-section of the reconstruction of a retaining wall on the left bank (Sequeiro, Santo Tirso 1904) (Fonte: APA).

it is due to physical features (morphological and hydrometric features of the drainage basins), or a greater population on the riverside areas is unknown. At the sources of the Ave and Vizela rivers, the high number of applications related to retaining walls derives essentially from geomorphological factors and the need to hold back the land on the banks against water erosion (Costa 2008).

Apart from their retaining and safeguarding functions, building walls is closely associated to regulating most of the watercourses. Here, we should also distinguish between the main types of operations—alterations of the river beds and canalisation.

Much of the work involved in altering the river beds was done in small stretches of the river, almost always less than one hundred metres, and included aligning the current with retaining walls built on the new banks (Fig. 13).

The applications were often restricted to the applicants' properties, intended to increase land for cultivation, by joining plots on each bank. Apart from the agricultural benefits of these interventions, aligning the currents at small, highly pronounced bends in the watercourses, facilitated water drainage, thus improving the river's flow (Fig. 14).

Supressing bends so as to improve drainage and lower the risk of flooding farm lands is also one of the main factors in this type of intervention. According to the regulations, the alignments should maintain the morphometric features of the abandoned channel, that is, the new stretch built should maintain the size of the supressed course,

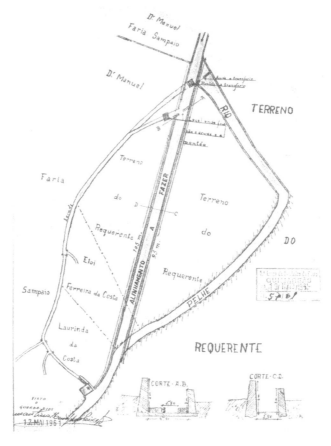

Fig. 13. Plan and cross-section of the process to align the course of the Pelhe River (Ribeira de Cima, Vale (São Cosme), Vila Nova de Famalicão 1961) (Source: APA).

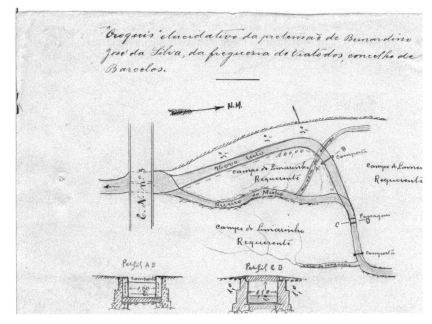

Fig. 14. Plan and cross-sections of the process of altering the course of the Micho Stream (Micho, Viatodos, Barcelos 1918) (Source: APA).

in terms of height, width, and depth. Consequently, the new alignments required side walls to be built, which also guaranteed protection of riverside lands and barriers to flooding (Fig. 15).

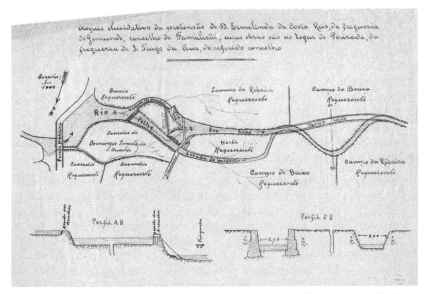

Fig. 15. Plan and cross-section of alignment works on the Pelhe River (Pousada, Cruz, Vila Nova de Famalicão 1918) (Source: APA).

Two consequences resulted directly from building walls on the new river bed: alterations to the river banks and abandonment of the rectified stretch. The change in river bed required action be taken with regard to the abandoned stretch. The simplest and most frequently used solution on farm land was to cover or fill up the old channel (Fig. 16).

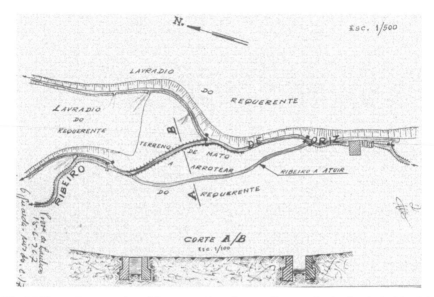

Fig. 16. Plan and cross-section of the construction of a new alignment for the river bed of the Oriz Stream (Ribeira, Gonça, Guimarães 1967) (Source: APA).

Apart from altering the river bed alignment on farm land, this operation was also used in situations related to urban expansion and, more significantly, industrial development. Many factories during initial installation or in times of expansion also had needed to alter watercourses crossing their industrial property. It is here that covering channels became more frequent since, on one hand, it made urban and industrial development possible over watercourses and, on the other, it resolved some public health issues related to bodies of water (Fig. 17).

With varying impacts from a hydrological and environmental perspective, these interventions were mostly local in scope and small scale. They often took place on industrial plots and involved different types of works:

- Expansion – the firm, Brito & Gomes, Lda., on the Passos Stream, in 1939;
- Channelling – the firm, Alfredo Correia da Silva, on the Abelha Stream in 1944; the Fábrica de Malhas de Silvares, Lda., on the Ave River in 1962; Varela Pinto & Companhia Lda., on the Passos Stream in 1963; Sampaio Ferreira & Companhia, on the same river, two years later, and MABOR, on the Reais Stream (Ave) in 1965 (Fig. 18);
- The canalisation accomplished by the Fábrica de Fiação e Tecidos de Pevidém, on the Selho River, in 1939, and the firm, Joaquim Oliveira & Filhos, Lda., on the Vilamão Stream, in 1964;

- Regulation of the banks and bed of the Este River by the firm, Henrique Buero e Costa, in 1923;
- Alteration of a stretch of the Passos Stream, by Varela Pinto & Companhia Lda., in 1973, and by MABOR, on a stretch of the Ave River, in 1943.

The use of different concepts raises some confusion since, as we have seen, they all fall under the same type of works, whether on the river bed or the river banks. One of the larger projects involved regulating the Este River in the city of Braga (Fig. 19).

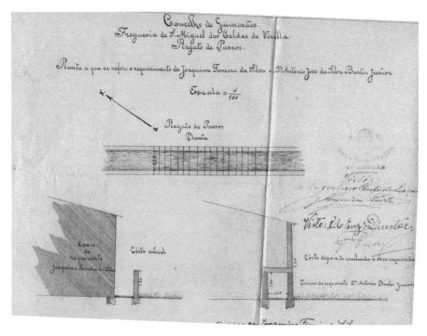

Fig. 17. Plan and cross-section of the process of covering the Passos Stream (rua Pereira Caldas, Caldas de Vizela (São Miguel), Guimarães 1919) (Source: APA).

Fig. 18. Cross-sections of the channelling process of the Reais Stream (Salgueiro, Lousado, Vila Nova de Famalicão 1965) (Source: APA).

This project was first planned at the beginning of the 1950s, in response to an application submitted to the Ministry of Public Works by the Braga City Hall, which was approved by Order on 11th August 1950. Heavy silting in the Este River downstream

Fig.19. Plan of the regulating project of the Este River in Braga (1959) (Source: APA).

from the São João Bridge caused deficient drainage because of its weak slope and direction, which caused frequent flooding of riverside lands. The works intended to improve drainage between the two existing agricultural plots, at each end of the stretch of river considered. The stretch would be altered into two straight alignments, joined by a wide bend in the river. This regulation project for the Este River was approved in 1959, and involved several ground levelling works, building retaining walls, and placing aprons on the river bed to prevent erosion, as well as raising some stretches of the river to increase their discharge section (Costa 2008). The Este River has had several other interventions on its passage through Braga's urban centre up to the present day.

In order to identify the possible effects resulting from the regulating works, namely canalisation of the river beds, and thus make sustained decisions with regard to licencing, the Directorate-General of the Water Services issued a circular in 1968 which identified the following elements as requirements to evaluate the respective projects (Costa 2008):

1. The delimitation of the river basin and the respective area;
2. Evaluation of the flood discharge foreseen;
3. The longitudinal profile of the current, at least 300 metres upstream and downstream; large-scale works or river beds with low gradients should cover a greater extension, up to 1000 metres;
4. The transversal profiles of the river bed and the stretch corresponding to the longitudinal profile mentioned in point 3. These profiles could be surveyed every 100 metres, at minimum widths of 5 metres on each bank.

Apart from this set of technical characteristics, these projects should also include information on the levels of flooding known in the location, the risks and inconveniences resulting from the floods, or the possibility of rising backwater areas, resulting from the works planned. The hydraulic studies accomplished provided a better understanding of the river dynamics of some of the small drainage basins and served to better justify the decisions made with regard to applications for canalisation licencing.

Related to this circular, we can give the example of the information provided by the Hydrology Division on the calculations of the flood discharge to be considered in the building of aqueducts on the Mouta Stream in 1972. It is a small basin of $1.6\ km^2$, located on the left bank of the Pelhe River (Vila Nova de Familicão). To determine maximum rainfall, the hydrology division, based on data from the pluviometric station of Barcelos, used the corresponding curve of pluviometric possibility to determine the maximum probable rainfall in the time interval corresponding to the time of concentration – 1.6 hours – extrapolating to 100, 50, 25, 10, and 5 years (Table 1):

Table 1. Discharge flows to the Mouta Stream (Vizela 1972) (Source: APA).

Frequency	m³/s	m³/s/km²
Once every 100 years	17	10,5
Once every 50 years	16	9,9
Once every 25 years	14	8,8
Once every 10 years	11	6,8
Once every 5 years	9	5,6

The crossing of a diversity of information (based essentially on meticulous fieldwork) with the several hydrographic and hydraulic models served to characterise the floods and to define and delimit the flood areas. By that time, Portuguese legislation established the delimitation of flood areas, on a case-by-case basis, which it called adjacent areas, subdivided into prohibited and conditioned built areas (decree-law No. 468/71 of 5th November). The identification of a flood area corresponding to the centennial discharge is usually the one that matters the most in legal terms to define a flood river bed. It is in this setting that hydraulic studies appear, particularly focused on the frequency of the main floods, as well as the references to the main morphometric indices.

It is possible to conclude that the options included mainly structural measures, which raises some questions about the sustainability of the river system in these locations.

Bridges—a potential problem for the drainage of water runoff

Among the different types of works carried out on river beds and river banks, we must also consider the construction of infrastructures, with particular focus on bridges and aqueducts. The building of pillars (Fig. 20) or spans (Fig. 21) on bridges often affected the discharge section and thus the river flow (Costa 2004a, 2008).

During flooding, the stability of bridge structures was often at risk. The great floods of 1909 left many enlightening reports on these situations, particularly with regard to the inspections carried out on the bridges affected (Costa 2008).

The structures' frailty and many of the bridges' poor designs in terms of spans were responsible for some situations in which they collapsed or were destroyed:

- In 1916, a four-span stone bridge fell on the Macieira Stream. The spans' covers hindered the current and the Barcelos City Hall had to raise the entire pavement 0.5 meters so as not be affected by regular flooding;
- In 1928, a wooden bridge over the Vizela River, a former public passageway, whose ruined state posed a risk to traffic, was repaired by the Santo Tirso City Hall;
- In 1937, an improvised bridge on the Ave River, in Póvoa de Lanhoso, collapsed, a situation which was to be repeated in years of flooding;
- In 1938, a project was submitted to repair the Reboto Bridge and its access ramps over the Selho River. Given its state of disrepair and insufficient discharge section, the fields to which it was connected became completely unusable in times of flooding.

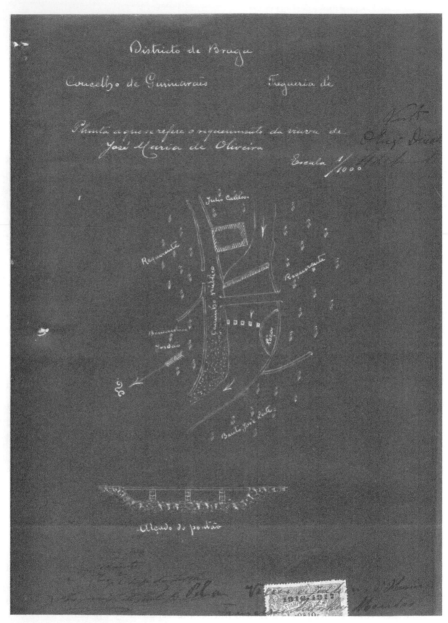

Fig. 20. Cross-section of the maintenance works on a stone bridge over the Couros Stream (Poça de Relhos, Guimarães (São Sebastião), Guimarães 1917) (Source: APA).

So as to mitigate these types of problems, it became essential to reinforce the existing structures and many applications were submitted to this end (Fig. 22).

The concern with floods and their impact on the bridges' structures meant that many projects had to include calculations regarding the flood discharge foreseen. Until 1968, the bridge projects did not require a hydraulic study, but it became mandatory when

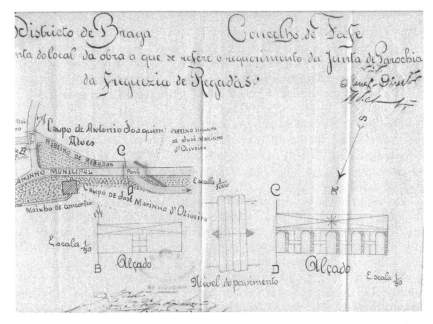

Fig. 21. Plan and cross-sections of the construction of two passageways over the Regadas Stream (Moinhos do Bairro do Rego, Regadas, Fafe 1917) (Source: APA).

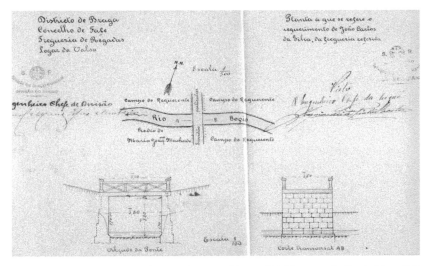

Fig. 22. Plan and cross-sections of the works to modify an old bridge over the Bugio River (Valsa, Regadas, Fafe 1920) (Source: APA).

Decree No. 48373 was passed on the 8th of May. Nevertheless, some projects already included calculations on the discharge section for larger-scale works. The project to build a pontoon over the Selho River submitted by the company Francisco Inácio da Cunha Guimarães & Filhos included a number of calculations in its brief and specifications:

"(…) the Brandão Bridge on Trunk Road No. 310 (…) has three aqueducts (…) with 9.2 m² of total usable section. Given the bridge has served its purpose and because upstream we have an added section of 13.22 m² in our favour (…) the characteristics given to the pontoon's deck means that, when exceptional flooding occurs, the structure will be able to resist the force of the water, seeing as a beamless deck has been used (…)". As we can see, there is in fact concern with the safety of the work with regard to flooding.

The Este River is undoubtedly the watercourse that has been the most studied in this regard (Costa 2008). In 1966, following an application from Grundig for several works related to its premises, a survey was conducted on the agricultural pontoons and easements along the Este River, between the company and the Torrão Bridge in Priscos (Fig. 23).

Following the calculations of the flood discharges, the DHD concluded that only four of the existing twenty-seven structures allowed for the reasonable discharge of the Este River and, consequently, the others required expansion so as to be served with sufficient discharge sections. On this matter, a report was issued on the 9th of October, 1968, so that qualified technicians could take note of the unitary discharge rates and take them into consideration in the study of all of the Este River's problems, particularly with regard to the possibility of substituting the bridges with insufficient discharge sections. This recommendation would become valid for all the rivers on which any motive hydrological studies similar to those of the Este River were to be conducted.

At the time, several hydrological studies were in fact conducted on the bridges which directly affected the public waters system. These studies gave an important contribution to knowledge of the hydrological systems at the local level, since they were intended to provide the most appropriate solutions: even when applying different

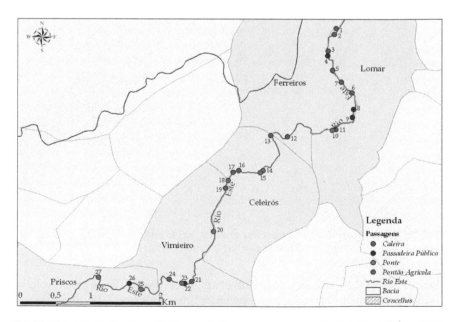

Fig. 23. Plan with the location of bridges and other passageways over the Este River (Braga 1966) (Source: APA).

methodologies in their implementation, the hydraulic studies serve to show that the decisions made implied safety works so as to minimise the risks associated with extreme hydrological events.

Generally speaking, the different works accomplished by the DPH were mostly small-scale and, consequently, their impact was local in scope. An exception was made of the regulation of the Este River in the urban part of the city of Braga. We cannot, however, neglect to mention that the conjugation of a high number of works involving the different types of intervention must have undoubtedly had detrimental consequences, which are still felt today, as well as explaining many of the hydrological and hydraulic problems that affect numerous watercourses of Ave River basin.

Conclusion

Archives are indispensable and are indubitably built from and are builders of historical memory and of human action over the centuries. In this context, researchers play a role in sharing responsibilities, given that they have to ensure not only access to the documental collections, but also, and above all, they have to ensure the necessary conditions for their conservation, storage, dissemination, and construction of collective memory (Veiga 2010). The mission to safeguard and transmit collective institutional memories bestows on historical archives a symbolism and a strategic role in the edification of the identity of a nation and of a region (Barros 2007).

The wealth of documentation in the APA archive and its potential for research can serve as the basis for a diversity of studies, not only in the field of geography, but in the historical sciences, heritage studies, planning and in some fields of hydraulics and hydrology connected to the spatial management of riverside areas.

The historical collection of the SHD is one of the richest local archives in the country, gathering thousands of documents dating from the 19th century to the present day. It represents a documental repository with unique characteristics and is a testimony to the identity, memory, and history of the Water Services (Costa and Cordeiro 2012a). To make this collection better known will mean that essential information on the coastal areas and their management is made available and processed.

As noted by Certeau (2002), we cannot forget that a reading of the past, no matter how controlled it is by the analysis of documents, is always directed at a reading of the present. The SHD archive has enormous potential for research, which means that we are also responsible for publicising this important heritage in the defence of a common memory which must be safeguarded (Costa 2012; Costa and Cordeiro 2012:9, 2012a, 2015:2; Costa et al. 2015:62).

Acknowledgments

This work is funded by FEDER funds through the Operational Programme Competitiveness Factors - COMPETE and national funds by FCT - Fundação para a Ciência e a Tecnologia under the project UID/GEO/04084.

Keywords: Ave river, Archives, Portuguese Environmental Agency, Channels Interventions.

References

Bacellar, C.A.P. 2005. Fontes documentais uso e mau uso dos arquivos. pp. 23–80. *In*: Pinsky, Carla B. (Org.) Fontes Históricas. São Paulo: Contexto.

Barbosa, J.N., Couto, A.F. and Valente, J.T. 1992. Os aproveitamentos hidroeléctricos da bacia hidrográfica do Rio Ave, Recursos Hídricos 13(3): 15–19.

Barros, F. 2007. Arquivos históricos nos dias de hoje: aliciantes desafios, múltiplos papéis, 9.º Congresso Nacional BAD, 2007.

Camargo, F. As fontes históricas. Consultado em 17 de abril de 2015.

http://filoinfo.net/disciplinasonline/pluginfile.php/3041/mod_resource/content/1/AS%20FONTES%20HIST%C3%93RICAS.pdf.

Campelo, A. 2011. Das hidráulicas aos recursos hídricos: história, sociedade e saber. Edições ARH do Norte, I.P. Porto, 2011, 139 p.

Carvalho, R. 2009. Historiador e as Fontes Históricas. Consultado em 15 de junho de 2015.

http://www.webartigos.com/artigos/o-historiador-e-as-fontes-historicas/22598/

Certeau, M. 2002. A escrita da História. 2 ed. Rio de Janeiro: Forense Universitária.

Cordeiro, J.M.L. 1995. Indústria e paisagem na bacia do Ave, Cadernos do Noroeste, Vol. 8, nº2, Braga, pp. 47–68.

Costa, F.S. 2003. O rio Ave no início do século XX: uma perspectiva segundo os aproveitamentos hidroeléctricos, Actas do II Simpósio dos Aproveitamentos Hidroeléctricos, Vila Real, 13 p.

Costa, F.S. 2004. Os aproveitamentos hidráulicos e hidroeléctricos do rio Ave no período 1902–1936, Actas do 7º Congresso da Água, Lisboa, 15 p.

Costa, F.S. 2004a. As águas públicas na bacia do Ave: Uma perspectiva do ordenamento do território no início do século XX, Actas do 7º Congresso da Água, Lisboa, 14 p.

Costa, F.S. 2008. A Gestão das Águas Públicas: o caso da bacia hidrográfica do rio Ave no período 1902–1973. Dissertação de doutoramento em Geografia, Universidade do Minho, Braga, 857 p.

Costa, F.S. 2008a. A Central Hidroeléctrica de Santa Rita – Um contributo para a história da sua implantação. Dom Fafes, revista cultural, Ano XIII, nº 13/14, Câmara Municipal de Fafe, Fafe, pp. 83–97.

Costa, F.S. 2008b. O papel dos moinhos no aproveitamento hidráulico das águas públicas do rio Ave – Um contributo na perspectiva do património ligado à água. Atas do VII Colóquio Ibérico de Estudos Rurais "Inovação e Território", 23 a 25 de Outubro, Escola Superior Agrária de Coimbra (ESAC), Coimbra, 23 p. Edição em CD-ROM.

Costa, F.S. 2008c. Hidro-conflitos na bacia hidrográfica do rio Ave – uma análise a partir das transgressões cometidas no período 1902–1973. Atas do XI Colóquio Ibérico de Geografia "A perspectiva Geográfica face aos novos desafios da sociedade e do ambiente no contexto ibérico", 1 a 4 de Outubro, Universidade de Alcalá de Henares, Espanha, 15 pp.

Costa, F.S. 2009. A indústria têxtil na bacia hidrográfica do rio Ave – uma perspectiva segundo as fábricas de fiação e tecidos, numa relação historicamente sustentada pelo Domínio Público Hídrico. XIX Encontro da APHES "Memória social, patrimónios e Identidades, Porto, 14 de Novembro, 16 pp.

Costa, F.S. 2010. Águas públicas e sua utilização no concelho de Fafe - Um contributo do ponto de vista histórico-geográfico. Câmara Municipal de Fafe, Fafe, 144 p.

Costa, F.S. 2010a. O ciclone de 1941 e os prejuízos causados na sua passagem pela bacia hidrográfica do rio Ave – Uma perspectiva a partir do relatório do chefe da 2ª Secção da 1ª Direcção Hidráulica do Douro. pp. 121–131. *In*: Moisés de Lemos Martins (org.), Caminhos nas Ciências Sociais Memória, mudança social e razão – Estudos em homenagem a Manuel da Silva Costa, Centro de Estudos Comunicação e Sociedade, Universidade do Minho, Grácio Editor, Coimbra,

Costa, F.S. 2010b. O património industrial no vale do Ave. O têxtil como chave de leitura territorial. pp. 349–368. *In*: Gonçalves, Eduardo C. (ed.). Dinâmicas de Rede no Turismo Cultural e Religioso. Maia: Ed. ISMAI e CEDTUR – CETRAD, Vol. II.

Costa, F.S. 2010c. Domínio Público Hídrico na bacia hidrográfica do rio Ave – uma breve perspectiva histórica. 1º Seminário sobre Gestão de Bacias Hidrográficas "As Regiões do Norte e as

Perspectivas Futuras de Gestão", Associação Portuguesa dos Recursos Hídricos – Núcleo Regional do Norte, Associação Portuguesa dos Recursos Hídricos – Núcleo Regional do Norte, pp. 111–116.

Costa, F.S. 2010d. Geopatrimónio ligado à água. O caso do património industrial na bacia hidrográfica do rio Ave. Atas do VI Seminário Latino-Americano, II Seminário Ibero-Americano de Geografia Física "Sustentabilidade da Gaia: ambiente, ordenamento e desenvolvimento", CEGOT, Universidade de Coimbra, 26 a 30 de Maio de 2010, 12 p.

Costa, F.S. 2011. Licenciamento em águas públicas e cartografia – O caso do rio Ave no início do século XX. pp. 593–602. *In*: Norberto Santos e Lúcio Cunha, Trunfos de uma geografia ativa, Imprensa da Universidade de Coimbra, Coimbra.

Costa, F.S. 2011a. Aproveitamentos hidráulicos no rio Ave – uma cascata de pequenas barragens. 2º Seminário sobre Gestão de Bacias Hidrográficas "Reabilitação e utilização da rede hidrográfica", Associação Portuguesa dos Recursos Hídricos – Núcleo Regional do Norte, Associação Portuguesa dos Recursos Hídricos – Núcleo Regional do Norte, pp. 1–6.

Costa, F.S., Nossa, P.N.S., Magalhães, S.C.M. and Magalhães, M.A. 2011. A legislação dos recursos hídricos em Portugal e no Brasil – Uma análise histórica comparativa. XIV Congresso Mundial da IWRA e 10º Simpósio de Hidráulica e Recursos Hídricos dos Países de Língua Portuguesa "Adaptative Water Management: Looking to the Future", Porto de Galinhas, Recife, Brasil, 6 p.

Costa, F.S. 2012. O arquivo da Administração da Região Hidrográfica do Norte. Roteiro metodológico. pp. 267–293. *In*: Manuela Martins, Isabel Vaz de Freitas, Maria Isabel Del Val Valdivieso (Coords.). Caminhos da água. Paisagens e usos na longa duração., CITCEM-Centro de Investigação Transdisciplinar "Cultura, Espaço e Memória", Braga.

Costa, F.S. 2012a. Poluição e Domínio Público Hídrico. Um contributo histórico para o estudo da bacia hidrográfica do rio Ave. 11º Congresso da água "Valorizar a água num contexto de incerteza", Porto, 6 a 9 de Fevereiro de 2012, Associação Portuguesa de Recursos Hídricos, 10 p.

Costa, F.S. and Cordeiro, J.M.L. 2012a. O CEDOCAVE - Centro de Documentação sobre Água no Cávado e Ave: um projeto para preservar a memória e divulgar o património e cultura da água. Atas das VIII Jornadas de Geografia e Planeamento "Cidades, criatividade(s) e sustentabilidade(s)", Coleção atas nº 2, Departamento de Geografia, Universidade do Minho, Guimarães, pp. 21–28.

Costa, F.S. and Cordeiro, J.M.L. 2012a. O CEDOCAVE - Centro de Documentação sobre Água no Cávado e Ave: um projeto para preservar a memória e divulgar o património e cultura da água. Atas das VIII Jornadas de Geografia e Planeamento "Cidades, criatividade(s) e sustentabilidade(s)", Cordeiro, J.M.L. (2001) – Indústria e energia na Bacia do Ave: [1845–1959], Cadernos do Noroeste, Série História, Nº1 (2001), Braga, pp. 57–174.

Costa, F.S. and Cordeiro, J.M.L. 2015. Archiv-AVE "Património documental da bacia do Ave. *In*: Costa, F.S., Cordeiro, J.M.L., Vieira, A.A.B. and Silva, C.C.S. (eds.), UMinhoDGEO, Departamento de Geografia da Universidade do Minho, Guimarães, 52 p.

Costa, F.S. and Cordeiro, J.M.L. 2015a. O Sistema de Informação Arquivística da Agência Portuguesa do Ambiente (SIAPA) – um projeto para recuperar a memória dos Serviços Hidráulicos. pp. 63–67. *In*: Membiela, P., Casado, N.C. and Cebreiros, M.A. (eds.). Panorámica interdisciplinar sobre el agua. Educación Editora, Ourense, Espanha.

Costa, F.S., Cordeiro, J.M.L., Vieira, A.A.B. and Silva, C.C.S. 2015. Archiv-Ave: um projeto para conservar e divulgar o património documental do rio Ave, pp. 50–63. *In*: António Vieira and Francisco Costa (eds.). II Simpósio de Pesquisa em Geografia, Universidade do Minho – Universidade Federal de Santfa Maria, 27 e 28 de maio de 2015, Guimarães, Coleção Atas, 4, UMinhoDGEO, Departamento de Geografia da Universidade do Minho, Guimarães.

Direcção Regional de Ambiente e do Ordenamento do Território-Norte. 2000. Plano de Bacia Hidrográfica do Rio do Ave, 1ª Fase, Volume I, Porto, 95 p.

Figueira, J.M. 2003. A importância da energia eléctrica para o surto da industrialização no Vale do Ave, pp. 196–217. *In*: Património e Indústria no Vale do Ave, um passado com futuro, Rota do Património Industrial do Vale do Ave, ADRAVE – Agência de Desenvolvimento Regional do Vale do Ave, S.A.

INE. 2004. Classificação Portuguesa das Construções (CC-PT), Lisboa, Instituto Nacional de Estatística, 22 p.

Ketelaar, E. 2004. Time future contained in time past. Archival science in the 21st century. Journal of the Japan Society for Archival Science 1: 20–35.

Martins, E. 1997. As fontes documentais: análise da vida quotidiana e elementos para a história social e educativa. Coleção - Cadernos do Projecto Museológico sobre Educação e Infância. Nº55. p. 5–20.

Milligan, J.D. 1979. The Treatment of an Historical Source. History and Theory 2: 177–196.

Pato, J. 2008. O Valor da Água como Bem Público. Tese de doutoramento em Ciências Sociais, Instituto de Ciências Sociais da Universidade de Lisboa.

Prado, E.L. 2010. A importância das fontes documentais para a pesquisa em História da Educação. InterMeio: revista do Programa de Pós-Graduação em Educação, Campo Grande, MS, v.16, n.31, jan./jun. 2010, pp. 124–133.

Providência, P. 2003. Um olhar sobre o Vale do Ave, in Património e Indústria no Vale do Ave, um passado com futuro, Rota do Património Industrial do Vale do Ave, ADRAVE – Agência de Desenvolvimento Regional do Vale do Ave, S.A., pp. 142–147.

Veiga, F.C. 2010. Os arquivos históricos na sala de aula: os documentos no processo ensino-aprendizagem. Atas do V Encontro de Pesquisa em Educação de Alagoas (Epeal), Programa de Pós-Graduação em Educação do Centro de Educação (PPGE) da Universidade Federal de Alagoas, 14 p.

Wasson, J.-G., Malavoi, J.-R., Maridet, L., Souchon, Y. and Paulin, L. 1998. Impacts écologiques de la chenalisation des rivières, Études Gestion des milieux aquatiques, Cemagref Ed., Lyon, 158 p.

CHAPTER 19

Sediment Yield in Different Scales in a Semiarid Basin

The Case of the Jaguaribe River, Brazil

José Carlos de Araújo,[1],* *Everton Alves Rodrigues Pinheiro,*[1]
Pedro Henrique Augusto Medeiros,[2] *José Vidal de Figueiredo*[2]
and *Axel Bronstert*[3]

INTRODUCTION

Erosion, sediment transport, and deposition are environmental issues that affect societies, especially due to agricultural productivity depletion, desertification, augmenting of flooding hazards, and reduction in water availability (Bormann et al. 2011; Baker et al. 2013; Navarro-Hevia et al. 2014). Regarding semiarid regions, these problems are even bigger, since dry spells make water resources very limited, nevertheless, most of the river basins in semiarid regions are ungauged and hence reliable data are scarce (Costa et al. 2013; Zende and Nagarajan 2014). For these environments, therefore, modelling can be a very useful tool (Medeiros et al. 2010), as long as models are supported by temporal representative monitoring programs (Bronstert et al. 2014). Despite the data deficiency of water-scarce basins, some relevant investigations based on field measurements can be identified (Lima Neto et al. 2011; Rhoton et al. 2011; Cavalcante and Cunha 2012; Nichols et al. 2013; Vanmaercke et al. 2015; Geeraert et al. 2015; de Figueiredo et al. 2016).

When sediment yield measurements of large basins are plotted against the respective drainage areas, the general trend is negative, that is, it decreases as the catchment area increases (Owens and Slaymaker 1992). According to Xu and Yan (2005), this pattern is mainly because (i) small upstream catchments are often highly erosive due to their

---

[1] Federal University of Ceará – UFC, Department of Agricultural Engineering, Fortaleza, CE 60.356-000, Brazil.
[2] Federal Institute of Education, Science and Technology of Ceará – IFCE, Brazil.
[3] University of Potsdam, Department of Hydrology and Climatology, 14476 Potsdam, Germany.
* Corresponding author: jcaraujo@ufc.br

steep relief; (ii) in small catchments, single intensive precipitation events can cover the whole area, whereas in large catchments this is unlikely to occur; and (iii) as the catchment area enlarges, the silting opportunities generally rise due to decreasing slope gradients and increasing floodplain sedimentation. However, Church and Slaymaker (1989) have shown, for a British Columbia (Canada) basin, that allometry was positive for catchments up to 30,000 km², decreasing afterwards. For river-erosion dominated basins, positive allometry has been identified throughout the world, considering that the deposition zones are less important. De Vente and Poesen (2005) identified positive allometry in Mediterranean basins, possibly caused by the increase of erosive processes on the hillslopes and on the river margins. Xu and Yan (2005) also found positive allometry for the Yellow River Basin, China, and according to the authors, a large amount of sediment has been deposited on the alluvial network in previous ages. This explains the present spatial behaviour of sediment yield in that basin. De Araújo and Knight (2005) analyzed the sediment yield of more than 300 basins and verified that no clear allometry could be identified when the data were clustered by continent, except for Africa, where negative allometry has been found. Vanmaercke et al. (2011) studied the phenomenon in Europe, finding a very weak relation between sediment yield and catchment area, including the Mediterranean mountainous basins. Despite the scientific relevance of this issue, such investigations are – to our best knowledge—inexistent for South American basins, especially in its tropical semiarid zones.

The objective of this manuscript is to analyze the sediment yield in the 75,000 km² semiarid basin of the Jaguaribe River (JRB) in the north-eastern Brazil, based on measurements from 10^{-4} to 10^{+4} km² using different methods. The main scientific challenges tackled in this research included establishing the spatial pattern of sediment yield in the basin, the behaviour of the sediment yield as a function of several attributes (geology, status of vegetation conservation and monitoring method), and the existence (or not) of positive allometry.

Material and Methods

Description of the study area

The focus area of this investigation is the 75,000 km² semiarid Jaguaribe River Basin, located in the northeast of Brazil (Fig. 1). The climate is tropical semiarid with average annual precipitation of 600 mm and average annual potential evaporation of 2,500 mm. The rainy season lasts five months (January to May), with average temperature of 25°C (Medeiros and de Araújo 2014). All its rivers are naturally intermittent. Runoff typically varies from 10 to 60 mm per year and hortonian events prevail, generating a transport-limited sediment environment (Medeiros et al. 2010). Geologically, the area is composed by a sedimentary basin in the south (Fontenele et al. 2013) and in the north (Cavalcante and Cunha 2012), with a large crystalline zone in the centre extending to the extreme west, where the prevailing soils are shallow with rock fragments (Creutzfeldt 2006). In the high sedimentary plateaus, however, soils are mostly deep and permeable. In the crystalline zone, there is the *Depressão Sertaneja*, a vast erosion surface generated by severely dry tropical conditions which was formed during the Cenozoic, more specifically during the Pleistocene. Its relatively flat surface has evolved from the lateral degradation of hillslopes by sheetflood. The Jaguaribe River Basin (JRB) is integrally

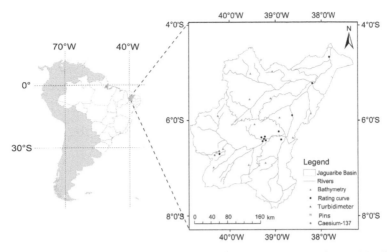

Fig. 1. Geographical location of the Jaguaribe River Basin (JRB), Brazil, as well as the spatial distribution of the 27 sediment-yield assessment sites analysed in this research.

inserted into the Caatinga biome, characterized by deciduous vegetation capable of adapting to the extreme conditions caused by the long dry season and by the recurrent droughts (Pinheiro et al. 2013a; Pinheiro et al. 2016).

Measurement sources

The sediment-yield analysis was based on 27 measurement sites using different methods and scales, which are explained below. Table 1 presents the main information concerning the measurements, which include surveys conducted in the context of this research, as well as secondary data.

Pins

As demonstrated by Hancock and Lowry (2015), erosion pins can provide a reliable data on hillslope erosion, producing data which are within the range measured using other established methods. A small upstream 68 m² hillslope, located within the Experimental Aiuaba Basin has been monitored since 2008. On the hillslope, which has steepness of 15% and is covered with preserved native Caatinga over the sedimentary geological basement, there are 5-mm accurate erosion pins at a rate of one pin per square meter. The measurements have been made seasonally, that is, before and after the rainy season.

Caesium-137

Sediment yield has been assessed on a 0.24 km² hillslope, on which soil samples were analyzed concerning the ¹³⁷Cs inventory (Medeiros et al. 2014). In turn, the ¹³⁷Cs reference inventory had been determined through a soil sample taken from an area approximately 5 km from the studied hillslope, with relatively high altitude, very low surface gradient, with negligible erosion or deposition since the peak of ¹³⁷Cs fallout

Table 1. Sediment-yield assessment sites in the Jaguaribe River Basin, with the respective catchment area, measurement method, sediment yield (SY), duration of the measurement representativeness (D) and data source.

Identification	Area (km^2)	Method	SY (t km^{-2} yr^{-1})	D (yr)	Source(s)
Small hillslope	6.80x10^{-5}	Pins	228	4	This research
Micro-basin PC[1]	1.15x10^{-2}	Rating curve	165	3	Santos (2012)
Micro-basin grass	1.20 x10^{-2}	Rating curve	198	3	Santos (2012)
Micro-basin TC[2]	2.06 x10^{-2}	Rating curve	124	3	Santos (2012)
Large hillslope	2.40x10^{-1}	Caesium-137	125	46	This research
Micro-basin corn	2.80x10^{-1}	Rating curve	150	3	Santos (2012)
Canabrava reservoir	2.85x10^{0}	Bathymetry	690	57	de Araújo (2003)
Aiuaba Exp Basin	1.20x10^{+1}	Rating curve	25	6	This research
Junco reservoir	1.82x10^{+1}	Bathymetry	364	52	COGERH
Ingarana reservoir	2.00x10^{+1}	Bathymetry	21	66	COGERH
S Joaquim reservoir	2.98x10^{+1}	Bathymetry	1,340	63	This research
Berilópolis reservoir	3.90x10^{+1}	Bathymetry	1,080	51	COGERH
Marengo reservoir	7.50x10^{+1}	Bathymetry	487	54	This research
Altamiro reservoir	7.56x10^{+1}	Bathymetry	120	61	COGERH
Quincoe reservoir	1.54x10^{+2}	Bathymetry	1,259	19	COGERH
N Floresta reservoir	1.60x10^{+2}	Bathymetry	227	82	COGERH
Cedro reservoir	2.20x10^{+2}	Bathymetry	1,277	95	de Araújo (2003)
Umbuzeiro River	8.00x10^{+2}	Turbidimeter	10[3]	2	This research
Poço Pedra reservoir	9.01x10^{+2}	Bathymetry	467	52	COGERH
Benguê reservoir	9.33x10^{+2}	Bathymetry	351	5	This research
V do Boi reservoir	1.22x10^{+3}	Bathymetry	122	47	de Araújo (2003)
Jaguaribe section I	1.19x10^{+4}	Rating curve	4.3	4	Cavalcante (2012)
Jaguaribe section II	2.26x10^{+4}	Rating curve	1.9	1	Cavalcante (2012)
Jaguaribe section III	2.46x10^{+4}	Rating curve[4]	148	25	Lima Neto et al. (2011)
Orós Reservoir	2.57x10^{+4}	Rating curve	206	49	Lima Neto et al. (2011)
Jaguaribe section IV	3.98x10^{+4}	Rating curve	1.1	8	Cavalcante (2012)
Jaguaribe section V	4.80x10^{+4}	Rating curve	1.5	9	Cavalcante (2012)

[1]Preserved Caatinga; [2]Thinned Caatinga; [3]The measurements have only been made in 2011 (meteorologically regular year) and in 2012 (severe drought), which led to a low sediment yield; [4]Suspended- and bed-load

from the atmosphere (year 1963). For the soil redistribution estimations, the hillslope had been divided into five zones according to the toposequence, and for each zone, a mixture of five randomly collected soil samples was taken as representative. The composite soil sampling approach had been used in order to minimize the effects of localized features (such as flow convergence or divergence) on the results. Erosion and sediment deposition had been quantified with the proportional model, which assumed that there is complete mixing of ^{137}Cs in the top soil layers, and that the lateral soil redistribution is directly proportional to the gain or loss of the isotope (Walling and He 1999).

Rating curves

The measurements are based on suspended-sediment rating curves (Table 1) that encompassed eleven sites, of which six are located in the Jaguaribe River, four in the Iguatu micro-basins, and one in the Aiuaba Experimental Basin. The Jaguaribe River measurements were accomplished in five sections (with contribution areas ranging from 25- to 48-thousand km^2 and monitoring period from one to 25 years), as well as in the Orós reservoir (Lima Neto et al. 2011; Cavalcante and Cunha 2012). One should note that Lima Neto et al. (2011) have considered the daily suspended load, as well as the daily bedload, for a 25-year period. The four experimental micro-basins located in Iguatu, close to the Orós reservoir, have different land uses and have been intensively monitored since 2009 (Santos 2012), whereas the Aiuaba Experimental Basin has been monitored since January 2003 (de Figueiredo et al. 2016). Both experimental areas were set to help understand the prevailing processes concerning water and sediment fluxes in the Caatinga biome.

Turbidimeter

A turbidimeter, together with automatic rainfall and river-discharge gauges, has been installed in the Aroeira section of the Umbuzeiro River (800 km^2), Upper Jaguaribe Basin. The turbidimeter was calibrated using both direct river sediment, as well as sediment deposited in a 6,000 m^3 pond located at the outlet. The measurements presented in this paper (Table 1) corresponded to the events of the years 2011 and 2012 (Pinheiro et al. 2013b).

Bathymetric surveys

Three bathymetric surveys were done in the context of this research (Benguê, Marengo, and São Joaquim reservoirs), whereas eight were performed by the Ceará State Water Management Company (COGERH), as shown in Table 1. In order to assess the sediment yield, Eq. (1) has been used:

$$SY = \frac{\Delta V \cdot \rho}{\eta \cdot \Delta t} \tag{1}$$

Where ΔV corresponds to the volume loss by silting between two bathymetric surveys, Δt is the period between the surveys, ρ is the dry bulk density of the deposited sediment and η is the reservoir trap efficiency, estimated using the Brune curve.

Attributes

The sediment yield assessed in the different sites was plotted against the respective catchment areas clustered by some relevant attributes, such as geology, conservation status of the vegetation (Caatinga), and measurement method.

In order to identify the prevailing geological stratum for each catchment area, the state geological map provided by FUNCEME was used. The vegetative preservation status was admitted as low (less than 40%), moderate (40%–80%), or high (higher than 80%), depending on the occurrence of preserved Caatinga in the respective catchment.

To identify the preserved Caatinga, satellite imagery was used. The calibration process consisted simply of the comparison of image standard with ground-truthing works done by Creutzfeldt (2006) in the Upper and by Cavalcante and Cunha (2012) in the Lower Jaguaribe River Basin.

Results and Discussion

Measurements

The sediment yield measurements in the semiarid Jaguaribe Basin, ranged from 10^0 to 10^3 t km^{-2} yr^{-1}. For the 68-m^2 hillslope monitored by pins, the sediment yield (SY) was 228 t km^{-2} yr^{-1} Boardman et al. (2015). Using erosion pins to estimate sediment yield of badlands had a significant relationship with measurements based on ^{137}Cs and on exported sediment, confirming the method's robustness. The sediment yield reported for the hillslope, which has high erodibility, is likely to be settled on the secondary banks of the local brook, not reaching the main stream. This feature might explain the lack of connectivity during the non-extreme events, although Harel and Mouche (2014) state that higher runoff does not necessarily imply higher connectivity. However, whenever floods cover the secondary bank—as in 2004—these sediments are expected to be washed out during the ascending limb, causing hysteresis.

Application of the ^{137}Cs technique indicated erosion rate of 500 t km^{-2} yr^{-1} in the large upper area of the 0.24-km^2 hillslope and deposition as high as 1,500 t km^{-2} yr^{-1} in the lowlands close to the stream. The areal weighted average resulted in a 125 t km^{-2} yr^{-1} sediment yield. Considering that the measurements of ^{137}Cs inventory reflect the accumulated erosion and sediment deposition since the peak of ^{137}Cs fallout (1963, according to Walling and He 1999), the estimated sediment yield may be taken as a long-term (almost five decades) average.

The results from the rating curves can be found in Table 1. A relevant aspect shown by Lima Neto et al. (2011) assessed both suspended and bedload in a Jaguaribe River section, whose catchment area is $2.40 \times 10^{+4}$ km^2. The authors op cit. observed that the first semester of 2004 alone yielded 75% of the sediment transferred from 1984 to 2009. This gives an insight of the temporal heterogeneity in terms of sediment yield.

The turbidity measured on site showed a very high correlation with the suspended sediment concentration, yielding a very low error (6%) for the total sediment load during the monitored events. This correlation has been particularly high for turbidity below 400 NTU (Nephelometric Turbidity Unit), but low for turbidity above 700 NTU. The average measured SY of the Umbuzeiro River was 10 t km^{-2} yr^{-1}; however, due to the severe drought observed in the year 2012 (average discharge less than 20% of the long-term average), this value cannot be interpreted as representative. In the dry year of 2012, the SY totaled only 0.4 t km^{-2} yr^{-1}. Despite this restrictive aspect of the monitoring period, the use of the turbidimeter has shown some relevant features of the sediment dynamics in the Umbuzeiro River. The most expressive rainfall event (107 mm) occurred late January 2012, when the soil was still very dry and initial abstractions were high, yielding maximum discharge below 9 m^3 s^{-1}. During this event, in which the SSC peak was observed earlier than the discharge peak, clockwise hysteresis occurred: the average SSC on the 12-h ascending limb was 438 mg L^{-1} (maximum of 580 mg L^{-1}), whereas the average SSC in the 160-h descending limb diminished to 62 mg L^{-1} (Fig. 2A). During the base flow, the SSC remained stable at 16 mg L^{-1}.

The most likely reason for the hysteresis is the previous deposition of highly erodible sediments originated from hillslopes and temporarily deposited on the river secondary banks, as noticed on the small hillslope monitored by pins. Mao et al. (2014) studied the occurrence of hysteresis in a mountainous river and concluded that clockwise hysteretic patterns prevail. However, in this case, the phenomenon is mainly explained by the snowmelt, whereas in the tropical semiarid basins it is associated with sediment availability. Figure 2B presents another relevant aspect of the measurements with the turbidimeter. Daily suspended sediment load estimated by the rating curve represented the average hourly values measured with the turbidimeter very well (Figure 2B). Nonetheless, contrarily to the rating curve, the turbidimeter has been able to identify the sediment dynamics (Mano et al. 2009) and, therefore, hysteresis. For example, for water discharge of 7 m³ s⁻¹, the rating curve suggests the suspended sediment load to be 6.5 t h⁻¹, whereas the turbidimeter showed that it varied from 11 to 1 t h⁻¹, for ascending and descending limbs, respectively.

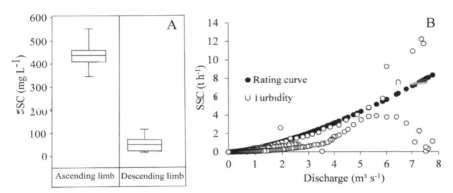

Fig. 2. Monitored data of sediment load for the 107 mm rainfall event that occurred from 22 to 28 January 2012 in the Arocira section of the Umbuzeiro River (800 km²), Upper Jaguaribe Basin. (A) Box-plot (average ± standard deviation) of the ascending and descending limbs, showing hysteresis. (B) Suspended sediment load (Qss) as a function of the water discharge for both daily rating-curve assessment and hourly turbidimeter measurements.

The results of the bathymetric surveys indicated that the Benguê reservoir, which controls a 933-km² area in the transition of sedimentary to crystalline geological formations, had a storage capacity of 21.7 hm³ in 2011. This represents a capacity reduction of 5.4% in five years, yielding 351 t km⁻² yr⁻¹. The Marengo reservoir (75 km² catchment), located in the crystalline geological basement, had a storage capacity of 15.3 hm³ in 2011. This is roughly 10% less than the original capacity measured in 1958, implying SY equal to 487 t km⁻² yr⁻¹. These values are in good agreement with the average SY for seven basins located in the Brazilian semiarid region (426 t km⁻² yr⁻¹: de Araújo 2003). The São Joaquim reservoir, whose 30-km² basin is entirely located on the crystalline geological basement, suffered a 1.87 hm³ capacity reduction in the last 63 years, showing SY of 1,340 t km⁻² yr⁻¹. Despite very high SY, this value is similar to the highest SY measured by de Araújo (2003) in the Jaguaribe River Basin: 1,277 t km⁻² yr⁻¹ in the Cedro-reservoir basin (Table 1), close to the São Joaquim basin. Vanmaercke et al. (2014) after compiling sediment yield measurement for the whole African continent found an average of 634 t km^{-2} yr^{-1}, but for some basins the highest values frequently exceeded 1000 t km^{-2} yr^{-1}.

Although in some cases bedload might be as low as 1% of the total load, it can reach 70% in some mountainous rivers. Lima Neto et al. (2011) concluded that bedload upstream the Orós dam corresponded to 30% of the total sediment load, a value up to five times higher than in other semiarid regions. One of the plausible explanations for this is the existence of a dense small-reservoir network in the Jaguaribe River Basin (Malveira et al. 2012; Mamede et al. 2012; Peter et al. 2014). Lima Neto et al. (2011) showed that, although the small-reservoir network does not effectively retain sediment load during extreme events, it traps a large fraction of sediments yielded in regular and dry years, doubling the life cycle of the strategic reservoirs, that is, those that can provide water during multi-year droughts. According to Geeraert et al. (2015), it is only for basins with autogenic processes (where river bed dynamic and bank erosion processes prevail) that dam construction does not appear to have greatly impacted sediment fluxes.

Spatial distribution of the sediment yield

The northwest zone of the Jaguaribe basin presents the highest sediment yield, possibly caused by the combination of geology (particularly the *Depressão Sertaneja* formation), relief, and long-term intensive land use, especially livestock and rainfed agriculture. The southern side of the basin, despite being the area where the relief is the steepest, does not present high sediment yield, which can be explained not only by the relatively well preserved vegetative status, but also by the combination of permeable deep soils located on top of a permeable sedimentary hydro-geological basin (Fontenele et al. 2013). As expected, the Middle Jaguaribe Basin showed moderate sediment yield, whereas the Lower Jaguaribe Basin yielded the least sediment annually.

The spatially average sediment yield in the JRB is 347 t km^{-2} yr^{-1}, which agrees with previous results for the region (426 t km^{-2} yr^{-1}: de Araújo 2003). According to Einsele and Hinderer (1997), the sediment yield of semiarid basins, whose area is 2x10^{+4} km^2, are expected to be between 300 and 2,000 t km^{-2} yr^{-1}, showing that the JRB is close to the lowest limit. From our results, the sediment yield in the JRB does not differ from other basins with similar catchment areas in the globe (de Araújo and Knight 2005). It is relevant to recall that the Caatinga is a deciduous dry forest and, in the dry season, the vegetation has no leaves (Pinheiro et al. 2013a). Beside the previous point, highly erosive events have been registered in the late dry season (Medeiros and de Araújo 2014). The combination of these two features led us to expect high rates of sediment yield compared with other regions; however, the data did not confirm this assumption. The main reason for this behavior in the Caatinga is assumed to be the effective action of the thick litter, which reduces direct runoff and, consequently, limits the capacity of sediment transport, as shown by Medeiros et al. (2010). Another aspect that might play a role is the existence of a dense reservoir network in the basin (Mamede et al. 2012; Peter et al. 2014), as discussed by Lima Neto et al. (2011).

Sediment yield and basin attributes

Figure 3 presents sediment yield against catchment area in the Jauaribe River Basin considering the prevailing geological basement, preservation status of the Caatinga vegetation, and the measurement methods. No clear trends could be identified: the small-area sites showed relatively high values, whereas the large-area sites showed mostly low values. The middle portion of the plot, however, provided a scatter of over two orders of magnitude.

Figure 3A showed that SY in the crystalline basins was 500 t km^{-2} yr^{-1}, a value four times higher than the average SY in the sedimentary basins (117 t km^{-2} yr^{-1}). After analyzing the sediment yield in a sedimentary basin in the Brazilian semiarid region, Medeiros et al. (2010) observed that the main constraint to sediment flux is the limited transport capacity due to low runoff (67 mm yr^{-1}), high interception losses (Pinheiro et al. 2016), and infiltration. If one considers only the sedimentary basins (Upper and Lower parts of the JRB), negative allometry is observed. Figure 3B shows the data for three conservation statuses of the preserved Caatinga. For drainage areas below 1 km^2, no evident trend can be identified. However, for areas above 10 km^2, it is clear that preserved watersheds deliver less sediment. This is also noticeable considering the average sediment yield for highly preserved (> 80% preservation), moderately preserved (40%–80%), and degraded (< 40%) areas: 139, 279, and 545 t km^{-2} yr^{-1}, respectively. This is in agreement with the results presented by López-Vicente et al. (2011). The Caatinga vegetation protects the soil degradation by means of its litter, as well as by enhancing the infiltration capacity during the dry season, which ceases the connectivity (see Harel and Mouche 2014) and, therefore, the sediment displacement. In fact, results from Pinheiro et al. (2013a) and de Figueiredo et al. (2016) suggest that the Caatinga forms macro-pores during the dry season (root shrinking and expansion), enhancing infiltration (therefore, reducing transport capacity of the runoff) during the periods in which the soil is most vulnerable (i.e., when the leaves have fallen due to low soil moisture). From Fig. 3C one can see that, for the meso-scale areas (10^{+1}–10^{+4} km^2), where SY is higher, the prevailing investigation method has been the bathymetric

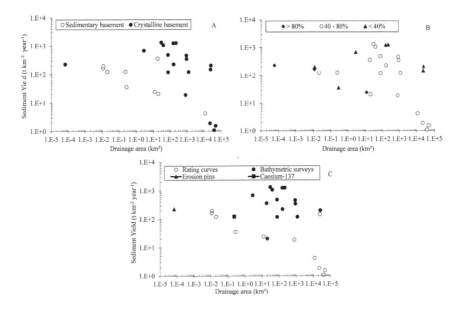

Fig. 3. Sediment yield in the Jaguaribe River Basin in the 27 measurement sites considering three attributes. (A) Prevailing geological basement. (B) Vegetative (Caatinga) preservation status. (C) Measurement method (SSC: rating curves).

survey because it coincides with the location of most of the relevant dams in the region. For the largest catchment areas ($> 10^{+4}$ km²), the rating curves notably preponderate, mainly because no dam has been constructed in the lower portion of the Jaguaribe River.

Allometry

Figure 4 presents the results of the four-term moving average of the sediment yield as a function of the drainage area (as in Xu and Yan 2005) for the Jaguaribe River Basin. It shows clearly that, for areas lower than 400 km², SY increases with the area (positive allometry), decreasing afterwards. This behavior differs from that described in Owens and Slaymaker (1992) and de Araújo and Knight (2005), but similar to those found by other authors. Church and Slaymaker (1989), who studied basins in British Columbia, observed positive allometry for watersheds up to 30,000 km² and explained that the phenomenon is caused by quaternary remobilization of sediment stored in the valleys and channels. Xu and Yan (2005) noted that, for the Yellow River Basin (ten times larger than the JRB), positive allometry was found for areas smaller than 2,000 km², which could be explained by three main factors: spatial distribution of the surface material, morphological adjustment in the macro-temporal scale, and the difference on the energy level release in the different spatial scales. Yan et al. (2011) studied the Yangtze River Basin, showing negative allometry in the sub-basins strongly influenced by agricultural activities. Nonetheless, the authors found positive allometry for the larger scales, mainly due to the remobilization of the main river sediment.

The explanations found for the positive allometry usually rely on the geomorphology, where the landscape units are still under formation. Besides, all these regions have strong influence of glacial areas whose sediment yield values are on average 5 to 10 times than in no-glaciated basins (Hinderer et al. 2013). In terms of geomorphology, the 400-km² areas in the JRB correspond to river lengths of the order of 50 km. The rivers of this order of magnitude are all inside the *Depressão Sertaneja*, a landscape unit that still suffers denudation from high hillslope erosive processes. The continuous erosion has generated surfaces, which seem to still be active. The larger sediment fractions tend to generate colluvial deposits, whereas the finer fractions reach the alluvial channels

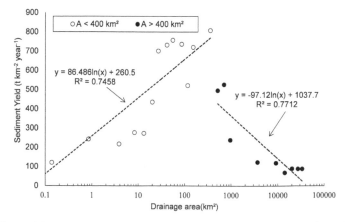

Fig. 4. Sediment yield versus catchment area in the Jaguaribe River Basin based on 27 site measurements. Each point represents the four-term moving average, as in Xu and Yan (2005).

(Cavalcante and Cunha 2012). It is, therefore, possible that the *Depressão Sertaneja*, the most relevant geomorphologic unit in the JRB, is still adapting from processes that started in the Cenozoic. In fact, Val et al. (2014) verified that watersheds in Central Amazonia, whose sizes are also of hundreds of square kilometers, still provide evidence of "landscape reorganization" and, therefore, evidence of sediment activeness. The small basins in the JRB (below 400 km²) have erodible material limitation and high sediment-transport potentiality, whereas the erodible material in the middle course of the river is abundant, resulting in high sediment yield. As the basin areas increase, the deposition processes—even on the river bed (Cavalcante and Cunha 2012) prevail, causing negative allometry that is the sediment delivery is transport limited and the sediment retention along the topography is the main process for sediment retention (Bronstert et al. 2014).

Conclusions

From the results gathered through this research, one can draw the following conclusions: Sediment concentration hysteresis has been observed by the hourly turbidimeter measurements at a river section, whose catchment area is $8.0 \times 10^{+2}$ km². Hysteresis can be explained by sediment (dis)-connectivity. The ascending limb disposes of highly erodible material, deposited during recent events, in contrast with the descending limb, which reduces its sediment load. The average sediment yield of the JRB is 347 t km^{-2} yr^{-1}, ranging from 10^{+2} to 10^{+3} t km^{-2} yr^{-1}. This value is in good agreement with other basins (semiarid or not) of the same catchment area throughout the globe. Despite the fact that the Caatinga vegetation is deciduous and that the rainfall is highly erosive, this moderate yield can be explained mainly by the litter, macro-pore flow, the low runoff sediment-transport capacity, and the dense reservoir network in the basin. The JRB presented positive allometry for areas up to 400 km², possibly caused by the *Depressão Sertaneja*, a still-active Cenozoic landscape unit located in the centre of the Jaguaribe River Basin.

Acknowledgements

The authors acknowledge DFG (*Deutsche Forschungsgemeinschaft*) for funding the research project entitled "Generation, transport and retention of water and suspended sediments in large dryland catchments: Monitoring and modelling of fluxes and integrated connectivity phenomena"; and the Brazilian CNPq, for the financial support to the field activities within the IPAS research project (Impact of small reservoirs on water availability in the semiarid - 474323/2011-0).

Keywords: Allometry, Caatinga, hysteresis, spatial scale.

References

Baker, T.J. and Miller, S.C. 2013. Using the soil and water assessment tool (SWAT) to assess land use impact on water resources in an East African watershed. Journal of Hydrology 486: 100–111, doi:10.1016/j.jhydrol.2013.01.041.

Boardman, J., Favis-Mortlock, D. and Foster, I. 2015. A 13-year record of erosion on badland sites in the Karoo, south Africa. Earth Surface Processes and Landforms 40: 1964–1981. doi:10.1002/esp.3775.

Bormann, H., Pinter, N. and Elfert, S. 2011. Hydrological signatures of flood trends on German rivers: flood frequencies, flood heights and specific stages. Journal of Hydrology 404: 50–66. doi:10.1016/j.jhydrol.2011.04.019.

Bronstert, A., de Araújo, J.C., Batalla, R., Costa, A.C., Francke, T., Förster, S., Güntner, A., Lopez-Tarazon, J.A., Mamede, G.L., Medeiros, P.H.A., Müller, E.N. and Vericat, D. 2014. Process-based modelling of erosion, sediment transport and reservoir siltation in mesoscale semi-arid catchments. Journal of Soils and Sediments 4: 2001–2018. doi:10.1007/s11368-014-0994-1.

Cavalcante, A.A. and Cunha, S.B. 2012. Morfodinâmica fluvial em áreas semiáridas: discutindo o vale do rio Jaguaribe-CE-Brasil (in Portuguese). Revista Brasileira de Geomorfologia 13(1): 39–49.

Church, M. and Slaymaker, O. 1989. Disequilibrium of Holocene sediment yield in glaciated British Columbia. Nature 337: 452–454. doi:10.1038/337452a0.

Costa, A.C., Foerster, S., de Araújo, J.C. and Bronstert, A. 2013. Analysis of channel transmission losses in a dryland river reach in north-eastern Brazil using streamflow series, groundwater level series and multitemporal satellite data. Hydrological Processes 27: 1046–1060. doi:10.1002/hyp.9243.

Creutzfeldt, B.N.A. 2006. Remote sensing based characterization of land cover and terrain properties for hydrological modeling in the semi-arid Northeast of Brazil. 104p. MSc. Thesis, University of Potsdam, Germany, 2006.

de Araújo, J.C. 2003. Assoreamento em reservatórios do semiárido: modelagem e validação (in Portuguese). Revista Brasileira de Recursos Hídricos 8(2): 39–56.

de Araújo, J.C. and Knight, D.W. 2005. A review of measurement of sediment yield in different scales. Revista da Escola de Minas 58(2): 257–266. doi:10.1590/S0370 44672005000300012.

de Figueiredo, J.V., de Araújo, J.C., Medeiros, P.H.A. and Costa, A.C. 2016. Runoff initiation in a preserved semiarid Caatinga small watershed, Northeastern Brazil. Hydrological Processes. doi:10.1002/hyp.10801.

de Vente, J. and Poesen, J. 2005. Predicting soil erosion and sediment yield at the basin scale: scale issues and semi-quantitative models. Earth-Science Reviews 71: 95–125. doi:10.1016/j.earscirev.2005.02.002.

Einsele, G. and Hinderer, M. 1997. Terrestrial sediment yield and the lifetimes of reservoirs, lakes, and larger basins. Geologische Rundschau 86: 288–310.

Fontenele, S.B., Mendonça, L.A.R., de Araújo, J.C., Santiago, M.M.F. and Gonçalves, J.Y.B. 2013. Relationship between hydrogeological parameters for data-scarce regions: the case of the Araripe sedimentary basin, Brazil. Environmental Earth Sciences 71(2): 885–894. doi:10.1007/s12665-013-2491-z.

Geeraert, N., Omengo, F.O., Tamooh, F., Paron, P., Bouillon, S. and Govers, G. 2015. Sediment yield of the lower Tana River, Kenya, is insensitive to dam construction: sediment mobilization processes in a semi-arid tropical river system. Earth Surface Processes and Landforms 40: 1827–1838. doi:10.1002/esp.376.

Hancock, G.R. and Lowry, J.B.C. 2015. Hillslope erosion measurement—a simple approach to a complex process. Hydrological Processes 29: 4809–4816. doi:10.1002/hyp.10608.

Harel, M.A. and Mouche, E. 2014. Is the connectivity function a good indicator of soil infiltrability distribution and runoff flow dimension? Earth Surface Processes and Landforms 39: 1514–1525. doi:10.1002/esp.3604.

Hinderer, M., Kastowski, M., Kamelger, A., Bartolini, C. and Schlunegger, F. 2013. River loads and modern denudation of the Alps—A review. Earth-Science Reviews 118: 11–44. doi:http://dx.doi.org/10.1016/j.earscirev.2013.01.001.

Lima Neto, I.E., Wiegand, M.C. and de Araújo, J.C. 2011. Sediment redistribution due to a dense reservoir network in a large semi-arid Brazilian basin. Hydrological Sciences Journal des Sciences Hydrologiques 56: 319–333. doi:10.1080/02626667.2011.553616.

López-Vicente, M., Poesen, J., Navas, A. and Gaspar, L. 2011. Predicting runoff and sediment connectivity and soil erosion by water for different land use scenarios in the Spanish Pre-Pyrenees. Catena 102: 62–73. doi:10.1016/j.catena.2011.01.001.

Malveira, V., de Araújo, J.C. and Güntner, A. 2012. Hydrological impact of a high-density reservoir network in semiarid northeastern Brazil. Journal of Hydrologic Engineering 17(1): 109–117. doi:10.1061/(ASCE)HE.1943–5584.0000404.

Mamede, G.L., Araújo, N., Schneider, C.M., de Araújo, J.C. and Herrmann, H.J. 2012. Overspill avalanching in a dense reservoir network. Proceedings of the National Academy of Sciences of the USA 109: 7191–7195. doi:10.1073/pnas.1200398109.

Mano, V., Némery, J., Belleudy, P. and Poirel, A. 2009. Assessment of suspended sediment transport in four alpine watersheds (France): influence of the climatic regime. Hydrological Processes 23(5): 777–792. doi:10.1002/hyp.7178.

Mao, L., Dell'agnese, A., Huincache, C., Penna, D., Engel, M., Niedrist, G. and Comiti, F. 2014. Bedload hysteresis in a glacier-fed mountain river. Earth Surface Processes and Landforms 39(7): 964–976. doi:10.1002/esp.3563.

Medeiros, P.H.A., Güntner, A., Francke, T., Mamede, G. and de Araújo, J.C. 2010. Modelling spatio-temporal patterns of sediment yield and connectivity in a semiarid catchment with the WASA-SED model. Hydrological Sciences Journal des Sciences Hydrologiques 55(4): 636–648. doi:10.1080/02626661003780409.

Medeiros, P.H.A. and de Araújo, J.C. 2014. Temporal variability of rainfall in a semiarid environment in Brazil and its effect on sediment transport processes. Journal of Soils and Sediments 14: 1216–1223. doi:10.1007/s11368-013-0809-9.

Medeiros, P.H.A., de Araújo, J.C. and Andrello, A.C. 2014. Uncertainties of the ^{137}Cs technique for validation of soil redistribution modelling in a semiarid meso-scale watershed. Engenharia Agrícola 34(2): 222–235. doi:http://dx.doi.org/10.1590/S0100–69162014000200004.

Navarro-Hevia, J., de Araújo, J.C. and Manso, J.M. 2014. Assessment of 80 years of ancient-badlands restoration in Saldaña, Spain. Earth Surface Processes and Landforms 39(12): 1563–1575. doi:10.1002/esp.3541.

Nichols, M.H., Nearing, M.A., Polyakov, V.O. and Stone, J.J. 2013. A sediment budget for a small semiarid watershed in southeastern Arizona, USA. Geomorphology 180-181: 137–145. doi:10.1016/j.geomorph.2012.10.002.

Owens, P. and Slaymaker, O. 1992. Late Holocene sediment yields in small alpine and subalpine drainage basins, British Columbia. IASH Publications 209: 147–154.

Peter, S., de Araújo, J.C., Araujo, N. and Herrmann, H.J. 2014. Flood avalanches in a semiarid basin with a dense reservoir network. Journal of Hydrology 512(408): 408–420. doi:10.1016/j.jhydrol.2014.03.001.

Pinheiro, E.A.R., Costa, C.A.G. and de Araújo, J.C. 2013a. Effective root depth of the Caatinga biome. Journal of Arid Environment 89: 1–4. doi:10.1016/j.jaridenv.2012.10.003.

Pinheiro, E.A.R., de Araújo, J.C., Fontenele, S.B. and Lopes, J.W.B. 2013b. Calibração de turbidímetro e análise de confiabilidade das estimativas de sedimento suspenso em bacia semiárida (in Portuguese). Water Resources and Irrigation Management 2(2): 103–110.

Pinheiro, E.A.R., Metselaar, K., de Jong Van Lier, Q. and de Araújo, J.C. 2016. Importance of soil-water to the Caatinga biome, Brazil. Ecohydrology. doi:10.1002/eco.1728.

Rhoton, F.E., Emmerich, W.E., Nearing, M.A., McChesney, D.S. and Ritchie, J.C. 2011. Sediment source identification in a semiarid watershed at soil mapping unit scales. Catena 87(2): 172–181. doi:10.1016/j.catena.2011.05.002.

Santos, J.C.N. 2012. Uso de terras agrícolas no semiárido e mensuração da erosão do solo em diferentes escalas espaciais (in Portuguese). MSc. Thesis, Universidade Federal do Ceará, Fortaleza, Brazil.

Val, P., Silva, C., Harbor, D., Morales, N., Amaral, F. and Maia, T. 2014. Erosion of an active fault scarp leads to drainage capture in the Amazon region, Brazil. Earth Surface Processes and Landforms 39(8): 1062–1074. Doi: 10.1002/esp.3507. doi:10.1002/esp.3507.

Vanmaercke, M., Poesen, J., Maetens, W., de Vente, J. and Verstraeten, G. 2011. Sediment yield as a desertification risk indicator. Science of the Total Environment 409(9): 1715–1725. doi:10.1016/j.scitotenv.2011.01.034.

Walling, D.E. and He, Q. 1999. Improved models for estimating soil erosion rates from Cesium-137 measurements. Journal of Environmental Quality 28(2): 611–622. doi:10.2134/jeq1999.00472425002800020027x.

Xu, J. and Yan, Y. 2005. Scale effects on specific sediment yield in the Yellow River basin and geomorphological explanations. Journal of Hydrology 307(1-9): 219–232. doi:10.1016/j.jhydrol.2004.10.011.

Yan, Y., Wang, S. and Chen, J. 2011. Spatial patterns of scale effect on specific sediment yield in Yangtze River Basin. Geomorphology 130(1-2): 29–39. doi:10.1016/j.geomorph.2011.02.024.

Zende, A.M. and Nagarajan, R. 2014. Sediment yield estimate of river basin using SWAT model in semi-arid region of Peninsular India. Engineering Geology for Society and Territory 3: 543–546. doi:10.1007/978-3-319-09054-2_110.

CHAPTER 20

Application of Remote Sensing and the GIS in Interpretation of River Geomorphic Response to Floods

Milan Lehotský,* Miloš Rusnák[a] and Anna Kidová[b]

INTRODUCTION

Floods modify channel landforms and affect processes taking place in the riparian zone. Extreme events increase the energy of streams, which manifest themselves by distinct geomorphological effects and increased erosion/accumulation processes in the channel. Geomorphological effects of floods are controlled by the discharge magnitude and physical properties of the channel and the floodplain (Miller 1990). Effects of such extreme events on the channel morphology, in terms of erosion or aggradation depend on their frequency and timing. The resulting effect is erosion of floodplain (bank, channel floor) on the one side, and accumulation (accretion) of sediments in the form of gravel bars accompanied by remodelling of the channel and floodplain morphology both in horizontal and vertical directions.

These morphological changes (except for vertical changes) are identifiable through remote sensing data, that is, from satellite and aerial images or maps (Hooke and Redmond 1989; Gilvear and Bryant 2003) and their processing in GIS (Geographic Information Systems). The outputs capture horizontal changes in channel pattern, dynamics of the stream, lateral migration caused by bank erosion and deposition of sediments in form of gravel bars, development of erosion corridor, or the trend of general planform change. The objective of this chapter is to give the general overview on the

Department of Physical Geography, Geomorphology and Natural Hazards, Institute of Geography, Slovak Academy of Sciences, Štefánikova 49, 814 73 Bratislava, Slovakia.

[a] E-mail: geogmilo@savba.sk

[b] E-mail: geogkido@savba.sk

* Corresponding author: geogleho@savba.sk

application of remote sensing and GIS data in the research of geomorphic response of rivers to floods, but also to present some results of using these procedures to determine how flood event periods have influenced the evolutionary trend of two rivers.

Remote Sensing and Maps in Fluvial Geomorphology

Historical maps and remote sensing data are the basic source of the spatial information in river research (Gurnell et al. 2003). There are a lot of suitable sources (different kind of archives) of historical maps as far as local or regional records of river courses are concerned. For example, in Central Europe, the military maps (from period 1764–1787, 1810–1869 or from 1875–1884) covering the entire territory of the former Austro-Hungarian Empire are well known. Remote sensing data can be divided based on sensor type and carrier into three main categories (Fig. 1): satellite images, aerial photographs, and UAV's (Unmanned Aerial Vehicle) photographs (Carbonneau and Piégay 2012).

Some of the most popular satellite data are from Landsat Program (launched in 1972). They provide images (currently, satellite Landsat 7 or 8) with pixel resolution 15 m (panchromatic) and 30 m (multispectral), usable for studying the large river systems. Data with the higher pixel accuracy are available from other satellite platforms such as Ikonos, GeoEye, QuickBird, WorldView, and others (Table 1). Spatial resolution of these images is approximately 0.5 m, which is sufficient for the fluvial geomorphic research.

Aerial photos and satellite images orthorectified in specialized software capture high-resolution details of the changing fluvial landscape (Gilvear et al. 2004; Bryant and Gilvear 1999). Milton et al. (1995) and Gilvear et al. (2004) point to the benefits of images from remote sensing for research of hydromorphological properties and channel changes. In addition to classic orthophotographs, the aircraft can carry many other sensors for example, multispectral cameras or LIDAR's.

The greatest advantage though is their temporal and spatial flexibility in mapping of changes. Their benefit lies in the accessibility of dense and high accuracy information (depends on resolution), which allows relatively precise quantification of morphological processes (bank erosion, planform changes, bar accumulation, etc.).

Fig. 1. Main sources of remote sensing data: Landsat 8 launched in 2013, as a representative satellite data platform (a)—copyright NASA, aircraft for airborne remote sensing (b)—copyright EUROSENSE, and the six-rotor platform for UAV's photography (c).

Table 1. Several common satellite platforms with available images of earth for river research.

Satellite platform	Operator	Resolution		Bands	Launched	Comment
		Panchromatic	Multispectral			
Landsat 7/8	USGS/NASA	15 m	30 m	8/11	1999/2013	
Sentinel 2	ESA		10 m	13	2015	
ALOS	JAXA	2,5 m	10 m	5	2006	
SPOT-5	AIRBUS Defence & Space	2,5 m	10 m	5	2002	
SPOT-6/7	AIRBUS Defence & Space	1,5 m	6 m	5	2012/2014	decommissioned 2015
Pleiades 1A/1B	AIRBUS Defence & Space	0,5 m	2 m	5	2011/2012	
KOMPSAT-3	KARI	0,7 m	2,8 m	5	2004	
KOMPSAT-3A	KARI	0,55 m	2,2 m	6	2015	
IKONOS	DigitalGlobe	0,82 m	3,2 m	5	1999	
QuickBird	DigitalGlobe	0,65 m	2,62 m	5	2001	
GeoEye 1	DigitalGlobe	0,46 m	1,84 m	5	2008	decommissioned 2015
WorldView-1	DigitalGlobe	0,5 m	-	1	2007	
WorldView-2	DigitalGlobe	0,46 m	1,84 m	9	2009	
WorldView-3	DigitalGlobe	0,31 m	1,24 m	30	2014	
SkySat 1/2	SkyBox Imaging	0,9 m	2,0 m	5	2013/2014	

Hooke and Redmond (1989) recognize the limit of the method, that is, the fact that the image captures the state of the channel in the moment of imaging/mapping and does not record the process of changes that take place between individual events. It often leads to a problem with the assessment and defining of changes and the identification of contribution of individual change attractors. Meanwhile, it is indispensable to use images with satisfactory spatial resolution and correct spatial referencing of study objects.

UAV technologies represent a new way of collecting spatial information and an interesting alternative to the conventional means (satellites, aircraft, helicopters, etc.) used in large-scale research (Sládek and Rusnák 2013). Carriers can be remotely controlled models of aircrafts or helicopters, balloons, kites, or special multi-rotor platforms (copters). These platforms can be loaded with different types of sensors depending on their technical parameters or objective of research. The UAV technology enables data collection with high resolution and accuracy (from 0.5 to 2 cm). By using an open source or commercial software (for instance Airphoto SE, Pix4D, Agisoft PhotoScan) can visualize the terrain from bird eye view, orthophoto mosaic, point clouds, 3D models of objects, or digital terrain models (Fig. 2). At present, micro-UAVs are mainly used in the study of lateral movements of river channels, granulometry of gravel bars, and bathymetry of channel or identification of large woody debris (Miřijovsky et al. 2015). The potential of UAV lies in high operability, which makes it possible to capture changes in landscape either in progress (flooding) or in time (for instance, spatial changes of vegetation during seasons by producing a chronological series of images). Unmanned Aerial Vehicles allow concentration on small areas, very accurate capturing of seasonal changes, and visualization or reconstruction of earth's surface or various objects by high precision photogrammetric techniques (Hervouet et al. 2011).

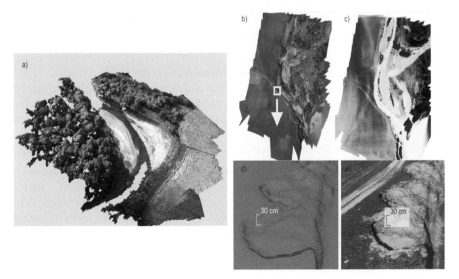

Fig. 2. Examples of riverine landscape visualisation by UAV. Dense point cloud of meander loop created by the UAV photogrammetry by Agisoft PhotoScan software with density 800 point on m² (a). Orthomosaic (b) and digital elevation model (c) of avulsion channel (formed by three flood events with recurrence interval from 2 to 5 years in 2010; River Ondava, Slovakia). Details show high precision of the solid TIN model (d) and 3D texture model (e) with vertical accuracy 20 mm.

GIS Approach Methodology

In a systemic view, a stream can be divided into various biophysical components, including the channel and floodplain landforms (Albert and Piégay 2011). Most of the methodological frameworks are based on the delineation and measurement of river geomorphic objects (in vectors or raster format) in the GIS environment.

The first step is the identification of the geomorphic object based on the automatic classification or visual interpretation (or a combination of both) of raw data (images). Boundary delimitation of the geomorphic component (polygons, lines, points, or raster) depends on the resolution of the raw data. Geomorphic objects are spatially continuous (e.g., a polygon delineating the water surface) or discontinuous (e.g., a polygon delineating the channel bars) along the stream. For example, the borderline of the channel is visually interpreted along the bank of the stream (bank line) either considering the bank morphology as the distinct terrain step between the channel and the floodplain or it is as the borderline between the channel and riparian vegetation. Selection and classification of geomorphic objects, which are identified by images, depend on the aim of research.

Once features have been extracted, they can be disaggregated into elementary segments using the "disaggregation processes". This spatial disaggregation is aimed to provide fluvial units at a higher resolution which has two functions: (i) to characterize objects with high spatial resolution continuously, and more generally (ii) to provide linear referencing systems for object attributes along the channel and over time, which are pertinent to river geomorphic responses to floods like channel shifting, widening/narrowing, erosion or deposition area, bar area and bar number, island area and island number and planform development, etc. Morphologic attributes (width, area, length, etc.) can be extracted by direct measurement of the geomorphic objects. Additional attributes can also be derived from spatial analysis.

Thus, this methodological framework in the GIS is subdivided into four main scales as follows (Fig. 3):

1. Raw data processing scale (orthorectification and correction).
2. Scale of extraction of geomorphic objects and delineating the features of interest.
3. Scale of disaggregation of the objects into high resolution spatial units providing a characterization of longitudinal changes along the channel and resulting into sets of attributes (GIS-tables).
4. Interpretation scale of the channel adjustment processes as responses to floods.

Individual remote sensing data are chosen based on the analysis of hydrographs in a way that captures the state of the river system after (and before as a reference state) a specific flood event, events, or periods with occurrence of specific N-year floods. What is important is that images do not capture individual flood events in longer study periods and the effects of small floods are negligible due to the resolution and rectification errors (Piégay et al. 2005). Extraction of geomorphic objects took place in GIS where a logical model of spatial geodatabase was produced.

Effect of Flood Events

One of the key factors affecting the behaviour of a river channel is extreme discharge, which increases the energy of the stream. Extreme discharge leads to intensification of

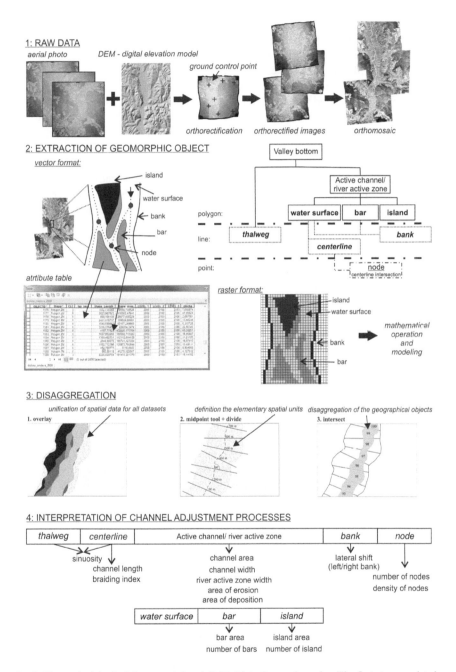

Fig. 3. The methodological framework is subdivided into four main scales. The first step consists in raw data (1) processing followed by extracting individual geomorphic objects that delineate the river features of interest (2). Then, the spatial disaggregation of individual geomorphic objects (3) results in disaggregated geomorphic objects allowing us to measure their basic metrics and interpret responses of geomorphic objects to flood periods (4).

erosion and deposition processes in the channel and floodplain. Long-term response to the series of flood events manifested itself first in the inner-channel organisation of landforms. The temporal distribution of flood events is also important. Temporal distribution of flood events reflects the overall climatic conditions of the region and respond to the precipitation regime. The basic characteristics of floods are: magnitude, frequency, duration, and timing. Geomorphological effectiveness depends mainly on the magnitude and frequency of flood events and is influenced by the catchment scale boundary conditions (Wolman and Miller 1960; Charlton 2008). However, similar floods in terms of magnitude and frequency can produce dissimilar morphological response (Fuller 2008).

High magnitude floods are more destructive to channel morphology, causing bank erosion, floodplain destruction, and total geomorphological changes in river systems. Normal or low magnitude floods (approximately 1–2 year recurrence interval, bankfull discharge) contribute to the accretion of sediments and system stabilization (Corenblit et al. 2007; Opperman et al. 2010; Ward et al. 2002). Bertoldi et al. (2009) on the river Tagliamento (Italy) show that the flow at a level higher than the bankfull discharge is responsible for significant erosion of the floodplain and edges of river islands. High levels of discharge lead to the formation of new avulsion channels. Surian et al. (2009) similarly highlight the effects of low-flow discharges as formative for channels and high magnitude floods for gravel bars transport and significant morphological changes. Phillips (2002) argues that the smaller discharge below bankfull level is responsible for the formation and shaping of a channel and larger flows are needed to transport sediments and bank erosion.

Ward (1978) emphasises the significance of flood frequency. Effects of two consecutive floods in a short period are significant and have a larger geomorphological effect when the interval between events is longer. Time gap allows vegetation to recover, which results in channel stabilization. When comparing two equal floods, the second exerts a lesser geomorphic impact, because the first flood adjusted the channel (Kochel 1988). The effectiveness of floods depends on the morphological state of the channel (sinuosity, pattern, gradient) during high flow events and geological and morphological (topographical) conditions of the river catchment (Hooke 2015). Further variables, which affect the response of the channel system to floods, are changes in vegetation cover (succession, seasonality, and land use), as they partially influence the sediment supply or flood capacity of channel. Spatial or local factors, such as the geotechnical characteristics of banks (geometry, moisture, etc.), geological settings, bedrock outcrop, sediment characteristic, sediment availability, or channel incision are important too. In river research, one must also bear in mind the impacts of humans: flood defence structures, flow regime modification, or direct intervention into the channel (gravel extraction, channelization).

The application of remote sensing and GIS in interpretation of river geomorphic response to floods is presented in two case studies (the braided and meandering gravel bed rivers) where the basic source for the extraction of geomorphic objects were orthophotographs and aerial photographs. In case of the braided stream (the River Belá, Slovakia), it was from seven time horizons (1949, 1961, 1973, 1986, 1992, 2003, and 2009) and in case of the gravel beds and meandering streams (the River Topľa, Slovakia), it was from six time horizons (1949, 1961, 1981, 1987, 2002, 2009) (Kidová and

Lehotský 2012; Kidová et al. 2016; Lehotský et al. 2013; Rusnák and Lehotský 2014). Time horizons of object extraction were chosen based on the analysis of hydrographs in a way that captured the state of the river system after periods with occurrence of specific N-year floods. Flood periods with different flood characteristics were identified based on the variability of maximum annual discharges. The borderline of the channel has been drawn along the bank of the stream (bank line) either considering the bank morphology as the distinct terrain step between the channel and the floodplain or it was identified as the borderline between the channel and riparian vegetation. Channel landforms were assessed visually. The compiled geodatabase represented the basic information, which entered individual analyses as the vector or raster layer. The raster layer with pixel size of 0.5 m and the function of aligning the individual raster cells with reference to the 2009 raster layer was used in time-spatial operations.

In the case of braided river changes during 1949–2009 in the river active zone area, the river active zone width and channel length, island area and number, bar area and number, accretion, erosion, and total reworked area exhibit the decreasing trend. Braiding index as a disaggregated value of braided pattern along the channel length varied from one year to another in seven time horizons, while the maximum values occurred in 1949. Low magnitude floods when their n-return period did not exceed the 10-year discharge lead to the decrease of the lateral channel mobility. Short interval of floods will likely have a more pronounced effect on river morphology than similar floods with more space in time. Incision of channels into the bedrock as an effect of flooding (Fig. 4) connected with land cover changes (reduction in catchment sediment supply) and human impact is evident in the study area. This process leads to the stabilisation of channels position, fixation of central bars, and their transformation into islands.

In the case of the meandering gravel bed river (Fig. 5), small discharges (recurrence interval on the level of 1- to 2-year flood) did not have any pronounced destroying effects in accord with Corenblit et al. (2007), Opperman et al. (2010), or Ward et al. (2002). Small discharge effects were constructive and contributed to accumulation of fine sediments and stabilisation of the system. On the contrary, extreme flood events with 5–10 year recurrence intervals led to a pronounced increase of channel dynamics. In accord with the results of Lehotský et al. (2013) or Rusnák and Lehotský (2014) obtained from other east-Slovakian River Ondava, the lower discharge is precisely responsible for the formation of the channel, ground plan pattern, and relatively slow bank erosion connected with the overall behaviour of the river.

Large floods lead to sudden and significant bank erosion, destruction of planforms, and partial or total rejuvenation of the stream followed by its widening, enlarged areas of gravel bars and destruction of old stands and creation of new stands for plants. It is the discharge with 10-year return period, which is significant for local rejuvenation of the stream.

The period with absence of floods led to the rapid growth of vegetation in the channel and especially on the newly formed gravel bars. Vegetation succession is also influenced by accumulation of dead wood and local topography. In this way, bars become stands for the development of vegetation, which gradually stabilizes them. Generally, the trend of morphological development of both rivers is in accordance with Corenblit et al. (2007) while major floods are rather destructive. Low-magnitude, high frequency floods are constructive and contribute to the accretion of sediments on the floodplain and islands and to stabilization of the river system.

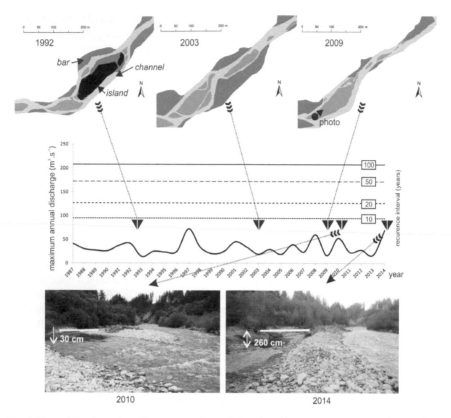

Fig. 4. Several flood events with recurrence interval less than 10 years cannot reverse the negative trend of channel evolution. Floods of such magnitude lead to channel narrowing and channel incision. These processes were accelerated in the last flood event in 2010 (69.9 $m^3.s^{-1}$, equal to recurrence interval 2–5 years).

Conclusions

Application remote sensing data and the GIS may provide relatively complex information about rivers focused on their geomorphic response to floods. Based on the determined geomorphic objects using remote sensed data and the GIS and flood periods, we were able to clarify their geomorphic evolution since the middle of the 20th century. Results have definitely showed that the studied rivers are subjected to distinct morphological changes. The long-term morphological dynamics identified from historical maps, aerial photographs, or satellite images possibly correlate with the hydrological conditions. Capturing of spatial-temporal changes and monitoring of destruction of the existing landforms, deposition of new ones, followed by succession of vegetation and overall land cover changes in the riparian zone is the most relevant advantage of these resources.

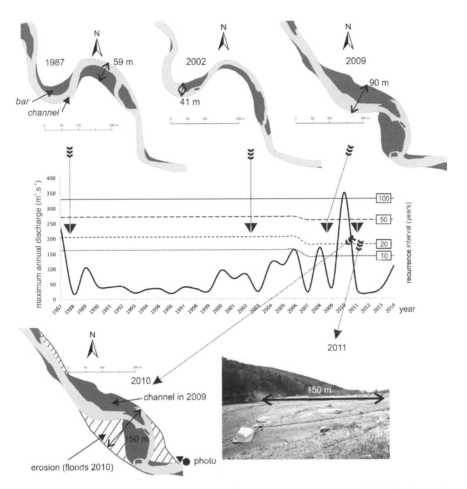

Fig. 5. Planform changes of the River Topľa (East Slovakia), where small magnitude floods (RI_{1-2}) between 1987 and 2002 led to channel stabilization and shaping of meandering channel (erosion only in meander loop). Intensive floods after 2002 (RI_{10} and higher) increased the channel erosion and caused widening of the channel and destruction of the meandering planform.

Acknowledgement

The research was supported by Science Grant Agency (VEGA) of the Ministry of Education of the Slovak Republic and the Slovak Academy of Sciences; 02/0020/15.

Keywords: River, remote sensing, GIS, geomorphic, floods.

References

Albert, A. and Piégay, H. 2011. Spatial disaggregation and aggregation procedures for characterizing fluvial features at the network-scale: application to the Rhône basin (France). Geomorphology 125: 343–360.

Bertoldi, W., Gurnell, A.M., Surian, N., Tockner, K., Zanoni, L., Ziliani, L. and Zolezzi, G. 2009. Understanding reference processes: linkages between river flows, sediment dynamics and vegetated landforms along the Tagliamento River, Italy. River Research and Applications 25: 501–516.

Bryant, R.G. and Gilvear, D.J. 1999. Quantifying geomorphic and riparian land cover changes either side of a large flood event using airborne remote sensing: River Tay, Scotland. Geomorphology 29: 307–321.

Carbonneau, P.E. and Piégay, H. 2012. Fluvial Remote Sensing for Science and Management. Wiley, Chichester.

Charlton, R. 2008. Fundamentals of Fluvial Geomorphology. Routledge, New York.

Corenblit, D., Tabacchi, E., Steiger, J. and Gurnell, A.M. 2007. Reciprocal interactions and adjustments between fluvial landforms and vegetation dynamics in river corridors: a review of complementary approaches. Earth-Science Reviews 84: 56–86.

Fuller, I.C. 2008. Geomorphic impacts of a 100-year flood: Kiwitea Stream, Manawatu catchment, New Zealand. Geomorphology 98: 84–95.

Gilvear, D. and Bryant, R. 2003. Analysis of aerial photography and other remotely sensed data. pp. 135–170. *In*: Kondolf, M. and Piégay, H. (eds.). Tools in Geomorphology. Wiley, Chichester, England.

Gilvear, D., Davids, C. and Tyler, A.N. 2004. The use of remotely sensed data to detect channel hydromorphology; River Tummel, Scotland. River Reserch and Applications 20: 795–811.

Gurnell, A.M., Peiry, J.L. and Petts, G.E. 2003. Using historical data in fluvial geomorphology. pp. 77–103. *In*: Kondolf, M. and Piégay, H. (eds.). Tools in Geomorphology. Wiley, Chichester, England.

Hervouet, A., Dunford, R., Piégay, H., Belletti, B. and Trémélo, M.L. 2011. Analysis of post-flood recruitment patterns in braided-channel rivers at multiple scales based on an image series collected by unmanned aerial vehicles, ultralight aerial vehicles, and satellites. GIScience & Remote Sensing 48: 50–73.

Hooke, J.M. and Redmond, C.E. 1989. Use of cartographic sources for analyzing river channel change with examples from Britain. pp. 79–93. *In*: Petts, G.E. (ed.). Historical Change of Large Alluvial Rivers: Western Europe. John Wiley and Sons, Chichester, England.

Hooke, J.M. 2015. Variations in flood magnitude–effect relations and the implications for flood risk assessment and river management. Gomorphology 251: 91–107.

Kidová, A. and Lehotský, M. 2012. Časovo-priestorová variabilita morfológie divočiaceho a migrujúceho vodného toku Belá [Spatio-temporal morphological variability of the braided-wandering river Belá]. Geografický časopis 64(4): 311–333.

Kidová, A., Lehotský, M. and Rusnák, M. 2016. Spatio-temporal geomorphic diversity in the braided-wandering Belá River, Slovak Carpathians, as a response to modern flood periods and environmental changes. Geomorphology 272: 132–149.

Kochel, R.C. 1988. Geomorphic impact of large floods: review and new perspectives on magnitude and frequency. pp. 169–188. *In*: Kochel, R.C., Patton, P.C. and Baker, V.R. (eds.). Flood Geomorphology. Wiley, New York, USA.

Lehotský, M., Frandofer, M., Novotný, J., Rusnák, M. and Szmańda, J.B. 2013. Geomorphic/Sedimentary responses of rivers to floods: case studies from Slovakia. pp. 37–52. *In*: Lóczy, D. (ed.). Geomorphological Impacts of Extreme Weather. Springer, Dordrecht, Germany.

Miller, A.J. 1990. Flood hydrology and geomorphic effectiveness in the Central Appalachians. Earth Surface Processes and Landforms 15: 119–134.

Milton, E.J., Gilvear, D.J. and Hooper, I.D. 1995. Investigating change in fluvial systems using remotely sensed data. pp. 276–301. *In*: Gurnell, A.M. and Petts, G.E. (eds.). Changing River Channels. John Wiley and Sons, Chichester, England.

Miřijovsky, J., Michalková, M.S., Petyniak, O., Máčka, Z. and Trizna, M. 2015. Spatiotemporal evolution of a unique preserved meandering system in Central Europe – The Morava River near Litovel. Catena 127: 300–311.

Opperman, J.J., Luster, R., Mckenney, B.A., Roberts, M. and Meadows, A.W. 2010. Ecologically functional floodplains: connectivity, flow regime, and scale. Journal of the American Water Resources Association 46(2): 211–226.

Phillips, J.D. 2002. Geomorphic impacts of flash flooding in a forested headwater basin. Journal of Hydrology 269: 236–250.

Piégay, H., Darby, S.E., Mosselman, E. and Surian, N. 2005. A review of techniques available for delimiting the erodible river corridor: a sustainable approach to managing bank erosion. River Research and Applications 21: 773–789.

Rusnák, M. and Lehotský, M. 2014. Time-focused investigation of river channel morphological changes due to extreme floods. Zeitschrift für Geomorphologie 58(2): 251–266.

Sládek, J. and Rusnák, M. 2013. Nízkonákladové mikro-UAV technológie v geografii (nová metóda zberu priestorových dát) [Low-cost micro UAV technologies in geography (a new method of spatial data collection)]. Geografický časopis 65(3): 269–285.

Surian, N., Mao, L., Giacomin, M. and Ziliani, L. 2009. Morphological effects of different channel-forming discharges in a gravel-bed river. Earth Surface Processes and Landforms 34: 1093–1107.

Ward, J.V., Tockner, K., Arscott, D.B. and Claret, C. 2002. Riverine landscape diversity. Freshwater Biology 47: 517–539.

Ward, R. 1978. Floods – A Geographical Perspective. MacMillan Press, London.

Wolman, M.G. and Miller, J.P. 1960. Magnitude and frequency of forces in geomorphic processes. The Journal of Geology 68: 54–74.

CHAPTER 21

Human Impact on Mountain Streams and Rivers

Joanna Korpak

INTRODUCTION

For thousands of years, people have chosen river valleys as places to settle due to ready access to freshwater. Initially, only the middle and lower sections of river valleys were populated, with the main forms of human activity being crop cultivation and animal husbandry. Mountain areas were viewed as relatively inaccessible and not very useful for the purpose of agriculture. As human populations rapidly expanded, existing settled lands could no longer sustain all residents. People had to learn how to farm and otherwise manage mountainous areas. By the end of the 20th century, approximately 10% of the world's population lived in mountain areas (Grôtzbach and Stadel 1997). The most popular areas for new settlements were areas along rivers and streams. On one hand, the close proximity to rivers and streams allowed for easy access to freshwater, but on the other hand, increased the risk of flooding. Humans soon learned to "tame" rivers and streams to some extent—this will be discussed later on.

Systematic research on the role of man in the evolution of fluvial systems began in the 1950s. This was quite late, considering the timeframe, scale, and scope of human impact on the natural environment (Gregory 2006). Papers on this subject usually focus on specific rivers that are subjected to specific types of human impact. A good summary of the state of knowledge on human impacts on mountain rivers in various parts of the world is that of Wohl (2006). This particular work provides a broad overview of different types of human impacts on mountain river systems (Table 1). It describes the response of mountain river channels to different forms of human impacts in terms of morphology, discharge, water chemistry, physical properties of water, and biological life. The publication also includes an attempt at evaluating the intensity of select forms of human impacts in different mountain massifs around the world.

Institute of Engineering and Water Management, Cracow University of Technology, Warszawska 24, 31-155 Cracow, Poland.
E-mail: joanna.korpak@iigw.pl

Table 1. Types of direct and indirect human impacts on mountain rivers and streams. Developed from Wohl (2006), with changes.

Direct impact	Indirect impact
Regulation of flow with use of dams or diversions	Deforestation and agricultural land use
In-channel structures (check dams, grade-control structures, culverts)	Afforestation
Bank stabilization	Lode or strip mining
Channelization (levees, wing dams, channel straightening)	Building of transportation corridors (railroads, roads)
Gravel mining from riverbed	Urbanization
Trapping beaver	Altered fire regime
Log driving	
Recreation (fishing, boating)	
Introducing of exotic species	
Riparian grazing	

This chapter focuses on the discussion of river channel response to the most frequently encountered forms of human impacts on mountain rivers around the world. The emphasis herein is on the role of specific types of mountain river regulation. The chapter also discusses alternatives that are often used today to adjust fluvial systems in a way that improves their ecological quality.

Types of Human Impacts

Human impacts on fluvial systems may be divided into direct impacts and indirect impacts. Indirect impacts are what often occur on slopes in a given catchment. This form of impact can alter the water cycle on slopes, which affects denudation and erosion processes that facilitate the supply of slope material to river channels. Direct impact occurs in river channels and contributes to changes in morphology, debris transport, and discharge (Wohl 2006).

The most important forms of human impact around the world include changes in land use (indirect), river-bed gravel mining (direct), and channel regulation (direct) (Soja 1977; Krzemień 1981, 1984, 2003; Punzet 1981; Klimek 1987; Wyżga 1992; Marston et al. 1995; Brookes 1997; Landon et al. 1998; Rinaldi and Simon 1998; Winterbottom 2000; Kondolf et al. 2002; Liebault et al. 2002; Surian and Rinaldi 2003; Simon and Rinaldi 2006). All of these forms of impact yield a change in the type and quantity of transported river material, which destabilizes river channels and alters their morphology and discharge.

The majority of mountain areas have experienced a number of forms of human impacts over centuries of human history. This impact was compounded by natural environmental change, including changes associated with climate shifts, producing variations in flood frequency and magnitude. It is difficult, and often impossible, to separate the effects of different factors on changes in river channel systems. This is particularly true of geographic regions with a long history of human colonization (Rinaldi 2003).

Every form of interference in a river's natural channel system disturbs its dynamic equilibrium. Rivers and streams then "pursue" a return to the state of equilibrium, which often leads to further changes in their morphology. Some of these additional changes are not planned and are not considered desirable from an engineering point of view.

Changes in Land Use

Deforestation and crop farming

As man began to migrate to the mountain regions, logging became an issue. Farming became a mainstay of the human economy and this new lifestyle required clearing of forests. Deforestation encroached upon increasingly larger and higher slope areas in different parts of the world. This process occurred at various points in time and with varying intensity. However, most mountain massifs in the world were affected, even those not experiencing any other human impact (Wohl 2006).

In the European Alps, the practice of deforestation and land cultivation started several thousand years ago and ended only in the 1960s (Marston et al. 2003; Wohl 2006). In the Western Carpathians, the first settlers began to fell trees in the 13th century and would systematically continue to do so until the 1950s (Dobrowolski 1931; Klimek 1987; Adamczyk 1997; Chwistek 2002; Guzik 2004). The largest rate of deforestation in this part of the Carpathians would occur in the 18th century when Wallachian settlers began to graze sheep in the region (Dobrowolski 1931; Škarpich et al. 2011). As late as the early 1900s, about 30% of slope areas found below 1,600 meters of elevation were still used for agriculture in the central part of the Spanish Pyrenees (Lasanta 1989). There are also examples of mountain massifs where people lived in harmony with the natural environment for quite a long period of time; these settlements lived in mountain massifs without making any significant changes in their functioning. One good example of such settlement living in harmony with mountain massifs is the Colorado Front Range in the western United States, which has been inhabited for about 12,000 years, but manmade changes in river channels have only been dated back to the 1850s (Eighmy 1984).

Sedimentological and archaeological studies have shown that changes in the supply of water and slope material in river channels are synchronous with the onset of agriculture (Wohl 2006). As forest coverage declines, precipitation and meltwater runoff increases and accelerates across slope surfaces (Chełmicki 1999). The effectiveness of the wash depends on land use. In forests, this process is relatively insignificant. Its intensity increases substantially across land that is cultivated. Slope wash in areas planted with potatoes occurs with an intensity that is several dozen to several hundred times greater than that in areas planted with cereal crops and grass (Gil and Słupik 1972; Słupik 1981).

As human settlements increased in number and agriculture flourished, the number of dirt roads increased as well. Roads, especially those that line rivers and streams in both woodland and non-woodland catchments, play the most important role in the supply of slope material (Krzemień 1976; Słupik 1981; Froehlich 1982, 1992; Reid and Dunne 1984; Gil 1999; Wałdykowski 2005). Research conducted in Carpathian river channels have shown that 70% to 80% of the suspended material carried by their waters is supplied to their channels through dirt roads during strong precipitation events and snowmelt events (Froehlich 1992).

Plowed slopes no longer protected by vegetation and segmented by roads began to supply fine-grained material to river channels. Research in the Blue Ridge Mountains in the eastern United States has shown that the higher the degree of catchment deforestation, the higher the share of suspended matter in river water (Price and Leigh 2006). The quantity of sediment available for transport exceeded the river capacity. Deposition became the dominant fluvial process. River and stream channels become shallow and broad. Channels often transition from narrow and meandering to broad and braided (Klimek and Trafas 1972; Szumański 1977; Krzemień 1981, 1984; Kukulak 1994). Such altered channels experience higher flood waves, but with a shorter duration (Klimek 1983). Fluctuations in river and stream water levels also increase. As a result, strong droughts alternate with larger and more abrupt flood events across valley floors (Kukulak 1994; Adamczyk 1997).

Afforestation

Forest succession has been observed in recent decades in many regions around the world (Landon et al. 1998; Rinaldi and Simon 1998; Marchetti 2002; Simon and Rinaldi 2006). This process was particularly intensive in the French Alps, especially in the late 19th century. The period is characterized by a mass migration of rural residents to cities. The French government's response was to plant forests on abandoned farmland, especially that found at higher elevations (Rinaldi and Simon 1998; Kondolf et al. 2002; Liebault et al. 2002). In other regions, the process of reforestation proceeded in a more natural manner, without any special planning. It also started at a later point in time. Agriculture in the central part of the Spanish Pyrenees declined in the 1950s and former farmland is now occupied by bushes and trees (Garcia-Ruiz et al. 1997). The 1950s also had a gradual decline of agriculture in the Western and Eastern Carpathians, as well as in the Sudety Mountains. Bushes, trees, and grass began to cover these three regions in the face of the decline of land cultivation (Guzik 2004; Latocha 2009; Rădoane et al. 2013). In general, forest cover in the Polish Carpathians increased by about 15% over the twentieth century (Kozak et al. 2007) (Fig. 1).

The emergence of turf and sometimes woodland helped change the water cycle in the catchment. According to Dynowska (1993), even small changes in forest coverage in a small mountain catchment strongly affect local water circulation patterns. This is a result of the strong water retention capacity of woodland areas. In addition, grassland is also characterized by a strong ability to retain water (Kopeć and Kurek 1975). Reductions in the area of agricultural land have resulted in decreased intensity of soil wash. Other factors have also contributed to a sediment deficit in rivers, including decrease in density of unused dirt roads (Bucała 2014), as well as the abandonment of plowing down the slope and slope terracing (Lach 1984).

As a result, rivers are not carrying as much load as their potential would suggest, which leads to physical changes in the horizontal and vertical profiles of their channels. Many braided river channels found in mountain areas have become meandering channels (Rinaldi 2003). This is particularly true of rivers supplied with fine material originating in catchment slope areas. Rivers supplied with coarse fraction material have often maintained their multi-threaded flow patterns, but their central bars have often become islands reinforced with vegetation (Wyżga 2001; Wyżga et al. 2012). At the same time, river channels have become deeper, often reaching down to solid rock. Studies on rivers

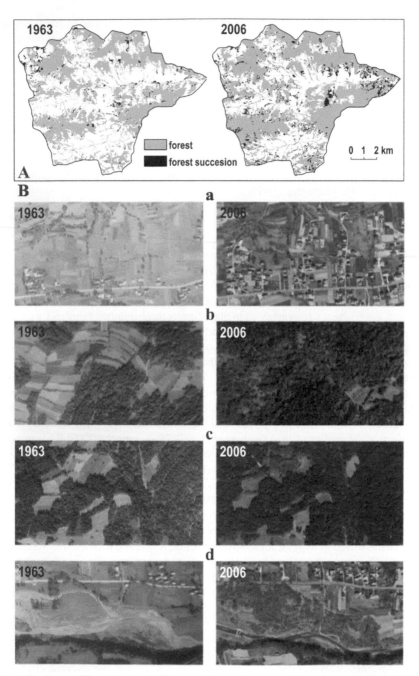

Fig. 1. Unplanned afforestation in the Krzczonówka catchment (Polish Western Carpathians) in the period 1963–2006: A – An increase in the surface area of forests and area experiencing forest succession of almost 14%, B – location of the largest increase in forest area: a – abandoned agricultural fields located across the valley bottom near villages as well as at higher elevations, b – areas situated along the former borders of woodland areas, c – forest clearings previously used to graze sheep, now unused, d – surface areas of formerly wide river channels made useless due to regulation.

in the Pyrenees have shown that river channels have become as much as 2 meters deeper due to reforestation in the last 20 years. At the same time, river channels have narrowed, leaving behind narrow terraces along both banks (Garcia-Ruiz et al. 1997). River and stream channels in the Sudety Mountains in southwestern Poland have become from 1 to 1.5 meters deeper in the last 80 years or so due to reforestation (Latocha 2009).

Channel Gravel Mining

The first papers on the effects of this form of intervention in fluvial systems began to appear in the 1990s (Kondolf 1994, 1997). Mountain river channels are particularly susceptible to this type of intervention due to the unique nature of the material carried by river water. Alluvial matter is sizable, well-rounded, and sorted in terms of both size and resistance. It is also more readily available than material from other sources (Kondolf 1994; Rinaldi et al. 2005).

The extraction of gravel from river channels occurred with varying intensity in different periods of time. In many European countries, this process accelerated after World War II when the destruction of European infrastructure led to a search for alternative sources of construction materials. Channel gravel was extracted on an industrial scale in Italy, which became the primary reason for changes in local river channels (Capelli et al. 1997). For example, more than 24 million cubic meters of gravel were extracted from the Tagliamento River and its tributaries in the period 1970–1991 (Rinaldi et al. 2005). In the eastern part of Romania, an average of 7 million cubic meters of material per year were extracted from the catchment of the Siret River (Rădoane et al. 2013). River debris extraction in the Ropa River in the Low Beskid Mountains of the Polish Carpathians led to a complete loss of alluvial matter from the river channel and the exposure of solid rock in the period 1941–1966. At least a million cubic meters of material were collected (Augustowski 1968). Even larger amounts of debris were extracted from the Wisłoka River in the Low Beskid, up to 2.1 million cubic meters over the period 1955–1964 (Osuch 1968). Calculations of debris transport rates have shown that it will take the river about 500 years to replace this quantity of debris (Osuch 1968). In the Romanian Carpathians, debris extraction has commonly occurred since the 1960s. Many European countries including France, Italy, and Poland have banned river debris extraction. In spite of this ban, some people continue to collect river debris for "home use" (Radecki-Pawlik 2002) (Fig. 2). Some debris is collected legally under the pretense of flood protection (Rinaldi et al. 2005; Wishart et al. 2008).

In addition to the collection of debris from the river channel floor, some material is also collected from the floodplain, which produces the illusion of being less harmful to the natural environment. However, sometimes floodwaters cross the river's threshold and inundate extraction pits, making them part of the active channel. In addition, some debris extraction pits found across the floodplain are deep enough to reach the groundwater table, which then leads to the pollution of groundwater (Rinaldi et al. 2005).

The collection of channel bed material upsets the balance between the amount of bedload being transported and river capacity (Rinaldi et al. 2005; Wyżga et al. 2010). The rate of mining and the quantity of extracted material are normally far larger than the resupply rate and quantity of bedload transported from upstream sections of river or from the river's catchment in general (Kondolf 1997; Kondolf et al. 2002; Radecki-Pawlik 2002). Implications for river channel morphology are described below.

Fig. 2. Selective (illegal) collection and use of river debris by local residents (Photo: J. Korpak).

Change in vertical stability of the channel

Rivers without an adequate amount of transportable material redirect their energy towards channel bed downcutting (Galarowski and Klimek 1991; Kondolf 1997; Radecki-Pawlik 2002; López 2004). Erosion proceeds both upstream and downstream. Headward erosion is induced by a locally increased bed gradient at sites where debris is extracted. Erosion proceeds upstream, covering up to several kilometers of river and encroaches upon tributaries (Kondolf 1997; Marston et al. 2003; Rinaldi et al. 2005). Material transported from sections located upstream is deposited in post-extraction pits. This creates a shortage of material downstream, which means that a river will then erode its channel bed and banks (Kondolf 1994; Marston et al. 2003). The channel deepening pattern described herein differs somewhat in the case of selective debris collection of the largest fractions, which is a common practice among local residents. Research in the Czarny Dunajec River channel have shown that headward erosion occurs in such cases, and no shortages of debris were observed downstream of the deepened section of river, as this section is continuously filled with finer material washed out of the deepened section (Zawiejska et al. 2015).

The magnitude of the deepening process varies from one mountain area to another, and this depends on many factors including extraction period, alluvial layer thickness, quantity of the material arriving from upstream sections of river, and the potential presence of several types of human impacts within a given section of river. Erosion continues for years following the cessation of the extraction process. The resulting depth of the river channel is produced by debris extraction itself as well as years of erosion due to a shortage of debris flowing downstream (Martín-Vide et al. 2010). For example, the channel of the Malnant River in the French Alps increased about 7 meters deeper (Marston et al. 2003), Stony Creek in California (USA) increased about 5 meters deeper (Kondolf and Swanson 1993), the Arno River in Italy increased 1 to 3 meters deeper (Rinaldi and Simon 1998), the Tagliamento River in Italy increased 2 to 3 meters deeper (Surian 2006), and the Gallego River in the Spanish Pyrenees increased up to 6 meters deeper (Martín-Vide et al. 2010). The degree of river deepening also depends on the predominant flow pattern present prior to human impact. In most cases, braided rivers erode to a lesser extent than do single-threaded rivers. For example, channel deepening in the Brenta River in Italy deepened up to 5 meters in braided sections of river and up to 8 meters in single-threaded sections (Rinaldi et al. 2005).

Channel pattern change and narrowing

Deepening is sometimes accompanied by a loss of horizontal channel stability. In most cases, channels become increasingly narrow and braided channel systems transition into single-threaded channels. This type of change was observed in the channels of Italian rivers such as the Tagliamento, Brenta, and Piave, whose width decreased between 53% and 63%. At the same time, the braiding index decreased for the same three rivers (Surian 2006). Some sections of river experience lateral migration or channel widening, as in the case of Cache Creek in California (USA) (Rinaldi et al. 2005).

Armouring

One additional human impact of debris extraction is armouring. Given the shortage of debris supply downstream of extraction pits, selective erosion washes out the finer fractions, leaving behind only coarser fractions, which may protect a river channel from intensive degradation (Begin et al. 1981).

Hydrological and ecological changes

Deeper incising rivers produce fewer floods, as floodwaters usually stay within channel boundaries. This results in the drying of floodplains and the loss of unique and valuable wetland habitats. The drying of floodplains found adjacent to debris extraction pits has been observed in the catchments of the Piava, Brenta, and Tagliamento rivers in Italy (Surian 2006). The same was true of Cache Creek in the United States (Kondolf 1997) and the Rhone River in France (Petit et al. 1996). In Poland, such observations were made for the Ropa and Wisłoka rivers (Augustowski 1968; Osuch 1968). On the other hand, flooding rivers can significantly damage bridges as well as river engineering structures

(Rinaldi et al. 2005). The risk of flooding increases downstream of deepened sections of the river (Rinaldi et al. 2005). Deepened sections also feature fewer fluvial forms, which does not favor the development of some waterborne species (Kondolf 1994).

Regulation of Mountain Rivers

People first began to regulate rivers centuries ago. Early human settlers learned to use the energy provided by fast-moving mountain rivers and streams to construct watermills. Rivers had to be deepened for the purpose of watermill construction. Water would collect in front of a barrier, and retention ponds with dikes were also needed sometimes. Canals used to redirect water flow were often constructed (Kaniecki and Brychcy 2010). Manmade changes in river channels included both vertical and horizontal changes. Wooden thresholds and drop structures were built to reduce gradient. Most of these projects did not form some larger system, but were designed to meet the needs of local populations. Large-scale river engineering projects began in the late 19th century on larger mountain rivers, and in the mid-20th century, on their tributaries.

Today, most river and stream channels in many mountainous regions are regulated to some extent. Most regulation work has been done on the downstream sections of rivers, which are characterized by high population density and intensive land use. The upstream sections of mountain rivers are usually formed in solid rock, which guarantees their stability. The main purpose of river regulation was to protect local communities from floods and the effects of river erosion. A secondary purpose consisted of the acquisition of new land for agriculture.

Hydrotechnical structures most often used in mountain river channels

The following sections describe the need for the use of certain hydrotechnical structures—both transversal and longitudinal—and how these structures function in mountain river channels.

Dams are classified as collection structures, and this includes all types of barriers used to maintain different water levels in a body of flowing water (Wołoszyn et al. 1994). Dams can reach several dozen meters in height. Large manmade reservoirs of water are created upstream of each given dam. Reservoirs are used for a number of purposes including protection from flooding and supplying of drinking water, as well as purely recreational uses.

Weirs also serve as collection structures, but are usually much lower than dams (Fig. 3A). Unlike dams, weirs do not produce upstream reservoirs of water and do not actually store water. The purpose of a weir is to collect water to some degree—and this degree varies depending on local human needs. Artificially elevated water levels can be used to feed irrigation systems, "fuel" hydropower stations, provide a supply of drinking water, and supply water for purely industrial use (Janson et al. 1979).

Debris dams resemble weirs in terms of size and their primary feature is a reservoir that collects bedload. The purpose of a debris dam is to stop the transport of bedload by containing it in a reservoir and to limit the amount of bedload being deposited along downstream sections of river, which helps to protect adjacent land areas from flooding. Debris dams were built along alluvial stretches of river characterized by the intensive transport of bedload and substantial deposition thereof.

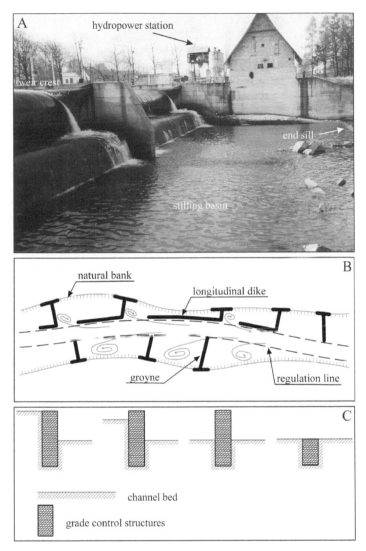

Fig. 3. Hydrotechnical structures often used in mountain river channels: A – weir (Photo: J. Korpak), B – groins and longitudinal dikes, C – grade control structures; B and C based on Dębski (1978) and Wołoszyn et al. (1994), with changes.

Groins and longitudinal dikes concentrate river flow (Fig. 3B). Groins are positioned transversally to a river's direction of flow, and longitudinal dikes—lengthwise. The crest of both structures rise usually as high as the annual average water level or lower. Hence, these structures halt flow at low water stages and decelerate flow velocity at high water stages within the bodies of water confined by them. This function leads to a gradual accumulation of sediment in the groin fields (Łapuszek and Lenar-Matyas 2013). The primary purpose of groins and longitudinal dikes is to reduce the width of a river channel and produce an engineered water flow pathway as well as to protect banks from erosion (Wołoszyn et al. 1994).

Grade control structures are barrier structures that halt water flow across the entire width of the river channel (Wołoszyn et al. 1994) (Fig. 3C). These include drop structures (or check dams), whose height usually ranges between 1 and 2 meters—as well as sills, whose height ranges between 0 and 1 meters. Drop structures are often constructed in groups over longer sections of river. They are often used along river stretches affected by strong channel bed down-cutting. Sills are usually not independent structures, but serve as structural elements of other hydrotechnical structures. The purpose of drop structures is to reduce channel gradient and stabilize the river channel bed. The main purpose of sills is to reinforce existing channel protection structures (Wołoszyn et al. 1994).

Riverbank reinforcements are constructed in the form of independent structures or in the form of associated structures (e.g., with drop structures). Banks are reinforced with a variety of materials, including rip-rap (a layer of various-sized rocks), rocks in netting, concrete, prefabricated elements, fascine, turf, and wicker. The main purpose of riverbank reinforcements is to protect banks from erosion and to prevent channel migration (Surian 1999).

Changes in channel morphology in sections with dams, debris dams, and weirs

From a geomorphologic perspective, the roles of dams, weirs, and debris dams are similar to some extent. The main effect produced by these types of structures is to interrupt the continuity of the fluvial system and bedload transport (Pickup 1977). These structures store bedload and starve the river downstream of material. In effect, the fluvial processes that affect channel upstream and downstream of the structures are different.

As barrier structures halt material transported by the river, its downstream section experiences less bedload and redirects its energy towards channel deepening (Cyberski 1984; Marston et al. 1995; Kondolf 1997). Deposition becomes the predominant process in the upstream section. The river's reaction is similar in the face of these manmade structures, but its scale varies based on the size of each given structure. Sections of river affected by dams are much longer and can even reach tens of kilometers. Sections affected by weirs and debris dams are usually no more than 1 kilometer long (Korpak 2007a). The magnitude of channel changes depends on channel characteristics, bed material size, and the sequence of flood events following structure erection.

Dams, weirs, and debris dams also yield hydrologic and ecological changes in the fluvial system. Dams, and to a lesser extent, weirs, stabilize water flow by increasing low-water levels and decreasing flood levels (Williams and Wolman 1984). Debris dams do not store water and do not regulate water flow. Areas adjacent to dams—both upstream and downstream—experience a change in the water cycle and habitats of living organisms also change. Flooding by reservoirs affects upstream sections. A reduction in flood rates and a lowering of the groundwater table caused by deepened channels are observed in downstream sections of the river (Bunn and Arthington 2002). Moreover, upstream habitats become characterized by increased moisture, while downstream habitats by decreased moisture. Dams also cause changes in river water temperature, dissolved oxygen content, nutrient content, and water chemistry (Wohl 2006). As a result, the plant and animal species composition changes in river channels and adjacent areas (Nilsson and Bergggren 2000). The construction of a dam, weir, or debris dam makes fish migration impossible, which most strongly affects salmon-type fish that migrate upriver to reproduce. It has been shown that populations of these types of fish have decreased substantially in mountain rivers and streams featuring dam-type structures (Mann and Plummer 2000).

Changes upstream of dams, weirs, and debris dams caused by interrupted material transport

The effects of dams, weirs, and debris dams on the morphology of upstream river channel sections are rarely covered in the literature. Existing publications usually focus on changes in the upstream sections of river with a fine-grained channel bed and high suspended sediment load (Xu 1990; Xu and Shi 1997). Few studies cover changes in gravel bed channels in mountain areas (Liro 2014, 2015). Changes upstream of debris dams and weirs are also part of just a few research papers (Ratomski 1991; Korpak 2007a,b; Korpak et al. 2008).

Most studies point to a predominance of deposition upstream of dams (Makkaveyev 1972; Książek 2006; Wiejaczka et al. 2014) as well as weirs and debris dams (Korpak 2007a,b; García-Ruiz et al. 2011). The rate of deposition is the highest in the period immediately following the erection of the structure. The thickness of deposited material and the magnitude of other key changes increases with decreasing distance to the given structure (Korpak 2007b; Liro 2014).

Liro (2014) states that change patterns upstream of a dam may vary depending on the relationship between river transport capacity and sediment supply in effect during the pre-dam period. The period of channel equilibration is the longest in the case of rivers with a shortage of debris in the pre-dam period.

Research upstream of the Czorsztyn Dam on the Dunajec mountain river in southern Poland has shown that channel widening affected the river during the first years (1994–2003) following dam construction (Liro 2015). This can be explained via intensive deposition during this period of time, which reduced the channel capacity. This led to riverbank erosion and the widening of the river channel (Xu 1990; Xu and Shi 1997). The braiding index for this section of river also increased during this time period. This initial period following dam construction was followed by a second (i.e., different) period of change (2003–2009) when the river channel began to narrow and its braiding index decreased (Liro 2015). This was caused by increasing flood deposition across the floodplain, which resulted in increases in riverbank height and increased resistance to erosion (Xu 1990; Xu and Shi 1997). Liro (2015) notes that the Dunajec river channel is likely to continue to narrow in the future, while its sinuosity is likely to increase, which would be consistent with observations for fine-grained river channels in China (Xu 1990; Xu and Shi 1997).

Similar observations were made for Carpathian rivers and streams featuring debris dams (Korpak 2007a,b; Korpak et al. 2008). In the initial reservoir fill period, channels found within the backwater zone became wider. Rivers and streams divided into several branches and water encircled newly formed bars. Once the resulting reservoir became relatively full, channels in the backwater zone became narrow and transitioned into single-threaded channels. A number of lateral sandbars then transitioned into a newly formed floodplain.

The fill rate for reservoirs of debris dams is usually larger than that assumed by river engineers at the dam planning stage (Ratomski 1991; Korpak 2007b). For example, a reservoir becomes filled with sediment only at high discharge (Ratomski

1991). Floodwaters sometimes extend the life of a reservoir completely filled with sediment by tearing away vegetation and reshaping its surface. Fast-moving water can unclog old side arms and can increase the active surface of a reservoir (Korpak 2007b).

According to Ratomski (1991), sedimentation patterns in manmade reservoirs behind debris dams depend on reservoir geometry, which in turn depend on the width of the valley floor at a given location. Wider reservoirs tend to accumulate coarser debris in their initial period of existence. Coarse debris is transported by the main river channel as far as the dam. Its deposition creates additional resistance to movement and a general change in flow patterns, which stops the downriver movement of subsequent waves of coarse debris. This material becomes accumulated in the form of a tongue in the rear part of the body of the reservoir. Intense deposition of suspended matter, also accelerated by permanent and seasonal vegetation, occurs in the main part of the reservoir, which further limits bedload movement. The pattern of sedimentation is different in the case of narrow reservoirs. These tend to capture mostly coarse debris consisting of sand, gravel, and rock (Ratomski 1991). These types of material fill in the reservoir evenly. From the perspective of dams designed to halt any bedload and allow suspended matter to pass, narrow reservoirs appear to be most appropriate for this purpose.

Case study. The Porębianka Stream is a Carpathian gravel-bed stream. It is about 15.5 km long, with a mean gradient of 0.035. Elevations in its catchment range from 400 meters to 1,276 meters. A debris dam was built at the 5.4th kilometer in 1960, with a height of 5.6 meters, capacity of 173.2 m^3, and surface area of 65,000 m^2. The expected lifespan of the reservoir was estimated to be 55 years. By 1976 the reservoir was already 68% full, 76% full by 1978, and 81% full by 1992. By 2009 the reservoir was completely filled with sediment and overgrown with vegetation, with only one functional narrow and sinuous channel (Fig. 4). The actual rate of deposition (7,361 m^3/yr) was more than twice the estimated rate of deposition (3,150 m^3/yr). Turf began to encroach upon the newly deposited debris—and wicker followed. The river began to pile up debris at the narrow end of the reservoir, as this is the site where it loses energy by dividing into several branches. As these branches became dry and filled in with vegetation, material was deposited in the main river channel increasingly close to the dam.

Changes downstream of dams, weirs, and debris dams caused by interrupted material transport

A transition from a multi-threaded channel to a single-threaded channel often occurs downstream of a dam. Such transitions were observed in Stony Creek in the United States (Kondolf and Swanson 1993), Ain River in France (Marston et al. 1995), Arno River in Italy (Rinaldi 2003), Mszanka and Porębianka Streams in Poland (Korpak 2007b), and Moravka River in the Czech Republic (Škarpich et al. 2013).

The most important fluvial process affecting a narrowed river channel is channel bed downcutting. The rate of erosion is the highest immediately after a structure is constructed (Williams and Wolman 1984). Most rivers analyzed by Brandt (2000) attained half of their additional depth within the first seven years of dam construction. The erosion rate peaks just below the structure (Williams and Wolman 1984). Erosion then declines substantially, especially if the river reaches bedrock or if/and armour has developed via selective washout of fine fractions out of alluvia (Kondolf 1997; Brandt 2000; Wierzbicki and Wicher 2002). Bed armouring, consisting of a coarse surface

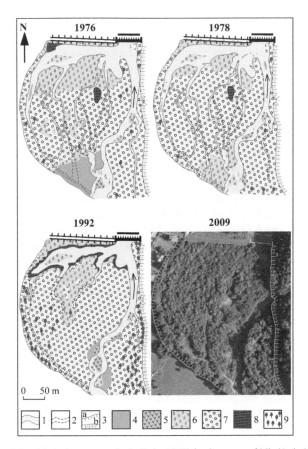

Fig. 4. Filling of the debris dam reservoir (built in 1960) in the town of Niedźwiedź, as shown on sketches made by geography students doing fieldwork in 1976 and 1978 as well as by Komędera (1993) in 1992: 1 – channel with flowing water, 2 – dry channel, active during flood events, 3 – riverbank scarps with the following heights: a – less than 5 m, b – more than 5 m, 4 – bars not reinforced with vegetation, 5 – bars reinforced with vegetation, 6 – bedload reinforced with turf, 7 – bedload reinforced with wicker, 8 – wetlands, 9 – areas with trees.

layer overlying finer sediment, makes the channel bed more resistant and helps delay mass transport of material (Pickup 1977; Wyżga 1992; Bartnik and Strużyński 1998; Rinaldi and Simon 1998). Armouring may be torn away at large discharge velocities, and subsequent stages of channel deepening then occur in the river. The leading edge of the erosion zone tends to shift downstream (Babiński 1992; Malarz 2002). Hence, the length of the eroded channel section increases over time. The degree to which river channels become deeper varies depending on many different factors. In most cases, erosion downstream of large dams is stronger than erosion downstream of weirs and debris dams. A river channel may become as much as 7 meters deeper downstream of a large dam (Williams and Wolman 1984). In the case of weirs and debris dams, river channels are usually about 2 meters deeper in the downstream section (Korpak 2007b).

As river channels become deeper, they also become narrower. The rate of this process is largest in the time immediately following the construction of a training structure. When a river cuts into the bedrock, its channel narrows, and thin rock shelves form along its course. Large areas of formerly active river channels become overgrown with vegetation and transform into floodplains.

Channel deepening and narrowing downstream of dams and weirs are facilitated not only by a lack of bedload from upstream sections of river, but also by a lack of material from areas immediately adjacent to river channels. Vegetation invades channel-adjacent areas, which limits the supply of slope material.

Case study. The processes and changes in channel morphology downstream of a debris dam described earlier are presented in this section in the context of the Mszanka River in the flysch Western Carpathians in southern Poland. The Mszanka River is 19.5 km long; its elevation range starts at 373 m at the mouth and reaches 1,311 m at the highest point in the catchment. The mean annual flow is 3.25 m³/s. The studied channel reach was braided and alluvial until 1959. Its width was 30 to 50 m. In 1961, a concrete debris dam (3.5 m high) was built at kilometer 7.835 to reduce further sediment transport and prevent flood risk in this reach. The surface area of the studied reservoir was 3.02 ha and its capacity was 90,700 m³. The lifespan of the reservoir or the time it takes to fill in its entire volume was calculated to be 25 years. This reservoir became filled in much faster than expected. It was largely full by 1975 or just 14 years after it was constructed. About 68% of its surface area had become fixed by vegetation (Fig. 5A). The Mszanka River had several channels and a number of islands within the reservoir area. Its channel was braided upstream of the reservoir and had a width of 35 to 50 meters (Fig. 5A). In 1981, the thickness of deposited material reached 4 meters in the reservoir and 1.5 meters near the bridge (Fig. 5C). In 1997 virtually the entire reservoir was filled with debris and overgrown with vegetation (Fig. 5B). At this point in time, the Mszanka River had a single-threaded channel with a maximum width of 35 meters.

The erection of the debris dam has starved the river channel downstream of material. The channel gradient increased from 0.010 to 0.012 downstream of the dam. This resulted in intensive river channel downcutting. By 1967, the river channel bed directly below the dam lowered by 1.5 meters. The bed was about 0.5 meters lower at a distance of about 200 meters downstream (Fig. 5C). The river reached solid rock at a distance of 7.59 km downstream of the dam (Fig. 5C). The channel width decreased from about 50 m to 9 m and the dominant process transforming this channel reach changed from re-deposition into downcutting (Korpak 2007a). In order to reduce further channel incision, a concrete drop structure (1.2 m high) downstream from the debris dam was built (Fig. 5A). However, the measures proved ineffective and channel bed degradation continued until bedrock was exposed. In 1997, many rocky outcrops were found in the channel (Fig. 5B). The previous floodplain, which was active several decades ago, has been transformed into a terrace. On either side of the channel, narrow rocky bands (1–2 m wide) developed which are initial floodplains today.

Changes in channel morphology along reaches regulated with groins and longitudinal dikes

The construction of groins and longitudinal dikes ultimately leads to a narrower river channel. The filling in of intra-groin basins results in the incorporation of channel sides into the floodplain (Liro 2012). The rate of intra-groin basin filling is usually highest immediately after regulation works, with a gradual reduction that follows (Citterio and Piégay 2009). For example, the mean rate of sedimentation in the Rodan

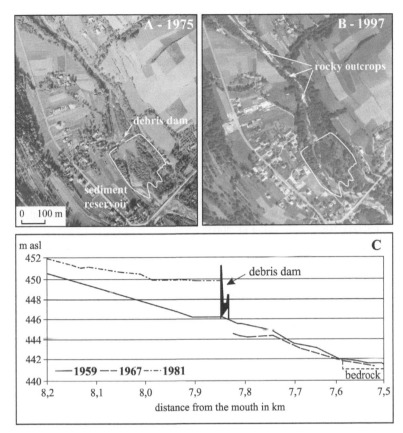

Fig. 5. Differences in the Mszanka River channel morphology upstream and downstream of the debris dam at Mszana Górna (built in 1961): A – deepened, narrow channel downstream of the dam and a wide, braided channel upstream of the dam, B – additional deepening and narrowing of the river channel downstream of the dam (thin rocky shelves along both riverbanks) as well as a sinuous depositional reach upstream of the dam (already filled with debris), C – changes in the channel bed level in the section with the debris dam resulting from different channel processes acting upstream and downstream of the structure (based on regulation projects no. 2196, 1682, 3375).

River channel in France was about 2.5 cm per year (Citterio and Piégay 2009). In the upstream sections of the Vistula and Odra River channels in Poland, the mean rate of sedimentation ranged between 1.7 and 6.0 cm per year (Czajka 2007). The analogous rate for the Dunajec River channel in Poland was about 5 cm per year (Liro 2012). Post-regulation river channels are usually shorter and less sinuous than natural channels. In the case of braided mountain rivers, side channels are cut off in the regulation process and a single-threaded channel is produced. Changes of this type have been observed in many mountain rivers regulated using longitudinal channel structures (Surian 1999; Rinaldi 2003; Korpak 2007a; Korpak et al. 2008; Škarpich et al. 2013).

The concentration of flow in this type of altered channel, which is also characterized by a larger gradient, leads to increased erosive power and a reduction in flow resistance. Transport capacity increases due to a reduction in the sediment supply to the channel caused by relatively ineffective erosion of newly reinforced riverbanks. As an effect, a rapid river incision begins. The rate of channel bed down-cutting is the largest immediately following the process of river regulation. As the channel bed becomes lower, riverbanks become higher. In effect, the previously active channel becomes a floodplain (or even terrace) and vegetation invades over time. This further reduces the amount of material being transferred to the river and reinforces its incision and narrowing tendencies (Marston et al. 1995; Bravard et al. 1999).

Once a river channel becomes significantly deeper, it is unlikely that floodwaters will affect adjacent areas. A reduction in the frequency of riverbank flooding limits opportunities for sediment deposition outside of river channels, on floodplains (Wyżga 2001). Most suspended material is transported to downstream reaches, which become shallower over time (Korpak 2007b; Kiss et al. 2011). This gradual process then increases flood risk in areas downstream of regulated sections of river (Bormann et al. 2011). This problem affects many piedmont areas. For example, rapid supply of material by Carpathian rivers in Poland has reduced the depth of the channel of the middle Vistula River (Łajczak 1997; Bojarski et al. 2005).

When a river channel becomes shorter and more narrow and its banks are reinforced, which prevents lateral migration, the only way for a river to reach a new state of equilibrium is to reduce gradient via headward erosion and accumulation of migrating alluvial material downstream of a deepened section of river (Parker and Andres 1976; Petit et al. 1996; Rinaldi and Simon 1998; Wharton 2000; Kondolf et al. 2002; Surian and Rinaldi 2003; Simon and Rinaldi 2006). Rivers regulated in this manner are characterized by much greater flow energy than natural rivers experiencing the same environmental conditions (Capelli et al. 1997).

The rate of channel adjustment depends on the resistance of the channel bed and riverbanks to erosion as well as on unit stream power—defined as stream energy per unit bed area, given a defined rate of water flow (Brookes 1987; Wharton 2000; Bojarski et al. 2005). The return to equilibrium for channels that used to be multi-threaded, and covered with a thick layer of alluvial material follows a very different pattern (including rate of change) than that for channels that are initially single-threaded, and covered with a thin layer of alluvial material (Korpak et al. 2008; Zawiejska and Wyżga 2010).

Morphological changes in channels that are initially multi-threaded and covered with a thick layer of alluvia

Channel bed down-cutting in river channels of this type occurs rapidly, especially immediately after regulation is completed, and substantially increases channel depth. An effect of channel deepening is the exposure of the foundations of groins or longitudinal dikes and gradual damage to these key structures. This process accelerates substantially following major flood events during which these training structures often become completely destroyed. Once groins are removed, affected riverbanks become highly susceptible to erosion, and a river begins to wander, increasing its length, sinuosity, and width. Lateral erosion leads to an increased supply of transportable material, which significantly reduces the intensity of the channel deepening process (Simon and Rinaldi 2006). However, the channel does not return to its previous width. This is impossible due to significant deepening and the establishment of at least some permanent riverbanks lined with young trees. The inability to return to original width by artificially narrowed rivers has been noted for Carpathian streams (Korpak 2007a; Korpak et al. 2008) and the Tummel River in Scotland (Winnterbottom 2000). In addition, rivers do not reproduce braided channels in existence prior to regulation works, but exhibit a tendency to meander. This is a fairly typical fluvial response designed to yield an equilibrium profile given an increased gradient and less sediment storage (Schumm 1968).

Case study. One example of the aforesaid patterns is that of the downstream section of the Biały Dunajec River, an example of a trained gravel-bedded mountain river. The Biały Dunajec River (35 kilometers long) begins in the Polish Tatra Mountains and joins the Dunajec River in the Orawsko-Nowotarska Basin located north of the Tatra Range. Some 35 percent of the catchment network is located above 1,000 m a.s.l., with the mouth at 580 m a.s.l. The mean annual flow near the mouth is 5.5 m³/s. The studied downstream section of the river channel was regulated using groins in 1971. The channel reach performed an accumulation function prior to river regulation and was a typical braided channel (Fig. 6A). River regulation reduced the width of the channel—reaching 350 m in places—to 30 m along a 3 km stretch (Fig. 6B). The new post-regulation route was to follow a nearly straight pattern, which made the new river channel 18%

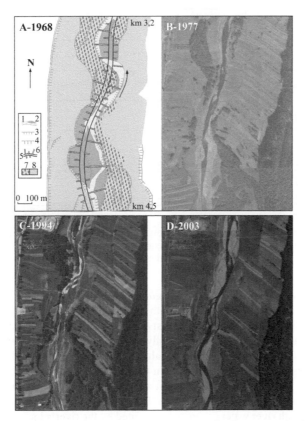

Fig. 6. Changes of the Biały Dunajec River channel at the town of Szaflary as a result of river regulation in 1971 using groins: A – braided channel prior to regulation (based on project no. 3964), B – straightened, narrowed, and deepened channel six years after regulation work, C – increased lateral erosion at sites previously occupied by groins, D – altered channel characterized by a greater tendency to meander after floods in 1997 and 2001; 1 – channel with flowing water, 2 – bar, 3, 4 – edges of riverbanks with the following heights: 3 – 0 to 2 m, 4 – 2 to 5 m, 5 – designed training path, 6 – groins, 7 – forest, 8 – agricultural land.

shorter. The gradient of the channel increased from 7.7 to 8.6‰, which helped produce stronger channel bed down-cutting. In the course of just six years after regulation, the channel became deeper by about two meters and today it is up to four meters deeper (Fig. 7). Channel deepening led to a gradual destruction of groins. The river began to undercut unreinforced banks at sites where groins had become damaged. By 1994, the length of the river channel had increased 100 m along this particular stretch (Fig. 6C). Floods in 1997 and 2001 destroyed all groins and the river channel changed completely (Fig. 6D). It gained another 60 m in length and its width became variable ranging from 43.4 to 141.9 m. The average width of the channel increased from 35.4 m in 1994 to 72.6 m in 2006. The sinuosity index increased from 1.02 to 1.04 over the first 17 years since regulation work. It further increased to 1.08 over the following nine years—the period after the destruction of river engineering structures. Today, the river tends to meander along this particular stretch.

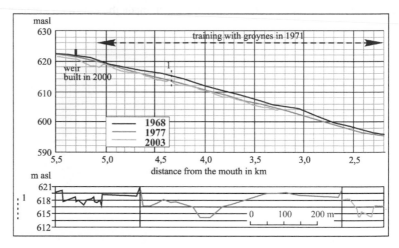

Fig. 7. Changes in the longitudinal profile and cross section of the Biały Dunajec River channel shown for a section regulated in 1971 using groins (on the basis of regulation projects no. 3964 from 1968, 101 from 1977, and 2064 from 2003).

Morphological changes in initially single-threaded channels with a thin layer of alluvial material. Once a channel floor with a thin layer of alluvial material is regulated using groins or longitudinal dikes, intensive channel bed downcutting leads to rapid disruption of alluvial cover and subsequently exposure of bedrock. The erosion rate then decreases significantly, but does not stop altogether. However, a river does not cut into rock across its entire width, but mainly in the middle of the channel (Korpak 2007a; Korpak et al. 2008). Rocky shelves form along the riverbanks. As the river channel become deeper, it also becomes narrower. Reinforcements are no longer damaged, as their contact with river water becomes interrupted by the river's incision into bedrock. Ultimately, reinforcements "hang" over the channel and no longer play any role in the life of the river.

Case study. This situation is illustrated well by the Mszanka Stream in the town of Mszana Górna. The studied Mszanka Stream reach (km 7.0–6.2) had an alluvial channel prior to regulation (Fig. 8A). The width of its active channel (with bars) varied between

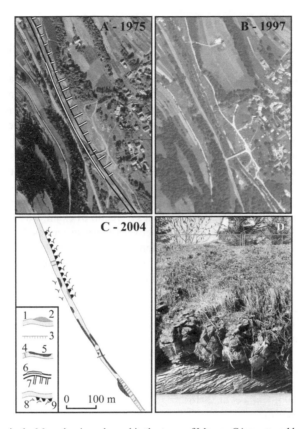

Fig. 8. Changes in the Mszanka river channel in the town of Mszana Górna caused by river regulation in 1978 using groins: A – multi-threaded alluvial channel prior to regulation, with the training design shown (based on project no. 114), B – single-threaded, narrow, deepened channel after regulation, C – morphology of the channel incised to bedrock (based on a field survey), D – longitudinal dike "hanging" over a deepened channel (Photo: J. Korpak); 1 – channel with flowing water, 2 – bar, 3 – cutbank, 4 – rocky step, 5 – rocky outcrop, 6 – longitudinal dike, 7 – groin, 8 – wicker, 9 – rip-rap.

30 and 90 m. The channel was regulated in 1978. The purpose of the regulation was to protect the left bank from erosion; thus, protecting the road that ran alongside. The new channel had a trapezoidal cross section and uniform width –16 m at bed level. The left bank, with its adjacent road, was reinforced with a longitudinal dike (Fig. 8A). The right bank of the Mszanka River was reinforced using groins. The substantially narrowing of the channel (by 32%) resulted in increase of the flowing water energy and intensification of channel bed down-cutting. As the river channel became deeper, it also became even narrower. The river began to cut alluvia apart, reaching bedrock, which also became incised to some extent. Thin rocky shelves appeared along both riverbanks (Fig. 8B,C). The channel became more than one meter deeper between 1978 and 2004 at some locations—the river cut into a layer of alluvial material and one meter of solid rock. The groin fields became filled and covered with vegetation. Today, these fields functions as a terrace. Longitudinal dikes securing the left bank are "hanging" over water level (Fig. 8D).

Changes in channel morphology in sections with drop structures

A staircase-type sequence of transverse structures is one of the most often used means of regulation in mountain rivers (Radecki-Pawlik 2013a). The use of drop structures requires that a natural river channel be transformed into a single-threaded regulation channel. Side arms of channels are cut off, which often leads to a transition from a braided to a single-threaded pattern. A river channel becomes straightened, shortened, and narrower. Riverbanks are most often reinforced along the entire length of the regulated stretch or along concave bends. The channel gradient is adjusted via drop structures in order to prevent channel deepening, which is quite intensive under such conditions (Galia et al. 2016). Lateral erosion is not very effective due to riverbank reinforcement. The river channel pattern becomes stable.

The river channel attempts to regain its vertical stability following an event that upset its equilibrium. The literature discusses a wealth of examples of parameters for drop structures and the channel gradient between these structures designed to assure equilibrium between potential erosion and deposition processes (Lenzi et al. 2003; Galia et al. 2016). Research has shown that the highest rates of success in river regulation are provided by designing drop structures that help mimic natural riffle-pool sequences (Marion et al. 2004).

A scour forms directly below a new drop structure (Lenzi and Comiti 2003; Galia et al. 2015). The maximum depth of a scour depends on specific critical flow energy and drop height (Lenzi et al. 2003) as well as sediment supply from upstream reaches (Marion et al. 2006). Increased debris supply helps decrease maximum scour depth. Local scour below drop structures threatens their stability. This problem may be solved by building an overflow basin and end sill right below the drop structure of interest (Lenzi et al. 2003; Galia et al. 2016).

Deposition is the predominant process upstream of drop structures, up to the point when sedimentary wedges become filled in (Galia et al. 2016). The rate of deposition depends on sediment supply from upstream sections of river and from river bank erosion (Korpak 2007a, 2015). Once sedimentary wedges become filled in, the channel regains transport connectivity, especially in the case of the fine grain fraction (Boix-Fayos et al. 2007; Korpak 2007b).

Studies in the Czarny Dunajec River channel have shown that a section of river about 8 km in length and regulated using drop structures between the early 1960s and late 1990s was fully filled with debris already by 1999 (Krzemień 2003). This fill material is characterized by coarse fractions and fairly good sorting, which indicates that fluvial transport was very important in this section of river during the period of interest (Zawiejska et al. 2015).

In the initial period of functioning, a sequence of drop structures plays a role similar to that of a dam by making material transport impossible, especially in the case of coarse fractions. A shortage of debris in the section located downstream of this sequence of drop structures leads to channel deepening and narrowing as well as bed armouring (Boix-Fayos et al. 2007; Castillo et al. 2007; Wohl et al. 2013; Korpak 2015). This type of process was observed in the Wear river channel in northern England, where drop structures were used upstream of material extraction areas. The purpose of these structures was to reduce gradient and retard headward erosion. These actions resulted in downcutting erosion downstream of the drop structures (Wishart et al. 2008). With a limited supply of debris from the section of river upstream of the described drop structures, the process of downcutting erosion is particularly intensive and long-lasting (Galia et al. 2015). This pattern was observed in the Mszanka Stream in the Polish Western Carpathians. A 5.7 km section of this stream had been regulated using a sequence of drop structures (Korpak 2015). An old debris dam is found upstream of the regulated stretch, which limits the supply of debris from the upstream section of the stream. The banks of the stream section with drop structures were reinforced, which makes it impossible for material to be supplied by lateral erosion. In effect, the Mszanka channel was deepened as far as solid rock not only downstream of the drop structures, but also within the section with drop structures. Outcrops and rock thresholds can be found between the drop structures, while the structures themselves appear to be damaged to some extent. This example illustrates that in an extreme case of material shortage in a channel, hydrodynamic equilibrium in a river channel regulated via drop structures may not be achieved (Korpak 2015).

River regulation using drop structures is not matched with variable water stages and discharge over the course of the year (Korpak et al. 2008; Korpak 2010). The channel is too wide at low water stages. Narrow lateral bars tend to form, which reduces the channel cross section. Vegetation sometimes invades the new bars and stabilizes them. During floods, the flow velocity increases in a straightened channel regulated with drop structures, which produces changes in channel bed morphology—old bars are washed away and new bars are formed. A river attempts to make its channel more sinuous by eroding its bed and banks. River engineering structures become destroyed—especially crests of drops, stilling basins, as well as bank reinforcements (Korpak 2010; Galia et al. 2016).

Case study. The Krzczonówka Stream is located in the Polish Western Carpathians, in the catchment of the Raba River. The region is built of Carpathian flysch, composed primarily of sandstone and shale. The area of the Krzczonówka Stream catchment is 92.2 km^2, the river channel is 17 km long. The highest peak in the catchment lies at an elevation of 866 m a.s.l. and the mouth of the river is located at an elevation of 329 m a.s.l. The mean annual flow is 1.66 m^3/s. The studied reach (km 4.6 to 10.3) was regulated using drop structures in the 1980s. Prior to regulation efforts, the river was braided and wide with numerous central and lateral bars of substantial size (Fig. 9A). The regulation project for the river created a new and single-threaded channel characterized by little sinuosity. Its cross section was trapezoidal in geometry and its width was uniform at about 20 meters (Fig. 9B). The mean width of the channel decreased from 66 meters to only 11 meters in the period 1963–2006 (Fig. 9C). A few small bars are found between drop structures on the river. A lack of material can be observed in the river channel downstream of the regulated stretch, as fluvial material is now halted by the new drop structures. Sediment trapping between the drops caused the transformation of the river channel below them from depositional-type to erosional-type. Downcutting became the dominant process since the period of regulation over a stretch of about 800 meters of river. The mean width of the channel (along this section) decreased from 87 meters in 1963 to 51 meters in 1977. A further reduction to about 16 meters occurred by 1993. The width of the regulated section has virtually not changed since. The primary fluvial forms in this section are rocky outcrops and rocky steps with a mean height of 0.5 m.

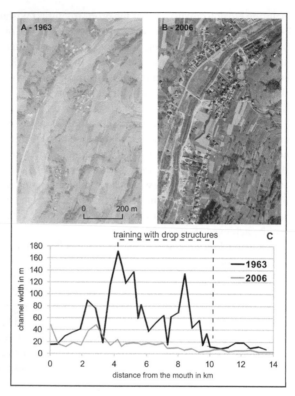

Fig. 9. Changes in channel morphology in the Krzczonówka Stream following regulation using drop structures along the 4.6 to 10.3 km stretch; regulation was completed in the 1980s. A – wide and braided reach of river channel prior to regulation, B – straightened and narrowed reach of regulated river channel, C – width reduction and consolidation of the Krzczonówka River channel in the period 1963–2006.

Assessment of Human Impact and Restoration Activities

The main outcome of all forms of human impact in mountain river catchments, both direct and indirect, has been a reduction in the supply of sediment that reaches river channels. The equilibrium between river storage and transport capacity has been upset. Rivers have attempted to regain a certain state of equilibrium by carving new dynamic profiles via vertical and horizontal changes in channel morphology (Chang 1988). The primary ways in which river channels adapt to new conditions—observed in many rivers in mountain areas in the 20th century—were channel deepening and narrowing. These two ways of adjustment were predominant in mountain area rivers in France (Liebault and Piegay 2001), Scotland (Gurnell 1997; Winterbottom 2000), Italy (Surian 1999; Rinaldi 2003), as well as Poland (Korpak 2007a; Korpak et al. 2008; Zawiejska and Wyżga 2010). The rate of change in channel morphology has been noted to be most rapid during the short period following an event upsetting the river's equilibrium; subsequently the rate tends to decrease (Williams and Wolman 1984). Rates of change and change patterns are also intensely affected by catastrophic flood events, which in most cases tend to accelerate ongoing change processes (Soja

1977). In some instances, catastrophic flood events can completely change the manner of river channel evolution (Hooke 1997).

Channel adjustments, due to lateral and vertical instability, has prompted a number of negative environmental and societal outcomes (Bravard et al. 1999; Rinaldi 2003). The first is increased flood risk that affects areas downstream of the deepened section of river (Pinter and Heine 2005). Bridge supports and regulation structures are damaged in the course of floods that stay within the banks of a deepened river channel. Areas adjacent to the river channel tend to experience excessively dry conditions due to a lower groundwater table triggered by channel bed incision (Arnaud-Fassetta 2003). At the same time, the newly higher riverbanks do not permit the river to flood surrounding areas in a cyclical manner. This yields implications for the human economy via the loss of fertile riverside land as well as for the realm of biodiversity, which was rather substantial in the now lost riparian forests. A channel that is deepened, and often reaches bedrock, is also morphologically homogeneous, which does not help organisms living in the water, especially fish and large benthic invertebrates (Kłonowska-Olejnik and Radecki-Pawlik 2002). Today many mountain rivers are characterized by poor ecological conditions (Wyżga et al. 2009). In order to improve the ecological condition of rivers, a brand new approach is needed in the way water resources are managed.

Restoration efforts have been underway in many parts of the world over the last 30 years promoting river channel and floodplain connectivity (Shields et al. 2011; Gumiero et al. 2013) and longitudinal connectivity of water flow and sediment transport (Konrad et al. 2011; Wohl et al. 2015) as well as ecological quality (Palmer et al. 2010; Lepori et al. 2005). Given the recent nature of these efforts, it is quite difficult to predict their long-term effects on the morphology and functioning of river channels, water quality, and biological communities. It appears that even well-planned projects can yield unexpected results (Wohl et al. 2015). For example, studies have shown that almost all of the 400 renaturalization projects in southern Appalachian river channels damaged by debris extraction produced disappointing results in terms of the improvement of the ecological quality of rivers (Palmer and Hondula 2014).

It must be clearly stated that existing traditional river engineering structures usually have not produced expected results and have damaged the ecological state of rivers. The least effective method appears to have been efforts to shorten and make narrow river channels in mountain areas, usually via the use of groins or longitudinal dikes. This method led to rapid channel deepening, greater flow energy, and increased flood risk in downstream areas.

Many researchers today are actively looking for ways to delay river discharge, which is supposed to reduce flood risk in river valleys (Poulard et al. 2005). A variety of solutions are being sought with respect to many different parts of flood-prone catchments. The gradual increase in the surface area of forest over the last few decades in many mountainous regions increases the retention capability of the catchments. The elimination of levees and riverbank reinforcements is proposed wherever feasible, especially since these structures do not protect anything of real value at many locations (Galia et al. 2015). An erodible channel corridor should be left for a river in broad, unsettled valley bottoms (Wyżga and Zawiejska 2012). This corridor would be a broad migration zone, where a river could freely evolve in terms of its course and channel morphology (Fig. 10). At low water stages, a river can flow using a fraction of the zone

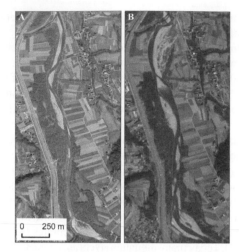

Fig. 10. A "free migration corridor" situated on the Raba River, formed unintentionally as a result of the destruction of a longitudinal regulation structure during the flood of 2010: A – regulated channel section prior to a flood, B – "free migration corridor" after a flood.

(Fig. 11A), while at high water stages, it can use an entire corridor width (Fig. 11B). The erodible channel corridor can also increase the channel storage of flood water and slow down the passage of flood waves (Laszek et al. 2014) (Fig. 11C). A decrease in the river flow velocity leads to a high rate of deposition of flood material (Fig. 11D).

Traditional dams, debris dams, weirs, and drop structures need to be rebuilt in a way they do not stop bedload transport, as the main problem in most mountain rivers and streams today is the lack of bedload. In addition, the ban on channel debris extraction needs to be strictly enforced.

Dams used in decades past produced large changes in mountain river regimes. Their basic flaw consisted of their ability to retain most debris including suspended matter whose transport is not to be limited (Żbikowski and Żelazo 1993). Dams built in recent years do not halt all bedload, but regulate the pattern and rate of bedload transport (Bartnik and Strużyński 1998).

Many projects have been undertaken around the world, especially in the United States, which involve the demolition or the lowering of existing dams of various size (Gleick et al. 2009) in order to restore fluvial system connectivity. Still, there are no answers to many questions regarding the reaction of a river channel to such an intervention. The problem that is still discussed the most is the amount and rate of transport of material deposited in a reservoir (Shuman 1995; Doyle et al. 2003) and the consequences of the lowering of the erosional base for a segment that lies upstream (Doyle et al. 2003). Long-term channel changes following the lowering of a dam are due to the performed works and flood events, which occur after the completion of the project (Stanley et al. 2002). A sudden influx of sediment after the lowering of the dam may quickly destroy the structures on the channel bottom, and bury riffles and pools (Lisle et al. 2001). Rivers usually compensate for reduced depth by widening their channel (Pizzuto 2002). Rivers upstream from the dam most frequently react through headward erosion. The time necessary for the gravel bed channel to regain its balance after such a disturbance would last several years, depending on the frequency of large

Fig. 11. Functioning of the "free migration corridor" situated on the Raba River at different water stages: A – at low water stages, B – at high water stages, C – during a flood event, D – deposition of material carried by the flood (Photo: J. Korpak).

flood events (Madej 2001). For example, a major flood occurred immediately after a debris dam on Krzczonówka Stream in the Polish Carpathians was lowered in 2014 (Lenar-Matyas et al. 2015) (Fig. 12A). The Krzczonówka channel was completely re-shaped. Almost all sediment was eroded out of the reservoir and transported to the channel segment downstream. It was not only fragments of pools, but also the riffles that were damaged and covered with sediment whose thickness ranged from 0.21 m to 0.91 m (Lenar-Matyas et al. 2015) (Fig. 12B). The dam reservoir had almost entirely eroded and all sediment was washed out of it in one flood event. The river channel bottom here was lowered by as much as 2 m (Fig. 12C). One year after the event, a primary role in the further development of the channel was played by the widening of the channel by lateral erosion or by landslides from the riverbanks, which had lost their stability due to the deepening of the channel. A similar case of river channel adjustment to weir removal was noted in the gravel-bed Monnow River in Wales (UK) (Thomas et al. 2015). The channel responded via degradation and widening upstream of the former weir as well as aggradation and narrowing downstream of the former weir.

Another means of improving the ecological state of a river with a dam is the addition of debris to the river section downstream of the dam. The primary purpose of such a step is to help restore fish habitats. This type of effort is pursued in many European countries, as well as in Japan and the United States (Kondolf et al. 2014).

Traditional grade control structures disrupt channel continuity by preventing sediment transport and establishing a barrier that fish from the salmon family cannot overcome. One potentially workable solution consists of the removal of grade control

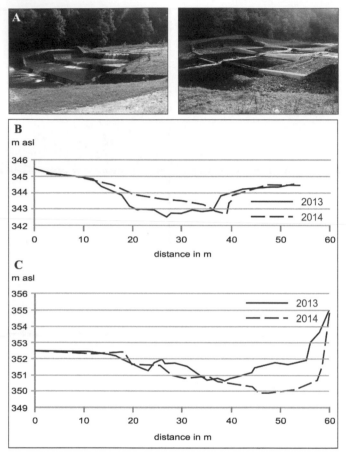

Fig. 12. Effects of debris dam lowering and flood occurrence on changes in channel morphology: A – debris dam prior to lowering (year 2013) and after lowering (year 2014) (Photo: J. Korpak), B – deposition downstream of the dam, C – headward erosion and lateral erosion upstream of the dam.

structures and bank reinforcements, as in the case of the Mareit River in Italy (Wohl et al. 2015). The river channel became much broader and began to follow a braided pattern. This type of solution is not workable in built-up areas situated along rivers. In such places it is still possible to improve te ecological state of rivers by replacing a sequence of traditional grade control structures with hydraulic structures resembling natural rapids (Lenzi 2002; Radecki-Pawlik 2013b). The new structures are constructed using natural features that increase the roughness of the river channel (Fig. 13). They play the role of traditional concrete drop structures by reducing gradient, but are more aesthetic in the mountain landscape. In addition, they bring morphologic diversity to streams and rivers by imitating naturally occurring riffles, which are a characteristic fluvial landform in mountain river channels (Radecki-Pawlik 2013b; Galia et al. 2015; Lenar-Matyas et al. 2015). Furthermore, rapid hydraulic structures provide sufficient connectivity for aquatic organisms and sediment transport (Lenzi 2002). The monitoring of artificial step-pools formed in 2013 from sandstone boulders in the Malá Ráztoka

Fig. 13. Rapid hydraulic structures – "nature-friendly" transversal structures used to decrease river channel gradient and increase river channel roughness: A – on the Porębianka River, B – on the Krzczonówka River (Photo: J. Korpak).

Stream in the flysch Carpathians has shown that these structures enable the connection of fine grain-size fractions during lower flows (Galia et al. 2015). Yet another benefit is more aerated water (Żbikowski and Żelazo 1993). They also provide organic matter retention and increased macroinvertebrate diversity and richness (Comiti et al. 2009).

The various efforts undertaken around the world to renaturalize rivers are likely to succeed if several conditions are met. First, it is not actually possible to restore a river to its original pre-regulation state due to environmental and economic changes in catchments that have happened since the start of regulation (Dufour and Piégay 2009). The monitoring of ongoing renaturalization projects have shown that effective river restoration must be preceded by a fundamental analysis of local physical and ecological processes, as well as interactions between them. In addition, it is important to investigate longitudinal, lateral, and vertical connectivity for each studied fluvial system (Dufour and Piégay 2009). It is also important to study the details of each river or stream and to investigate its past development. Studies of this type should cover not just a section of river or stream, but the entire catchment of interest. In summary, a comprehensive approach to river channel renaturalization requires the collaboration of river scientists and restoration practitioners (Wohl et al. 2015).

Final Remarks

The scientific literature includes a discussion of whether changes in fluvial systems over the course of centuries past were mostly caused by humans or climate change associated with the so-called Little Ice Age, as well as present-day climate warming (Macklin and Lewin 1997; Surian 1999; Winterbottom 2000; Starkel 2002; Surian and Rinaldi 2003; Rădoane et al. 2013). While scientific views on this issue are divided, even opponents of the view that human impact has been the dominant factor agree that human made changes have played a very important role.

Human made changes in one section of river channel contribute to changes in other sections of river channel throughout a fluvial system. River channels adapt to new conditions not only in the months and years after a given human made change, but over decades and even centuries (Surian 1999; Wohl 2006).

It is always better to prevent harmful human impact rather than have to proceed with costly restoration projects later. It is simpler and less expensive to maintain any existing stretches of river characterized by a good ecological state. There is no universal solution to all river-related problems. Different rivers are characterized by different natural and anthropogenic issues and every river reach is unique, while being part of an entire fluvial system. It is not possible to go ahead and fix problems without understanding the evolution and contemporary role of a given reach of river in the context of other sections of the same river. The right solutions to local problems can be found by looking at the entire fluvial system and the catchment as a whole.

Acknowledgment

Sincere thanks to Dr. Anna Lenar-Matyas for her helpful comments and critique of the chapter.

Keywords: River channel regulation, river-bed gravel mining, land use change, river restoration.

References

Adamczyk, M.J. 1997. Gorce w latach 1670–1870 i zmiany w ich krajobrazie. Wierchy 62: 93–118.

Arnaud-Fassetta, G. 2003. River channel changes in the Rhone Delta (France) since the end of the Little Ice Age: geomorphological adjustment to hydroclimatic change and natural resource management. Catena 51: 141–172.

Augustowski, B. 1968. Spostrzeżenia nad zmianami antropogenicznymi w korycie Ropy w Karpatach w okolicy Biecza. Zeszyty Naukowe WSP w Gdańsku 10: 161–168.

Babiński, Z. 1992. Współczesne procesy korytowe dolnej Wisły. Prace Geograficzne IGiPZ PAN 157: 1–171.

Bartnik, W. and Strużynski, A. 1998. Deformation of river bed after flood at Tenczyński stream. Science Conference in Cracow 7–9 May 1998. Powódź w dorzeczu górnej Wisly w lipcu 1997. Ed. Polish Academy of Science, Cracow: 155–167.

Begin, Z.B., Meyer, D.F. and Schumm, S.A. 1981. Development of longitudinal profiles of alluvial channels in response to base-level lowering. Earth Surface Processes and Landforms 6: 49–68.

Boix-Fayos, C., Barberá, G.G., López-Bermúdez, F. and Castillo, V.M. 2007. Effect of check-dams, reforestation and land-use changes on river channel morphology: case study of the Rogativa catchment (Murcia, Spain). Geomorphology 91: 103–123.

Bojarski, A., Jeleński, J., Jelonek, M., Litewka, T., Wyżga, B. and Zalewski, J. 2005. Zasady dobrej praktyki w utrzymaniu rzek i potoków górskich. Ministerstwo Środowiska, Departament Zasobów Wodnych, Warszawa.

Bormann, H., Pinter, N. and Elfert, S. 2011. Hydrological signatures of flood trends on German rivers: flood frequencies, flood heights and specific stages. Journal of Hydrology 404(1-2): 50–66.

Brandt, S.A. 2000. Classification of geomorphological effects downstream of dams. Catena 40: 375–401.

Bravard, J.P., Kondolf, G.M. and Piegay, H. 1999. Environmental and societal effects of channel incision and remedial strategies. pp. 303–341. *In*: Darby, S.E. and Simon, A. (eds.). Incised River Channels: Processes, Forms, Engineering and Management. Wiley. Chichester, New York.

Brookes, A. 1997. River dynamics and channel maintenance. pp. 293–307. *In*: Thorne, C.R., Hey, R.D. and Newson, M.D. (eds.). Applied Fluvial Geomorphology for River Engineering and Management. John Wiley, Chichester.

Bucała, A. 2014. The impact of human activities on land use and land cover changes and environmental processes in the Gorce Mountains (Western Polish Carpathians) in the past 50 years. Journal of Environmental Management 138: 4–14.

Bunn, S.E. and Arthington, A.H. 2002. Basic principles and ecological consequences of altered flow regimes for aquatic biodiversity. Environmental Management 30: 492–507.

Capelli, G., Miccadei, E. and Raffi, R. 1997. Fluvial dynamics in the Castel di Sangro plain: morphological changes and human impact from 1875 to 1992. Catena 30: 295–309.

Castillo, V.M., Mosh, W.M., Conesa-García, C., Barbéra, G.G., Vavarro Cano, J.A. and López-Bermudez, F. 2007. Effectiveness and geomorphological impacts of check dams for soil erosion control in a semiarid Mediterranean catchment: El Carvano (Murcia, Spain). Catena 70: 416–427.

Chang, H.H. 1988. Fluvial Processes in River Engineering. John Wiley, New York.

Chełmicki, W. 1999. Degradacja i ochrona wód. Część druga – Zasoby Inst Geogr UJ, Kraków.

Chwistek, K. 2002. Historia lasów i leśnictwa w Gorcach. Wierchy 68: 135–160.

Citterio, A. and Piégay, H. 2009. Overbank sedimentation rates in former channel lakes: characterization and control factors. Sedimentology 56: 461–482.

Comiti, F., Mao, L., Lenzi, M.A. and Siligardi, M. 2009. Artificial steps to stabilize mountain rivers: a post-project ecological assessment. River Research and Applications 25(5): 639–659.

Cyberski, J. 1984. Zjawiska akumulacyjno-erozyjne w rzekach objętych oddziaływaniem budowli piętrzących. Czasopismo Geograficzne 55: 355–363.

Czajka, A. 2007. Środowisko sedymentacji osadów przykorytowych rzek uregulowanych na przykładzie górnej Odry i górnej Wisły. Wydawnictwo Uniwersytetu Śląskiego. Katowice.

Dębski, K. 1978. Regulacja rzek. PWN, Warszawa.

Dobrowolski, K. 1931. Dzieje wsi Niedźwiedzia w powiecie limanowskim do schyłku dawnej Rzeczypospolitej. Studia z historii społecznej i gospodarczej poświęcone dr Franciszkowi Bujakowi. Lwów.

Doyle, M.W., Stanley, E.H. and Harbor, J.M. 2003. Channel adjustments following two dam removals in Wisconsin. Water Resources Research 39: 1011–1026.

Dufour, S. and Piégay, H. 2009. From the myth of a lost paradise to targeted river restoration: forget natural references and focus on human benefits. River Research and Applications 25(5): 568–581.

Dynowska, I. 1993. Przemiany stosunków wodnych w Polsce w wyniku procesów naturalnych i antropogenicznych. Kraków.

Fighmy, J.L. 1984. Colorado Plains Prehistoric Context for Management of Prehistoric Resources of the Colorado Plains. State Historical Society of Colorado, Denver.

Froehlich, W. 1982. Mechanizm transportu fluwialnego i dostawy zwietrzelin do koryta w górskiej zlewni fliszowej. Prace Geograficzne IGiPZ PAN 143.

Froehlich, W. 1992. Mechanizm erozji i transportu fluwialnego w zlewniach beskidzkich. System denudacyjny Polski. Prace Geograficzne IGiPZ PAN 155: 171–189.

Galarowski, T. and Klimek, K. 1991. Funkcjonowanie koryt rzecznych w warunkach zagospodarowania. pp. 231–259. In: Dynowska, I. and Maciejewski, M. (eds.). Dorzecze górnej Wisły, cz. I. PWN, Warszawa-Kraków.

Galia, T., Hradecký, J. and Škarpich, V. 2015. Sediment transport in headwater streams of the Carpathian Flysh belt: its nature and recent effects of human interventions. pp. 13–26. In: Heininger, P. and Cullmann, J. (eds.). Sediment Matters. Springer. Heidelberg.

Galia, T., Hradecký, J., Škarpich, V. and Přibyla, Z. 2016. Effect of grade-control structures at various stages of their destruction on bed sediments and local channel parameters. Geomorphology 253: 305–317.

García-Ruiz, J.M., White, S.M., Lasanta, T., Marti, C., Gonzalez, C., Paz Errea, M., Valero, B. and Ortigosa, L. 1997. Assessing the effects of land-use changes on sediment yield and channel dynamics in the central Spanish Pyrenees. Human Impact on Erosion and Sedimentation. Proceedings of Rabat Symposium S6: 151–158.

García-Ruiz, J.M., Luis, C., Alatorre, A., Gómez-Villar, A. and Beguería, S. 2011. Upstream and downstream effects of check dams in braided rivers, Central Pyrenees. pp. 307–322. In: Conesa-Garcia, C. and Lenzi, M.A. (eds.). Check Dams, Morphological Adjustments and Erosion Control in Torrential Streams. Nova Science Publishers Inc. New York.

Gil, E. and Słupik, J. 1972. The influence of plant cover and land use on the surface run-off and wash down during heavy rain. Studia Geomorphologica Carpatho - Balcanica 6: 181–190.

Gil, E. 1999. Obieg wody i spłukiwanie na fliszowych stokach użytkowanych rolniczo w latach 1980–1990. Zeszyty IGiPZ PAN 60: 1–78.

Gleick, P.H., Cooley, H., Cohen, M.J., Marikawa, M., Morrison, J. and Palaniappan, M. 2009. Dams removed or decommissioned in the United States, 1912 to present. The World's Water 2008–2009. Pacific Institute for Studies in Development, Environment, and Security, Island Press, Washington, DC.

Gregory, K.J. 2006. The human role in changing river channels. Geomorphology 79: 172–191.

Grôtzbach, E. and Stadel, C. 1997. Mountain peoples and cultures. pp. 24–35. *In*: Messerli, B. and Ives, J.D. (eds.). Mountains of the World: A Global Priority. Parthenon Publishing Group, New York and Carnforth, UK.

Gumiero, B., Mant, J., Hein, T., Elso, J. and Boz, B. 2013. Linking the restoration of river and riparian zones/wetlands in Europe: sharing knowledge through case studies. Ecological Engineering 56: 36–50.

Gurnell, A.M. 1997. Channel change on the River Dee, 1946–1992, from the analysis of air photographs. Regulated Rivers: Research and Management 13: 13–26.

Guzik, C. 2004. Wieś podhalańska i jej gospodarcze przeobrażenia w historii regionu. pp. 209–218. *In*: Izmaiłow, B. (ed.). Przyroda – Człowiek – Bóg. Cracow, Poland.

Hooke, J.M. 1997. Styles of channel change. pp. 237–268. *In*: Thorne, C.R., Hey, R.D. and Newson, M.D. (eds.). Applied Fluvial Geomorphology for River Engineering and Management. John Wiley, Chichester.

Janson, P.P., van Bendegom, L., van den Berg, J., de Vries, M. and Zanen, A. 1979. Principles of River Engineering. The Non-Tidal Alluvial River. Pitman. London, San Francisco, Melbourne.

Kaniecki, A. and Brychcy, D. 2010. Średniowieczne młyny wodne i ich wpływ na przemiany stosunków wodnych na przykładzie zlewni Obry Skwierzyńskiej. Badania Fizjograficzne 61: 145–156.

Kiss, T., Andrási, G. and Hernesz, P. 2011. Morphological alteration of Dráva as the result of human impact. AGD Landscape & Environment 5(2): 58–75.

Klimek, K. and Trafas, K. 1972. Young-Holocene changes in the course of the Dunajec river in the Beskid Sądecki Mts (Western Carpathians). Studia Geomorphologica Carpatho - Balcanica 6: 85–91.

Klimek, K. 1983. Erozja wgłębna dopływów Wisły na przedpolu Karpat. pp. 97–108. *In*: Kajak, Z. (ed.). Ekologiczne podstawy zagospodarowania Wisły i jej dorzecza. PWN, Warszawa-Łódź.

Klimek, K. 1987. Man's impact on fluvial processes in the Polish Western Carpathians. Geografiska Annaler 69A: 221–226.

Kłonowska-Olejnik, M. and Radecki-Pawlik, A. 2002. Ocena jakości wody i warunków siedliskowych potoku Białego na podstawie metod biologicznych i parametrów hydrologicznych. Polskie parki narodowe – ich rola w rozwoju nauk przyrodniczych. Konferencja Jubileuszowa z okazji 80-lecia Białowieskiego Parku Narodowego: 15.

Komędera, M. 1993. Zmiany systemu korytowego Mszanki. Master's Thesis, Jagiellonian University, Cracow, Poland.

Kondolf, G.M. and Swanson, M.L. 1993. Channel adjustments to reservoir construction and gravel extraction along Stony Creek, California. Environmental Geology and Water Sciences 25: 256–269.

Kondolf, G.M. 1994. Geomorphic and environmental effects of instream gravel mining. Landscape and Urban Planning 28: 225–243.

Kondolf, G.M. 1997. Hungry water: effects of dam and gravel mining on river channels. Environmental Management 21(4): 533–551.

Kondolf, G.M., Piégay, H. and Landon, N. 2002. Channel response to increased and decreased bedload supply from land use change: contrasts between two catchments. Geomorphology 45: 35–51.

Kondolf, G.M., Gao, Y., Annandale, G.W., Morris, G., Jiang, E., Zhang, J., Cao, Y., Carling, P., Fu, K., Guo, Q., Hotchkiss, R., Peteuil, C., Sumi, T.,Wang, H.-W., Wang, Z., Wei, Z., Wu, B., Wu, C. and Yang, C.T. 2014. Sustainable sediment management in reservoirs and regulated rivers: experiences from five continents. Earth's Future 2: 256–280.

Konrad, C.P., Olden, J.D., Lytle, D.A., Melis, T.S., Schmidt, J.C., Bray, E.N., Freeman, M.C., Gido, K.B., Hemphill, N.P., Kennard, M.J., McMullen, L.E., Mims, M.C., Pyron, M., Robinson, C.T. and Williams, J.G. 2011. Large-scale flow experiments for managing river systems. BioScience 61(12): 948–959.

Kopeć, S. and Kurek, S. 1975. Wpływ szaty roślinnej na odpływ i retencję w małych zlewniach górskich na przykładzie dorzecza górnego Grajcarka. Zeszyty Problemowe Postępów Nauk Rolniczych 162: 337–353.

Korpak, J. 2007a. The influence of river training on mountain channel changes (Polish Carpathian Mountains). Geomorphology 92: 166–181.

Korpak, J. 2007b. Morfologiczna rola budowli regulacyjnych w górskich systemach fluwialnych. Ph.D. Thesis, Jagiellonian University, Cracow, Poland.

Korpak, J., Krzemień, K. and Radecki-Pawlik, A. 2008. Wpływ czynników antropogenicznych na zmiany koryt cieków karpackich. Infrastruktura i Ekologia Terenów Wiejskich, Monografia 4. PAN, Kraków.

Korpak, J. 2010. Geomorphologic effects of river engineering structures in Carpathian fluvial systems. Landform Analysis 14: 34–44.

Korpak, J. 2015. Evolution of the lower Mszanka Channel section after training using stage correction method. Infrastruktura i ekologia terenów wiejskich 4: 1285–1302.

Kozak, J., Estreguil, C. and Troll, M. 2007. Forest cover changes in the northern Carpathians in the 20th century: a slow transition. Journal of Land Use Science 2: 127–146.

Krzemień, K. 1976. Współczesna dynamika koryta potoku Konina w Gorcach. Folia Geographica, ser. Geographica - Physica 10: 87–122.

Krzemień, K. 1981. Zmienność subsystemu korytowego Czarnego Dunajca. Zeszyty Naukowe UJ, Prace Geograficzne 53: 123–137.

Krzemień, K. 1984. Współczesne zmiany modelowania koryt potoków w Gorcach. Zeszyty Naukowe UJ, Prace Geograficzne 59: 83–96.

Krzemień, K. 2003. The Czarny Dunajec River, Poland, as an example of human-induced development tendencies in a mountain river channel. Landform Analysis 4: 57–64.

Książek, L. 2006. Morfologia koryta rzeki Skawy w zasięgu cofki zbiornika Świnna Poręba. Infrastruktura i Ekologia Terenów Wiejskich 4/1: 249–267.

Kukulak, J. 1994. Antropogeniczne przemiany w środowisku przyrodniczym Podhala w latach 1931–1988. pp. 265–285. *In*: Górz, B. (ed.). Studia nad przemianami Podhala. Wydawnictwo Naukowe WSP, Kraków.

Lach, J. 1984. Geomorfologiczne skutki antropopresji rolniczej w wybranych częściach Karpat i ich Przedgórza. Prace Monograficzne WSP w Krakowie 66: 1–142.

Łajczak, A. 1997. Anthropogenic changes in the suspended load transportation by and sedimentation rates of the river Vistula, Poland. *In*: Maruszczak, H. and Starkel, L. (eds.). Anthropogenic Impact on Water Conditions (Vistula and Oder River basins). Geographia Polonica 68: 7–30.

Landon, N., Piégay, H. and Bravard, J.P. 1998. The Drôme river incision (France): from assessment to management. Landscape and Urban Planning 43: 119–131.

Łapuszek, M. and Lenar-Matyas, A. 2013. Utrzymanie i zagospodarowanie rzek górskich. Wydawnictwo Politechniki Krakowskiej, Kraków.

Lasanta, T. 1989. The process of desertion of cultivated areas in the central Spanish Pyrenees. Pirineos 132: 15–36.

Laszek, W., Radecki-Pawlik, A., Wyżga, B. and Hajdukiewicz, H. 2014. Modelling hydraulic parameters of flood flows for a Polish Carpathian river. Local Responses to Global Challenges. Proceedings of Forum Carpaticum 2014: 108–111.

Latocha, A. 2009. Land-use changes and longer-term human-environment interactions in a mountain region (Sudetes Mountains, Poland). Geomorphology 108: 48–57.

Lenar-Matyas, A., Korpak, J. and Mączałowski, A. 2015. Influence of extreme discharge on restoration works in mountains river—a case study of the Krzczonówka River (southern Poland). Journal of Ecological Engineering 16(3): 83–96.

Lenzi, M.A. 2002. Stream bed stabilization using boulder check dams that mimic step-pool morphology features in Northern Italy. Geomorphology 45: 243–260.

Lenzi, M.A. and Comiti, F. 2003. Local scouring and morphological adjustments in steep channels with check-dam sequences. Geomorphology 55: 97–109.

Lenzi, M.A., Marion, A. and Comiti, F. 2003. Local scouring at grade-control structures in alluvial mountain. Water Resources Research 39(7): 1176.

Lepori, F., Palm, D., Brannas, E. and Malmqvist, B. 2005. Does restoration of structural heterogeneity in streams enhance fish and macroinvertebrate diversity? Ecological Applications 15: 2060–2071.

Liébault, F. and Piégay, H. 2001. Assessment of channel changes due to long-term bedload supply decrease, Roubion River, France. Geomorphology 36: 167–186.

Liébault, F., Clément, P., Piégay, H., Rogers, C.F., Kondolf, G.M. and Landon, N. 2002. Contemporary channel changes in the Eygues basin, southern French Prealps: the relationship of subbasin variability to watershed characteristics. Geomorphology 45: 53–66.

Liro, M. 2012. Wpływ regulacji koryta na warunki sedymentacji osadów na równinie zalewowej dolnego Dunajca. Przegląd Geologiczny 60/7: 380–386.

Liro, M. 2014. Conceptual model for assessing the channel changes upstream from dam reservoir. Quaestiones Geographicae 33/1: 61–74.

Liro, M. 2015. Gravel-bed channel changes upstream of a reservoir: the case of the Dunajec River upstream of the Czorsztyn Reservoir, southern Poland. Geomorphology 228: 694–702.

Lisle, T.E., Cui, Y., Parker, G., Pizzuto, J.E. and Dodd, A.M. 2001. The dominance of dispersion in the evolution of bed material waves in gravel-bed rivers. Earth Surface Processes and Landforms 26: 1409–1420.

López, J.L. 2004. Channel response to gravel mining activities in mountain rivers. Journal of Mountain Science 1: 264–269.

Macklin, M.G. and Lewin, J. 1997. Channel, floodplain and drainage basin response to environmental changes. pp. 15–45. *In*: Thorne, C.R., Hey, R.D. and Newson, M.D. (eds.). Applied Fluvial Geomorphology for River Engineering and Management. Wiley, Chichester, UK.

Madej, M.A. 2001. Development of channel organization and roughness following sediment pulses in singlethread, gravel bed rivers. Water Resources Research 37: 2259–2272.

Makkaveyev, N.I. 1972. The impact of large water engineering projects on geomorphic processes in stream valleys. Soviet Geography 13/6: 387–393.

Malarz, R. 2002. Powodziowa transformacja gruboklastycznych aluwiów w żwirodennych rzekach Zachodnich Karpatach fliszowych (na przykładzie Soły i Skawy). Wydawnictwo Naukowe AP, Kraków.

Mann, C.C. and Plummer, M.L. 2000. Can science rescue salmon? Science 289: 716–719.

Marchetti, M. 2002. Environmental changes in the central Po Plain (Northern Italy) due to fluvial modifications and anthropogenic activities. Geomorphology 44: 361–373.

Marion, A., Lenzi, M.A. and Comiti, F. 2004. Effect of sill spacing and sediment size grading on scouring at grade control structures. Earth Surface Processes and Landforms 29: 983–993.

Marion, A., Tregnaghi, M. and Tait, M. 2006. Sediment supply and local scouring at bed sills in high-gradient streams. Water Resources Research 42: W06416.

Marston, R.A., Girel, J., Pautou, G., Piegay, H., Bravard, J.P. and Arneson, C. 1995. Channel metamorphosis, floodplain disturbance and vegetation development: Ain River, France. Geomorphology 13: 121–131.

Marston, R.A., Bravard, J.-P. and Green, T. 2003. Impacts of reforestation and gravel mining on the Malnant River, Haute-Savoie, French Alps. Geomorphology 55: 65–74.

Martín-Vide, J.P., Ferrer-Boix, C. and Ollero, A. 2010. Incision due to gravel mining: modeling a case study from the Gállego River, Spain. Geomorphology 117: 261–271.

Nilsson, C. and Berggren, K. 2000. Alterations of riparian ecosystems caused by river regulation. Bioscience 50(9): 783–792.

Osuch, B. 1968. Problemy wynikające z nadmiernej eksploatacji kruszywa rzecznego na przykładzie rzeki Wisłoki. Zeszyty Naukowe. Akademii Górniczo-Hutniczej 219: 283–301.

Palmer, M.A., Menninger, H.L. and Bernhardt, E. 2010. River restoration, habitat heterogeneity and biodiversity: a failure of theory or practice? Freshwater Biology 55, suppl. 1: 205–222.

Palmer, M.A. and Hondula, K.L. 2014. Restoration as mitigation: analysis of stream mitigation for coal mining impacts in southern Appalachia. Environmental Science and Technology 48: 10552–10560.

Parker, G. and Andres, D. 1976. Detrimental effects of river channelization. Proceedings of the Conference Rivers '76. American Society of Civil Engineers: 1248–1266.

Petit, F., Poinsart, D. and Bravard, J.-P. 1996. Channel incision, gravel mining and bedload transport in the Rhône river upstream of Lyon, France ("canal de Miribel"). Catena 26: 209–226.

Pickup, G. 1977. Simulation modelling of river channel erosion. pp. 47–60. *In*: Gregory, K.J. (ed.). River Channel Changes. John Wiley & Sons.

Pinter, A. and Heine, R.A. 2005. Hydrodynamic and morphodynamic response to river engineering documented by fixed-discharge analysis, Lower Missouri River, USA. Journal of Hydrology 302: 70–91.

Pizzuto, J. 2002. Effects of dam removal on river form and process. Bioscience 52(8): 683–691.

Poulard, C., Szczęsny, J., Witkowsk, H. and Radzicki, K. 2005. Dynamic SlowDown: a flood mitigation strategy complying with the integrated management concept – implementation in a small mountainous catchment. Journal of River Basin Management 3(2): 75–85.

Price, K. and Leigh, D.S. 2006. Morphological and sedimentological responses of streams to human impact in the southern Blue Ridge Mountains, USA. Geomorphology 78: 142–160.

Punzet, J. 1981. Zmiany w przebiegu stanów wody w dorzeczu górnej Wisły na przestrzeni 100 lat (1871–1970). Folia Geographica, ser. Geographica - Physica 14: 5–28.

Radecki-Pawlik, A. 2002. Pobór żwiru i otoczaków z dna potoków górskich. Aura 3: 17–19.

Radecki-Pawlik, A. 2013a. The influence of a drop-hydraulic structure on the mountain stream channel regime-case study from the Polish Carpathians. Georeview 23: 46–57.

Radecki-Pawlik, A. 2013b. On using artificial rapid hydraulic structures (RHS) within mountain stream channels: some exploitation and hydraulic problems. pp. 101–115. *In*: Rowiński, P. (ed.). Experimental and Computational Solutions of Hydraulic Problems. Springer, Berlin, Heidelberg.

Rădoane, M., Obreja, F., Cristea, I. and Mihailă, D. 2013. Changes in the channel-bed level of the eastern Carpathian rivers: climatic vs. human control over the last 50 years. Geomorphology 193: 91–111.

Ratomski, J. 1991. Sedymentacja rumowiska w zbiornikach przeciwrumowiskowych na obszarze Karpat fliszowych. Politechnika Krakowska, Monografia 123: 1–131.

Reid, L.M. and Dunne, T. 1984. Sediment production from forest road surfaces. Water Resources Research 20: 1753–1761.

Rinaldi, M. and Simon, A. 1998. Bed-level adjustments in the Arno River, central Italy. Geomorphology 22: 57–71.

Rinaldi, M. 2003. Recent channel adjustments in alluvial rivers of Tuscany, central Italy. Earth Surface Processes and Landforms 28: 587–608.

Rinaldi, M., Wyżga, B. and Surian, N. 2005. Sediment mining in alluvial channels: physical effects and management perspectives. River Research and Applications 21: 805–828.

Schumm, S.A. 1968. River adjustment to altered hydrologic regimen – Murrumbidgee River and paleochannels, Australia. Geological Survey Professional Paper 598: 1–65.

Shields, F.D., Knight, Jr., S.S., Lizotte, Jr., R. and Wren, D.G. 2011. Connectivity an variability: metrix for riverine floodplain backwater rehabilitation. pp. 233–246. *In*: Simon, A. (ed.). Stream Restoration in Dynamic Fluvial Systems: Scientific Approaches, Analyses, and Tools. Geophysical Monograph Series 194, AGU, Washington, DC.

Shuman, J.R. 1995. Environmental considerations for assessing dam removal alternatives for river restoration. Regulated Rivers: Research and Management 11: 249–261.

Simon, A. and Rinaldi, M. 2006. Disturbance, stream incision, and channel evolution: the roles of excess transport capacity and boundary materials in controlling channel response. Geomorphology 79: 361–383.

Škarpich, V., Hradecký, J. and Tábořík, P. 2011. Structure and genesis of the quaternary filling of the Slavíč River valley (Moravskoslezské Beskydy Mts., Czech Republic). Moravian Geographical Reports 19: 30–38.

Škarpich, V., Hradecký, J. and Dušek, R. 2013. Complex transformation of the geomorphic regime of channels in the forefield of the Moravskoslezské Beskydy Mts.: Case study of the Morávka River (Czech Republic). Catena 111: 25–40.

Słupik, J. 1981. Rola stoku w kształtowaniu odpływu w Karpatach fliszowych. Prace Geograficzne IGiPZ PAN 142.

Soja, R. 1977. Deepening of river channel in the light of the cross profile analysis (Carpathian river as example). Studia Geomorhologica Carpatho - Balcanica 11: 127–138.

Stanley, E.H., Luebke, M.A., Doyle, M.W. and Marshall, D.W. 2002. Short-term changes in channel form and macroinvertebrate communities following low-head dam removal. Journal of the North American Benthological Society 21: 172–187.

Starkel, L. 2002. Change in the frequency of extreme events as the indicator of climatic change in the Holocene (in fluvial systems). Quaternary International 91: 25–32.

Surian, N. 1999. Channel changes due to River regulation: the case of the Piave River, Italy. Earth Surface Processes and Landforms 24: 1135–1151.

Surian, N. and Rinaldi, M. 2003. Morphological response to river engineering and management in alluvial channels in Italy. Geomorphology 50: 307–326.

Surian, N. 2006. Effects of human impact on braided river morphology: examples from Northern Italy. pp. 327–338. *In*: Sambrook Smith, G.H., Best, J.L., Bristow, C. and Petts, G.E. (eds.). Braided Rivers: Process, Deposits, Ecology and Management. IAS Special Publication 36. Blackwell Science, Oxford.

Szumański, A. 1977. Zmiany układu koryta dolnego Sanu w XIX i XX wieku oraz ich wpływ na morfogenezę tarasu łęgowego. Studia Geomorhologica Carpatho - Balcanica. 9: 139–154.

Thomas, R.J., Constantine, J.A., Gough, P. and Fussell, B. 2015. Rapid channel widening following weir removal due to bed-material wave dispersion on the river Monnow, Wales. River Research and Applications 31: 1017–1027.

Wałdykowski, P. 2005. Rola sieci drogowej w przekształcaniu stoków i den dolin w rejonie Turbacza (Gorczański Park Narodowy). pp. 495–500. *In*: Kotarba, A., Krzemień, K. and Święchowicz, J. (eds.). Współczesna ewolucja rzeźby Polski. Kraków.

Wharton, G. 2000. Managing River Environments. Cambridge University Press, Cambridge.

Wiejaczka, Ł., Kiszka, K. and Bochenek, W. 2014. Changes of the Morphology of the Ropa River —Upstream and Downstream of the Klimkowka Water Reservoir. Studia Geomorphologica Carpatho-Balcanica 48(1): 61–76.

Wierzbicki, M. and Wicher, J. 2002. Wpływ erozji poniżej zbiornika Jeziorsko na zmiany zachodzące w korycie rzeki Warty, Erozja gleb i transport rumowiska rzecznego. Materiały Sympozjum Naukowego, Zakopane.

Williams, G.P. and Wolman, M.G. 1984. Downstream effects of dams on alluvial rivers. Geological Survey Professional Paper 1286: 83.

Winnterbottom, S.J. 2000. Medium and short-term channel planform changes on the Rivers Tay and Tummel, Scotland. Geomorphology 34: 195–208.

Wishart, D., Warburton, J. and Bracken, L. 2008. Gravel extraction and planform change in a wandering gravel-bed river: the River Wear, Northern England. Geomorphology 94: 131–152.

Wohl, E. 2006. Human impacts to mountain streams. Geomorphology 79: 217–248.

Wohl, E., Chin, A., Haltiner, J.P. and Kondolf, G.M. 2013. Managing stream morphology with check-dams. pp. 135–150. *In*: García, C.C. and Lenzi, M.A. (eds.). Check Dams, Morphological Adjustments and Erosion Control in Torrential Streams. Nova Science Publishers, New York.

Wohl, E., Lane, S.N. and Wilcox, A.C. 2015. The science and practice of river restoration. Water Resources Research 51(8): 5974–5997.

Wołoszyn, J., Czamara, W., Eliasiewicz, R. and Krężel, J. 1994. Regulacja rzek i potoków. Wydawnictwo. Akademii Rolniczej we Wrocławiu, Wrocław.

Wyżga, B. 1992. Zmiany w geometrii koryta i układzie facji jako odzwierciedlenie transformacji reżimu hydrologicznego Raby w ciągu ostatnich dwustu lat. Czasopismo Geograficzne 63: 279–294.

Wyżga, B. 2001. Impact of the channelization-induced incision of the Skawa and Wisłoka Rivers, Southern Poland, on the conditions of overbank deposition. Regulated Rivers: Research and Management 17: 85–100.

Wyżga, B., Amirowicz, A., Radecki-Pawlik, A. and Zawiejska, J. 2009. Hydromorphological conditions, potential fish habitats and the fish community in a mountain river subjected to variable human impacts, the Czarny Dunajec, Polish Carpathians. River Research and Applications 30: 517–536.

Wyżga, B., Hajdukiewicz, H., Radecki-Pawlik, A. and Zawiejska, J. 2010. Eksploatacja osadów z koryt rzek górskich – skutki środowiskowe i procedury oceny. Gospodarka Wodna 6: 243–249.

Wyżga, B. and Zawiejska, J. 2012. Hydromorphological quality as a key element of the ecological status of Polish Carpathian rivers. Georeview 21(1): 56–67.

Wyżga, B., Zawiejska, J., Radecki-Pawlik, A. and Hajdukiewicz, H. 2012. Environmental change, hydromorphological reference conditions and the restoration of Polish Carpathian rivers. Earth Surface Processes and Landforms 37: 1213–1226.

Xu, J. 1990. Complex response in adjustment of Weihe channel to the construction of the Sanmenxia Reservoir. Zeitschrift für Geomorhologie 34: 233–245.

Xu, J. and Shi, C. 1997. The river channel pattern change as influenced by the floodplain geoecosystem: an example from the Honshan Reservoir. Zeitschrift für Geomorhologie 41: 97–113.

Zawiejska, J. and Wyżga, B. 2010. Twentieth-century channel change on the Dunajec River, southern Poland: patterns, causes and controls. Geomorphology 117: 234–246.

Zawiejska, J., Wyżga, B. and Radecki-Pawlik, A. 2015. Variation in surface bed material along a mountain river modified by gravel extraction and channelization, the Czarny Dunajec, Polish Carpathians. Geomorphology 231: 353–366.

Żbikowski, A. and Żelazo, J. 1993. Ochrona środowiska w budownictwie wodnym. Materiały Informacyjne, Ministerstwo Ochrony Środowiska i Zasobów Naturalnych, Warszawa.

Technical Designs

2196: Projekt koncepcyjnego rozwiązania zabudowy potoków Mszanka i Łostówka od km 8.165 do km 5.140. Hydroprojekt Oddział w Krakowie. 1959.

1682: Projekt techniczno-roboczy zabudowy potoku Mszanka w km 7.834–7.500. Okręgowy Zarząd Wodny w Krakowie. 1967.

3964: Projekt zabudowy potoku Biały Dunajec w km 2.182-5.101. Okręgowy Zarząd Wody w Krakowie. 1968.

101: Regulacja Białego Dunajca w odcinku 0.000-3.260. Projekt techniczny. 1976.
114: Zabezpieczenie brzegu potoku Mszanki w km 6.224–6.981 w Mszanie Górnej. Przedsiębiorstwo Budownictwa Wodnego w Krakowie, Pracownia Projektowa. 1976.
3375: Analiza stosunków wodnych w węźle potoku Mszanka – ujście potoku Łętowego. Hydroprojekt Oddział Kraków. 1981.
2064: Usuwanie skutków powodzi z lipca 2001. Projekt budowlany regulacji koryta potoku Biały Dunajec w km 3.260–6.500 w miejscowości Szaflary. Hydroprojekt Oddział Kraków. 2003.

CHAPTER 22

Introduction to Floods
Analysis and Modelling

Fabian Rivera-Trejo,[1,*] *Juan Barajas-Fernández,*[1]
Gabriel Soto-Cortés[2,a] *and Alejandro Mendoza-Reséndiz*[2,b]

INTRODUCTION

Hydro-meteorological events from the last decades have resulted in human and economic losses. It is estimated that more than three quarters of all disasters reported in the last 10 years are caused by weather events linked with floods (Table 1). The floods have become more frequent and more severe due to factors that had not been considered before. Disordered urban growth (Sayama et al. 2012), the lack of legislation that defines risk zone (Rivera et al. 2010), the short-term memory of people who live in endangered areas (Adina 2014), and the climate variability (Matishov 2014; Schröter et al. 2014) are some factors that play a significant role in floods. Flooding occurs when the natural river capacity is exceeded, as a consequence of excessive rainfall (Fig. 1) or caused by failures in the flood control infrastructure (Fig. 2). Identifying the variables involved during flood events is an important matter for governments around the world. Protective actions and works are designed, and risk scenarios are simulated from an analysis of the involved variables. However, despite numerical modelling capacities having improved over time and there being efficient simulation models of floods which exist nowadays, numerical models are unable to predict the increasing frequency in floods all over the world. Floods are dynamic and two identical floods rarely occur. This is because natural conditions of runoff change by cause of climate variation or anthropogenic alterations to the landscape. As a result of economic losses (damage to infrastructure, cattle raising, agriculture, and economic activities) and loss of human life caused by floods, proper predictions are required. The first step for establishing mechanisms for protection and defense against flooding is to understand how floods occur.

[1] Professor of Hydraulic Engineering, Juarez Autonomous University of Tabasco, Tabasco, México.
   E-mail: jbarajasf@gmail.com
[2] Professor of Hydraulic Engineering, Autonomous Metropolitan University Campus Lerma, México.
[a] E-mail: gsoto@ler.uam.mx
[b] E-mail: a.mendoza@ler.uam.mx
* Corresponding author: jgfabianrivera@gmail.com

Table 1. Total number of reported disasters, by type of phenomenon and year (Adapted from WDS 2015).

	2005	2006	2007	2008	2009	2010	2011	2012	2013	2014	TOTAL
Droughts	28	20	13	21	31	27	24	31	13	15	223
Dry mass movements	n.d.r	1	n.d.r	3	1	n.d.r	n.d.r	1	1	n.d.r	7
Earthquakes	25	24	21	23	22	25	30	29	28	26	253
Extreme temperatures	29	32	25	11	26	34	19	52	17	17	262
Floods	193	232	219	175	160	190	160	141	149	132	1 751
Insect infestations	n.d.r.	1	n.d.r.	n.d.r.	1	1	1	n.d.r.	n.d.r.	n.d.r.	4
Landslides	12	20	10	12	28	32	17	13	11	15	170
Storms	132	77	105	111	87	95	86	90	106	99	988
Volcanic activities	8	12	6	7	3	6	6	1	4	8	61
Wildfires	13	10	18	5	9	7	8	7	10	3	90
Subtotal climato-, hydro- and meteorological disasters	**407**	**392**	**390**	**335**	**342**	**386**	**315**	**334**	**306**	**281**	**3 488**
Subtotal geophysical disasters	33	37	27	33	26	31	36	31	33	34	321
Total natural disasters	**440**	**429**	**417**	**368**	**368**	**417**	**351**	**365**	**339**	**315**	**3 809**

*n.d.r. No disaster reported

Fig. 1. Flooding in Tabasco, México (October 2007), through heavy precipitation on a large scale.

Fig. 2. Flooding in New Orleans, USA (August 2005) by failure of defense against floods.

Classification of Floods

Although flood events may be similar, they have different origins. For engineering purposes, it is necessary to identify the source of the flood. The International Glossary of Hydrology, from UNESCO (WMO 2012), defines a flooding as: "(1) Overflowing by water of the normal confines of a watercourse or other body of water, and (2)

Accumulation of drainage water over areas which are not normally submerged", where normal confines are the water surface elevations.

Several causes can produce elevations of water surface above its normal level. Figure 3 shows the flooding classification according to their source and the response time.

Pluvial floods are result of precipitation; they occur when the soil is saturated and the rainwater surplus begins to accumulate. The causes of excessive precipitation are: tropical storms, cyclones, hurricanes, and monsoons (Knabb et al. 2011, 2012; Sayama et al. 2012). All of them can produce a long heavy storm. Orographic rainfall is caused by humid air currents that hit the mountain streams (Fox and Roentree 2013; Hicks et al. 2005), winter rains or cold fronts, caused by the movement of cold air from the poles to warmer areas (Rivera et al. 2010; Delrieu et al. 2005), and convective rains, due to the warming of the earth surface, particularly in urban areas (Silvestro et al. 2012; Sto. Domingo et al. 2010; Llasat 2001). Pluvial-river–floods (Salas and Jiménez 2014) are generated when the water transported by the current exceeds the river banks (Rivera et al. 2010; Schröter et al. 2014) or have high sedimentation rates (George et al. 2007). Coastal flooding occurs when the sea level rises by effect of meteorological tides or tsunamis and penetrates inland (Matishov et al. 2014; Mori et al. 2007; Shentsis et al. 2012). Hydraulic infrastructure failures are often the most severe since neither populations nor governments are prepared when a system fails (Adina 2014), in that scenario, a flash flood occurs and causes severe damage (Escuder-Bueno et al. 2012; Schröter et al. 2014). Failures in hydraulic infrastructure can be caused by: bad design, inadequate operation, poor or lack of maintenance, and at the end of the period of life of the system. On the other hand, physiographic characteristics affect the response time of a flood by slowing or hastening a flood's response time. Flat zones generally

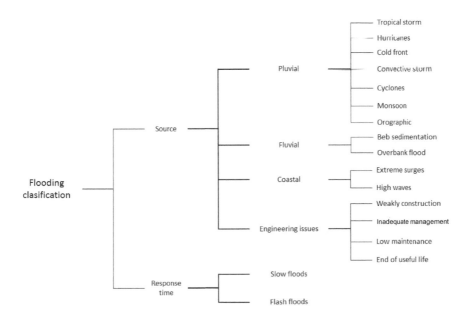

Fig. 3. Classification of floods.

have slow floods that develop during days (Fox and Rowntree 2013; Matishov 2013; Schröter et al. 2014; Jorge et al. 2007; Rivera et al. 2010); meanwhile, mountainous zones have flash floods that can be developed within hours (Costache and Pravalie 2013; Delrieu et al. 2005; Sayama et al. 2012; Escuder-Bueno et al. 2012; Silvestro et al. 2012; Adina 2014).

A factor to consider is the anthropic influence. While a valley with low population density and poor infrastructure is flooded, it is not considered an emergency; however, the same zone with high population and serious modification in their natural drainage will undergo severe effects of flooding. This is a result of lack of preventive measures and inappropriate studies about the nature of runoff, valley drainage, and consequences of modifications generated when the rivers are altered.

The flood analysis is often made after the event happened. The purpose of flood analysis is to characterize the causes and origins of the flooding and generate strategies to control and manage runoff. However, these analyses should be developed as priority and included in urban development plans.

Flood Analysis

Forensic Disaster Analysis (FDA) establishes the flood analysis criteria, gathers information concerning a particular flood, and describes the origin, evolution, and consequences of that particular flood (Schröter et al. 2014). Actions to mitigate future flooding events will be derived from these results. In this chapter, we only focus on the diagnostic. The remedial actions—structural and nonstructural—are determined by the physical characteristics of the study area.

The main post-flooding stage analysis is gathering information, basically, four types of information: (a) topographic, (b) hydrological, (c) climatological, and (d) demographical.

Topographical information. The analysis begins by limiting the geographical area and the physical characteristics of the basin by using topographical data (Salinas and Jiménez 2014). It consists of collecting information about: cartographic maps and/or digital elevation models of the influenced area, hydraulic structures, topographic detail of high-risk zones, soil, and vegetation characteristics and field observations to find out marks left by the maximum levels of water reached during the flood.

Hydrologic and hydraulic information. Allows for the reconstruction of the evolution of flooding events from rain and runoff (Rivera et al. 2010; Costache and Pravalie 2013). This stage consists of collecting the data available before, during and after the event, particularly the location of measuring stations (hydrometric and rainfall), rainfall intensities, flow discharge, levels of the free surface water, sediment discharge, and dam management. Sometimes it is possible to find out information from different sources (news reports, bulletins, and alarms). Then, it is necessary to choose the most reliable source of information. We recommend international or national meteorological services by example the Goddard Earth Science Data and Information Service Centre's web tool (Fox and Rowntree 2013).

Climatological information. Floods are largely linked to weather conditions, particularly those that result in high rainfall events (Silvestro et al. 2012), for example: hurricanes, monsoons, cyclones, cold fronts, tropical storms, etc. The National Oceanic and Atmospheric Association (NOAA) monitors on a daily basis the conditions that may cause flooding events (Meintel 2015). In this category, the information that is required beforehand, during and after the event are the atmospheric conditions. This information is reported in bulletins and alerts of extreme phenomena.

Demographical information. Floods are particularly important when they damage infrastructure and cause deaths. In this sense, the social aspect plays a key role in the development of a flood, so it is important to collect information about communication ways, existence of alert systems, coordination between office governments-emergency plans- and the public, and educational programs to face floods. Usually this type of information is the most difficult to collect, and for this reason, is often not considered in the analysis.

Flood modelling

Flood analysis begins with a diagnosis and it is complemented with numerical models. Numerical models are useful tools for designing defense plans against floods (Moussa and Cheviron 2015). These models are supplied and calibrated with data collected from a flood event. The state of the art in numerical modeling is based on hydrological models of precipitation (Fiorentino et al. 2006), coupled with hydraulic models of runoff (Fig. 4). In this sense, the runoff is linked to the precipitation, where the weather forecasts play a key role to determine rainfall patterns (Mendez et al. 2011). The hydrological models have grown over the past three decades (Gayathri et al. 2015; Fiorentino et al. 2006). Models are developed to physically represent the hydrological processes in a watershed through analogies and mathematical simplifications. The precipitation has a strong impact over results of the hydrological modeling, in the sense that density of the pluviographic network is important to improve the quality of results (Faures et al. 1995; Morin 1995). Another key point is the alert system, which has greatly been improved with the development of Geographic Information Systems (GIS) and the Internet (Vieux

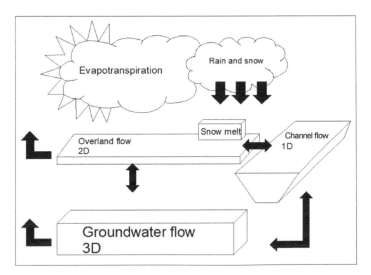

Fig. 4. Hydrological models + hydraulic models (Adapted from Sto. Domingo et al. 2010).

2003). In this sense, remote sensing instruments, such as radars and weather satellites, are able to estimate the spatial variation of precipitation to provide data in real time, and are ideal instruments to use in hydrological modelling. Data obtained from these instruments feed hydrological models to estimate the response of the basin in the presence of an extreme precipitation event and to improve the quality of the results.

Hydraulic modeling of runoff is based on Saint Venant shallow water equations (Hunter et al. 2007; Soto and Berezowsky 2003). Conservation of mass (1), and momentum (2) and (3), consider transient flow (Sayama et al. 2012)

$$\frac{\partial h}{\partial t} + \frac{\partial q_x}{\partial x} + \frac{\partial q_y}{\partial y} = r \tag{1}$$

$$\frac{\partial q_x}{\partial t} + \frac{\partial u q_x}{\partial x} + \frac{\partial v q_x}{\partial y} = -gh\frac{\partial H}{\partial x} - \frac{\tau_x}{\rho_w} \tag{2}$$

$$\frac{\partial q_y}{\partial t} + \frac{\partial u q_y}{\partial x} + \frac{\partial v q_y}{\partial y} = -gh\frac{\partial H}{\partial y} - \frac{\tau_y}{\rho_w} \tag{3}$$

where h is the height of water from local surface, q_x and q_y are unit width discharges, u and v are flow velocities, r is the rainfall intensity, H is the elevation of water surface, ρ_w is density of water, g is gravitational acceleration, τ_x and τ_y are shear stresses, and x and y are the directions. The second terms of the right-hand side of Eqs. (2) and (3) may be calculated with coefficient of Manning

$$\frac{\tau_x}{\rho_w} = \frac{gn^2 u\sqrt{u^2+v^2}}{h^{1/3}} \tag{4}$$

$$\frac{\tau_y}{\rho_w} = \frac{gn^2 v\sqrt{u^2+v^2}}{h^{1/3}} \tag{5}$$

where n is Manning's roughness parameter.

The domain of integration is divided into a grid that contains the topographic information of the system. The slope utilized corresponds to the average bed slope of the channel (1D) and to the slope of the floodplain (2D), so a coupled 1D-2D system (Fig. 5) is solved. Many existing commercial and free models employ these schemes, such as the free software Hec-Ras (Hicks et al. 2005; George et al. 2007), Iber (Cea and Blade 2015), Telemac (Horritt and Bates 2002), and Storm Water Management Model (Rossman 2015). Although the numerical simulation is very attractive and has made great strides in computer terms, is not acceptable to use models without adequate validation. Physical parameters used to calibrate numerical models are: detailed topography, levels of free surface water, flow discharge, roughness coefficient (Manning n), and aerial photographs, among others.

Case Study

Below is shown as an example, the post event investigation, about the most significant flood that has taken place in Tabasco, Mexico (Rivera et al. 2010), and for the country, the worst flood ever recorded up to now. The study zone is located in a delta below a mountainous system (Fig. 6). As a result of heavy rainfall over the region, the system of rivers overflowed with catastrophic effects, affecting the city of Villahermosa, capital of Tabasco, with more than 850,000 inhabitants. The information used in the analysis

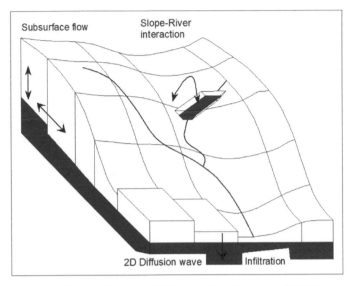

Fig. 5. Scheme diagram model (Adapted from Sayama et al. 2012).

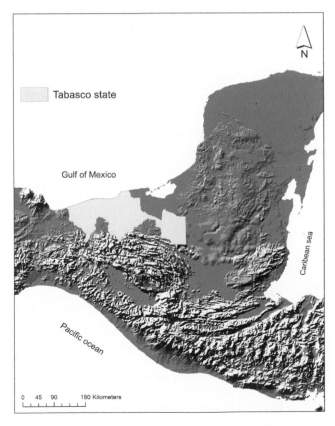

Fig. 6. Map of the drainage areas and basin in the study area.

consisted on the hydrometric, climate, and dam data published in daily basis by the National Water Commission (CONAGUA) which is a Government office of Mexico in charge of water management.

Topographic description

Topographic data was necessary to delimit the area, also, maps that define the river system, location of dams, and hydrometric stations (Fig. 7). Maps were obtained from the web site of the National Institute of Geography and Statistics of Mexico (INEGI

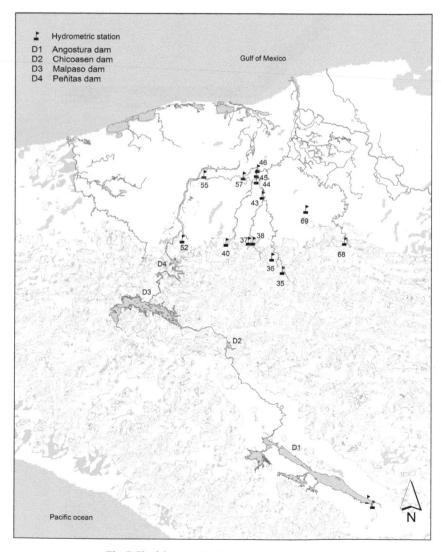

Fig. 7. Fluvial system, hydrometric stations, and dams.

2015), but it is possible to find word maps from the web site of National Aeronautic and Space Administration (NASA 2015) or web site ASTER Global Digital Elevation Model (ASTER 2016).

We analyzed the data from 16 hydrometric gauge stations (Table 2) and four dams (Table 3) located in the area.

Table 2. Hydrometric stations and its critical water levels.

Id.	Name	Critical level (masl*)	Id.	Name	Critical level (masl*)
52	Platanar	32.28	40	San Joaquín	23.12
55	Samaria	14.90	44	Gaviotas	5.42
57	González	8.77	45	Muelle	5.24
35	Oxolotán	39.53	46	Porvenir	4.74
36	Tapijulapa	24.63	68	Salto de Agua	10.99
38	Puyacatengo	29.65	69	Macuspana	9.45
37	Teapa	37.71	59	Boca del Cerro	18.65
43	Pueblo Nuevo	7.49	63	San Pedro	9.01

*masl, meters above sea level

Table 3. Characteristic of the dam system.

Dam	Angostura	Chicoasen	Malpaso	Peñitas
Maximum extraordinary water level (MEWL) [masl]	539.50	395.00	188.00	93.50
Maximum ordinary water level (MOWL) [masl]	533.00	392.00	182.50	87.40
Minimum ordinary water level (MINOWL) [masl]	500.00	380.00	144.00	85.00
Top of the gate [masl]	538.60	394.00	183.50	91.13
Crest of spillway [masl]	519.60	373.00	167.60	76.50
Total capacity [10^6m^3]	19,736.00	1,443.00	14,058.00	1,485.00
Conservation capacity [10^6m^3]	15,548.00	1,372.83	12,373.00	1,091.00
Minimum capacity [10^6m^3]	2,379.50	1,168.81	3,055.73	960.99

The forecast (Meteorological information)

The data was collected from weather bulletins from the National Meteorological Service of Mexico from dates October, 18 to 24, 2007. The scenario started on October 18 when the cold front No. 4 associated with an air mass travelled across the USA. It was predicted three days in advance as an event that would produce severe rainfall throughout the Gulf of Mexico (Fig. 8).

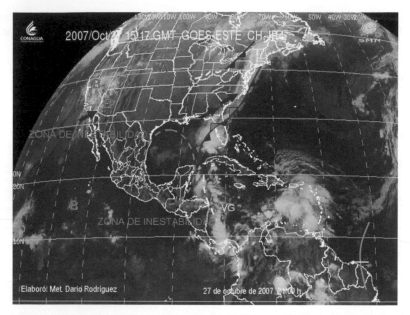

Fig. 8. Image from satellite GOES-E from October 27, 21:00 hours. Cold Front No. 4 (Adapted from Hernández 2007).

Precipitation analysis

The hyetograph for the region in the period from October 1 to November 3, 2007 was built from pluviographic records (Fig. 9).

Hydrometric and hydraulic Information

Curves of water surface elevation in the river system from October 1 to November 17 were obtained from bulletins of hydrometric stations and were compared against the critical levels of rivers (C.L.). We observed that from October 28 to November 5 (Fig. 10) all rivers overflowed.

In addition to the analysis of the river system, the operation of the last dam located downstream was inferred. This dam storage is directly linked to the levels of the rivers in Villahermosa (Fig. 11).
Where: MOWL—maximum ordinary water level; MINOWL—minimum operating water level, and MEWL—maximum extraordinary water level.

Demographic information

In this case, demographic information was not collected.

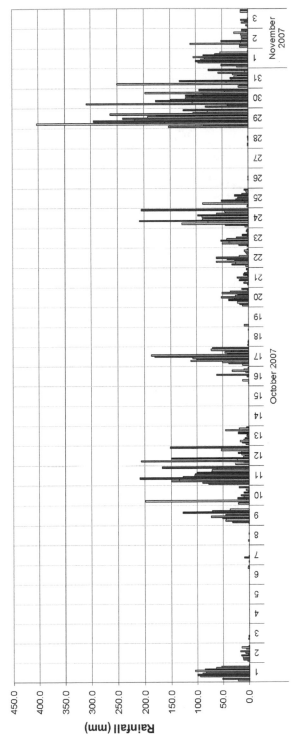

Fig. 9. Daily Rainfall over the storm period October 1 to November 3, 2007 (Adapted from Rivera et al. 2010).

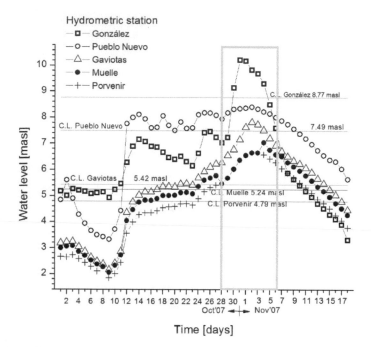

Fig. 10. Water levels in hydrometric stations over the period from October 1 to november 17 (adapted from Rivera et al. 2010).

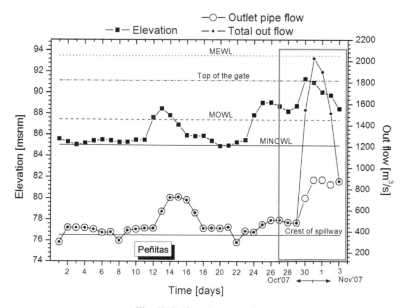

Fig. 11. Peñitas Dam operation.

Integral description of the phenomenon (Hydro-meteorological forecasting chain)

The particular phenomenon analyzed was severe and the result of a conjunction of the following factors: periods of successive rains (meteorological analysis), very intense rains (hyetograph analysis), soils that were saturated and partly saturated by rains from the previous days, rivers over critical levels (hydrometric analysis) and an extraordinary output of water from a dam (hydraulic analysis).

The estimated losses include: 1,487,000 people affected (74% of the population), 1,526,601 flooded hectares (62% of the land), 1,456 affected localities (58% of localities), 3,505 schools (66% of total schools in the area), and adding affected roads and hydraulic infrastructure turned this event into one of the most catastrophic to national level.

Conclusions and Recommendations

In this chapter, we introduced general aspects about flooding events. Flooding event analysis involves two main aspects: technical and social. From the technical side, challenges are presented in data collection and its analysis. However, evaluation and analysis of the demographic part is not easy work. In our experience, the government and agencies involved in these phenomena are disorganized, improvise actions, and are without adequate prevention plans. Few governments are prepared to deal with floods; they pay poor attention and usually are overwhelmed during its development, and try to repair the damages in the later stage. We can see this in the reported losses of human lives during the floods. As engineers and technicians, the expected result of a flood analysis must be objective, without under or overestimating the actions of the agencies. Analysis should be particularly careful to make recommendations and all of them must be supported by rigorous analysis.

Although there is widespread literature in the subject of flooding, we wanted to summarize about the phenomena, some useful tools, and practice aspects for conducting the analysis. Also for those who wish to read about the subject further, we have added some references.

Acknowledgments

The authors acknowledge the support of the CB-2001-1660168 Basic Science project, financed by CONACYT.

Keywords: Forensic Disaster Analysis, Floodplain, Numerical models, practical tools, integral analysis.

References

Adina, M.G. 2014. A case study on the diagnosis and consequences of flash floods in south – western Romania: the upper basin of desnatui river. Journal of the Geographical Institute Jovan Cvijic 62(4): 161–176.
ASTER. 2015. Web Site [http://www.jspacesystems.or.jp/ersdac/GDEM/E/4.html].

Cea, L. and Bladé, E. 2015. A simple and efficient unstructured finite volume scheme for solving the shallow water equations in overland flow applications. Water Resources Research 51: 5464–5486, doi:10.1002/2014WR016547.

Costache, R. and Pravalie, R. 2013. The analysis of May 29, 2012 flood phenomena in the lower sector of Slanic drainage basin (case of Cernatesti locality area). GEOREVIEW, 78–87.

Delrieu, G., Ducrocq, V., Gaume, E., Nicol, J., Payrastre, O., Yates, E. and Wobrock, W. 2005. The Catastrophic Flash-Flood Event of 8–9 September 2002 in the Gard Region, France: A First Case Study for the Cévennes–Vivarais, Mediterranean Hydrometeorological Observatory. Journal of Hydrometeorology 6(1): 34–52.

Escuder-Bueno, I.E., Castillo-Rodríguez, J.T., Zechner, S., Jöbstl, C., Perales-Monparler, S. and Petaccia, G. 2012. A quantitative flood risk analysis methodology for urban areas with integration of social research data. Natural Hazards and Earth System Sciences, 2843–2863.

Faures, J.M. Goodrich, D.C., Woolisher, D.A. and Sorooshian, S. 1995. Impact of small-scale rainfall variability on runoff Modeling. Journal of Hydrology 173: 309–326.

Fiorentino, M., Gioia, A., Iacobellis, V. and Manfreda, S. 2006. Analysis on flood generation processes by means of a continuous simulation model. Advances in Geosciences 7: 231–236.

Fox, R. and Rowntree, K. 2013. Extreme weather events in the Sneeuberg, Karoo, South Africa: a case study of the floods of 9 and 12 February 2011. Hydrology and Earth System Sciences, 10809–10844.

Gayathri, K.D., Ganasri, B.P. and Dwarakish, G.S. 2015. A review on hydrological models. Aquatic Procedia 4: 1001–1007.

Hernández, U.A. 2007. Review of the intense cold front No. 4 based on notifications issued by the National Weather Service (Spanish). Storm Warning No. 866, National Weather Service – CONAGUA, 27 pp.

Hicks, N.S., Smith, J.A., Miller, A.J. and Nelson, P.A. 2005. Catastrophic flooding from an orographic thunderstorm in the central Appalachians, Water Resources Research 41: W12428, doi:10.1029/2005WR004129.

Horritt, M.S. and Bates, P.D. 2002. Evaluation of 1D and 2D numerical models for predicting river flood inundation. Journal of Hydrology 268(1): 87–99.

Hunter, N.M., Bates, P.D., Horritt, M.S. and Wilson, M.D. 2007. Simple spatially-distributed models for predicting flood inundation: a review. Geomorphology 90(3): 208–225.

INEGI. 2015. Web Site [http://antares.inegi.org.mx/analisis/red_hidro/SIATL/#].

Jorge, M.C., De León, N.M., Rodríguez, J.M., González, Y.M. and López, Y.R. 2007. Análisis de inundaciones. Estudio de casos. (Spanish). Ingenieria Hidraulica y Ambiental 28(1): 21–29.

Knabb, R.D., Rhome, J.R. and Daniel, B.P. 2011. Tropical Cyclone Report Hurricane Katrina 23–30 August 2005. E.E.U.U.: National Hurricane Center.

llasat, M.C. 2001. An objective classification of rainfall events on the basis of their convective features: application to rainfall intensity in the northeast of Spain. International Journal of Climatology, 21(11): 1385–1400.

Matishov, G., Chikin, A., Berdnikov, S. and Sheverdyaev, I. 2014. The extreme flood in the don river delta, March 23–24, 2013, and determining factors. Doklady Earth Sciences 455(1): 360–363. doi:10.1134/S1028334X14030295.

Meintel, J. 2015. Rising Waters. Mobility Forum: The Journal of the Air Mobility Command's Magazine 24(2): 28–29.

Méndez-Antonio, B., Domínguez, R., Soto-Cortés, G., Rivera-Trejo, F., Magaña, V. and Caetano, E. 2011. Radars, an alternative in hydrological modeling: Lumped model. Atmósfera 24(2): 157–171.

Mori, J., Mooney, W.D., Kurniawan, S., Anaya, A.I. and Widiyantoro, S. 2007. The 17 July 2006 tsunami earthquake in West Java, Indonesia. Seismological Research Letters 78(2): 201–207, doi:10.1785/gssrl.78.2.201.

Morin, J., Rosenfeld, D. and Amitai, E. 1995. Radar rain field evaluation and possible use of its high temporal and spatial resolution for hydrological purposes. Journal of Hydrology 172: 275–292.

Moussa, R., & Cheviron, B. (2015). Modeling of floods—State of the art and research challenges. *In*: Rivers–Physical, Fluvial and Environmental Processes (pp. 169-192). Springer International Publishing, Switzerland.

NASA. 2015. Web Site [http://reverb.echo.nasa.gov/].

Rivera-Trejo, F., Soto-Cortés, G. and Méndez-Antonio, B. 2010. The 2007 flood in Tabasco, Mexico: an integral analysis of a devastating phenomenon. International Journal of River Basin Management 8(3-4): 255–267.

Rossman, L. 2015. Storm Water Management Model, Reference Manual, Volume I – Hydrology. National Risk Management Laboratory, Cincinnati, OH, pp. 235.

Salinas, M.A. and Jiménez, M. 2014. Inundaciones, Serie Fascículos. Centro Nacional de Prevención de Desastres, México.

Sayama, T., Ozawa, G., Kawakami, T., Nabesaka, S. and Fukami, K. 2012. Rainfall–runoff–inundation analysis of the 2010 Pakistan flood in the Kabul River basin. Hydrological Sciences Journal/Journal Des Sciences Hydrologiques 57(2): 298–312. doi:10.1080/02626667.2011.644245.

Schröter, K., Kunz, M., Elmer, F., Mühr, B. and Merz, B. 2014. What made the June 2013 flood in Germany an exceptional event? A hydro-meteorological evaluation. Hydrology and Earth System Sciences 19(1): 8125–8166.

Shentsis, I., Laronne, J.B. and Alpert, P. 2012. Red Sea Trough flood events in the Negev, Israel (1964–2007). Hydrological Sciences Journal/Journal Des Sciences Hydrologiques 57(1): 42–51. doi:10.1080/02626667.2011.636922.

Silvestro, F., Gabellani, S., Giannoni, F., Parodi, A., Rebora, N., Rudari, R. and Siccardi, F. 2012. A hydrological analysis of the 4 November 2011 event in Genoa. Natural Hazards and Earth System Sciences 12(9): 2743–2752.

Soto Cortés, G. and Berezowsky, M. 2003. Numerical simulation of wetting and drying in shallow waters using a curvilinear adaptive scheme (in Spanish), JUL-SEP; XVIII, pp. 29–44.

Sto. Domingo, N.D., Refsgaard, A., Mark, O. and Paludan, B. 2010. Flood analysis in mixed-urban areas reflecting interactions with the complete water cycle through coupled hydrologic-hydraulic modelling. Water Science & Technology 62(6): 1386–1392.

Vieux, E.B. 2001. Distributed Hydrologic Model Using GIS. Water Science and Technology Library, Vol. 38. Kluwer Academic Publisher. The Netherlands.

WDC. 2105. Word Disaster Reports 2015, The International Federation of Red Cross and Red Crescent Societies [http://ifrc-media.org/interactive/wp-content/uploads/2015/09/1293600-World-Disasters-Report-2015_en.pdf].

WMO. 2012. International Glossary of Hydrology WMO-No. 385, World Meteorological Organization, Switzerland.

CHAPTER 23

Mountain Slopes Protection and Stabilization after Forest Fires in Mediterranean Areas

Research Developed in Mountain Areas in Portugal

António Avelino Batista Vieira,[1,*] *António José Bento Gonçalves,*[1,a] *Francisco da Silva Costa,*[1,b] *Luís Miguel da Vinha*[2] and *Flora Carina Ferreira Leite*[1,c]

INTRODUCTION

Fire is an integral part of many ecosystems (Bento-Gonçalves et al. 2012), and it has served to clear forests throughout time, both for agricultural use and grazing, thus taking on the role of crucial "ecological factor" for the development or regression of forest systems around the world (Ferreira-Leite et al. 2012).

Although wildland fires are characteristic of certain regions and seasons, vegetation fires (including grass fires, forest fires, and scrub fires) are a global phenomenon which occur in the tropical, temperate, and boreal regions (Fig. 1) (González-Pérez et al. 2004) and they can occur in the most unfavorable places.

It has been estimated that more than 30% of the global land surface is subject to a considerable frequency of fires (Chuvieco et al. 2008). For millennia, fires were a natural phenomenon (Pausas et al. 2008) and there is unambiguous evidence that

---

[1] Departamento de Geografia, Centro de Estudos em Geografia e Ordenamento do Território-UMinho, Universidade do Minho. Campus de Azurém, 4800 Guimarães (Portugal).
[a] E-mail: bento@geografia.uminho
[b] E-mail: costafs@geografia.uminho.pt
[c] E-mail: floraferreiraleite@gmail.com
[2] Valley City State University, Valley City, North Dakota 58072 (USA). E-mail: luis.davinha@vcsu.edu
* Corresponding authors: vieira@geografia.uminho.pt

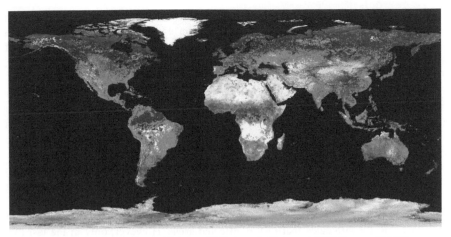

Fig. 1. Forest fires in the world, 1 June 2005 through 31 August 2005 (Source: NASA).

wildfires go back to the Famennian Age (Devonian/Carboniferous periods) (Jones and Rowe 1999). Pausas and Keeley (2009) argue that the appearance of wildfires is concomitant with the origin of terrestrial plants. Even today, in some climates, fires are a natural phenomenon, but most of the fires are caused by human activity and only a negligible number are produced by environmental factors (FAO 2001). Equally, only a relatively small number of forest fires are responsible for the very high proportion of the total damage (Strauss et al. 1989).

Although fire has been a natural and fundamental factor influencing the evolution of the landscape throughout time (Naveh 1975; Pyne 1982; Pausas et al. 2008; Pausas and Keeley 2009), recent decades have been marked by a rising concern with the issues related to the ongoing global changes, such as land-use (Meyer and Turner 1994; IPCC 2000; Dolman et al. 2003; Lambin and Meyfroidt 2011) and climate change (IPCC 1990; Viner et al. 1995; Hulme et al. 1999; IPCC 2007a; Hitz and Smith 2004; Venkata Raman et al. 2012) and its direct and indirect effects on society, namely those inducing abrupt changes in the community (Kazanis and Arianotsou 2004; Rodrigo et al. 2004; Luís et al. 2006; Arnan et al. 2007; Shakesby 2011).

Parallel to this tendency, research seeking to identify causes, effects, and mitigation measures has also multiplied (Woodward 1992; IPCC 2000; IPCC 2007b; IPCC 2007c; NAS 2001; Hitz and Smith 2004; Easterbrook 2011; Holmes 2011; Humlum et al. 2011; Venkata Raman et al. 2012). The interest in the effects of fires in the ecosystems, in all its different dimensions, has been the subject of extensive research, especially related to soil erosion and degradation (for more extensive references, see Bento-Gonçalves et al. 2012).

In fact, the destruction of vegetation by forest fires makes soils vulnerable to erosion by promoting the removal of nutrients and mineral components. This significant and continuous degradation of soil, especially recognized in the Mediterranean areas, makes the implementation of slope protection measures urgent in order to mitigate the effects of forest fires and reduce the loss of soil and nutrients (Shakesby et al. 1993; Bento-Gonçalves and Coelho 1995; Shakesby et al. 1996; Walsh et al. 1998; Bento-Gonçalves and Lourenço 2010; Vega et al. 2010).

Forest recovery intervention measures after wildland fires have long been implemented, particularly in the "Mediterranean world" where fires have functioned as a natural and fundamental factor in determining the evolution of the landscape throughout history (Naveh 1975; Pyne 1982; Pausas et al. 2008; Mataix-Solera and Cerdà 2009; Pausas and Keeley 2009; Shakesby 2011).

In the United States, post-fire intervention actions have been implemented for some decades (since 1930s, according to Robichaud et al. 2005). These interventions count on specialized multidisciplinary teams that evaluate the needs and the treatments specifically required for each burned area (Robichaud 2009) and which apply specific programs to evaluate the interventions in areas of risk (BAER – Burnt Areas Emergency Response) (Napper 2006). Also in other countries affected by forest fires, such as Australia and Canada, large campaigns and post-fire rehabilitation plans are being promoted (Pike and Ussery 2006; Robichaud 2009).

In the Mediterranean countries, the efforts of the authorities have been directed mainly towards strategies that restore affected areas. However, only in the last two decades have some emergency stabilization measures been implemented—but on a small scale. The importance of this problem in the Mediterranean countries of the European Union alerted the authorities to the urgent need of financing for the development of scientific research projects, such as EUFIRELAB, which systematically report on the adequate tools and methodologies for restoring burnt areas (Vallejo 2006). In the meantime, some initiatives were developed in Spain (Bautista et al. 1996; Pinaya et al. 2000; Carballas et al. 2009; Vega 2011) and in Greece (Raftoyannis and Spanos 2005).

1. Forest Fires in Portugal and Soil Erosion Mitigation Techniques

The evolution of the forest cover in Portugal has followed, over the years, a pattern which is shared by the entire Mediterranean region. This area is characterized by the destruction of original forestland through frequent fires which make way for grazing grounds, the use of the best soils for cereal culture, and the use of wood as fuel source and as a raw material for construction (Andrada e Silva 1815; Rego 2001). However, the phenomenon described is not that of a contemporary perspective of forest fires, but that of slash-and-burn practice carried out by societies to make way for their daily activities or for their own protection.

In the Mediterranean, the role of fires has been particularly relevant due to a combination of very special characteristics that make the Mediterranean ecosystems different from others around the world. This specificity is the result of its particular weather characteristics and the prolonged and intense human presence with its consequent influence on fires (Pausas and Vallejo 1999).

Although fire has shaped the Mediterranean ecosystems, fire occurrence schemes, meaning their frequency and intensity, have changed in the last 50 years. The natural cycle of fire has been reduced (Pereira et al. 2006), fires have become recurrent (Ferreira-Leite et al. 2011), their intensity and expansion have increased, and they have taken on catastrophic proportions and have lost their role as catalysts for the renewal of the ecosystems (Noss et al. 2006).

A set of socio-economic changes which took place during the second half of the twentieth century in Mediterranean countries, seems to have contributed to a scenario where fires are not only more plausible, but also more difficult to extinguish due to the considerable amounts of biomass accumulated over the years (Lourenço 1991;

Vélez 1993; Moreno et al. 1998; Rego 2001). When combined with very unfavorable weather conditions (Lourenço 1988; Pyne 2006), this accumulation of biomass fuels catastrophic fires, resulting in an increase in burned areas (Ferreira-Leite et al. 2013).

Even if the number of large forest fires is statistically irrelevant in Portugal when compared to the total number of occurrences, it is still the main factor behind most of the annual burned area. Despite there being a statistically relevant increase in the number of large forest fires over the last ten years, there is a slight trend toward the increase in the expansion of large forest fires, that is, expansion which is proportional to the size of the fire (Ferreira-Leite et al. 2013).

Portugal has a predominantly warm and temperate Mediterranean climate, characterized by hot, dry summers and fresh, wet winters. Areas of rugged terrain are common and the natural vegetation is typically evergreen, which is resistant to drought and fire-prone (Ferreira-Leite et al. 2011).

Since the last quarter century, forest fires have registered a significant increase in both the number of occurrences and the size of the scorched areas. This phenomenon is due mainly to the profound social changes that began to be felt in the population residing in the forest areas since the second half of the twentieth century, especially in the Portuguese interior.

With the disintegration of the rural world (Bento-Gonçalves et al. 2010), especially after the 1970s, we have witnessed an increase in forest fires and burned area in Portugal (Fig. 2).

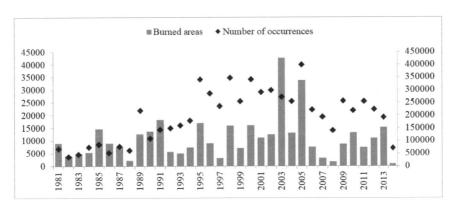

Fig. 2. Evolution of the number of occurrences and burned area, between 1981 and 2014 (Source: ICNF).

The observation of the graph is elucidative of the evolution of the number of occurrences and scorched areas over the period under consideration. More precisely, it highlights several differentiated trends.

In terms of the number of occurrences, the values are highly determined by human factors. In fact, human behavior, either intentionally or unintentionally, is mostly responsible for the growing number of occurrences over the years.

When analyzing the extent of the scorched areas, we can conclude that there is an irregular pattern that is directly linked to the different weather conditions throughout the years, the lack of land and forest management, and, finally, to some inefficiencies in the firefighting operations.

Contrasting with the slow evolution of forest cover growth over the centuries, recent years have demonstrated very rapid changes, mainly due to the high incidence of forest fires, which resulted, among other things, in a profound change in tree species, particularly of the autochthonous species.

At present, the two most representative species of forests in northern and central Portugal are the pine and eucalyptus forests, whereas previously, oak and chestnut trees were predominant.

Although the holm and the cork oak still continue to dominate, eucalyptus and pine trees have begun to gain ground rapidly and occupy an increasingly significant area. This transformation has also contributed to the increase in the number of ignitions, and especially in recent years, to the occurrence of large fires south of the Tagus River, and in particular, in the hills of the Algarve region.

Taking into consideration that a significant amount of Portuguese forests are located in mountainous areas, their destruction by forest fires contributes to the development of conditions for soil degradation.

As we stated before, soil erosion is one of the major effects of forest fires due to the intense and continuous degradation of the quality and quantity of the top soil layer where the nutrients in the majority of Portuguese soils of mountain areas are concentrated (Bento-Gonçalves et al. 2008).

In a climate with Mediterranean characteristics such as the one in Portugal, the removal of sediments and nutrients occurs in the first four to six months after a fire. Therefore, it's fundamental to study and implement effective solutions that reduce those losses and prevent the destruction of this valuable resource (Shakesby et al. 1993; Bento-Gonçalves and Coelho 1995; Shakesby et al. 1996; Walsh 1998; Bento-Gonçalves and Lourenço 2010; Vega et al. 2010).

Nevertheless, this process is intimately dependent on fire recurrence, intensity, severity, and spatial variability of soil hydrophobicity (Jungerius and DeJong 1989; Ritsema and Dekker 1994; Coelho et al. 2004; Bento-Gonçalves et al. 2012), as well as the physical characteristics of the affected area (slope, exposure, climate, geological composition...) as some pioneer studies conducted in Portugal have demonstrated (Lourenço 1989; Lourenço and Bento-Gonçalves 1990; Lourenço et al. 1991). Therefore, is important to take all these factors into consideration and to adapt the different mitigation measures and techniques to each specific context.

There are several mitigation and rehabilitation strategies that may be implemented based on degradation risk and management objectives. In fact, forestall management can have multiple objectives. However, if what is intended is the reduction of the impacts of forest fires, a minimum set of priority objectives must be defined in the majority of the cases (Vallejo 2006): soil protection and hydrological regulation; fire risk reduction; increase in the resilience of ecosystems and landscapes against forest fires; development of adult, diversified, and productive stands.

Currently in the Mediterranean, suppressing all fires is one of the dominant management policies. Nevertheless, an alternative to this practice is the use of prescribed fires as an important management tool (Fireparadox – http://www.fireparadox.org/). The main advantage of prescribed fires is the reduction of fuel load, promoting the discontinuity of fuel and, therefore, limiting the severity of forest fires. However, in the long-term, preventive forestall management measures must be applied (Shakesby 2011).

In this context, is important to implement strategies that increase the heterogeneity and fragmentation of the vegetation and increase structural diversity in order to promote an increase in the resistance of landscapes to fire (Moreira et al. 2009).

There are other options with the objective of creating and spreading the scientific basis and intervention techniques for the management of burnt areas which may be implemented immediately after a fire in order to reduce erosion and, at the same time, contribute to the conservation of the ecosystem.

As previously referred to, in the United States of America, the experience regarding the implementation of intervention measures to burnt areas is considerable and there is a systematization of measures, which take into account the specific strategy to implement and the time frame for its implementation (USA General Accounting Office 2006; Robichaud 2009).

In general, they are grouped into three categories. The first one, corresponds to the "emergency stabilization" procedures, which are implemented immediately after the fire (sometimes before the fire is completely controlled) and within one year. The main objectives of these measures are to control and reduce soil erosion and to protect life, property, and resources. The main emergency stabilization procedures are mulching and seeding, contour-felled logs, and check dams among others (Neary et al. 2005; Napper 2006; Foltz et al. 2009; Robichaud et al. 2010).

In the second category, we may find the "rehabilitation" procedures which are implemented over a longer period of time—that is, about three years after the fire—and which encompass tasks such as reparation of facilities or the mitigation of land damage lacking auto-recovery ability. Finally, the "restoration" strategies, considered long-term actions, are implemented along with the other strategies, but have a longer temporal implementation. These strategies seek to promote the restoration of the quality and productivity of the habitat and increase its resilience (Robichaud 2009).

These measures are widely implemented (especially in the United States) in order to promote the effective rehabilitation of burned areas and to mitigate the effects of fires on soil and vegetation. However, before deciding which measures are adequate for a specific situation, we must decide which areas need soil protection measures.

Although the implementation of post-fire mitigation treatments undoubtedly promote soil protection from erosion and help vegetation recovery, for some areas it is preferable not to apply any kind of treatment in the burnt areas (Robichaud 2009; Bautista et al. 2009). Furthermore, its effectiveness must continue to be evaluated, as well as its impacts in the short and long term, particularly on soil, water, and plants (Kruse et al. 2004; Robichaud 2009; Neary 2009).

The implementation of these measures is, however, difficult. The majority of the post-fire soil protection measures are expensive and are difficult to implement. For these reasons, land owners are often reluctant to invest in soil protection because of the low profit they receive and the high risk involved in forestall activities.

2. Implementation of Soil Mitigation Measures after Forest Fires in an Experimental Field in the Northwestern Mountains of Portugal

2.1 Study area

The municipality of Terras de Bouro, located in the northwest of Portugal, has been affected continuously by forest fires. In the period between 1996 and 2010, the number of forest fires was above 100 per year, except in 1997, 2008, and 2010. Although registering a reduced number of fires, the year of 2010 witnessed the two major forest fires occurring in the period of consideration, with more than 1000 ha of burned area for each one. The natural characteristics of the burned areas, with steep slopes and poor soils, reveal a high risk of soil erosion. These characteristics lead us to implement different techniques for erosion mitigation in the experimental area, in order to evaluate not only their effectiveness, but also their cost/benefit relation.

The area identified for implementing the techniques was affected by the fires that broke out on August 2010 in the municipality of Terras do Bouro (Fig. 3), in the heart of the Peneda-Geres National Park (NW Portugal). These fires produced a continuous burnt area which was subject to different fire intensities and severity. It is a large area, occupied by scrublands and stands of *Pinus pinaster*. The lithology is mainly granite and the soils (cambisols) are generally thin and stony. The land cover in the last 50 years has been essentially composed of woodlands that are favored by the climate, which is characterized by high amounts of precipitation.

In order to evaluate the severity of the fire on the affected area, we analyzed the satellite images from before and after the forest fires. Based on Landsat 5, images of July 30, 2010, and April 28, 2011, a NBR algorithm (normalized burnt ratio) was implemented with the definition of five classes of severity: very high, high, moderate, low severity, and not burned (Fig. 4).

The result was validated in the field (Photos 1 and 2), where fire severity was evaluated using the BAER methodology presented by Parsons et al. (2010) and the methodology proposed by Lampin et al. (2003).

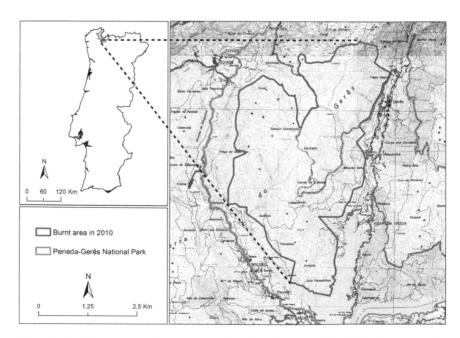

Fig. 3. Limits of the burnt area in the two large forest fires that occurred in 2010 in the municipality of Terras de Bouro.

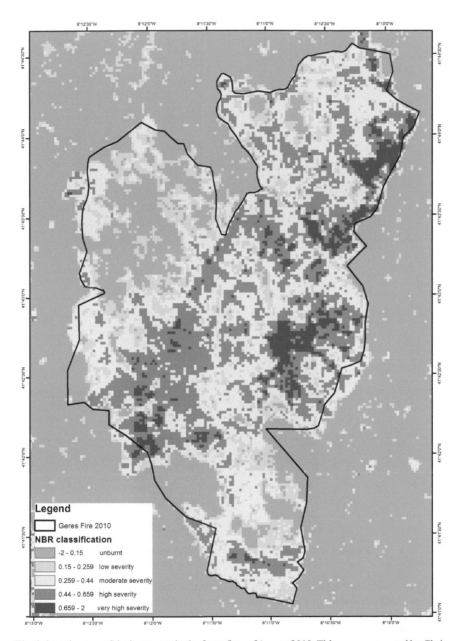

Fig. 4. Severity map of the burnt area in the forest fires of August 2010. This map was created by Chris Schafer, from the Sydney Catchment Authority – Australia.

Based on the distribution of the severity of fire and on the erosion risks identified in the field and topographic conditions, two experimental areas were defined, in order to implement the erosion mitigation techniques. Through the "Soil Protec" project (Emergency measures for post-fire soil protection), financed by the CEGOT, different

Photos 1 and 2. Junceda – Medium and High severity burnt areas.

low cost emergency measures were tested. The techniques were implemented in slopes and in channels immediately after the fires in areas of low to medium fire severity. These areas were essentially composed of *Pinus pinaster* stands in Serra do Gerês (Bento-Gonçalves et al. 2011).

The objectives of the research were: to test the role of pine needles in slopes affected by low to medium severity fires as protection for soil against erosion and compare it with the use of straw; to test a set of measures in channels, where there is a concentration of runoff in order to reduce the gully processes and the removal and transport of soil. This can be achieved by implementing structures, materials, and techniques which favor the retention of sediments and the possible consolidation of ridges and pre-existing gullies.

The measures were assessed in terms of their effectiveness in mitigating erosion (especially throughout time), as well as their cost/benefit.

2.2 Measures applied to slopes

Following the forest fires that broke out in the study area, we installed six plots,[1] measuring 10 meters in length and 2.5 meters in width,[2] in an area of a *Pinus pinaster* stand affected by medium severity fire and with an average slope of 15% (Fig. 5).

We then applied the different proposed treatments selected for evaluation, corresponding to straw (2, 4, and 8 kg) and pine needles (2 and 4 kg), spread over five plots. One plot was left over as a control plot (Fig. 6).

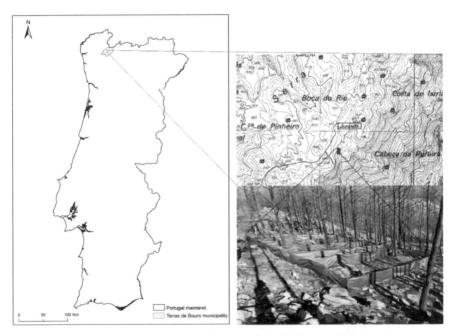

Fig. 5. Area of installation of the plots (Junceda, Terras de Bouro).

2.3 Measures applied to channels

After the field recognition, a small catchment revealing high erosion risk and the development of gullies was identified and some critical spots were selected and prepared for intervention. We proceeded with the installation of structural measures in the

---

[1] The plots we implemented in the study area were based on the design used in Galicia. Our objective was to establish a comparison with the results obtained by the "Instituto de Investigaciones Agrobiológicas de Galicia". Nevertheless, we concluded that this type of plot isn't the most appropriate for the steep and rocky slopes present in our study area.

[2] Although the total area of each plot is about 25 m², the use of the geotextile limited the useful area available and, therefore, for this analysis we calculated the effective area of each plot to be about 16 m².

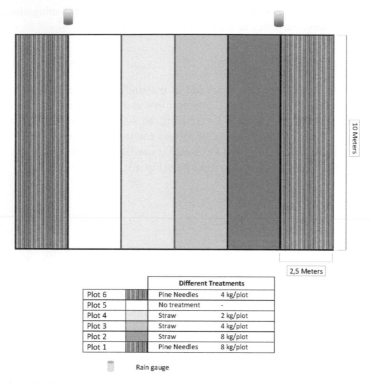

		Different Treatments	
Plot 6		Pine Needles	4 kg/plot
Plot 5		No treatment	-
Plot 4		Straw	2 kg/plot
Plot 3		Straw	4 kg/plot
Plot 2		Straw	8 kg/plot
Plot 1		Pine Needles	8 kg/plot

Rain gauge

Fig. 6. Experimental design for testing post-fire emergency measures for soil protection.

channels in order to evaluate its effectiveness on soil erosion mitigation. The selected area was subject to significant disturbance and increased risk of erosion after the fire due to the use of heavy machinery to remove burned timber.

Photo 3. Selected area for the implementation of channel erosion mitigation measures (the arrows indicate the gullies that were intervened).

The selected mitigation measures were particularly focused on the channels which served to drain the water from the slopes in the study area. The objective of these measures was to alter the flow of water and sediments in order to decrease the amount of soil spilling into the water lines and, consequently, the damaging effect of debris torrents on human infrastructures or agricultural production.

The techniques that we implemented consisted in log check dams, straw bale barriers, and barriers made from the remains of the cutting of the burnt pine trees.

The first dams (log check dams) required the use of mechanical means (tractors and electric saws) for their implementation. Accordingly, this technique is more demanding in terms of costs since it involves more man power for its implementation (Photos 4 and 5).

Photos 4 and 5. Construction of log check dams.

The overarching objective of these measures is to apply locally collected trunks and, therefore, decrease the inherent costs of transporting materials from outside of the area. The trunks are placed perpendicular to the water flow in order to create a barrier to the sediments transported by water and the runoff from the slope. The accumulation of this material will decrease the speed of the seepage and, eventually, lessening the peak periods of seepage. The accumulation of sediments ("sediment pools") may ultimately catalyze the regeneration of vegetation.

The straw bale barriers are significantly easier to implement. Their objective is the same as that of the log check dams. However, their use should be limited to areas with slopes that are not as steep and in smaller gullies. The placement of the straw bales —preferably between three and five bales—should also be perpendicular to the water way. It is common to use rocks (or logs) in order to support the bales. In the barrier implemented in the study area (Photos 6 and 7) we placed three straw bales which were secured by iron rods and supported by granite blocks.

Photos 6 and 7. Straw bale barrier implemented.

This method involves taking into consideration the costs of the straw bales and their transport. The manual labor involved is significantly less demanding and the handling of the bales is relatively easy.

The barriers made from the remains of the cutting of the burnt pine trees are made from materials available in the area. This strategy is one of the least expensive methods and the easiest to implement. For this reason, it is the measure that logging companies may systematically implement after extracting the wood. This is a variant of a technique applied in the US (Napper 2006) in which the canopies of the trees that have been cut are placed along the water ways.

In the study area, the removal of wood (essentially trunks with some economic potential) left large quantities of accumulated materials, such as pine branches, twigs, and other left over materials. This allowed us to implement this strategy in many other locales along the water ways and complement the measures initially planned. This sort of intervention requires very limited labor, taking into consideration that all the necessary resources are locally available.

3. Discussion

3.1 Measures applied to slopes

The analysis of the results of soil loss indicates that the annual erosion rates in the plots (with and without treatments) are not very significant (Fig. 7). Accordingly, the erosion can be considered tolerable.

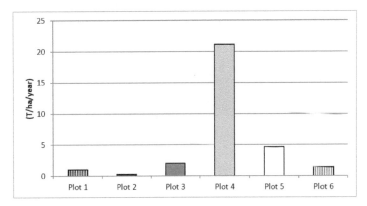

Fig. 7. Erosion rates (T/ha/year).

In fact, in similar conditions to those verified in our study area, Diaz Fierros et al. (1982) identified three thresholds of erosion (for the Galiza region). The three levels correspond to 11, 30, and 100 ton/ha/year. As a result, up to 11 ton/ha/year erosion is considered tolerable. Between 11 and 30 ton/ha/year it is classified as consisting of slight erosion. When the levels of erosion are between 30 and 100 ton/ha/year they are considered moderate and anything over 100 ton/ha/year is grave.

FAO-PNUMA-UNESCO (1980) also presented a classification with three levels of severity for soil degradation, which correspond to 10, 50, and 100 ton/ha/year. In this classification, values below 10 ton/ha/year are considered low or inexistent. Between 10 and 50 ton/ha/year degradation is low, and between 50 and 200 ton/ha/year it is moderate. Above 200 ton/ha/year, it is considered to be high.

In our experiment, only plot 4 was an exception to the tolerable degradation. This plot was subject to a treatment of 1.25 tons of straw per hectare, and it clearly surpassed 10–11 ton/ha/year by claiming more than 21 ton/ha/year. Nevertheless, this value is due to the systemic reception of materials running off from plot 5 (control parcel). The impossibility of totally isolating the plots from one another was responsible for this outlier.

Therefore, one of the first results obtained was the verification of inadequacy of these types of plots in slopes with shallow soils and an irregular and rocky relief (Photo 8). More precisely, the influence of a micro-relief with small "pools of sediments", which promotes the retention of a large amount of sediments, seems to overcome the influence of the implemented treatments.

Photo 8. Detailed image of rocky terrain in one of the trial plots.

The second conclusion from the results demonstrates the high efficacy in protecting the soil derived from all the types of soil covers applied (straw or pine needles). However, that efficacy tends to progressively diminish after one year (Fig. 6). Although the first significant peak of erosion, characterized by a high volume of precipitation (613.33 mm), occurred between 18 December 2010 and 15 January 2011, the largest peak only occurred nine months later, between 14 September and 28 October 2011. While the precipitation registered during this period was about half of that of the prior peak (290.75 mm) and the plots contained less material for mobilization, the covering material was already significantly damaged and thus favoring the transport of materials.

On the other hand, the low levels of precipitation in the months prior to this period also hindered the transport of materials and allowed for accumulation. Therefore, with a greater accumulation of materials and less retention capacity due to the damaged cover layer, it is natural that with a greater volume of rain and greater intensity, the quantity of transported material was the greatest in the period under analysis.

Comparing the efficacy of each of the materials in terms of the amount of material applied, we verified that, for the case of 0.5 kg/m^2 (5 ton/ha), the use of straw is more effective than that of the pine needles. This is due to the larger area covered by straw when compared with the pine needles and also to the fact that straw creates a higher aggregation of elements, producing a more effective protection—except in very exceptional situations (Fig. 7).

In the cases where the cover is of 0.25 kg/m^2 (2,5 ton/ha), the situation is the opposite. During the first year, the pine needles provided greater soil protection (Fig. 8).

Although the straw cover has demonstrated less durability, the results indicate that in the first year, straw is more effective for cover densities of 5 ton/ha. In contrast, pine needle cover revealed greater efficacy for covers of 2.5 ton/ha.

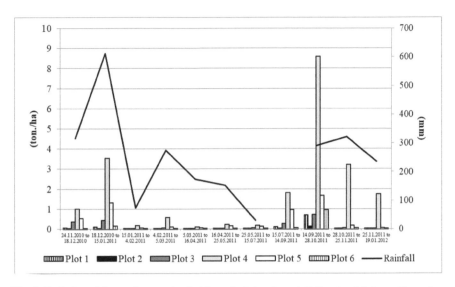

Fig. 8. Evolution of the erosion rate (ton/ha) in each plot and precipitation (mm) between December 2010 and January 2012.

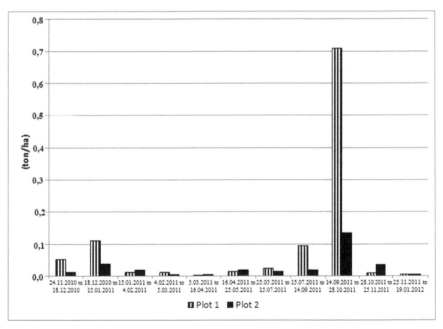

Fig. 9. Evolution of the erosion rate (ton/ha) in the plots with cover densities of 0.5 kg/m² (5 ton/ha) of pine needles and straw, between December 2010 and January 2012.

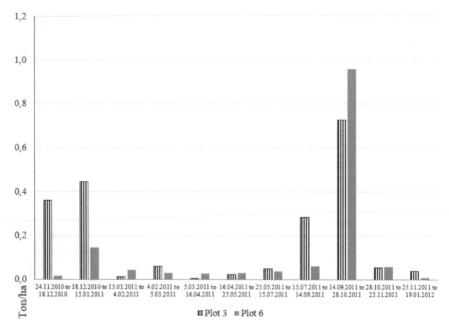

Fig. 10. Evolution of the erosion rate (ton/ha) in the plots with cover densities of 0.25 kg/m^2 (2.5 ton/ha) of pine needles and straw, between December 2010 and January 2012.

However, if we take into consideration that straw is an exogenous element in forest ecosystems and it carries seeds, which may alter the vegetation, pine needles may be the better option. Moreover, pine needles may be available locally and thus reduce the need to transport them from far away.

In addition, straw also presents a greater cost. Its purchase value is about €4 for a bale of 12 kg. Besides these costs, we must also add the costs involving its transportation and application in the field.

Taking these values into consideration, for an intervention using 0.25 kg/m^2 (2.5 ton/ha), a straw bale (20 kg) can cover an area of 80 m^2—that is, about 5 cents per square meter or € 500 per hectare (plus transportation and labor costs). Therefore, while pine needles seem to be the best option, we must take some precautions in their use. More precisely, we must not forget that if we transfer a large amount of pine needles from an area that has not been burned to another that has been scorched, we risk altering the area and leaving the original area unprotected. Thus, by removing the pine needles we are reducing the protective layer and organic materials in the soil. The removal may also imply moving the upper levels of the soil cover and risk depleting it, exposing it to the erosive agents.

Therefore, in case this solution is selected, we should always be careful not to remove the entire pine needle layer. We should limit ourselves to the upper level and avoid depleting the soil, particularly in areas with steep slopes.

3.2 Measures applied to channels

The techniques applied in order to mitigate erosion in channels include log check dams, straw bale barriers, and barriers made from the remains of cutting of the burnt pine trees. These different techniques were implemented in the early winter months of 2011 and the objective was to evaluate their efficacy in controlling erosion and fixating sediments, particularly in terms of cost/benefit.

The heavy rainfall that occurred immediately after the installation of these measures produced considerable erosion on the slopes of the intervened drainage basin (the average annual rainfall in this region is about 2500 mm). The rainfall caused an intense removal of sediments, which were deposited in the waterways that were part of the experiment. Accordingly, by verifying the accumulation of sediments, this phenomenon allowed us to attest to the efficacy of these techniques.

As a result, we verify that all the techniques applied were capable of retaining sediments. Of the two log check dams installed, one effectively served as a flux absorber and facilitated the accumulation of sediments upstream (Photo 9).

Photo 9. Accumulation of sediments in the log check dam (when it was installed and at the last survey).

When assessing the role of the straw bale barriers, we verified their positive effect in sediment retention (Photo 10). Its efficacy is, in our analysis, highly significant. Its main advantage consists in allowing for the adequate flow of water and avoiding situations of rupture of the system due to the retention of high volumes of water. Nevertheless, this technique is less robust than the barriers constructed from logs and trunks and has a much shorter durability cycle—probably no more than one year.

The barriers constructed from the remains of burnt pine trees and pine needles produced the most significant results. Regardless of the simplicity of this approach,

Photo 10. Accumulation of sediments in a straw bale barrier.

its efficacy in retaining sediments is high. In all the barriers where this technique was implemented, we verified the retention of sediments. In many of these plots, the quantity of sediments retained was very high considering the type of structure used (Photos 11, 12, and 13).

Photo 11. Accumulation of sediments in barriers made from the remains of burnt pines and pine needles.

Photos 12 and 13. Accumulation of sediments in barriers made from the remains of burnt pines and pine needles.

In sum, the observations allow us to conclude that the techniques applied possess a significant efficacy in retaining sediments transported by the gullies and, thus, contributing to the creation of sediment "pools" which may function as privileged locales for the recovery of plant life. Therefore, it is imperative to create conditions in these mountain areas, which favor the fixation of the existing soil cover and avoid the transport and deposit of sediments downstream, where it can affect human infrastructures and populations. On the other hand, the results do reveal some differences between the techniques, particularly when assessing their cost/benefit. More precisely, the barriers made of the remains of the removal of burnt pine trees and pine needles are a low cost solution, which produces very satisfactory results in terms of mitigating the effects of erosion.

4. Conclusions

The analysis of the results of the research carried out allows us to conclude that the techniques applied both in channels and in slopes, have a significant degree of effectiveness in soil erosion mitigation.

Although the measures applied to slopes presented different behavior, taking into account the amount of material used in each plot, both methods revealed efficacy in the mitigation of soil erosion. Straw is more effective when we applied 0.5 kg/m² (5 ton/ha), because of the larger area covered by straw when compared with the pine needles and because it creates a higher aggregation of elements, producing a more effective protection. Nevertheless, straw cover has demonstrated less durability, providing effective protection only in the first year. On the other hand, pine needles provided greater soil protection when the cover is of 0.25 kg/m² (2,5 ton/ha). If we consider the relation cost/benefit, pine needles can be considered the most adequate technique to implement.

Regarding the measures applied to channels, all techniques showed similar efficiency in trapping sediments carried by drainage channels of the water overflow, contributing to the creation of sediment "pools". Indeed, it is imperative to create conditions in these mountainous areas so that the small quantity of soil that is remaining is maintained, while simultaneously avoiding its transport and deposit in unwanted areas, namely downstream, where human settlements are located, where there is a dam, and other human infrastructures. Nevertheless, the results suggest a differentiation of techniques regarding the cost/benefit ratio. More precisely, the barriers made from the cut debris, pine rot, and needles revealed satisfactory erosion mitigation capacity at relatively low costs.

Although the results were encouraging, the generalized implementation of these techniques is difficult. Once most parts of forest properties are private and the fire risk is very high, it is necessary to promote awareness campaigns for land owners and national forest authorities, in order to show them that soil protection is fundamental, especially after forest fires, to assure long term productivity.

Acknowledgments

This research was funded by the Foundation for Science and Technology (FCT) through the scholarship SFRH/BSAB/113587/2015, co-financed by the European Social Fund and national funding by MEC.

Keywords: Post-fire soil erosion; erosion mitigation techniques.

References

Andrada e Silva, J. 1815. Memoria sobre a Necessidade e Utilidades do Plantio de Novos Bosques em Portugal, particularmente de pinhaes nos areaes de beria-mar; seu methodo de sementeira, costeamento, e administração, Typografia da Academia Real das Sciencias, Lisboa.
Arnan, X., Rodrigo, A. and Retana, J. 2007. Vegetation type and dryness drive the post-fire regeneration of Mediterranean plant communities at a regional scale. Journal of Vegetation Science 18: 111–122.
Bautista, S., Bellot, J. and Vallejo, V.R. 1996. Mulching treatment for post-fire soil conservation in a semiarid ecosystem. Arid Soil Research and Rehabilitation 10: 235–242.

Bautista, S., Robichaud, P.R. and Bladé, C. 2009. Post-fire mulching. pp. 353–372. *In*: Cerdá, A. and Robichaud, P. (eds.). Fire Effects on Soils and Restoration Strategies. Science Publishers, Enfield, New Hampshire.

Bento-Gonçalves, A. and Coelho, C. 1995. Wildfire impacts on soil loss and runoff in dry mediterranean forest, Tejo basin, Portugal: preliminary results. Proceedings of Course on Desertification in a European Context, Physical and Socio-Economic Aspects, Bruxelles, pp. 361–369.

Bento-Gonçalves, A., Vieira, A., Ferreira, A. and Coelho, C. 2008. Caracterização geomorfológica e implementação de um sistema integrado de informação, em ambiente SIG, no âmbito do projecto RECOVER (Estratégias de remediação de solos imediatamente após incêndios florestais). Revista Geografia Ensino & Pesquisa 12(1).

Bento-Gonçalves, A. and Lourenço, L. 2010. The study and measurement of overland flow and soil erosion on slopes affected by forest fires in Lousã mountain – main results. In Actas das Jornadas Internacionales – Investigación y gestión para la proteccion del suelo y restauración de los ecossistemas forestales affectados por incêndios forestales, Santiago de Compostela.

Bento-Gonçalves, A., Vieira, A., Martins, C., Ferreira-Leite, F. and Costa, F. 2010. A desestruturação do mundo rural e o uso do fogo – o caso da serra da Cabreira (Vieira do Minho), Caminhos nas Ciências Sociais. Memória, Mudança Social e Razão – Estudos em Homenagem a Manuel da Silva Costa, Universidade do Minho, Braga, pp. 87–104.

Bento-Gonçalves, A., Vieira, A., Lourenço, L., Salgado, J., Mendes, L., Castro, A. and Ferreira-Leite, F. 2011. The importance of pine needles in reducing soil erosion following a low/medium intensity wildfire in Junceda (Portugal)—an experimental design. pp. 181–185. *In*: Bento-Gonçalves, A. and Vieira, A. (eds.). Proceedings of the 3rd International Meeting of Fire Effects on Soil Properties. University of Minho, Guimarães, Portugal.

Bento-Gonçalves, A., Vieira, A., Úbeda, X. and Martin, D. 2012. Fire and soils: key concepts and recent advances. Geoderma 191: 3–13.

Carballas, T., Martín, A., González-Prieto, S.J. and Díaz-Raviña, M. 2009. Restauración de ecosistemas quemados de Galicia (N.O. España): Aplicación de residuos orgánicos e impacto de los retardantes de llama. pp. 49–72. *In*: Gallardo, J.F. (ed.). Emisiones de gases con efecto invernadero en ecossistemas iberoamericanos. Red Iberoamericana de Física y Química Ambiental, Salamanca.

Chuvieco, E., Giglio, L. and Justice, C. 2008. Global characterization of fire activity: toward defining fire regimes from Earth observation data. Global Change Biology 14(7): 1488–1502.

Coelho, C., Ferreira, A., Boulet, A. and Keizer, J. 2004. Overland flow generation processes, erosion yields and solute loss following different intensity fires. Quarterly Journal of Engineering Geology and Hydrogeology 37(3): 233–240.

Diaz-Fierros, V., Gil Sotres, F., Cabaneiro, A., Arballas, T., Leiros de la Peña, M.C. and Villar Celorio, M.C. 1982. Efectos erosivos de los incêndios forestales en suelos de Galicia. Anales de Edafología y Agrob XLI(3-4): 627–639.

Dolman, A., Verhagen, A. and Rovers, C.A. 2003. Global Environmental Change and Land Use. Kluwer Academic Publishers, Dordrecht.

Easterbrook, D. 2011. Geologic evidence of recurring climate cycles and their implications for the cause of global climate changes—The past is the key to the future. pp. 3–51. *In*: Easterbrook, D. (ed.). Evidence-Based Climate Science, Data Opposing CO2 Emissions as the Primary Source of Global Warming. Elsevier. Amsterdam.

FAO-PNUMA-UNESCO. 1980. Metodología provisional para la evaluación de la degradación de los suelos. Organización de las Naciones Unidas para el Desarrollo de la Agricultura y la Alimentación (FAO), Programa de las Naciones Unidas para el Medio Ambiente (PNUMA), Organización de las Naciones para el Medio Ambiente (UNESCO). Roma, FAO.

FAO. 2001. Global forest fire assessment 1990–2000. Forest Resources Assessment Programme, Working Paper 55. FAO, Rome, 495 p.

Ferreira-Leite, F., Bento-Gonçalves, A. and Vieira, A. 2011. The recurrence interval of forest fires in Cabeço da Vaca (Cabreira Mountain – northwest of Portugal). Environmental Research 111: 215–221.

Ferreira-Leite, F., Bento-Gonçalves, A. and Lourenço, L. 2012. Grandes incêndios florestais em Portugal. Da história recente à atualidade, Cadernos de Geografia, 30/31: 81–86.

Ferreira-Leite, F., Lourenço, L. and Bento-Gonçalves, A. 2013. Large forest fires in mainland Portugal, brief characterization. Méditerranée 121: 53–66.

Foltz, R.B., Robichaud, P.R. and Rhee, H. 2009. A synthesis of postfire road treatments for BAER teams: methods, treatment effectiveness, and decision making tools for rehabilitation. Gen Tech Rep RMRS-GTR-228, U.S.D.A., Forest Service, Rocky Mountain Research Station.

González-Pérez, J.A., González-Vila, F.J., Almendros, G. and Knicker, H. 2004. The effect of fire on soil organic matter-a review. Environment International 30(6): 855–870.

Hitz, S. and Smith, J. 2004. Estimating global impacts from climate change. Global Environmental Change 14(3): 201–218.

Holmes, J., Lowe, J., Wolff, E. and Srokosz, M. 2011. Rapid climate change: lessons from the recent geological past. Global and Planetary Change 79: 157–162.

Hulme, M., Mitchell, J., Ingram, W., Lowe, J., Johns, T., New, M. and Viner, D. 1999. Climate change scenarios for global impacts studies. Global Environmental Change 9: 3–19.

Humlum, O., Solheim, J. and Stordahl, K. 2011. Identifying natural contributions to late Holocene climate change. Global and Planetary Change 79: 145–156.

Intergovernmental Panel on Climate Change. 1990. Climate Change: The IPCC Scientific Assessment. Report for Intergovernmental Panel on Climate Change by Working Group I. Great Britain, New York and Melbourne.

Intergovernmental Panel on Climate Change. 2000. Land Use, Land-Use Change and Forestry. Special Report of the Intergovernmental Panel on Climate Change. United Kingdom.

Intergovernmental Panel on Climate Change. 2007a. Climate Change 2007: The Physical Science Basis. Contribution of Working Group I to the Fourth Assessment Report of the Intergovernmental Panel on Climate Change. United Kingdom and New York.

Intergovernmental Panel on Climate Change. 2007b. Climate Change 2007: Impacts, Adaptation and Vulnerability. Contribution of Working Group II to the Fourth Assessment Report of the Intergovernmental Panel on Climate Change. United Kingdom.

Intergovernmental Panel on Climate Change. 2007c. Summary for Policymakers. Climate Change 2007: Mitigation. Contribution of Working Group III to the Fourth Assessment Report of the Intergovernmental Panel on Climate Change. United Kingdom and New York.

Jungerius, P.D. and Dejong, J.H. 1989. Variability of water repellency in the dunes along the Dutch coast. Catena 16: 491–497.

Jones, T.P. and Rowe, N.P. 1999. The earliest known occurrences of charcoal in the fossil record. 9th Annual V.M. Goldschmidt Conference, August 22–27, 1999. Harvard University, Cambridge, MA, USA. Houston, TX, USA: Lunar and Planetary Institute. 143 p.

Kazanis, D. and Arianotsou, M. 2004. Factors determining low Mediterranean ecosystems resilience to fire: the case of Pinus halepensis forests. p. 13. *In*: Arianotsou, M. and Papanatasis, V.P. (eds.). Proceedings of the Tenth MEDECOS Conference. Rhodes, Greece.

Kruse, R., Bend, E. and Bierzychudek, P. 2004. Native plant regeneration and introduction of non-natives following post-fire rehabilitation with straw mulch and barley seeding. Forest Ecology and Management 196: 299–310.

Lambin, E.F. and Meyfroidt, P. 2011. Global land use change, economic globalization, and the looming land scarcity. PNAS 108(9): 3465–3472.

Lampin-Cabaret, C., Jappiot, M., Alibert, N. and Manlay, R. 2003. Une échelle d'intensité pour le phénomène Incendie de forêts, SIRNAT – JPRN Orléans, 2003.

Lourenço, L. 1988. Tipos de tempo correspondentes aos grandes incêndios florestais ocorridos em 1986 no Centro de Portugal. Finisterra XXIII(46): 251–270.

Lourenço, L. 1989. Erosion of agro-forester soil in mountains affected by fire in Central Portugal. Pirineos. A Journal on Mountain Ecology 133: 55–76.

Lourenço, L. and Bento-Gonçalves, A. 1990. The study and measurement of surface flow and soil erosion on slopes affected by forest fires in the Serra da Lousã. *In*: Proceedings, International Conference on Forest Fire Research, Coimbra, C.05–1 a 13.

Lourenço, L. 1991. Aspectos sócio-económicos dos incêndios florestais em Portugal, Biblos LXVII: 373–385.

Lourenço, L., Bento-Gonçalves, A. and Monteiro, R. 1991. Avaliação da erosão dos solos produzida na sequência de incêndios florestais. In Comunicações, II Congresso Florestal Nacional, Porto, vol. II, pp. 834–844.

Luís, M., Raventós, J. and Gonzalez-Hidalgo, J.C. 2006. Post-fire vegetation succession in Mediterranean gorse shrublands. Acta Oecologica 30: 54–61.

Mataix-Solera, J. and Cerdà, A. 2009. Incendios forestales en España. Ecosistemas terrestres y suelos. pp. 27–53. *In*: Cerdà, A. and Mataix-Solera, J. (eds.). Efectos de los incendios forestales sobre los suelos en España. El estado de la cuestión visto por los científicos españoles. FUEGORED, Cátedra Divulgación de la Ciencia, Universitat de Valencia, Spain.

Meyer, W. and Turner, I.I.B. 1994. Changes in Land Use and Land Cover: A Global Perspective. Cambridge University Press, Cambridge.

Moreira, F., Vaz, P., Catry, F. and Silva, J.S. 2009. Regional variations in wildfire preference for land cover types in Portugal: implications for landscape management to minimize fire hazard. International Journal of Wildland Fire 18(5): 563–574.

Moreno, J., Vázquez, A. and Vélez, R. 1998. Recent history of forest fires in Spain. pp. 159–185. *In*: Moreno, J.M. (ed.). Large Fires, Backhuys Publishers, Leiden, the Netherlands.

Napper, C. 2006. Burned Area Emergency Response treatments catalog. USDA Forest Service.

National Academy of Sciences. 2001. Climate Change Science: An Analysis of Some Key Questions. National Academies Press. Washington, D.C.

Naveh, Z. 1975. The evolutionary significance of fire in the Mediterranean region. Vegetatio 29: 199–208.

Neary, D.G., Ryan, K.C. and Debano, L.F. (eds.). 2005. Wildland Fire in Ecosystems: Effects of Fire on Soils and Water. Gen. Tech. Rep. RMRS-GTR-42-vol.4. Ogden, UT: U.S. Department of Agriculture, Forest Service, Rocky Mountain Research Station (revised 2008).

Neary, D.G. 2009. Post-wildland fire desertification: can rehabilitation treatments make a difference? Fire Ecology 5(1): 129–144.

Noss, R., Franklin, J., Baker, W., Schoennagel, T. and Moyle, P. 2006. Managing fire-prone forests in the western United States. Frontiers in Ecology and the Environment 4: 481–487.

Parsons, A., Robichaud, P., Lewis, S., Apper, C. and Clark, J. 2010. Field guide for mapping postfire soil burn severity. Gen. Tech. Rep. RMRS-GTR-243. Fort Collins, CO: U.S. Department of Agriculture, Forest Service, Rocky Mountain Research Station.

Pausas, J.G. and Vallejo, R. 1999. The role of fire in European Mediterranean ecosystems. pp. 3–16. *In*: Chuvieco, E. (ed.). Remote Sensing of Large Wildfires in the European Mediterranean Basin. Springer-Verlag.

Pausas, J.G., Llovet, J., Rodrigo, A. and Vallejo, V.R. 2008. Are wildfires a disaster in the Mediterranean basin? A review. International Journal of Wildland Fire 17: 713–723.

Pausas, J.G. and Keeley, J.E. 2009. A burning story: the role of fire in the history of life. BioScience 59: 593–601.

Pereira, J., Carreiras, J., Silva, J. and Vasconcelos, M. 2006. Alguns conceitos básicos sobre os fogos rurais em Portugal. pp. 134–161. *In*: Pereira, J.S., Cardoso Pereira, J.M, Rego, F.C., Silva, J.M. and Silva, T.P. (eds.). Incêndios Florestais em Portugal: caracterização, impactes e prevenção, ISA Press, Lisboa.

Pinaya, I., Soto, B., Arias, M. and Díaz-Fierros, F. 2000. Revegetation of burnt areas: relative effectiveness of native and commercial seed mixtures. Land Degradation and Development 11: 93–98.

Pike, R.G. and Ussery, J.G. 2006. Key points to consider when pre-planning for post-wildfire rehabilitation. FORREX Forest Res. Extension Partnership, FORREX Series 19, Kamloops, Canada.

Pyne, S.J. 1982. Fire in America: A Cultural History of Wildland and Rural Fire. Princeton University Press, Princeton.

Pyne, S. 2006. Fogo no jardim: Compreensão do contexto dos incêndios em Portugal. pp. 115–131. *In*: Pereira, J.S., Cardoso Pereira, J.M, Rego, F.C., Silva, J.M. and Silva, T.P. (eds.). Incêndios florestais em Portugal: caracterização, impactes e prevenção, ISA Press, Lisboa.

Raftoyannis, Y. and Spanos, I. 2005. Evaluation of log and branch barriers as post-fire rehabilitation treatments in a Mediterranean pine forest in Greece. International Journal of Wildland Fire 14: 183–188.

Rego, F.C. 2001. Florestas públicas, Direção Geral das Florestas e Comissão Nacional Especializada de Fogos Florestais.

Ritsema, C.J. and Dekker, L.W. 1994. How water moves in a water-repellent sandy soil. Dynamics of fingered flow. Water Resources Research 30: 2519–2531.

Robichaud, P. 2009. Post-fire stabilization and rehabilitation. pp. 299–320. *In*: Cerdá, A. and Robichaud, P. (eds.). Fire Effects on Soils and Restoration Strategies. Science Publishers, Enfield, New Hampshire.

Robichaud, P.R., Beyers, J.L. and Neary, D.G. 2005. Watershed rehabilitation. pp. 42–44. *In*: Wildland Fire in Ecosystems. Effects of Fire on Soil and Water. USDA Forest Serv, Gen Tech Rep RMRSGTR.

Robichaud, P.R., Ashmun, L.E. and Sims, B.D. 2010. Post-fire treatment effectiveness for hillslope stabilization. Gen. Tech. Rep. RMRS-GTR-240. Fort Collins, CO: U.S. Department of Agriculture, Forest Service, Rocky Mountain Research Station. 62 p.

Rodrigo, A., Retana, J. and Picó, X. 2004. Direct regeneration is not the only response of Mediterranean forests to large fires. Ecology 85: 716–729.

Shakesby, R., Boakes, D., Coelho, C., Bento-Gonçalves, A. and Walsh, R. 1993. Limiting the erosional effect of forest fires: background to the IBERLIM research programme in Águeda and Tejo basins, Portugal. Swansea Geographer 30: 132–154.

Shakesby, R., Boakes, D., Coelho, C., Bento-Gonçalves, A. and Walsh, R. 1996. Limiting the soil degradation impacts of wildfire in pine and eucalyptus forests, Portugal: comparison of alternative post-fire management practices. Applied Geography 16(4): 337–355.

Shakesby, R.A. 2011. Post-wildfire soil erosion in the Mediterranean: review and future research directions. Earth-Science Reviews 105: 71–100.

Strauss, D., Bednar, L. and Mees, R. 1989. Do one percent of the forest fires cause ninety-nine percent of the damage? Forest Science 35(2): 319–328.

USA General Accounting Office. 2006. Wildland fire rehabilitation and restoration: Forest Service and BLM could benefit from improved information on status of needed work. GAO-06-670. Washington DC.

Vallejo, R. (ed.). 2006. Ferramentas e metodologias para o restauro de áreas ardidas. EUFIRELAB, EVR1-CT-2020-40028, Report D-04-08.

Vega, J., Serrada, R., Hernando, C., Rincón, A., Ocaña, L., Madrigal, J., Fontúrbel, M., Pueyo, J., Aguilar, V., Guijarro, M., Carrillo, A., Fernández, C. and Marino, E. 2010. Actuaciones técnicas post-incendio y severidad del fuego: Proyecto Rodenal. In Actas das Jornadas Internacionales – Investigación y gestión para la proteccion del suelo y restauración de los ecossistemas forestales afectados por incêndios forestales, Santiago de Compostela, 305–308.

Vega, J.A. 2011. Criteria to develop protocols for post-wildfire soil rehabilitation: current experience in Galicia (NW Spain). pp. 99–103. *In*: Bento-Gonçalves, A. and Vieira, A. (eds.). Proceedings of the 3rd International Meeting of Fire Effects on Soil Properties. University of Minho, Guimarães, Portugal.

Vélez, R. 1993. High intensity forest fires in the Mediterranean Basin: natural and socioeconomic causes. Disaster Management 5: 16–21.

Venkata Raman, S.V., Iniyan, S. and Goic, R. 2012. A review of climate change, mitigation and adaptation. Renewable and Sustainable Energy Reviews 16: 878–897.

Viner, D., Hulme, M. and Raper, S. 1995. Climate change scenarios for the assessments of the climate change on regional ecosystems. Journal of Thermal Biology 20(1-2): 175–190.

Walsh, R., Coelho, C., Elmes, A., Ferreira, A., Bento-Gonçalves, A., Shakesby, R., Ternan, J. and Williams, A. 1998. Rainfall simulation plot experiments as a tool in overland flow and soil erosion assessment, North-Central Portugal. Geookodynamik XIX(3-4): 139–152.

Woodward, F.I. 1992. A review of the effects of climate on vegetation: ranges, competition and composition. pp. 105–123. *In*: Peters, R.L. and Lovejoy, T.E. (eds.). Global Warming and Biological Diversity. Yale University Press, New Haven.

CHAPTER 24

Ecohydrological Measures for Sustainable Catchment Management

An Outline

Maciej Zalewski[1,2]

According to International Council of Scientific Unions (ICSU), science of the 21st century has to be integrative, problem solving and policy oriented. Ecohydrology provides the methodological framework on how to use ecosystem processes, that is water-biota interplay, as a management tool. The background for this approach is based on the understanding about how evolution has been shaping geochemical processes, nutrients circulation, and energy flow in the river ecosystems.

In this chapter, the author synthesizes his broad range of papers to highlight the theoretical framework provided by ecohydrology for river sciences, as well as he provides some case studies of implementation of ecohydrology-based solutions for the improvement of ecological status of the rivers and reservoirs.

The author is convinced that ecohydrology and related nature-based solutions, which both have been in the phase of dynamic development, can to a great extent advance integration of the knowledge presented in the other chapters of this book, to the mutual benefit of all the involved sciences.

INTRODUCTION

In the Anthropocene Era, Humanity has, due to demographic processes and exploitation of natural resources, approached the carrying capacity of the biosphere. That is why it is very urgent to change the interaction between Man and the Biosphere towards sustainable development (compare: Sustainable Development Goals of the United Nations, SDG UN). Water as a key driver of biological processes and common

[1] European Regional Center for Ecohydrology of the Polish Academy of Sciences, Lodz, Poland.
[2] Department of Applied Ecology, University of Lodz, Poland.
  E-mail: m.zalewski@erce.unesco.lodz.pl; maciej.zalewski@biol.uni.lodz.pl

denominator of all ecological and socioeconomic processes in the face of predicted climate changes becomes the crucial factor for the achievement of SDG. Sustainable management of river basins should be looked at in the context of global conditions as gradual degradation of environment and resources, for example, soil erosion and decreasing water resources availability, relates to demography, economy, and climate changes. This creates a system of negative feedbacks that severely influences economy and negatively impacts social processes.

An intensive change in the mutual relations between humans and the environment begun with the advent of the industrial era. It was fuelled by the conviction of an unlimited potential of nature that can be used for any recognized needs of the humanity (UNESCO 2012; Fig. 1). After the phase of a wild exploitation of the natural resources there came a reflection of the need to protect nature, and later on, a phase concerned with its restoration. However, there are two important aspects of the current relations between humans and the environment.

The first, is a limited understanding among researchers of the integrity of ecological processes, that is, circulation of water, nutrients, and energy flow in ecosystems are under constant evolution and also under constant modification by man. Consequently, human actions based on incomplete information and inadequate understanding of the consequences of the planned actions, with no control form the outside lead to degradation of the environment; for example cage aquaculture in lakes drastically accelerates eutrophication (Penczak et al. 1981) and the loss of related ecosystem services (Mass et al. 2016), which is rarely considered in the water resources management.

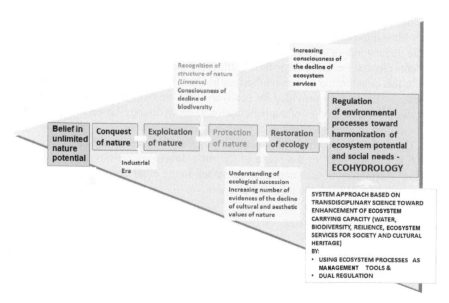

Fig. 1. Evolution of the human approach towards the use of natural resources, starting from the belief in the unlimited potential of nature to recent awareness of the necessity for regulating of ecological and hydrological processes for the enhancement of the ecosystem carrying capacity (Zalewski, courtesy of UNESCO 2012, modified).

The second aspect is an awareness of the complexity of a catchment that is a template for water resource management, including elaboration of river basin management plans (e.g., Piniewski et al. 2015, study on Pilca River nutrients modelling). Primarily, every catchment possesses a specific hierarchy of water cycle drivers related to the unique geomorphology, climate, plant cover, and so on. To various extents these drivers are being modified due to social development and population growth, combined with a variety of economy driven activities such as deforestation, urbanization, industrialization, and transportation. Also, all of the mentioned forms of human interventions in the water cycle are changing the catchment's heat budget, wind speed, and leed to the direct reduction of permeable areas; river channelization accelerates the water outflow rate from the catchments. This, in turn, has caused an increase, by orders of magnitude, of the transfer of mineral and organic matter, nutrients, and pollutants from land to rivers, reservoirs, lakes, and costal zones (Meybeck 2003; Wolanski 2007; Chicharo and Zalewski 2012; Kiedrzyńska et al. 2014). Consequently, researchers encounter that the landscapes across all continents become not only drier but also less fertile due to the loss of nutrients and organic matter from soils.

Therefore, while fundamental ecological processes such as water and nutrient cycling and energy flow become deeply modified, in order to increase the sustainability potential of catchments there arises an urgent need to expand the range of available compensatory measures by those aiming at regulation of the processes related to the water-biota interplay. This idea was formulated in the framework of UNESCO International Hydrological Programme (IHP) as Ecohydrology, which the key word is "dual regulation"—regulating the hydrology by shaping biocenoses and vice versa—shaping biocenoses through the regulation of hydrology (Fig. 1) (Zalewski 2014a, 2014b, 2015).

Water As a Driver of Ecosystem Structure and Dynamics: Background to Ecohydrology Theory

The two key interconnected issues that need to be addressed in the context of ecohydrology are (1) water cycle, and (2) its linkages with biocenoses. Firstly, the hydrological cycle must be considered as a primary regulator of ecological potential bioproductivity and biodiversity (for example, Zalewski 2003; Wojtal et al. 2008). Secondly, understanding of the role of biocenoses in shaping the water and nutrient cycling is fundamental to reversing the declining potential of the biogeosphere (Tilman 1999; Vorosmarty and Sahagian 2000; Rodrigues-Iturbe 2000). In the subsequent paragraphs these issues will be discussed in more detail.

Water is the primary factor limiting and regulating the ability of ecosystems to accumulate carbon, nitrogen, and phosphorus. When water is available *ad libitum*, then temperature determines the metabolic rates of microbial communities, plants, and poikilothermic animals. Thus, the assimilation of available nutrients is governed by stoichiometric relationships with limitation effect expressed by Liebig's law of the minimum. Moreover, biodiversity, a fundamental cumulative indicator of the human well-being and the prospects for sustainable future (Millenium Ecosystem Assessment 2005), is driven by water availability that determines plant yield (Visser 1971; Rodriguez-Iturbe 2000; Eamus et al. 2006), and solar radiation related to biomass increase (Kowalik and Eckersten 1984; Kędziora 1996). Hence, in given geomorphological conditions, water and temperature are the two major determinants of biodiversity and bioproductivity (Fig. 2). This is because the amount of water in a given temperature range determines the amount of carbon possible to accumulate in an ecosystem (in the form of living and decaying organic matter), while the temperature determines the allocation of carbon between the plant's biomass and soil organic matter. For example, when moving from the boreal zone southward to the tropics, with an increasing temperature there is a significant shift in allocation of organic matter, and therefore carbon, from soil into biomass. This is explained by the Van Hoff's law which explains the acceleration of organic matter decomposition with temperature increase. Thus, decomposition of organic matter in the soil can be up to 40 times faster in the tropics than in the boreal zones. High nutrients circulation rate and energy supply create good conditions for diversification of microbial, plant, and animal communities, and persistence of favorable mutations through natural selection and adaptation processes. Thus, researchers can assume, that opportunities for diversification of genomes, thus, adaptative capability of species, can be much higher in the tropics (biotic) then in the boreal zone (abiotic ecosystem regulation). What is more, under the harsh conditions of boreal zones, the short growing period for plants provides a very limited flow of energy and nutrients, and in consequence, catastrophic events may randomly eliminate emerging new genomes. Thus boreal and high mountainous ecosystems pose lower potential for regeneration and compensation of human impact. A similar phenomenon is observed in deserts where scarce biodiversity and bioproductivity is driven by low water availability and high temperature leading as a consequence to low carbon accumulation in the soils [abiotic ecosystem regulation, i.e., per Zalewski (2002a, 2010); Fig. 3].

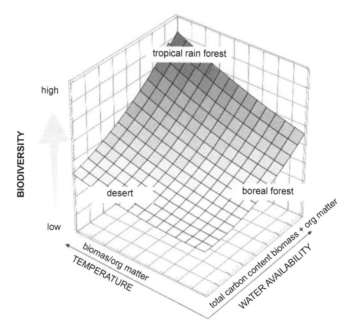

Fig. 2. Deductive background of ecohydrology theory: drivers of biodiversity in the terrestrial phase of water cycle. The amount of water determines, the amount of carbon accumulated in an ecosystem, while temperature determines the carbon allocation between biomass and soil organic matter. The maximum biodiversity and bioproductivity is achieved at highest water availability and highest temperatures (Zalewski 2002a).

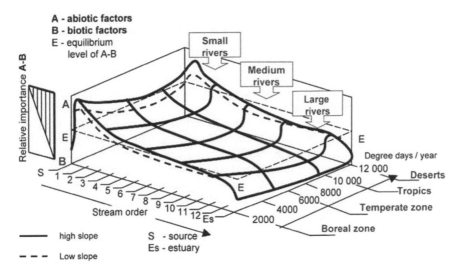

A - abiotic factors
B - biotic factors
E - equilibrium level of A-B

high slope
Low slope

S - source
Es - estuary

Fig. 3. Deductive background of ecohydrology theory, aquatic phase, the abiotic-biotic regulation concept; structure and dynamics of riverine fish communities are determined by a hierarchy of abiotic and biotic factors in specific ratios depending on the hydrology parameters determined by slope and stream order and on the energy budget of a given climatic zone determined by temperature; only when abiotic factors (hydrology) become stable and predictable the biotic drivers start to manifest themselves (adapted from Zalewski and Naiman 1985).

An example of the complexity of the interrelations between hydrology and biota, thus justifying the necessity of its profound understanding, is the abiotic-biotic regulatory concept (ABRC; Zalewski and Naiman 1985; Zalewski et al. 1986; Fig. 3). It was inspired by the river continuum (RC) concept (Vannote et al. 1980), which initiated the process-oriented thinking in river ecology, by referring to the observed shift in production/respiration ratio in streams from upstream to downstream (Newbold et al. 1982). The ABRC, as a background for ecohydrology, was also inspired by a scientific debate of ecologists concerning density-dependent (biotic drivers) and density-independent (abiotic drivers) determination of ecosystems structure and functioning and further by ecological bioenergetics (Grodzinski et al. 1975), where ecological processes were reduced to thermodynamics laws, thus increasing their quantification and predictability. The ABRC novelty was to underline the hierarchy of abiotic and biotic drivers in shaping the riverine ecosystems; only when abiotic (water) factors become stable and predictable the biotic factors start to manifest themselves (confirmed empirically by Wagner-Łotkowska, Zalewski 2016). Extrapolating this notion into all types of freshwater ecosystems, researchers can expect analogical adaptations to different abiotic stress factors not only in the rivers, but also in lakes and reservoirs. In the boreal zone, riverine organisms adapt to harsh conditions by fat accumulation, to compensate for temperature-limited food assimilation during long winters followed by a period of high energy expenditures due to hydrological stress during the snowmelt period. On the other hand, in hot and dry freshwater ecosystems [for example, Naiman and Soltz (1981)], high metabolic rates of organisms and consequential high oxygen demands clash with the low water oxygen solubility due to high temperature, and low oxygen availability during nights resulting from high respiration rates of the biotic community, thereby compromising the efficiency of Krebs cycle metabolic pathways.

Understanding of the relationships expressed by the models described above (Figs. 2 and 3) is fundamental to undertake informed and effective actions to reverse such negative processes as the worldwidely observed loss of biodiversity and bioproductivity, which are the results of such processes as the loss of organic matter in soils (UNEP 2008), as well as siltation and eutrophication in rivers, lakes, reservoirs (Hillbrich-Ilkowska 1993), and coastal zones (Wolanski et al. 2008; Chicharo et al. 1998; Chicharo Zalewski 2012).

This understanding requires an interdisciplinary knowledge of rudimentary physics, chemistry, plant physiology, biochemistry, geography, ecology, and evolution, and should be applied during hydroengineering projects. What is more, a broad understanding of the previously mentioned processes provides scientific background for the development of ecohydrological biotechnologies adaptable for different geographic regions (discussed in subsequent parts of this chapter).

Ecohydrology Multidimensional Goal for Enhanced Sustainability of Catchment

As mentioned at the beginning, the reductionist conception of nature, followed by the sectorial organization of science, led to mechanistic and deterministic perception of the environment, and had their consequences in the rally for maximizing resource use with minimal outlays (Fig. 4). However, it is postulated that to achieve sustainability the society needs to optimize their resource use instead of maximizing it by exploding consumption. Bioeconomy and circular economy concepts can help in achieving this goal.

The holistic conception of nature and its reflection in the transdisciplinary science needs to employ evolutionary and systematic approach to solving environmental

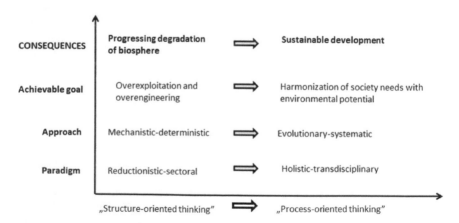

Fig. 4. Expected direction of a shifts in environmental science paradigm from the currently prevailing structure-oriented paradigm (conservation and restoration actions, focused on the assessment of the effect of various human activities on biota) to process-oriented approach (regulation of ecosystem processes, i.e., circulation of water and nutrients, and energy flow) to create a background for harmonization of societal needs with ecosystem potential, and ultimately the sustainable development; the arrows represent the expected shifts in the paradigm at three levels: (1) scientific, (2) operational/practical, (3) political, and the expected results.

problems and optimise ecosystem services (Zalewski 2000, 2014). Until the early 1970s, the ecology concentrated on the analysis of the structure of nature. Eugene P. Odum (Odum 1971) set fundamentals to break down this paradigm by focusing on the functions of ecosystems and their relation to economy; this has been recently developed under the bioeconomy concept. The further progress in understanding of ecological processes in rivers (Heynes 1970; Vannote et al. 1980; Junk et al. 1989; Acreman et al. 2011; Acreman 2014), reservoirs (Ward and Stanford 1983, 1995; Petts 1984, 1995; Tundisi and Tundisi 2008) and lakes (Gulati et al. 1990; Hillbricht-Ilkowska et al. 2000) created a background for process-oriented approach in ecology (Wilkinson 2006) and a development of a problem-solving science - Ecohydrology (Zalewski et al. 1997; Zalewski 2000, 2002a,b; Wolanski et al. 2002; Chicharo et al. 1998; Harper et al. 2012; McClain 2001; Rodriguez-Iturbe 2000; Zalewski 2000, 2013; Tundisi and Tundisi 2008). Hence, a paradigm shift to processes-oriented thinking in water and environment resources management becomes necessary for an effective implementation of system solutions in the holistically perceived catchment environment.

Process-oriented thinking in the integrative environmental science has to be driven by physics and integration of knowledge. For example, thermodynamics is applied in bioenergetics of organisms (Grodzinski et al. 1975) expressed by the Van Hoff law, and changes in heat budget of the landscape are quantified by geophysics (Olejnik and Kędziora 1990). Furthermore, the process-oriented approach needs quantification of the processes. For example, if due to deforestation of the landscape, temperature increases, consequently, growth rate of plants may be positively affected and generate increased yields, only if a sufficient, increased amount of water is available. However, a predicted temperature rise may affect water availability, thus limiting the expected positive effects.

Every strategy for success needs to be based on two elements as follows: (1) elimination of risks, and (2) amplification of opportunities in order to guarantee reaching of its goals. For example, aiming at curbing the ever growing social and economic aspirations, exacerbated by the global demographic growth, energy use, and overexploitation of resources, a Factor Four Concept of von Weizsäcker et al. (1997) could be a viable alternative, recently expanded by Circular Economy Strategy of the European Commission. The former refers to the fact that humans are able to double the wealth of humanity while halving the energy and other resource use, through setting a new direction for technological progress. This can be achieved only when the strategy of economic growth is shifted from competition for resources to competition in their efficient use. On the other hand, the enhancement of ecological potential of the highly-modified ecosystems by dual regulation from molecular to landscape scale, proposed by ecohydrology, is an important alternative to simplistic "silver bullet" approach dominating among engineering solutions.

Therefore, efficient large scale and long term actions, such as catchment management, have to be based on a vision expressed by clearly formulated goals (e.g., SDG UN). Moreover, the words, acronyms, and expressions in use determine the way of thinking of the engaged actors, and the efficiency of the actions being undertaken, to a great extent. Therefore, to translate ecohydrology principles into a viable catchment governance approach, Multidimensional Goal for Enhanced Sustainability (MGES) of catchment is being proposed Zalewski 2015.

Since amount of water (W) determines accumulation of carbon in catchment, and in consequence nutrients dynamics, and proportionally to temperature increase bioproductivity and biodiversity (B) also increases, both water and biodiversity are drivers of all related ecosystem services for society (S) which we rely on. As far as one of the major challenges in Anthropocene is adaptation to climate change, the goal of enhancement of catchment's resilience (R) using ecohydrology "dual regulation" should be considered equally important. All the interlinked goals must be considered in relation to society and its cultural heritage in a given river basin (C) (Wantzen et al. 2015). Thus, Multidimensional Goal for Enhanced Sustainability of catchment is defined by the five parameters (WBRS+C), which parallel enhancement always have to be considered in management plans.

Fundamentals of Ecohydrology—A Framework for Scientific Investigation and Problem-solving Implementation

The fundamental assumption of ecohydrology is that water is the major driver of biogeochemical evolution and thus of biodiversity and bioproductivity. Terrestrial and aquatic organisms, through evolution, have adopted certain life strategies to match with the prevailing water quantity and quality dynamics in the catchment (Zalewski 2000, 2002a; Janauer 2000, 2006; Harper et al. 2008). Therefore, biocenotic processes are shaped by hydrology and vice versa, hydrological processes by biocenotic structure and functions. Thus such modified two-way regulation of water-biota interplay has been called "dual regulation".

The novelty of ecohydrology is that it does not only aim to understand the complexity of water-biota interplay, but also develops a methodology how to use the ecosystem properties and the underlying processes as a management tool, often complementary to other water resources management measures (Zalewski 2002a; Zalewski et al. 2003, 2004). Ecohydrology expands the available management measures (conservation and restoration of ecosystems, aimed basically at maintaining their structure) with those of regulation of ecological processes towards enhancement of catchment sustainability potential (WBRS+C) (Zalewski 2013). The regulation measures should be applied primarily in the anthropogenically highly modified parts of river basins, the so-called novel ecosystems (Hobbs et al. 2006), and involve the processes from molecular to landscape scales. As far as the regulation of processes focuses on both the sides of the water-biota interplay the term 'dual regulation' also applies for these intentional actions (Zalewski 2000, 2006b).

Understanding of the processes needs to be primarily focused on the understanding of the hydrology-biota interplay, and the hierarchy of importance of the abiotic and biotic factors that drive the ecosystems structure and functions from molecular to landscape scales. In accordance with that shift, the catchment and hydrological mesocycles should become the template for quantification of the processes and for the strategic spatial planning and environmental resources management. The use of the catchment template for the management of existing resources, for example, in the framework of IWRM, as well as the water cycle template for precise quantification of not only water budgets, but also nutrients, pollutants, ecosystem performance, and socioeconomic processes, provides background for development of systemic transdisciplinary solutions.

Ecohydrology is a science that integrates knowledge in hydrology and ecology to elaborate methods that focus particularly on ecosystem biotechnologies (Zalewski and Wiśniewski 1997) and system solutions where both hydrological and ecological processes are applied, which appear in molecular scale and in the river basin scale.

The three key principles of ecohydrology (EHP) when applying the properties of ecosystems as a new and complementary tool to hydrotechnical methods in water management are as follows:

1. 'Dual regulation'—aiming at regulating biocenoses through hydrology and vice versa, regulating hydrology through shaping fauna and flora or controlling interactions between them (Zalewski 2000; Zalewski et al. 1990).
2. Synergism of activities that is aimed at integrating various types of biocenotic and hydrological solutions in the catchment, such as ecosystem biotechnologies and nature-based solutions, striving to achieve synergy in improving the quality of water, increasing the resources of freshwater and biological variety.
3. Integration—aiming at harmonising ecohydrological methods and ecosystem biotechnologies with hydrotechnological solutions such as dams, irrigation systems, or sewage treatment plants (Zalewski 2014).

Biotechnology has been defined (Zalewski and Wiśniewski 1997) as a process of converting one matter form into another using living organisms (a classic example is converting sugar into alcohol by using yeast). In the context of ecohydrology, the fundamental knowledge to apply such biotechnologies stems from the understanding of the dynamics of the water cycle and the biota (plant or vice versa bacteria) potential to convert the mineral forms of nutrients into organic ones. Such natural resource management applications can be called **ecohydrological biotechnologies**.

Activities in catchments management that would include the three above assumptions of ecohydrology serve to improve the quality of water, the environment and they are also complementary to technological solutions and thus socially accepted.

Despite the fact, that environmental problems occurring in a catchment are highly complex water management often tends to solve them by simple methods and disconnected actions, which usually turn to be ineffective. Consequently, to encourage decision makers to apply holistic catchment perspective and process-oriented thinking the three implementation principles of ecohydrology (EHIP) were formulated [see below, per Zalewski (2000, 2006a)]. They are the three steps and the three dimensions of analysis leading to understanding of the underlying ecohydrological processes in a catchment and to application of informed solutions. They also provide a systemic framework for integration of ecohydrology into Integrated Water Resources Management (IWRM) framework.

Whereas IWRM was defined by Global Water Partnership as "a process which promotes the coordinated development and management of water, land, and related resources to maximise economic and social welfare in an equitable manner without compromising the sustainability of vital ecosystems" (GlobalWater Partnership Technical Advisory Committee 2000), ecohydrology aims at harmonizing society needs with enhanced ecosystem potential through the increasing of carrying capacity of ecosystems. Therefore, instead of balancing social and economic needs with those of the ecosystems (IWRM), ecohydrology opts for harmonizing the socioeconomic requirements with enhanced ecosystem potential, and instead of/and apart from protecting pristine ecosystems it calls for regulating processes in the "novel ecosystems" in order to increase their ecological potential in terms of water resources, biodiversity, ecosystem services, and resilience to global change and anthropogenic stress (WBRS+C). As such, ecohydrology is compliant with IWRM concept but gives novel potent tools to achieve sustainability.

The first principle of ecohydrology (Fig. 5), the hydrological principle, implies quantification of hydrological processes at the basin scale and the entire hydrological cycle as a template for quantification of ecological processes. The quantification covers the patterns of hydrological pulses along the river continuum and identification of various forms of human impacts, for example, point and nonpoint sources of pollution. This principle is based in the assumption of superiority of abiotic factors over biotic interactions (Zalewski and Naiman 1985).

The second, ecological principle of ecohydrology (Fig. 6), implies the need for understanding of the evolutionary-established water-biota interplay, and thus quantification of nutrient flows and energy fluxes dynamics within the water cycle and catchment templates defined in the first step. It also calls for analysis of the spatial distribution of different types of ecosystems, that is, pristine, degraded, and modified, in order to identify the "novel ecosystems" which ecological potential could be enhanced through application of 'dual regulation'.

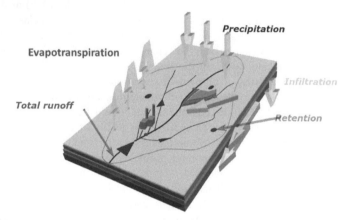

Fig. 5. The first principle of ecohydrology: hydrology. Quantification of hydrological cycle and its spatiotemporal dynamics; analysis from the perspective of socio-economy and various forms of human impact (Zalewski 2009).

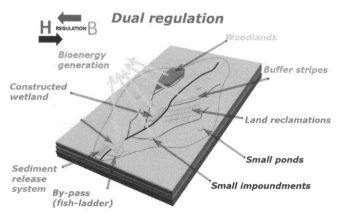

Fig. 6. The second principle of ecohydrology: ecology. Identification of the distribution of various types of biocenoses and their potential to enhance the resilience and absorbing capacity of human impact (Zalewski 2009).

The ecological principle has been focused on understanding the crucial role of vegetation in water cycling processes (Rodrigues-Iturbe 2000; Vorosmarty and Sahagian 2000).

Finally, the ecological engineering principle (Fig. 7) defines the way ecosystem properties/processes identified in the framework of the first and the second principles should be used as management tools. These tools and individual solutions developed based on them, currently dubbed nature-based solutions (NbS), should be considered as complementary to the already used hydrotechnical solutions in water management. The use of the ecosystem properties is compliant with the rules defined for ecological engineering (Mitsch 1993, 2012; Mitsch and Jorgensen 2003).

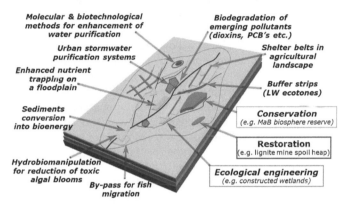

Fig. 7. The third principle of ecohydrology: ecological engineering. Managing biota to control hydrological processes and vice versa, using hydrology to regulate biota, integrated with conservation, restoration and ecological engineering measures, as well as hydrotechnical infrastructure at a basin scale (Zalewski 2013).

The effect of synergy on the basin scale, as well as other regulatory measures, have a potential to significantly reduce nutrient and pollutant loads into aquatic systems (Zalewski 2000). The restoration of an eutrophic, polluted reservoir by reducing the flow and changing the allocation of excessive nutrients and contaminants into the water through different ecosystem biotechnologies is an example of an ecohydrological approach at the river basin scale. Starting from the top of the catchment, the first stage has to be the enhancement of nutrient retention within the catchment by reforestation, creation of ecotone buffering zones, and optimisation of agricultural practices. Buffering zones (shelterbelts) at the land–water interface reduce the rate of groundwater flux due to evapotranspiration along the river valley gradient (Ryszkowski and Kedziora 1999; Izydorczyk et al. 2013a). This process may increase nutrient uptake on cultivated land up to 30% (Statzner and Sperling 1993). Nutrient transformation into plant biomass in ecotone zones may further reduce supply into rivers. Wetlands in a river valley form a buffering zone: they reduce mineral sediments, organic matter, and nutrient load transported by rivers during flood periods through sedimentation (Carling and Petts 1992; Kiedrzyńska et al. 2008; Mitch et al. 1995). In most natural and artificial wetlands, nitrogen load can be reduced significantly by regulating the water level to stimulate anaerobic denitrification processes (Bednarek et al. 2010).

Although these synergics between different regulatory ecohydrological measures have been significantly reducing the negative effect of excessive pollutants in aquatic systems, as with every method there are limitations. Their efficiency, in general, can be described by a parabolic curve: until the ecosystem absorbing capacity exceeds nutrient/pollutant load the efficiency of the ecohydrological regulatory measures increases. However, when the load exceeds the ecosystem potential absorbing capacity the potential of application of ecohydrology methods can be seriously reduced. This situation may happen for example when serious degradation of the biotic structure is accompanied by drastic modification of abiotic conditions, such as decreasing transparency of a water column or depleting water oxygen concentrations (Zalewski and Wagner-Łotkowska 2004).

An example of harmonisation of solutions with societal needs is illustrated in Fig. 8. Sewage treatment plants often are characterized by low efficiency. That is why should be supplemented by a constructed wetland in order to increase the efficiency of nutrient absorption from the effluent (Kiedrzyńska et al. 2008). The wetland was constructed based on local species of willows distributed according to their tolerance to flooding to enhance their optimal growth. Conventional environmental technology has been integrated with ecological engineering solution, thus creating a "hybrid system". The biological conversion of phosphorus into biomass, which instead of enriching the reservoir water stimulates bioenergy production (willow biomass), reduces carbon dioxide emission and creates employment opportunities; an example of an ecohydrological systemic solution. A broad range of such solutions have been implemented on different continents within the framework of UNESCO-IHP, Phase VII (http://ecohydrology-ihp.org/demosites/).

Due to the complexity of the applied knowledge, the application of the geographic information system (GIS) and development of mathematical models for high complexity hypotheses testing and decision support systems should be encouraged. These could be

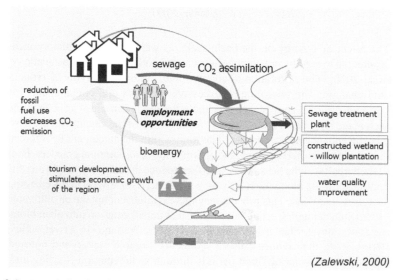

(Zalewski, 2000)

Fig. 8. System solutions based on three principles of EH – conversion of nutrients from partially purified sewage into bioenergy for improvement of water quality, human health, environmental opportunities, and quality of life (modified from Zalewski 2002b).

EDUCATION for sustainable future

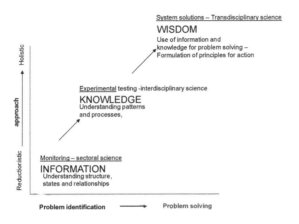

Fig. 9. Methodological background of ecohydrology as a problem-solving science. From information through knowledge to wisdom to develop a transdisciplinary system solution, a template for three principles of ecohydrology [drawing based on the concepts described in Zalewski (2000), (2002a,b); graph based on the concepts described in Zalewski (1999); images based on the concepts and data from Zalewski (1999), toxic algal blooms and molecular biology research from Mankiewicz-Boczek et al. (2002), and Mankiewicz-Boczek et al. (2006); figure adapted from Zalewski (2011)].

useful tools to test alternative scenarios and the implementation of the ecohydrology methodology for sustainable use of water and ecosystems by the societies (Fig. 9).

Mathematical and GIS-supported modelling are cross verified with field experiments. Such an interdisciplinary analysis creates a background for integration of the social and economic processes occurring in the catchment into a problem-solving exercise, which creates a background for a transition from multidisciplinary and interdisciplinary science to transdisciplinary science.

The ecohydrology concept embraces the entire hydrological cycle, focusing on terrestrial, and aquatic phases, and in the entire cycle, biological components of the environment play an important role as moderators of water quality and quantity. In the terrestrial phase, vegetation moderates water quantity and quality, its availability for plants [water-soil-plants interactions, per Baird and Wilby (1999)] and dynamics in the atmosphere (Vorosmarty and Sahagian 2000). In the aquatic phase, aquatic and riparian vegetation can modify nutrient fluxes, allocation and circulation affecting water quality and the related symptoms of eutrophication (e.g., toxic algal blooms) as well as surface water hydrological dynamics (Zalewski et al. 1990; Zalewski 2000). In this case, aquatic ecohydrology will investigate how water biota can modify nutrient loads into aquatic ecosystems.

The Role of Reservoirs in the Context of Climate Change

Global climate change is expected to affect the quantity and quality of water resources. It will affect certain elements of the hydrological cycle, changing river discharges, and hence, water retention time in reservoirs and water levels in lakes. It is predicted that the timing and intensity of floods and droughts will also change, leading to serious economic and sociological effects. Since water is the main medium responsible for the export of nutrients and pollutants from catchments into freshwaters and sees, the above processes will also alter the physical and chemical parameters of water. Due to predicted air temperature increases, water temperatures and the number of ice-free days will also change. The rate of all physical, chemical, and biological processes could be accelerated. Some species may disappear or the boundaries of their range could be shifted. All the above processes may seriously affect the ecosystem's structure and functioning, especially in the case of degraded ecosystems.

The most negative form of impact by traditionally constructed dams on river ecosystems is disruption of river flow continuity (river continuum, compare Vannote et al. 1980) directly connected to the physical barrier in a river. As an effect, the river biodiversity and fisheries decline, especially that of populations of migratory fish. Disturbance of nutrient composition results in an increased probability of the appearance of toxic algal blooms. Furthermore, a decline in mineral and organic matter deposition in coastal zones reduces land build-up processes as well as fertilisation, which in turn modifies habitats and trophy status, with negative consequences on coastal fisheries and ecosystems Chicharo et al. 2006.

However, artificial dam lakes could be an important positive element of a hydrological cycle in a river basin, provided they are properly designed and managed. Despite, they ensure renewable water resources they are much more prone to eutrophication than lakes because the size of their catchment is relatively significantly larger than that of lakes, thus, the area from which mineral and organic matter is transported to the river. Moreover, due to the 'open' character of the artificial dam lakes, there is a possibility of using integrated hydrological and ecological processes knowledge (ecohydrology) to increase their resistance to pollutant loads and occurrence of algal bloom; depending on a reservoir's retention time organic matter (nitrates and phosphates) is converted into algal biomass in favourable conditions high temperature and N/P ratio.

The traditional view of the role of dams in shaping basin-scale ecological processes (e.g., Lillehammer and Saltveit 1984; Petts 1984; Power et al. 1996; Ward and Stanford 1979), thus, has to be reconsidered. This is especially important in a global climate change context, when dams may also play an increasingly important role in protecting water resources and thus, contributing to human development by providing reliable sources of drinking and irrigation water, hydropower, recreation, navigation, "biodiversity refuge" income, and other important benefits (WCD 2000). As reported by Milly et al. (2002), there is the evidence of an increasing risk of significant floods in a warming climate as a result of intensification of the global water cycle. Chiew and McMahon (2002) reported that changes in rainfall result in greater percentage changes in runoff. A good example is provided by climatic scenarios of the South Baltic Sea catchment, where annual precipitation is 650 mm and 60% of the area is agricultural land (Helcom 2007). According to these predictions, summer river flow may decrease by up to 50%, whereas it may increase up to 70% in winter. If such dramatic changes occur, extreme floods and droughts—especially in agricultural and urbanised catchments—will constitute a severe threat to sustainable development. As a consequence, landcover in catchments should be modified following the first principle of EH and quantification of the water budget in forested and deforested parts of catchments should be conducted as a means of moving towards restitution of landcover and buffering strips at critical points of the landscape (Zalewski et al. 2003).

The other important role of dams is the reduction of the transfer of micropollutants (e.g., persistent organic pollutants or heavy metals) from land to coastal/mouth regions and consequently prevention of their accumulation in fish. Reservoirs are considered to be an efficient trap for sediments and, in consequence, associated chemical toxicants including nitrogen, phosphorus, and persistent organic pollutants (Devault et al. 2009). A decrease in flow velocity and increase in flocculent settling in constructed reservoirs create perfect conditions for sedimentation and pollutant deposition. It has been estimated that up to 97% of the released dioxins and dioxin-like compounds in a water column are retained in sediments (DiPinto et al. 1993), which serve as storage compartments for micropollutants (Knezovich et al. 1987). In the comparative studies of the three lowland reservoirs in Poland (Sulejow, Wloclawek, and Jeziorsko), it was demonstrated that the total toxicity of dioxins, along the reservoir has been reduced on the level 30% (Urbaniak and Zalewski 2010). Reservoirs thus, act as a sink for micropollutants and are therefore, important in pollution studies and the monitoring of ecosystem stress.

New dams thus have to be constructed with pre-reservoirs with enhanced resilience for sedimentation and eutrophication; as an example, the reservoir constructed on the Zala River reversed the eutrophication of Lake Balaton (Ta'trai et al. 2000). Moreover, the construction of bypass systems for maintaining the migration and restoration of rheophilic fish (Zalewski 2006; Zalewski and Welcomme 2001) should minimise the negative effects of impoundments on fish migration processes and biodiversity, and increase river basin resilience (Fig. 10). A proposed modification towards an

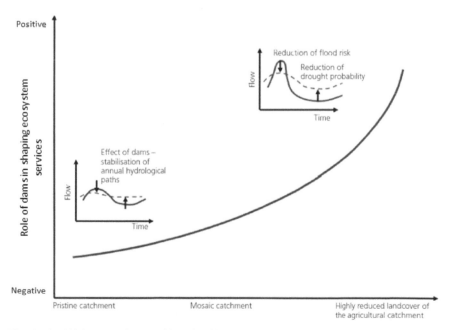

Fig. 10. The shift from negative to positive role of dams along the gradient of anthropogenic modification of catchments. Together with catchment's modification and ecosystem degradation increases the reservoirs sustainability potential (WBRS+C) (modified from Zalewski 2011).

environmentally friendly dam in Poland can be exemplified by the Wloclawek dam on the Vistula River. Here, migration of fish and aquatic organisms can be significantly enhanced not only by a traditional fish ladder but also by a bypass channel that will permanently transfer up to 2% of the Lower Vistula flow, which is approximately 8 m^3. The new, especially lowland reservoirs should be constructed with such by-pass, and additionally, to enhance biodiversity some areas of the banks and islands should be formed as a gradient of aquatic and terrestrial habitats containing diversity of vegetation communities.

Shaping reservoir hydrodynamics towards the creation of sedimentation zones near the banks allows for the periodic transfer of concentrated organic matter in sedimentary zones from the bottom of a reservoir to bioenergy plantations where nutrients stimulate the production of biofuels, and pollutants are fixed in the biomass and then burned.

Case Studies of Implementation

Water Quality Management: Applying Synergy between Ecohydrological Biotechnologies and Hydrological Engineering Measures

Eutrophication of artificial dam lakes and toxic cyanobacterial blooms (Fig. 11), which results from that, is a cumulative effect of the increasing intensification of various forms of using the river basin, such as farming, urbanisation, and transport. What is intrinsically connected with these forms of land use is unconstrained race for unification of rural areas by elimination of buffer zones between the land and water, and eradication of woodlots or wetlands, that serve as spots of intensive denitrification processes in the landscape.

Fig. 11. The process of cyanobacteria bloom formation as a result of great concentrations of phosphates in the groundwater and their input to the human-transformed littoral zone (photo by M. Zalewski).

Therefore, the transformation of nitrogen compounds into atmospheric nitrogen and accumulation of phosphorus compounds in the so-called hard-to-reach pool is going is disturbed. This leads to a massive transport of both organic matter and fertilizers to inland waters. Unification of rural areas causes further drying out of soil, and loss of organic matter and biogenic substances (P, N) as a result of intensified aeolian (wind) and water erosion (Fig. 12). It needs to be emphasised that so far, these detrimental

Fig. 12. Negative effects of extending the cultivation areas up to the stream banks, and unification of landscape for the quality of soil, water in rivers, reservoirs and in the Baltic Sea; case of the Pilica River basin (Zalewski 2014a).

processes have not been sufficiently included in the strategies for reclamation and catchment management.

Until now the problem of pollution and its reduction has been focusing majority of efforts on point pollution sources control; they can be easily dealt with through advanced technological solutions. However, for example in Poland, over 60% of the phosphorus and nitrogen load to the Baltic Sea originates from diffuse (nonpoint) source pollution. Also, in Europe, agricultural land covers up to 70% of the landscape, and reduction of nutrient loads (stimuli of eutrophication and toxic algal blooms in reservoirs, lakes, and coastal zones) from a catchment is one of the key challenges in implementing the European Union's (EU) Water Framework Directive (WFD). In this case, only widely applied, low-cost, and efficient solutions integrating engineering and biological methods (Statzner and Sperling 1993), for example, ecohydrological biotechnologies, will be capable enough to handle the problem.

Creation of land-water ecotones has proven to be an effective tool for reducing the impacts of nutrients originating from a landscape, which enter freshwater ecosystems (Ryszkowski and Kędziora 1993). However, very often shoreline zones are too narrow for these ecotones to work effectively. That is why the goal of the EU-funded EKOROB – Ecotones for the reduction of diffuse pollutions (LIFE08 ENV/PL/000518) project was

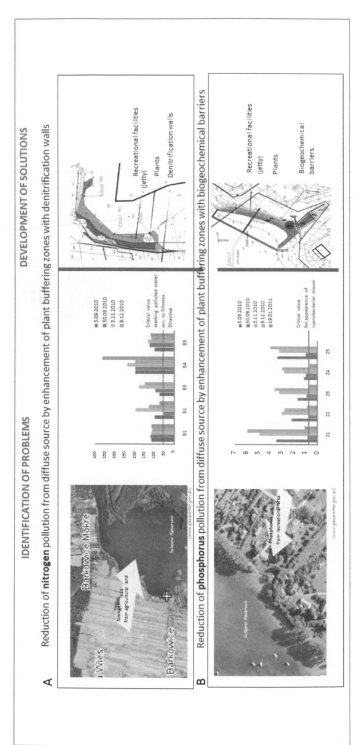

Fig. 13. (a) Reduction of nitrogen pollution generated from agriculture area by plant buffering zones enhanced with denitrification walls, Barkowice, Poland; (b) reduction of phosphorus pollution, which generates toxic cyanobacteria blooms (visible on the photo as lighter stripe along the shoreline), from groundwater in the recreational neighbourhood with plant buffering zones and biogeochemical and denitrification barriers, Zarzęcin, Poland [satellite images DigitalGlobe/European Space Imaging, distributor SmallGIS; data for nutrients concentrations and developed solutions from EKOROB LIFE08 ENV/PL/000519 (2011) project by the European Regional Center for Ecohydrology of Polish Academy of Sciences].

reduction of diffused pollution using enhanced ecotone zones within a limited space to reduce nitrogen and phosphorus fluxes into reservoirs and the Baltic Sea (Fig. 13). Within this project highly effective demonstrative buffer zones were designed and created in the areas threatened with permanent flow of biogenic compounds into inland waters from the direct catchment of the Sulejów reservoir (Figs. 14, 15). The biogeochemical barrier for reduction of contamination from diffuse sources was elaborated at the European Regional Ecohydrological Centre of the Polish Academy of Sciences (ERCE patent). This project also provides an important step towards complying with WFD in achieving good ecological status and reversing the eutrophication of inland waters and the coastal zone (Izydorczyk et al. 2013).

Some other examples of biotechnologies include microbial activity stimulation by addition of carbon to enhance denitrification processes for nitrogen polluted groundwater (Bednarek et al. 2010; Fig. 16), or regulation of excessive nutrients allocation in the aquatic trophy pyramid through hydrobiomanipulation (Zalewski et al. 1990; Wojtal-Frankiewicz and Frankiewicz 2010). These processes applied deliberately in the landscape have been described as landscape-scale biotechnologies. The fundamental knowledge to apply such biotechnologies in the framework of ecohydrology stems from the understanding of the dynamics of surface runoff in relation to reservoir, lake,

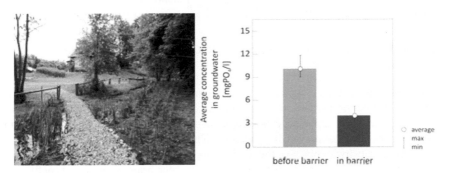

Fig. 14. Application of the biogeochemical barrier and phytotechnology in reducing the very high concentration of phosphates in inland waters, which are the result of flows from leaking septic tanks in the area of recreational urbanisation-Zarzęcin (Izydorczyk et al. 2013a).

Fig. 15. The rim of the Sulejowski reservoir in Zarzęcin (Fig. 15) before and after reconstructing the denitrification barrier and plant zone within the LIFE 08 ENV/PL/000519 project (EKOROB) – Frątczak et al. 2013.

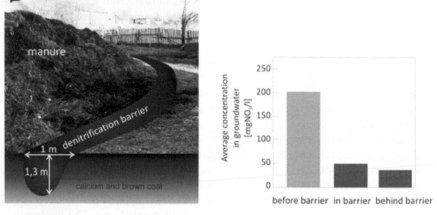

Fig. 16. Construction of denitrification barrier on the premises of a pigs farm so as to limit the concentration of nitrates in inland waters (Bednarek et al. 2010, modified).

or river level oscillations, and plant or bacterial potentials for converting the mineral forms of nutrients into biomass or gas.

Urban Storm Water Management: Low-Cost Advanced-Technology Approaches

The classic civil engineering paradigm concerning urban storm water management is to transfer the runoff water out of the city as soon as possible to avoid local floods. However, during the last decade, development of best management practices for storm water management has proved that reduction of impermeable areas and their detrimental effect on the city hydrology, together with enlargement of green areas and their potential in enhancement of the runoff infiltration to groundwater, brings significant benefit. Two research projects, which were conducted within the framework of EU-programmes, that is, (1) the Sustainable Water Management Improves Tomorrow's Cities' Health (SWITCH), and (2) "Innovative resources and effective methods of safety improvement and durability of buildings and transport infrastructure in the sustainable development", (POIG.01.01.02-10-106/09-04) in the city of Łódź expanded this approach through the analysis of the possibilities to improve the quality of the post-industrial city's landscape, primarily by introducing surface water bodies with enhanced purifying capacity, thus increasing the retentiveness of the landscape.

This was achieved by the construction of a cascade of small impoundments along an urban river (Sokolowka River) valley for retention and purification of storm water. The major challenge to implement this idea was a limited space for construction of a purifying wetland in the valley, a typical ecological engineering solution. Translating the understanding of the self-purification processes in natural rivers, such as importance of light and availability of calcium compounds in water (Zalewski et al. 1998) into technology, it was possible to conceive an innovative prototype of a sequential sedimentation biofiltration system (SBS; Fig. 17).

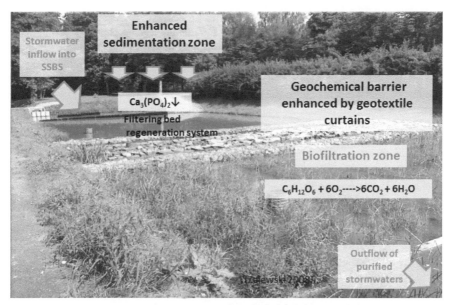

Fig. 17. Sequential biofiltration system (SBS) as a prototype for experiments on further enhancement of storm water purification. Arrangement of different zones in the biofiltration system based on concepts and data from Zalewski et al. (2012); image by M. Zalewski.

Typically, sequential biofiltration systems consist of (1) sedimentation zone, (2) geotextiles for prevention of colmatation and enhancement of PO_4 absorption, (3) biogeochemical, and (4) constructed wetland zones enhanced by bed regeneration system. The SSBS at Sokołówka River (SWITCH project, Zalewski et al. 2012) reduced concentrations of suspended matter up to 90% and total nitrogen and phosphorus by over 50% during the first experimental years of operation. It is worth mentioning that the surface of the SBS is only 0.3% of catchment area whereas, in a traditionally constructed wetland for urban storm water purification it should occupy over 2 to 3% of catchment area to achieve similar efficiency. Stabilization of the river flows by constructing a detention basin upstream of the system further reduced the stochastic character of the process. Additional work included shaping of the wetland plant structure, introducing biodegradable geotextiles, and monitoring of microbial activity. In the latter, the genetic method, that is, trinucleotide repeat sequences/polymerase chain reaction (TRS-PCR) using the common presence of TRS in the microbial genomes (Wojtasik et al. 2012; Adamus-Bialek et al. 2009), combined with partial sequencing of 16S recombinant deoxyribonucleic acid (rDNA) genes, showed qualitative dynamic changes in the microbial population in each zone. This demonstrated the diversity of the purification processes at each stage. Analysing these results relative to the hydrological dynamics of the system provided a baseline for further enhancing the purification efficiency by microbial process enhancement and ecohydrological regulation of flows.

The second challenge in the context of urbanised areas, the groundwater depletion, and a need for its recharge, has its roots in alterations of water cycle in the urban areas. There, infiltration and evapotranspiration, which naturally make up about 50 and 40% of the rainfall, respectively, are reduced to less than 5 and 30%, respectively, in many cases. Most storm water (up to 80%) is drained from cities by surface runoff channels and via highly efficient drainage systems. This process has far-reaching consequences for city inhabitants. The urban heat island phenomenon, characterized by increased temperatures, decreased humidity, and high amounts of dust and pollution in the air of the densely-urbanized areas, not only further affects precipitation, but also increases threefold occurrence of allergies and asthma compared to suburban areas (Kuprys-Lipinska et al. 2009). Declining comfort and growing health risks result in urban-suburban migrations, as humans are leaving cities in search for a better quality of life. Consequently, cities sprawl and induce larger investments in development and maintenance of infrastructure. Daily commuting further increases traffic and the associated emission of pollutants.

In contrast, in the new paradigm, ecohydrology-wise catchment management of the storm water, especially in the context of drought vulnerable cities, water should be treated as a valuable resource not an environmental risk. In such cities, the storm water ought to be consciously retained after prior effective treatment with ecohydrological biotechnologies (see Fig. 17). Expected results include improvement of microclimate and groundwater recharge within the city landscape (Zalewski and Wagner 2005; Wagner and Zalewski 2009).

The development of efficient and innovative purification biotechnologies (Fig. 17) is a foundation for the urban spatial planning concept, that is, the blue-green network concept (BGNC, Zalewski 2009, 2015). The network of blue-green corridors of water related ecosystems is weaved into the urban landscape and watered with purified storm water. It assumes that connected river valleys and green spaces create a network which not only reduces costs of storm water management infrastructures, but also improves microclimate, encourages healthy lifestyles, attracts developers, and make the city resilient to global climate change. This concept has been officially adopted by the City of Łódź as a part of its strategy for integrated development: Łódź 2020.

Ecohydrological rehabilitation of three recreational reservoirs in the urban catchment of the upper Bzura River in Łódź is yet another example of a practical implementation of the blue-green network with enhanced ecohydrological biotechnologies (EHREK, LIFE 08 ENV/000517, Jurczak et al. 2012). Whereas, the human impact in the urbanised areas increases proportionally to the population density, thus increases the risk of eutrophication of the urban waterbodies. The greatest impacts in such areas are stormwater and septic tanks. In order to limit their influence, several sequential sedimentation-biofiltration systems (SSBS) have been constructed just on the stormwater outlets to the waterbodies (Figs. 18, 19). Integrated knowledge of phytotechnology, geochemistry, hydrology, ecohydrology, and ecological and environmental engineering was used to design individual solutions suited to the characteristics of a given spot and the type of pollution. More details on the project, application of the elaborated methods and their efficiency for the reclamation of Upper Bzura and the enhancement of recreational function of the Arturówek reservoirs could be found at http://www.arturowek.pl/, or in Jurczak et al. (2012) (Fig. 20).

The outcomes of the project allowed to formulate the assumptions for achieving a proper balance of river systems and reservoirs. When implementing an ecohydrological rehabilitation scheme of a reservoir, the process must begin with the following:

1. analysis of the threats and the diagnosis of the causes and the state of the system according to ecohydrological principles;

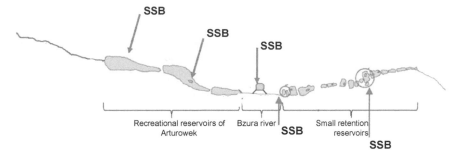

Fig. 18. The upper reach of Bzura River and the tree recreational reservoirs in Arturówek park, Łódź, Poland. Red arrows indicate major stormwater inflows mitigated by SSBS (EHREK project, LIFE 08 ENV/000517 – EHREK, Jurczak et al. 2012).

Fig. 19. The construction phases of the hybrid sequential sequential sedimentation-biofiltration system (SSBS) combine with separathors at the upper Bzura River, above the cascade of the Arturówek recreational reservoirs, suited to receive stormwater from a crossing street. Technological solutions used to separate oil fractions of the stormwater (EHREK project, LIFE 08 ENV/000517 – EHREK, Jurczak et al. 2012).

Fig. 20. Quality of water on a summer day in the Arturówek reservoir before the project in 2008 (A), and after the rehabilitation activities were implementation with the use of various ecohydrological biotechnologies (2014) (B) (EHREK project, LIFE 08 ENV/PL/000517 – EHREK, Jurczak et al. 2015).

2. elaboration of a concept of ecohydrological rehabilitation measures with the use of ecosystem biotechnologies;
3. plan for an adaptive implementation (Holling and Gunderson 2002), that is based on monitoring of the effectiveness of the iteratively implemented individual solutions, and their harmonisation to achieve a synergistic effect.

Currently, the applied methods and system solutions were implemented and are in operation in the two river basins, the Pilica River basin (LIFE 08 ENV/PL/00519 – EKOROB) and the upper Bzura River basin (LIFE 08 ENV/PL/00517 – EHREK). The effectiveness of the applied solutions that were to reduce the areal pollution and that of stormwater remains within the range of 60–90%. In the Upper Bzura system, for example, the water quality has remained good for the past three years. The adaptive assessment and management of the solutions and measures implemented in these projects is done under the two on-going projects: RADOMKLIMA (LIFE14CCA/PL/000101) and AMBER (Horyzon 2020, No. 689682).

Due to the complexity of socio-economic interactions with the environment's sustainability potential vis-à-vis implementation of such ecohydrological projects as introduced above, the whole society of a given river basin, including citizens, firms, government departments, associations, and institutions should be included in the creation of a sustainable natural-social-legal-economic and technological system.

Acknowledgements

The author would like to express a special thanks for stimulating and inspiring long term cooperation in development of the concept and testing the implementation of ecohydrology concept, to the team of European Regional Centre for Ecohydrology of the Polish Academy of Sciences: dr Iwona Wagner, dr hab. Katarzyna Izydorczyk, dr hab. Edyta Kiedrzyńska, prof. Joanna Mankiewicz-Boczek, dr hab. Magdalena Urbaniak, dr Kinga Krauze; the team of Department of Applied Ecology, University of Lodz: Prof. Piotr Frankiewicz, mgr Bogusława Brewińska-Zaraś, dr Małgorzata Łapińska, dr Tomasz Jurczak, dr Agnieszka Bednarek, dr Zbigniew Kaczkowski, prof. Adrianna Wojtal-Frankiewicz and colleagues from the UNESCO International Hydrological Programme, especially Luis Chicharo, Georg Janauer, David Harper, Michael McClain, Giuseppe Arduino, Shahbaz Khan, and many others.

Some of the case studies, project results and solutions presented herein were realized as part of the EU funded project, POIG.01.01.02-10-106/09-04, "Innovative resources and effective methods of safety improvement and durability of buildings and transport infrastructure in the sustainable development", financed from the European Fund of Regional Development, based on the Operational Program of the Innovative Economy; EKOROB – Ecotones for the reduction of diffuse pollutions (LIFE08 ENV/PL/000518), supported by EU EC LIFE+ Environment Policy and Governance programme, National Fund for Environmental protection and Water Management; SWITCH: Sustainable Water management Improves Tomorrow's Cities' Health; Integrated Project, 6 FP EU, G0CE018530; 2005–2011; EHREK: Ecohydrologic rehabilitation of recreational reservoirs "Arturówek" (Łódź) as a model approach to rehabilitation of urban reservoirs (LIFE PLUS, EU Project, LIFE08 ENV/PL/000517, 2010–2015). AMBER (Horyzon 2020, No. 689682).

Keywords: Ecohydrology, River basin, Catchment management, Ecosystem biotechnologies.

References

Acreman, M.C., Harding, R.J., Lloyd, C., McNamara, N.P., Mountford, J.O., Mould, D.J., Purse, B.V., Heard, M.S., Stratford, C.J. and Dury, S.J. 2011. Trade-off in ecosystem services of the Somerset Levels and Moors wetlands. Hydrological Sciences Journal 56(8): 1543–1565.

Acreman, M.C., Arthington, A.H., Colloff, M.J., Couch, C., Crossman, N.D., Dyer, F., Overton, I., Pollino, C.A., Stewardson, M.J. and Young, W. 2014. Environmental flows for natural, hybrid, and novel riverine ecosystems in a changing world. Front Ecol Environ 12(8): 466–473.

Adamus-Białek, W., Wojtasik, A., Majchrzak, M., Sosnowski, M. and Parniewski, P. 2009. (CGG)4-based PCR as a novel tool for discrimination of uropathogenic *Escherichia coli* strains: comparison with enterobacterial repetitive intergenic consensus-PCR. J Clin Microbiol 47(12): 3937–3944.

Baird, A.J. and Wilby, R.L. (eds.). 1999. Eco-Hydrology. Plants and Water in Terrestrial and Aquatic Environments. Routledge, London.

Bednarek, A., Stolarska, M., Ubraniak, M. and Zalewski, M. 2010a. Application of permeable reactive barrier for reduction of nitrogen load in the agricultural areas—preliminary results. Ecohydrology & Hydrobiology 10(2-4): 355–362.

Bednarek, A., Stolarska, M., Urbaniak, M. and Zalewski, M. 2010b. Ecohydrology for water ecosystems and society in Ethiopia—Application of permeable reactive barrier for reduction of nitrogen load in the agricultural areas—Preliminary results. Ecohydrol Hydrobiol 10(2-4): 355–361.

Bednarek, A., Szklarek, S. and Zalewski, M. 2014. Nitrogen pollution removal from area of intensive farming – comparison of various denitrification biotechnologies. Ecohydrology & Hydrobiology 14(2014): 132–141.

Carling, G.A. and Petts, G.E. 1992. Lowland Floodplain Rivers. Geomorphological Perspectives. Wiley, Chichester.

Chicharo, L.M.Z., Chicharo, M.A. and Ben-Hamadou, R. 2006. Use of a hydrotechnical infrastructure (Alqueva Dam) to regulate planktonic assemblages in the Guadiana estuary: basis for sustainable water and ecosystem services management, Estuarine Coastal and Shelf Science, 70, 1-2, 3-18.

Chicharo, L. and Zalewski, M. 2012. Introduction to ecohydrology and restoration of estuaries and coastal science. pp. 1–6. *In*: Wolanski, E. and McLusky, D. (eds.). Vol. 10, Academic Press, Waltham.

Chicharo, M., Chicharo, L., Valdés, L., López-Jamar, E. and Ré, P. 1998. Estimation of starvation and diel variation of the RNA/DNA ratios in field-caught Sardina pilchardus larvae off the north of Spain. Marine Ecology Progress Series 164: 273–283.

Chiew, F. and McMahon, T. 2002. Modelling the impacts of climate change on Australian stream flow. Hydrology Processes 16(6): 1235–1245.

Devault, D.A., Gerino, M., Laplanche Ch. et al. 2009. Herbicide accumulation and evolution in reservoir sediments. Science of the Total Environment 407(8): 2659–2665.

DiPinto, L.M., Coull, B.C. and Chandler, G. 1993. Lethal and sublethal effects of a sediment-associated PCB Aroclor 1254 on a meiobenthic copepod. Environmental Toxicology and Chemistry 12(10): 1909–1918.

Eamus, D., Hatton, T., Cook, P. and Colvin, Ch. 2006. Ecohydrology – Vegetation function, water and resource management. Common-wealth Scientific and Industrial Research Organization, Collingwood, Australia.

Frątczak, W., Izydorczyk, K. and Zalewski, M. 2013. Wysokoefektywne strefy buforowe dla zwiększenia potencjału ekologicznego i turystycznego Zbiornika Sulejowskiego. Gospodarka Wodna 12: 479–483.

Gulati, R.D., Lammens, E.H.R.R., Meijer, M.-L. and van Donk, E. (eds). 1990. Biomanipulation, Tool for Water Management. Proc., Int. Conf., Amsterdam, Netherlands.

Grodzinski, W., Klekowski, R.Z. and Duncan, A. (eds.). 1975. Methods for Ecological Bioenergetics. (I.B.P. Handbook 24). Blackwells, Oxford.

Harper, D.M. and Zalewski, M. (eds). 2001. Ecohydrology. Science and the sustainable management of tropical waters. A summary of the projects presented to the Conference Naivasha, Kenya, 11–16 April 1999. UNESCO-IHP. Technical Documents in Hydrology No. 46, Paris.

Harper, D.M., Zalewski, M. and Pacini, N. (eds.). 2008. Ecohydrology: Processes, Models and Case Studies. An Approach to the Sustainable Management of Water Resources. CABI, UK.

Helcom (Helsinki Commission). 2007. Climate Change in the Baltic Sea Area. Baltic Sea Environmental Proceedings No. 111. Helcom, Helsinki.

Heynes, H.B.N. 1970. The Ecology of Running Waters. University Toronto Press, Toronto, ON, Canada.

Hillbrich-Ilkowska, A. 1993. The dynamics and retention of phosphorus in lentic and lotic patches of two river-lake systems. Hydrobiologia 251(1-3): 257–268.

Hillbrich-Ilkowska, A., Rybak, J. and Rzepecki, M. 2000. Ecohydrological research of lake-watershed relations in a diversified landscape (Musurian Lakeland, Poland). Ecol Eng 16(1): 91–98.

Hobbs, Richard J., Arico, Salvatore, Aronson, James, Baron, Jill S., Bridgewater, Peter, Cramer, Viki A., Epstein, Paul R., Ewel, John J., Klink, Carlos A., Lugo, Ariel E., Norton, David, Ojima, Dennis, Richardson David M., Sanderson, Eric W., Valladares, Fernando, Vilà, Montserrat, Zamora, Regino and Zobel, Martin. 2006. Novel ecosystems: Theoretical and management aspects of the new ecological world order. Global Ecol Biogeogr 15(1): 1–7.

Izydorczyk, K., Fratczak, W., Drobniewska, A., Cichowicz, E., Michalska-Hejduk, D., Gross, R. and Zalewski, M. 2013a. A biogeochemical barrier to enhance a buffer zone for reducing diffuse phosphorus pollution – preliminary results. Ecohydrology & Hydrobiology 13: 104–112.

Izydroczyk, K., Fratczak, W. and Zalewski, M. 2013b. Poprawa jakości wody w obszarach użytkowanych rolniczo w wyniku zastosowania biotechnologii ekohydrologicznych. Panorama PAN 3: 2–4.

Janauer, G.A. 2000. Ecohydrology: Fusing concepts and scales. Ecol Eng 16(1): 9–16.

Janauer, G.A. 2006. Ecohydrological control of macrophytes in floodplain lakes. Ecohydrol Hydrobiol 6(1-4): 19–24.

Junk, W.J., Bayley, P.B. and Sparks, R.B. 1989. The flood pulse concept in river-floodplain systems. pp. 110–127. *In*: Dodge, D.P. (ed.). Proc, Int Large River Symp, Fisheries and Aquatic Sciences, Ottawa, Canada.

Jurczak, T., Wagner, I. and Zalewski, M. 2012. Urban Aquatic Ecosystems Management. Public Service Review. Europe 24, pp. 178.

Jurczak, T., Wagner, I. and Zalewski, M. (red.) 2012. Ekohydrologiczna rekultywacja zbiorników rekreacyjnych Arturówek (Łódź) jako modelowe podejście do rekultywacji zbiorników miejskich (EH-REK). Analiza zagrożeń i szans. Łódź, 115 pp.

Jurczak, T., Wagner, I., Mirosław-Świątek, D., Jaglewicz, M., Kaczkowski, Z., Oleksińska, Z. and Łapińska, M. 2015. System wspierania decyzji w rekultywacji małych zbiorników wodnych. Agent PR. Uniwersytet Łódzki.

Kędziora, A. 1996. Hydrological cycle in agricultural landscapes. Dynamics of an agricultural landscape. pp. 65–78. *In*: Ryszkowski, L., French, N. and Kędziora, A. (eds.). Państwowe Wydawnictwo Rolnicze I Leśne (PWRiL), Poznań, Poland.

Kiedrzyńska, E., Wagner-Łotkowska, I. and Zalewski, M. 2008. Quantification of phosphorus retention efficiency by floodplain vegetation and a management strategy for a eutrophic reservoir restoration. Ecological Engineering 33(1): 15–25.

Kiedrzyńska, E., Kiedrzyński, M., Urbaniak, M., Magnuszewski, A., Skłodowski, M., Wyrwicka, A. and Zalewski, M. 2014. Point sources of nutrient pollution in the lowland river catchment in the context of the Baltic Sea eutrophication. Ecol Eng 70: 337–348.

Kowalik, P. and Eckersten, H. 1984. Water transfer from soil through plants in the atmosphere in willow energy forest. Ecol Modell 26(3-4): 251–284.

Kuprys-Lipińska, I., Elgalal, A. and Kuna, P. 2009. Urban-rural differences in the prevalence of atopic diseases in the general population in Lodz province (Poland). Post Dermatol Alergol 26(5): 249–256.

LIFE+EKOROB: Ecotones for reduction of diffuse pollutions. 2011. The Parliament Magazine. Issue 328.

Lillehammer, A. and Saltveit, S.J. 1984. Regulated Rivers. Unversiteitsforlaget, Oslo.

Maass, M., Balvanera, P., Bourgeron, P., Equihua, M., Baudry, J., Dick, J., Forsius, M., Halada, L., Krauze, K., Nakaoka, M., Orenstein, D.E., Parr, T.W., Redman, C.L., Rozzi, R., Santos-Reis, M., Swemmer, A.M. and Vădineanu, A. 2016. Changes in biodiversity and trade-offs among ecosystem services, stakeholders, and components of well-being: the contribution of the International Long-Term Ecological Research network (ILTER) to Programme on Ecosystem Change and Society (PECS). Ecology and Society 21(3): 31.

Mankiewicz, J., Walter, Z., Tarczyńska, M., Palyvoda, O., Wojtysiak-Staniaszczyk, M. and Zalewski, M. 2002. Genotoxicity of cyanobacterial extracts with microcystins from Polish water reservoirs as determined by the SOS chromotest and comet assay. Environ Toxicol 17(4): 341–350.

Mankiewicz-Boczek, J., Urbaniak, M., Romanowska-Duda, Z. and Izydorczyk, K. 2006. Toxic Cyanobacteria strains in lowland dam resevoir (Sulejów Res, central Poland): amplification of mcy genes for detection and identification. Pol J Ecol 54(2): 171–180.

McClain, M. and Zalewski, M. (eds.). 2001. Ecohydrology. Hydrological and Geochemical Processes in Large River Basins. A summary of the projects presented to the International Symposium Manaus, Brazil, 15–19 Nov. 1999. UNESCO-IHP. Technical Documents in Hydrology No. 47, Paris.

Meybeck, M. 2003. Global analysis of river systems: From Earth system controls to Antropocene syndromes. Phil Trans R Soc London 358(1440): 1935–1955.

Milly, P.C.D., Wetherald, R.T., Dune, K.A. and Deworth, T.L. 2002. Increasing risk of great floods in a changing climate. Nature 415: 514–517.

Mitch, W.J., Cronk, J.K., Wu, X. and Narin, R.W. 1995. Phosphorus retention in constructed freshwater riparian marches. Ecological Applications 5(3): 830–845.

Mitsch, W.J. 1993. Ecological engineering—a co-operative role with the planetary life-support system. Environ Sci Technol 27(3): 438–445.

Mitsch, W.J. and Jorgensen, S.E. 2003. Ecological engineering: a field whose time has come. Ecol Eng 20(5): 363–377.

Mitsch, W.J. 2012. What is ecological engineering? Ecol Eng 45: 5 -12.

Naiman, R.J. and Soltz, D.L. 1981. An ecosystem overview: desert fishes and their habitats pp. 493–531. *In*: Naiman, R.J. and Soltz, D.L. (eds.). Fishes in North American Deserts. Wiley, New York.

Newbold, J.D., O'Neill, R., V., Elwood J.W., Van Winkel W. 1982. Nutrient spiraling in streams. Implications for nutrient limitation in invertebrate activity. Am Nat 120(5): 628–652.

Odum, E.P. 1971. Fundamentals of Ecology. W. B. Saunders Company, Philadelphia.

Penczak, T., Zalewski, M., Suszycka, E. and Molinski, M. 1981. Estimation of the density, biomass and growth rate of fish populations in two small lowland rivers. Polish Journal of Ecology.

Petts, G.E. (ed.). 1984. Impounded Rivers: Perspectives for Ecological Management. Wiley, Chichester, U.K.

Petts, G.E. (ed.). 1995. Man's Influence on Freshwater Ecosystems and Water Use. Proc, Boulder Symp, Int. Association of Hydrological Sciences Publication, Wallingford, U.K.

Piniewski, M., Marcinkowski, P., Kardel, I., Giełczewski, M., Izydorczyk, K. and Frątczak, W. 2015. Spatial Quantification of Non-Point Source Pollution in a MesoScale Catchment for an Assessment of Buffer Zones Efficiency Water 7: 1889–1920.

Power, M.E., Dietrich, W.E. and Finlay, J.C. 1996. Dams and downstream aquatic biodiversity: potential food web consequences of hydrologic and geomorphic change. Environmental Management 20(6): 887–895.

Rodriguez-Iturbe, I. 2000. Ecohydrology: a hydrological perspective of climate-soil-vegetation dynamic. Water Resour Res 36(1): 3–9.

Ryszkowski, L. and Kędziora, A. 1993. Energy control of matter fluxes through land-water ecotones in an agricultural landscape. Hydrobiologia 251: 239–248.

Ryszkowski, L. and Kedziora, A. 1999. State of the art in the appraisal of global climate change phenomena. Geograph Pol 72(2): 5–8.

Ryszkowski, L. and Kędziora, A. 2004. Energetics of ecosystem and landscape changes. Ecological Questions 5/2004: 9–21.

Statzner, B. and Sperling, F. 1993. Potential contribution of system-specific knowledge (SSK) to stream management decisions: Ecological and economic aspects. Freshwater Biol 29(2): 313–342.

Ta'trai, K., Korponai, M.J., Paulovits, G. and Pomogyi, P. 2000. The role of the Kis–Balaton water protection system in the control of water quality of Lake Balaton. Ecological Engineering 16(1): 73–78.

Tilman, D. 1999. The ecological consequences of changes in biodiversity: a search for general principles. J Ecol 80(5): 145501474.

Tomaszek, J.A. and Czerwieniec, E. 1995. Importance of denitrification process for nitrogen balance in water ecosystems. pp. 91–101. *In*: Zalewski, M. (ed.). Biological Processes in Protection and Recultivation of Lowland Reservoirs. Biblioteka Monitoringu Środowiska, Łódź.

Tundisi, J.G. and Tundisi, T.M. 2008. Limnologia. Oficina de Textos.

Tundisi, J.G. and Tundisi, T.M. 2016. Integrating ecohydrology, water management, and watershed economy: case studies from Brazil. Ecohydrology & Hydrobiology 16(2): 83–91.

Urbaniak, M. and Zalewski, M. 2010. Large dams as purification systems for toxic PCDD/PCDF and dl-PCB congeners. Hydrocomplexity: New Tools for Solving Wicked Water Problems (Hydrocomplexite´: Nouveaux Outils pour Solutionner des Proble`mes de l'eau Complexes). IAHS, Xxxxxxxxxxxxx, Report 338.

Vannote, R., L., Minshall G., W., Cummins K. W., Sedell J., R., Cushing C., E. 1980. The river continuum concept. Can J Fish Aquat Sci 37(1): 130–137.

Visser, W.C. 1971. Mathematical models in soil productivity studies, exemplified by the response to nitrogen. Plant Soil 30(2): 161–182.

von Weizsacker, E., Lovins, A.B. and Lovins, L.H. 1997. Factor Four: Doubling Wealth, Halving Resource Use. Earthscan, London.

Vorosmarty, C.J. and Sahagian, D. 2000. Antropocentric disturbance of the terrestrial water cycle. Bioscience 50(9): 753–765.

Wagner, I. and Zalewski, M. 2009. Ecohydrology as a basis for the sustainable city strategic planning—Focus on Lodz, Poland. Rev Environ Sci Biotechnol 8: 209–217.

Wagner I., Zalewski M. 2016. Temporal changes in the abiotic/biotic drivers of selfpurification in a temperate river. Ecological Engineering 94: 275–285.

Wantzen, K.M., Ballouche, A., Longuet, I., Bao, I., Bocum, H., Cisse, L., Chauhan, M., Girard, P., Gopal, B., Kane, A., Marchese, M.R., Nautiyal, P., Teixeira, P. and Zalewski, M. 2016. River culture: an eco-social approach to mitigate the biological and cultural diversity crisis in riverscapes. Ecohydrology & Hydrobiology 16(1): 7–18.

Ward, J.V. and Stanford, J.A. (eds.). 1979. The Ecology of Regulated Steams. Plenum, New York.

Ward, J.V. and Standford, J.A. 1983. The intermediate disturbance hypothesis: an explanation for biotic diversity patterns in lotic ecosystems. pp. 347–356. In: Fintaine, T.D. and Bartell, S.M. (eds.). Dynamic of Lotic Ecosystems. Ann Arbor Science, Ann Arbor, MI.

Wilkinson, D.M. 2006. Fundamental Processes in Ecology: An Earth Systems Approach. Oxford University Press, Oxford, U.K.

Wojtal, A., Bogusz, D., Menshutkin, V., Izydorczyk, K., Frankiewicz, P., Wagner-Lotkowska, I. and Zalewski, M. 2008. A study of Daphnia-Leptodora-juvenile Percids interactions using a mathematical model in the biomanipulated Sulejow Reservoir. Ann Limnol – Int J Lim 44(1): 7–23.

Wojtal-Frankiewicz, A. and Frankiewicz, P. 2010. Mathematical modelling as a tool for predicting the intensity of eutrophication symptoms based on zooplankton and fish density. Ecohydrol Hydrobiol 10(2-4): 247–257.

Wojtasik, A., Kubiak, A.B., Krzyżanowska, A., Majchrzak, M., Augustynowicz-Kopeć, E. and Parniewski, P. 2012. Comparison of the (CCG) 4-based PCR and MIRU-VNTR for molecular typing of Mycobacterium avium strains. Mol Biol Rep 39(7): 7681–7686.

Wolański, E. 2007. Estuarine Ecohydrology. Elsevier, Amsterdam, Netherlands.

Wolanski, E., Chicharo, L. and Chicharo, M.A. 2008. Estuarine ecohydrology. pp. 1413–1422. In: Jorgensen, Sven Erik and Fath, Brian D. (eds.). Encyclopedia of Ecology. Elsevier, Oxford.

Xia, J., Zhai, X., Zeng, S. and Zhang, Y. 2014. Systematic solutions and modeling on eco-water and its allocation applied to urban river restoration: case study in Beijing, China. Original Research Article, Ecohydrology & Hydrobiology 14(1): 39–54.

Zalewski, M. and Naiman, R.J. 1985. The regulation of riverine fish communities by a continuum of abiotic-biotic factors. Hab Modif Freshw Fish, Alabaster, J.S. (ed.). Butterworth, London, 3–9.

Zalewski, M., Frankiewicz, P. and Brewinska-Zaras, B. 1986. The production of brown trout (*Salmo trutta* L.) introduced to streams of various orders in an upland watershed. Pol Arch Hydrobiol 33(3-4): 411–422.

Zalewski, M., Brewińska-Zaraś, B., Frankiewicz, P. and Kalinowski, S. 1990. The potential for biomanipulation using fry communities in a lowland reservoir: concordance between water quality and optimal recruitment. Hydrobiologia 200/201: 549–556.

Zalewski, M. and Wiśniewski, R. (eds.). 1997. Zastosowanie biotechnologii ekosystemowych do poprawy jakości wód. Instytut Ekologii PAN, Zeszyty Naukowe nr 18.

Zalewski, M. 2000. Ecohydrology – the scientific background to use ecosystem properties as management tools toward sustainability of water resources. Guest Editorial Ecological Engineering 16: 1–8.

Zalewski, M. and Welcomme, R. 2001. Restoration of sustainability of physically degraded fish habitats—the model of intermediate restoration. Ecohydrology and Hydrobiology 1(3): 279–282.

Zalewski, M. 2002a. Ecohydrology—The use of ecological and hydrological processes for sustainable management of water resources. J Hydrol Sco 47(5): 823–832.

Zalewsi, M. 2002b. Ecohydrology—integrative science for sustainable water, environment and society. Ecohydrol Hydrobiol 2 (1-4): 3–10.

Zalewski, M. and Wagner-Lotkowska, I. (eds.). 2004. Integrated Watershed Management – Ecohydrology and Phytotechnology. Unesco IHP, Unesco ROSTE, UNEP–DTIE–IETC, ICE PAS, University of Łódź, Venice, Osaka, Shiga, Warsaw, Łódź.

Zalewski, M. and Wagner, I. 2005. Ecohydrology—The use of water and ecosystem processes for healthy urban environments. Ecohydrol Hydrobiol 5(4): 263–266.

Zalewski, M. 2009. Ecohydrology: a framework for reversing the degradation of the Baltic Sea. Baltex News 13: 7–10.

Zalewski, M., Negussie, J. and Urbaniak, M. 2010. Ecohydrology for Ethiopia—regulation of water biota interactions for sustainable water resources and ecosystem services for societies. Ecohydrology & Hydrobiology 10(2): 4101–106.

Zalewski, M. 2011. Ecohydrology for implementation of the EU Water Framework Directive. Proceedings of the Institute of Civil Engineering—Water Management 164(8): 375–385. doi:10.1680/wama.1000030.

Zalewski, M. 2012. Ecohydrology —process oriented thinking for sustainability of river basins. Ecohydrology & Hydrobiology, Selected papers from the 2nd Conference on Healthy Rivers and Sustainable Water Resource Management, 20–22 October 2011. Chongquing, China 12(2).

Zalewski, M., Wagner, I., Frątczak, W., Mankiewicz-Boczek, J. and Parniewski, P. 2012. Blue-Green City for compensating Global Climate Change. The Parliament magazine, Issue 350: 2–3.

Zalewski, M. 2013. Ecohydrology: process-oriented thinking towards sustainable river basins. Ecohydrol Hydrobiol 13(2): 97–103.

Zalewski, M. 2014a. Ecohydrology, biotechnology and engineering for cost efficiency in reaching the sustainability of biosphere. Ecology & Hydrobiology 14: 14–20.

Zalewski, M. 2014b. Ecohydrology for engineering harmony in the changing world. pp. 79–96. *In*: Eslamian, S. (ed.). Handbook of Engineering Hydrology. Fundamentals and Applications. Taylor & Francis.

Zalewski, M. 2015. Ecohydrology and hydrologic engineering: regulation of hydrology-biota Interactions for sustainability. J Hydrol Engineering 20. Special Issue: Grand Challenges in Hydrology.

Zhou, D., Zhang, H. and Liu, Ch. 2016. Wetland ecohydrology and its challenges. Ecohydrology & Hydrobiology 16(1): 26–32.

Index